PROCEEDINGS OF THE INTERNATIONAL CONFERENCE

ELECTRORHEOLOGICAL FLUIDS

PROCEEDINGS OF THE INTERNATIONAL CONFERENCE

ELECTRORHEOLOGICAL FLUIDS

15 – 16 October 1991
Carbondale, Illinois, USA

MECHANISMS, PROPERTIES, STRUCTURE, TECHNOLOGY, AND APPLICATIONS

edited by R. Tao

Department of Physics
Southern Illinois University at Carbondale

World Scientific
Singapore • New Jersey • London • Hong Kong

Published by

World Scientific Publishing Co Pte Ltd
5 Toh Tuck Link, Singapore 596224
USA office: 27 Warren Street, Suite 401-402, Hackensack, NJ 07601
UK office: 57 Shelton Street, Covent Garden, London WC2H 9HE

ELECTRORHEOLOGICAL FLUIDS: MECHANISM, PROPERTIES, STRUCTURE, TECHNOLOGY AND APPLICATIONS

ISBN-13 9789810209025
ISBN-10 9810209029

FOREWORD TO THE CONFERENCE

Electrorheological fluids (ER fluids) often called smart fluids or intelligent fluids are potential candidates for a wide variety of applications. These could range from the control of viscous damping such as in brakes, to the control of viscosity of fuels which results in a phase change. Implications of this are that electrorheological fuels may be stored as solid fuels and combusted as liquid fuels. These have both commercial and defense impacts. Successful application of this challenging, interdisciplinary technology would be possible if the fundamental mechanisms controlling the ER properties are understood, and production techniques are made available.

The Electrorheological Conference has provided one step to achieve the goal of understanding, controlling and applying ER phenomena. The meeting was organized to bring together the experts in this field to explore the state-of-the-art of this technology, and to disseminate the most recent research developments. Technical sessions included Materials Technology, Physical Mechanisms, Properties, and Applications, together with a poster session. The response from the scientific and engineering community was overwhelmingly positive. The meeting provided a forum for collaboration and interaction of industry and academia, which is required for the success of any emerging technology. Participation from UK, Japan, USSR, China, Germany, France and USA is evidence of the truly international interest in this technology area. It is evident from the presentations that considerable advances have been made in understanding the physics of ER fluids, and that the technology is applied in several components.

It gives me great pleasure to be involved in this scientific and engineering endeavor. I congratulate Professor Rongjia Tao for organizing a commendable meeting. I wish success to those who are involved in this area, and hope to see great growth in the science and engineering aspects of ER fluids, providing progress in commercial and industrial applications, and to the nation and to humanity at large.

January 1992

Gabriel D. Roy
Office of Naval Research
Arlington, Virginia

PREFACE

The International Conference on Electrorheological (ER) Fluids was held at Southern Illinois University at Carbondale (SIUC), Illinois, USA, October 15–16, 1991. Scientists and engineers from all over the world came together to make the conference successful and stimulating.

During the past several years there has been a remarkable increase in the interest and activity in the field of ER fluids. As evidenced by the papers presented at this conference and collected in this volume, significant progress has been made in both research and applications, since the last conference on ER fluids in 1989.

We are deeply indebted to the Office of Naval Research, Materials Technology Center, and College of Science of SIUC for their sponsorship which made the Conference possible. We especially would like to thank

Dr. Gabriel D. Roy, Office of Naval Research

Dr. Maurice A. Wright, Director of Materials Technology Center, SIUC

Dr. Russell R. Dutcher, Dean of College of Science, SIUC

whose continued support was the most essential ingredient for the success of the Conference. We express our special thanks to Dr. Jane Evers for her valuable assistance in the coordination of the Conference.

January 1992 Rongjia Tao

CONTENTS

III. PROPERTIES OF ER FLUIDS

IV. APPLICATIONS OF ER FLUIDS

I. PHYSICAL MECHANISMS OF ER FLUIDS

ORDER PARAMETERS AND PHASE TRANSITIONS IN ELECTRORHEOLOGICAL FLUIDS

R. TAO
Department of Physics, Southern Illinois University
Carbondale, Illinois 62901, USA

ABSTRACT

The ground state of an induced electrorheological (ER) solid is a body-centered tetragonal (bct) lattice with conventional Bravais vectors $\vec{a}_1 = \sqrt{6}a\hat{x}$, $\vec{a}_2 = \sqrt{6}a\hat{y}$, and $\vec{a}_3 = 2a\hat{z}$ where a is the radius of dielectric spheres and $\hat{z}$ is the field direction. The reciprocal lattice vectors are $\vec{b}_1 = 2\pi\hat{x}/\sqrt{6}a - \pi\hat{z}/a$, $\vec{b}_2 = 2\pi\hat{y}/\sqrt{6}a - \pi\hat{z}/a$, and $\vec{b}_3 = 2\pi\hat{z}/a$. Three order parameters are defined as $\rho_j = \sum_{i=1}^{N} \exp(i\vec{b}_j\vec{r}_i)/N, (j = 1, 2, 3)$, where N is the total number of particles. Among them, ρ_3 characterizes the formation of chains in the z direction, while ρ_1 and ρ_2 characterize the structure in the x-y plane. Monte Carlo simulations of canonical ensemble have shown that three different phases and two phase transitions exist in the ER fluid. At a fixed temperature, there are two critical electric fields $E_{c2} < E_{c1}$. When the applied electric field $E < E_{c2}$, ρ_j $(j = 1, 2, 3)$ are all vanishing and the ER fluid is a liquid. When $E_{c2} < E < E_{c1}$, $\rho_1 = \rho_2 = 0$, but ρ_3 is not vanishing, indicating that the ER fluid begins to form chains between two electrodes, but the distribution of these chains is random. This state is similar to nematic liquid crystal. The second phase transition occurs as E exceeds E_{c1}. Then ρ_1, ρ_2, and ρ_3 are all non-vanishing, indicating that the chains aggregate together to form thick columns which have the bct lattice structure. These two phase transitions can also be realized at a fixed field by lowering the temperature.

1. Introduction

An electrorheological (ER) fluid consists of a suspension of fine dielectric particles in a liquid of low dielectric constant and low viscosity.[1-9] Its apparent viscosity increases dramatically in the presence of an applied electric field. If the electric field exceeds a critical value, the ER fluid turns into a solid whose yield stress increases as the field is further strengthened. The phenomenon is completely reversible. Upon electric field cutoff, the system immediately resumes its original liquid state. The time scale for the transition is of the order of millisecond. This novel property makes ER fluids attractive for many futuristic technologies. The other papers in this volume have clearly illustrated potential applications of ER fluids in industries and technologies, indeed.

Since the discovery of ER fluids, a crucial question has been raised: What is the physical mechanism of the ER phenomenon? This is what the present paper is going to discuss. The fundamental physics of ER fluids has just begun to be understood.

Let us consider a system of small spherical dielectric particles of dielectric constant ϵ_p suspended in a fluid of dielectric constant ϵ_f, $\epsilon_f < \epsilon_p$. The whole system is placed between two parallel plates, upon which a voltage is applied to produce an electric field. When the dielectric particles have a density close to the liquid density, the buoyant force neutralizes the gravity and creates a low-gravity environment. Then, the thermal motion makes the particles randomly distributed in the space and form a uniform liquid suspension. As the electric field is applied, the particles obtain an induced dipole moment,

$$\vec{p} = \alpha a^3 \epsilon_f \vec{E}_{loc} \tag{1.1}$$

where a is the particle radius, $\alpha = (\epsilon_p - \epsilon_f)/(\epsilon_p + 2\epsilon_f)$ and $\vec{E}_{loc}$ is the local effective field acting on the particles. In a dilute case, $\vec{E}_{loc}$ can be approximated by the applied electric field $\vec{E}$. We take the field direction as the z direction. Two dipoles at $\vec{r}_i$ and $\vec{r}_j$ have an interaction

$$U(\vec{r}_{ij}) = \nu(1 - 3\cos^2\theta_{ij})/r_{ij}^3 = -\nu(2 + \rho\frac{\partial}{\partial\rho})\frac{1}{r_{ij}^3} \tag{1.2}$$

where $\nu = (\vec{p})^2/\epsilon_f$, $\rho = \sqrt{(x_i - x_j)^2 + (y_i - y_j)^2}$, $r_{ij} = \sqrt{\rho^2 + (z_i - z_j)^2}$, and $0 \le \theta_{ij} \le \pi/2$ is the angle between the joint line of the two dipoles and the z direction. As dipoles cannot overlap, we also add a hard-core interaction to the above dipolar interaction. The higher the field, the stronger the dipolar interaction. The competition between the Coulomb interaction and the thermal motion determines the state of the ER system.

We can estimate the critical field for the phase transition in the following way. The thermal energy is given by $3NkT/2$ where T is the temperature, k is the Boltzmann constant, and N is the total number of particles. The dipolar interaction is about $Np^2n/\epsilon_f = N\alpha^2a^6E_{loc}^2n\epsilon_f$ which is proportional to E_{loc}^2. We use n to denote the particle number density. In a low field, because $3NkT/2 \gg N\alpha^2a^6E_{loc}^2n\epsilon_f$, the thermal motion is dominate and the system remains a liquid. When the dipolar interaction overcomes the thermal motion as the field increases, the dielectric particles suddenly stop ignoring each other and begin to align in the field direction to form structures between the two electrodes. The nature of ER fluids lies in the electric field-induced solidification. From the above discussion, the critical electric field is about $\sqrt{3kT/(2\alpha^2na^6\epsilon_f)}$. A more detailed calculation[3,4], based on consideration about the free energy of the ER system, finds the critical electric field is approximately given by

$$E_c = \sqrt{\frac{8\pi kT(1-\phi)}{\alpha^2\bar{v}\phi\epsilon_f}} \tag{1.3}$$

where $\bar{v} = 4\pi a^3/3$ and $\phi = n\bar{v}$ is the volume fraction of the dielectric particles in the system. When $E > E_c$, the ER system becomes a solid. The system remains a liquid if $E < E_c$. Experimental results about E_c agree with Eq.(1.3) reasonably well.[4]

In materials technology of ER fluids, it is not easy to make dielectric particles have the same density as the base liquid. Petroleum oils, the most popular base liquid, have a density about 1 g/cm^3. Dielectric particles, such as alumina powders, have a density close to 3 g/cm^3. Sedimentation effects considerably weaken or even eliminate the ER effect. To overcome the sedimentation effects, before applying an electric field, experimenters often stir ER suspensions and create a strong liquid flow which forces the dielectric particles to move and spread into the space. Then, the solidification is a competition between the dipolar interaction and the high kinetic energy of the dielectric particles moving with the flow. In such a case, the electric field required to solidify the ER system is greatly increased,

$$E_c = \sqrt{\frac{8\pi m v^2(1-\phi)}{3\alpha^2\bar{v}\phi\epsilon_f}} \tag{1.4}$$

where v is the average velocity of the particles.

The above discussion shows that ER fluids will experience a field-induced phase transition in an electric field. However, the argument does not tell us what is the structure of the induced solid and how the phase transition occurs. In addition, E_c in Eq.(1.3) or Eq.(1.4) is only an estimation.

It is well known that a crucial step in study of phase transitions from a disordered state to an ordered state is to define order parameters.[10] For example, in a ferromagnet this quantity is the spontaneous magnetization. In superconductivity the order parameter is the number of electron pairs. When the system is in the disordered state, the order parameters are zero. They are non-vanishing when the system is in the ordered state. Therefore, further study of the phase transition in ER fluids must also employ the concept of order parameters which will be discussed in the following sections.

2. Ground State

The next question is naturally about the structure of the induced ER solid.[11] Because the physical properties of ER fluids depend on the induced structure, the issue is highly significant.

As mentioned early, at a sufficiently strong electric field or low temperature, an ER fluid becomes a solid. This is because the Coulomb interaction becomes dominate. The induced solid structure, therefore, must be the configuration which minimizes the Coulomb energy. The total Coulomb energy includes both dipolar interaction and the interaction between dipoles and the external field. Since the total Coulomb energy is minimized when the dipolar interaction is minimized, we only need to consider how to minimize the dipolar interaction energy.[5,12,13]

We have found that the ideal structure of the induced ER solid is a body-centered tetragonal (bct) lattice (Fig.1a). The applied electric field makes the space anisotropic. The dipolar force is also anisotropic, which leads to additional effects. The bct structure is the ground state benefited from these anisotropic interactions.

In the experiments and molecular dynamics simulations, it has been found that upon application of an electric field, the dielectric particles in ER fluids first form chains between two plates.[2,8] Chains then aggregate to form thick columns.[2,11] Halsey and Toor claimed that each thick column has a width $\sim a(L/a)^{2/3}$ where L is the spacing between two electrodes.[6]

The formation of chains is easy to understand because the dipolar energy has its minimum when two dipoles join together and align in the field direction. The interaction between chains needs to be examined here.

A dipole $\vec{p}$ inside the capacitor at (x, y, z) produces an infinite number of images which also interact with the dipole itself and other dipoles. All of its images have the same dipole moment $\vec{p}$ and are at $(x, y, -z)$ and $(x, y, 2Lj \pm z)$ for $j = \pm 1, \pm 2, \cdots$. The interaction between a dipole with an image produced by another dipole has the same form as Eq.(1.2). However, the interaction of a dipole at $\vec{r}_i$ with one of its own images at $\vec{r}_j$ is given by $\frac{1}{2}U(\vec{r}_{ij})$.

A chain, i.e. a line of particles between the two plates, combining with its images, becomes an infinite chain. Without loss of generality, we consider a chain having particles at $\vec{r}_j = 2aj\hat{z}$ $(j = 0, \pm 1, \pm 2, \cdots)$. The self-energy of the chain is given by

$$\sum_{i<j} U(\vec{r}_i - \vec{r}_j) = \frac{1}{2}\sum_{i \neq j} U(\vec{r}_i - \vec{r}_j). \tag{2.1}$$

As $\vec{r}_i - \vec{r}_j = 2a(i-j)\hat{z}$, the self-energy per particle in an infinite chain, u_0, is

$$u_0 = -\frac{\nu}{4a^3}\sum_{s=1}^{\infty}\frac{1}{s^3} = -0.300514\frac{\nu}{a^3}. \tag{2.2}$$

Consider the interaction between this chain and a dipole $\vec{p}$ at $\vec{r} = \vec{\rho} + z\hat{z}$,

$$U_i(\rho, z) = -\nu(2 + \rho\frac{\partial}{\partial\rho})\sum_{j=-\infty}^{\infty}\frac{1}{[\rho^2 + (z - 2ja)^2]^{3/2}}. \tag{2.3}$$

We introduce a function,

$$f(\rho, z) = \sum_{j=-\infty}^{\infty}\frac{1}{[\rho^2 + (z - 2ja)^2]^{3/2}}. \tag{2.4}$$

Since the periodicity, $f(\rho, z) = f(\rho, z + 2a)$, we can expand f into the form

$$f(\rho, z) = \sum_{s=-\infty}^{\infty} f_s(\rho)e^{-is\pi z/a}. \tag{2.5}$$

Then we have

$$f_s(\rho) = \frac{1}{2a}\int_0^{2a}\sum_{j=-\infty}^{\infty}\frac{dz e^{is\pi z/a}}{[\rho^2 + (z - 2ja)^2]^{3/2}} = \frac{\pi s}{a^2\rho}K_1(\frac{s\pi\rho}{a}) \tag{2.6}$$

where $K_1(x)$ is a modified Bessel function. From Eqs.(2.5) and (2.6), we have

$$f(\rho, z) = \frac{1}{a\rho^2} + \sum_{s=1}^{\infty} \frac{2\pi s}{a^2\rho} K_1(s\pi\rho/a)\cos(s\pi z/a). \tag{2.7}$$

The formula $\frac{d}{dx}(xK_1(x)) = -xK_0(x)$ enables us to write Eq.(2.3) into

$$U_i(\rho, z) = \frac{\nu}{a^3}\sum_{s=1}^{\infty} 2\pi^2 s^2 K_0(\frac{s\pi\rho}{a})\cos(\frac{s\pi z}{a}). \tag{2.8}$$

The property of K_0 makes Eq.(2.8) converge exponentially. Due to the hard-core interaction, the minimum value of ρ is $\sqrt{3}a$ at $z = (2j+1)a$. Then $s\pi\rho/a \geq \pi\sqrt{3}s \geq 5.4413$. The asymptotic behavior of $K_0(x) \to \sqrt{\frac{\pi}{2x}}e^{-x}$ (for $x \gg 1$) gives the leading term of Eq.(2.8), $U_i(\rho, z) \approx \frac{\nu}{a^3}\pi^2\sqrt{\frac{2a}{\rho}}e^{-\pi\rho/a}\cos(\pi z/a)$.

From $F_\rho = -\partial U_i/\partial\rho$, we note that the interaction U_i is attractive when $(2j+1/2)a < z < (2j+3/2)a$; otherwise, it is repulsive. $F_z = -\partial U_i/\partial z$ always pulls the dipole to a position of $z = (2j+1)a$. At $\rho = \sqrt{3}a$ and $z = (2j+1)a$, U_i is minimized to $-0.04442565\nu/a^3$.

When there is a second infinite chain around, the dipolar interaction energy per particle, u, is $u_0 + \frac{1}{2}U_i(\rho, z)$, because all dipoles in the second chain have coordinates (ρ, $z + 2aj$) ($j = 0, \pm1, \pm2, \cdots$). The factor $\frac{1}{2}$ comes from the fact that the interaction energy is shared by both dipoles. If these chains are far apart, $\rho \to \infty$, $u = u_0$. When $\rho = 2a$ and $z = 0$, two chains are repulsive, $u = -.291405\nu/a^3$. The minimum of u, $-.3227268\nu/a^3$, is obtained when the second chain has coordinates $\rho = \sqrt{3}a$ and $z = a$.

We define infinite chains with their dipoles at $z_j = 2ja$ ($j = 0, \pm1, \pm2, \cdots$) as chains of class A and infinite chains with $z_j = (2j+1)a$ as chains of class B. To benefit the dipolar interaction, the ER solid must have chains of both A and B. Two chains of the same class are repulsive, otherwise they are attractive, namely

$$U_{AA}(\rho) = U_{BB}(\rho) = \frac{\nu}{a^3}\sum_{s=1}^{\infty} 2\pi^2 s^2 K_0(\frac{s\pi\rho}{a}) \tag{2.9}$$

and

$$U_{AB}(\rho) = \frac{\nu}{a^3}\sum_{s=1}^{\infty} 2\pi^2 s^2 K_0(\frac{s\pi\rho}{a})\cos(s\pi). \tag{2.10}$$

These interactions are short-ranged and bear some similarity to the Yukawa potential. The problem of structure of ER fluids is equivalent to finding the ground state for a two-dimensional system which has two classes of particles, A and B, via the interactions in Eqs.(2.9) and (2.10). The distance between two chains of different class is $\rho \geq \sqrt{3}a$, but the distance between two chains of the same class is $\rho \geq 2a$.

To minimize the potential energy, the ground state must have such a structure that an A (or B) chain has as many B (or A) chains around at the minimum distance $\sqrt{3}a$ as possible and is as far away from other A (or B) chains as possible. Our bct lattice with $\vec{a}_1 = \sqrt{6}a\hat{x}$, $\vec{a}_2 = \sqrt{6}a\hat{y}$, and $\vec{a}_3 = 2a\hat{z}$ (Fig.1a) is such a structure. It can be described by a two-dimensional crystal in Fig.1b. In this bct structure, an A (or B) chain has four B (or A) chains as its nearest neighbor at $\rho = \sqrt{3}a$, while the four next nearest neighbor A (or B) chains are at $\rho = \sqrt{6}a$. The dipolar interaction energy per particle is $u = -.381268\nu/a^3$.

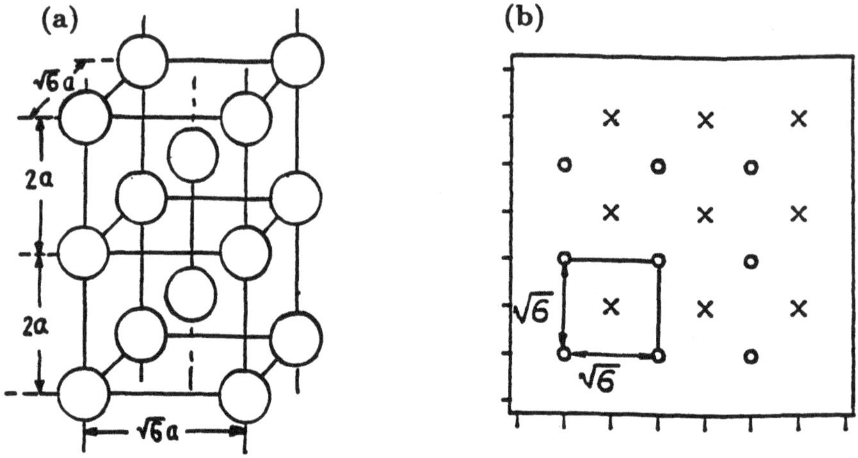

Fig.1. (a). Three-dimensional body-centered tetragonal (bct) structure (the particles have radius a and are not shown to scale). (b). The ideal bct structure is represented by a two-dimensional crystal in which o for A chains and × for B chains. The unit is a.

We have compared our bct structure with other structures.[5] The closed-packing has two structures (Fig.2a). The $ABCABC\cdots$ series is a fcc lattice which is equivalent to a two dimensional crystal in Fig.2b. It has Bravais lattice of $\vec{a}_1 = 2a\hat{x}$ and $\vec{a}_2 = 2\sqrt{2}a\hat{y}$ and a basis with one A chain at a corner and one B chain at the center. Though an A (or B) chain also has four B chains at $\rho = \sqrt{3}a$, it has two next nearest A (or B) chains at $\rho = 2a$. As a result, u is higher, $-.3702402\nu/a^3$ for the fcc. This value can also be obtained by Lorentz's method[13] which gives $u = \frac{1}{2}(-\frac{4\pi\vec{p}n}{3\epsilon_f})\vec{p}$ in the fcc structure. As the particle density, $n = 1/(4\sqrt{2}a^3)$, for the fcc lattice, $u = -(\pi/6\sqrt{2})\nu/a^3 = -.3702402\nu/a^3$

We note that there is another fcc structure suggested by Halsey and Toor which does not have chains in the field direction.[6] Instead, their fcc structure has three conventional Bravais vectors $2\sqrt{2}a\hat{x}$, $2\sqrt{2}a\hat{y}$, and $2\sqrt{2}a\hat{z}$. Two neighbor dipoles in the field direction in their fcc structure are separated by $2\sqrt{2}a$. From Lorentz's method, their fcc structure must have the same dipolar interaction energy as our fcc structure.

The $ABAB\cdots$ series is hexagonal closed-packing (hcp) structure which is described by a two-dimensional crystal in Fig.2c with Bravais vectors $\vec{a}_1 = 2\sqrt{3}a\hat{x}$ and $\vec{a}_2 = 4\sqrt{2/3}a\hat{y}$.

The basis has two A chains at $(0,0)$ and $(4a/\sqrt{3}, 2a\sqrt{2/3})$ and two B chains at $(\sqrt{3}a, 0)$ and $(a/\sqrt{3}, 2a\sqrt{2/3})$. Similar to fcc, around an A (or B) chain, hcp has two next nearest A (or B) chains at $\rho = 2a$. Thus, u is higher, $-.3700289\nu/a^3$.

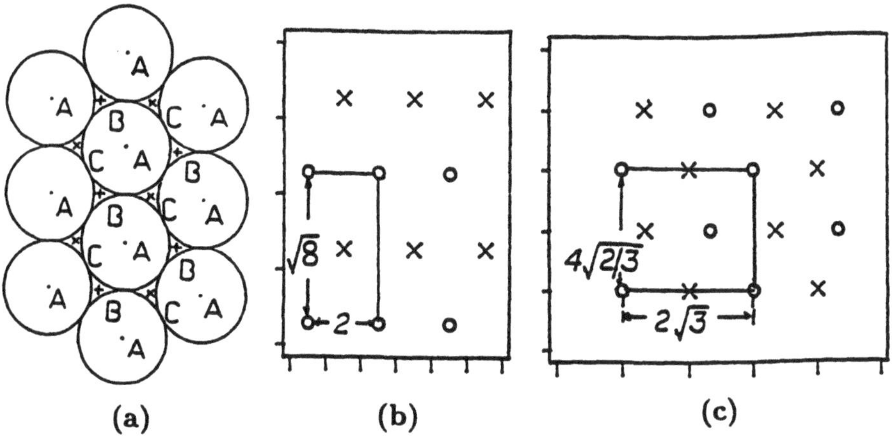

Fig.2. (a). A closed-packed layer of spheres with centers marked A. The second layer of spheres can be placed on top of this, above and parallel to the plane with centers at points B. When the third layer goes in over C, the sequence is $ABCABC\cdots$, and the structure is face-centered cubic (fcc). When the third layer goes in over A, the sequence is $ABAB\cdots$, the structure is hexagonal close-packed (hcp). (b). The fcc structure is represented by a two-dimensional crystal in which o for A chains and × for B chains. (c). The hcp structure is represented by a two-dimensional crystal in which o for A chains and × for B chains. The unit is a in (b) and (c).

We have also compared the bct structure with some other structures which is not formed by chains of A and B. For example, a three dimensional cubic lattice with lattice spacing $2a$ can be regarded as a structure formed by chains of A only. Using Lorentz's method, we have the dipolar interaction energy per particle $\frac{1}{2}(-\frac{4\pi\vec{p}n}{3\epsilon_f})\vec{p} = -\frac{\pi}{12}\nu/a^3 = -0.261799\nu/a^3$ where $n = 1/(2a)^3$ is the particle density. All the above results are summarized in Table I which shows that the bct structure has the lowest energy.

Table I. Dipolar Interaction Energy Per Particle for Various Structures

Structure	u (unit $[p^2/a^3\epsilon_f]$)
bct lattice	-.381268
fcc lattice	-.370240
hcp structure	-.370029
separated chains	-.300514
cubic lattice	-.261799

Though the above discussion about the bct lattice structure, as usual in many-body problems, can hardly be called a rigorous proof, the results clearly indicate that the bct lattice is the ground state.

3. Monte Carlo Simulations and Laser Diffraction Experiments

To verify the above conclusion, we have carried out Monte Carlo simulations to see if the bct lattice is the ground state for a small ER system.[7]

Since the present problem belongs to a class of problems for which the energy has many local minima, we choose the method of simulated annealing which has been proved to be very useful for optimization problems of large scale. At the heart of the method is an analogy with thermodynamics.[14,15] Our simulations have found that the above bct lattice is the ground state of the induced ER solid and this bct structure can be formed at room temperature for typical ER systems.

The present computing problem also belongs to a class of $NP-$*complete* problems, whose computation time for an exact solution increases with N as exp(const. $\times N$), becoming rapidly prohibitive in cost as N increases. Therefore, we take a system of $N = 16$ infinite chains via interactions in Eqs.(2.9) and (2.10). We label half of them as A chains and the other half as B chains. The x-y plane has size $24a \times 24a$ with a periodic boundary condition in both directions. Initially, the chains are randomly distributed in space. We calculate the total dipolar interaction energy, U_0, for the initial configuration and assign a temperature $T = 0.0075$ in the unit ν/a^3k. For a typical ER system, this unit ν/a^3k is quite large. For example, if $\epsilon_f = 2.2$, $\epsilon_p \approx 10$, $a = 1\mu m$, $E_{loc} = 4kV/cm$, this initial temperature is about 6237K. Then we randomly pick a chain and move it to a trial position in an arbitrary direction. We calculate the total dipolar interaction energy U_t for the trial configuration. If $\delta U = U_t - U_0 \leq 0$, the trial state is accepted. If $\delta U > 0$, the probability to accept the trial state is $\exp(-\delta U/T)$. If the trial state is accepted, the system has a successful move, then the next trail move starts from the new configuration. Otherwise, the next trial move is still from the initial state. We make $2000N$ random trial moves for temperature T. Then, we lower the temperature by a factor of 0.95 and repeat the above process for the new temperature. After another $2000N$ random trial moves, we reduce the temperature by a factor of 0.95 again, and so on. As a result of lowering the temperature, the number of successful moves among $2000N$ trial moves decreases. However, as long as there is any successful move for any temperature, we would continue the process and further lower the temperature. The Monte Carlo process is terminated until there is no successful move at a certain temperature. In our simulations, this final state is reached at $T = .3625 \times 10^{-3}$. For the early example, this final temperature is about 301K, room temperature. The simulations take a lot of cpu time on our vectorized IBM-3090 computer, but produce an excellent result.

We plot the final configuration in Figure 3. The final ER state is phase-separated. All infinite chains move into a small region to form a square crystal. A comparison between Figure 3 and Figure 1(b) confirms that the structure derived from our Monte Carlo simulations is the bct lattice. In Fig.3, an A (or B) chain has four B (or A) chains as its nearest neighbor at a distance very close to $\sqrt{3}a$, while the four next nearest neighbor A (or B) chains are at a distance very close to $\sqrt{6}a$. The deviation of these distances from the results of the exact bct lattice is within 1%. The dipolar interaction energy per particle is $-.361627\nu/a^3$ in our final configuration. The dipolar interaction energy per particle for the exact bct lattice of 16 infinite chains in $24a \times 24a$ square with a periodic boundary condition is $-0.362591\nu/a^3$, slightly lower than the result from our simulations. However, the difference between them is very tiny. Even if we exclude the self-energy u_0, their difference is only about 1.5%. If we want our state to approach the bct lattice a little further, we have to increase the number of Monte Carlo steps greatly. We note that the dipolar interaction energy per particle in our infinite bct lattice is $-.381268\nu/a^3$. Since our ER system on computer has only 16 infinite chains, its energy is slightly higher than that of the infinite lattice.

We have also carried out another simulation which allows, in addition to the above random moves, random switches between A chains and B chains. We have found that these additional random switches do not change our conclusion: the final ground state is still the bct lattice.

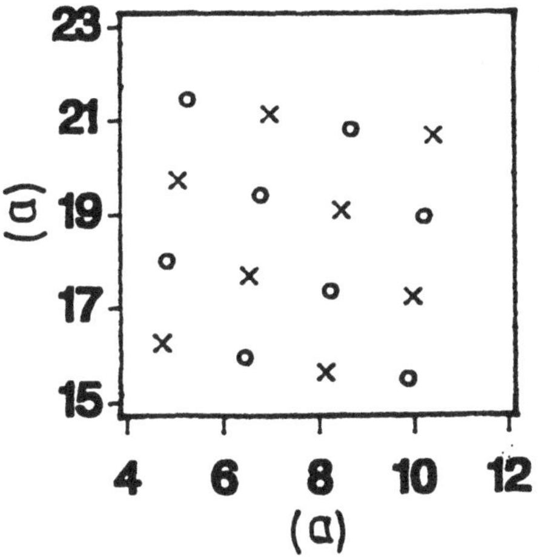

Fig.3. The final configuration from Monte Carlo simulations in which o for A chains and × for B chains. The unit a is the radius of particles. All infinite chains finally concentrate in a small region to form a square crystal. An A (or B) chain has four B (or A) chains as its nearest neighbor at a distance close to $\sqrt{3}a$, while the four next nearest neighbor A (or B) chains at a distance close to $\sqrt{6}a$.

Our simulations clearly demonstrate that the bct ground state is observable at room temperature. This encourages us to perform laser diffraction experiments to characterize the column structure.

The detail of the experiments and the new mechanism to produce laser diffraction patterns by a close-packed structure can be found in another paper in the present volume.[11] For several different samples, the structure constants measured from the experiments all match the theoretical prediction extremely well. The data is entirely consistent with the proposed bct lattice. Accordingly, the experiments have unambiguously confirmed the bct lattice as the ground state of the ER fluids.

4. Order Parameters and Phase Transitions

After clarifying the ideal structure of the induced ER solid, we are now in a position to define order parameters for ER fluids and examine the phase transition. The reciprocal lattice of the above bct lattice is a face-centered-tetragonal lattice with three conventional vectors $4\pi\hat{x}/\sqrt{6}a$, $4\pi\hat{y}/\sqrt{6}a$, and $2\pi\hat{z}/a$. We take three primitive vectors, $\vec{b}_1 = 2\pi\hat{x}/\sqrt{6}a - \pi\hat{z}/a$, $\vec{b}_2 = 2\pi\hat{y}/\sqrt{6}a - \pi\hat{z}/a$, and $\vec{b}_3 = 2\pi\hat{z}/a$. The order parameters are then defined as

$$\rho_j = \frac{1}{N}\sum_{i=1}^{N} \exp(i\vec{b}_j\vec{r}_i), \qquad (j = 1,2,3). \tag{4.1}$$

All these three order parameters are unity if the ER system is the ideal bct lattice. When the dielectric particles are randomly distributed, i.e., the ER system is in a liquid state, ρ_j $(j = 1,2,3)$ are all vanishing.

As mentioned early, the bct lattice can be regarded as a compound of chains of class A and class B where a chain of class A has their dipoles at $z_j = 2ja$ $(j = 1,2,\cdots)$ and a chain of class B has their dipoles at $z_j = (2j-1)a$ $(j = 1,2,\cdots)$. The order parameter ρ_3 characterizes the formation of chains in the z direction because chains of both class A and class B have a coherent contribution to ρ_3. The other two parameters ρ_1 and ρ_2 characterize the structure in the x-y plane for the ER system.

Let us introduce a dimensionless quantity

$$\theta = kTd^3\epsilon_f/(\vec{p})^2 \tag{4.2}$$

which characterizes the competition between the dipolar interaction and the thermal energy. We use $d = 2a$ to denote the diameter of the spherical particles.

To study the phase transition, we apply the Metropolis algorithm[16] to calculate the canonical ensemble average of ρ_j $(j = 1,2,3)$,

$$\langle\rho_j\rangle = \sum(e^{-\beta U}\rho_j)/\sum(e^{-\beta U}) \tag{4.3}$$

where $\beta = 1/kT$ and U is the Coulomb energy of the system. The summations are taken over all configurations of the ER system. The Coulomb energy of the system is given by

$$U = -\frac{1}{2}\sum(\vec{E} + \Delta\vec{E})\cdot\vec{p} \tag{4.4}$$

where $\vec{E}$ is the applied external field, $\Delta\vec{E}$ is the field produced by dipoles, and $\vec{E}_{loc} = \vec{E} + \Delta\vec{E}$. The quantity $U_d = -\frac{1}{2}\sum\Delta\vec{E}\cdot\vec{p}$ is just the dipolar interaction energy. When the applied external field is uniform, Eq.(4.3) can be written as

$$\langle \rho_j \rangle = \sum (e^{-\beta U_d} \rho_j) / \sum (e^{-\beta U_d}). \tag{4.5}$$

Consider how to calculate the dipolar interaction for the many-body system. We first note that the interaction between a dipole inside the capacitor at (x, y, z) with its infinite number of images at $(x, y, 2Lj \pm z)$ $(j = \pm 1, \pm 2, \cdots)$, is given by

$$U_1(z) = -\nu\zeta(3)/(4L^3) - \nu/(8z^3) - \nu \sum_{n=1}^{\infty} [1/(nL-z)^3 + 1/(nL+z)^3]/8, \tag{4.6}$$

where the constant $\zeta(3) = \sum_{n=1}^{\infty} 1/n^3 = 1.2020569\cdots$. U_1 can be considered as the self-energy of the dipole. Using the method which leads to Eq.(2.8), the interaction of a dipole at (x, y, z) with another dipole and its infinite number of images at $(x', y', 2Lj \pm z')$ $(j = 0, \pm 1, \pm 2, \cdots)$, $U_2(\rho, z, z')$, is given by

$$U_2(\rho, z, z') = \frac{\nu}{L^3} \sum_{s=1}^{\infty} 4\pi^2 s^2 K_0(\frac{s\pi\rho}{L}) \cos(\frac{s\pi z}{L}) \cos(\frac{s\pi z'}{L}) \tag{4.7}$$

where $\rho = \sqrt{(x-x')^2 + (y-y')^2}$. The special case in which $\rho = 0$ has the interaction

$$U_2(0, z, z') = -2\nu \sum_{j=-\infty}^{\infty} \left(\frac{1}{|z - z' - 2Lj|^3} + \frac{1}{|z + z' - 2Lj|^3} \right). \tag{4.8}$$

The total dipolar interaction energy of the system is given by

$$U_d = \sum_j U_1(z_j) + \frac{1}{2} \sum_{j \neq k} U_2(\rho_{jk}, z_j, z_k). \tag{4.9}$$

Our simulations take $N = 178$ dielectric particles inside a capacitor with spacing $L = 28a$. In the x and y directions, the system has $360a \times 360a$ and a periodic boundary condition is employed. Our simulations start at $\theta = 0$ then increase θ slowly. The state of $\theta = 0$ represents either zero temperature or extremely strong electric field. As discussed in the previous sections, at $\theta = 0$ the ER system must have the bct lattice structure.

At a fixed θ, the simulations repeat $500N = 89000$ Monte Carlo steps. In each step, we pick a particle in the initial configuration to make a random trial move, then calculate the change of dipolar interaction of the system due to the random move, δU_d. If $\delta U_d \leq 0$, we accept the new configuration. If $\delta U_d > 0$, the probability to accept the new configuration is $\exp(-\delta U_d/kT)$. When the new configuration is accepted, we calculate new ρ_j $(j = 1, 2, 3)$

for statistical average. If the new configuration is rejected, the previous configuration is retained and the previous values of ρ_j are counted again in statistical average. Then, the next random trial will be either on the newly accepted configuration or the retained previous configuration. After $500N$ Monte Carlo steps are completed at a fixed θ, we calculate the average values of ρ_j. Then, we increase θ by 0.01 and repeat the above procedure. To reduce the cpu time, we take the final configuration at θ in the simulations as the initial configuration at $\theta + 0.01$.

Our results of $|\rho_j|$ $(j = 1, 2, 3)$ are plotted in Fig.4. It is clear that $|\rho_1|$ and $|\rho_2|$ have almost identical behavior, while $|\rho_3|$ is noticeably different. At $\theta = 0$, all the three order parameters equal to unity, indicating the ER system has the ideal bct lattice structure. As θ increases, $|\rho_j|$ (j=1,2,3) remains unity for a while. Then, near $\theta = 0.16$, $|\rho_1|$ and $|\rho_2|$ drop quickly, while $|\rho_3|$ does not drop quickly until θ reaches 0.22. Afterwards, $|\rho_1|$ and $|\rho_2|$ tend to zero much faster than $|\rho_3|$. There are two important quantities, $\theta_1 = 0.264$ and $\theta_2 = 0.69$ When θ reaches θ_1, $|\rho_1|$ and $|\rho_2|$ are essentially vanishing and their fluctuation at $\theta > \theta_1$ is due to the finite size effect. However, $|\rho_3|$ does not vanish until θ reaches θ_2.

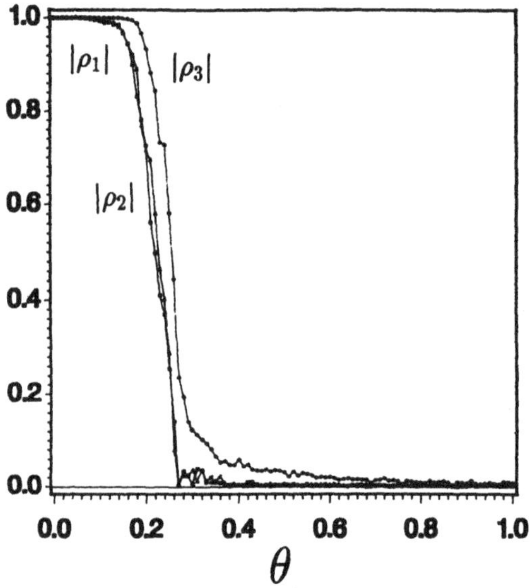

Fig.4 The change of $|\rho_1|$, $|\rho_2|$, and $|\rho_3|$ as vs θ. The curves of $|\rho_1|$ and $|\rho_2|$ are almost identical, while the curve of $|\rho_3|$ decreases much slower than them as θ increases.

The above results indicate three regions of θ. For $0 \le \theta < \theta_1$, all three $|\rho_j|$ are non-vanishing; the ER system is a solid which has the bct lattice structure. When $\theta_1 < \theta < \theta_2$, $|\rho_1| = |\rho_2| = 0$ while $|\rho_3| > 0$. Since ρ_3 characterizes the formation of chains in the z direction, in this state the ER system has ordering in the z direction, i.e. chains in the field direction, but the distribution of these chains is random, no ordering in the x and y directions. It is clear that this state is similar to nematic liquid crystal.[17] As $\theta > \theta_2$, $\rho_1 = \rho_2 = \rho_3 = 0$, the ER system is a liquid which has no ordering at all.

We can consider to induce the phase transitions in the ER fluid by the following two different approaches: (1) At a fixed temperature, increase the applied electric field; (2) at a constant applied electric field, lower the temperature of the ER system.

When the ER fluid has a fixed temperature, we have two critical electric fields which are given by

$$E_{c1} = \sqrt{\frac{8kT}{\alpha^2 \epsilon_f a^3 \theta_1}} \quad \text{and} \quad E_{c2} = \sqrt{\frac{8kT}{\alpha^2 \epsilon_f a^3 \theta_2}}. \tag{4.10}$$

When the applied electric field $E < E_{c2}$, the ER system is in a liquid state. When $E_{c2} < E < E_{c1}$, the ER system is similar to nematic liquid crystal which has ordering in the field direction but no ordering in the directions perpendicular to the field. When $E_{c1} < E$, the ER system is in the solid state of the bct lattice structure.

When the ER fluid is under a constant electric field, we have two critical temperatures,

$$T_{c1} = \frac{\alpha^2 a^3 \epsilon_f E^2 \theta_1}{8k} \quad \text{and} \quad T_{c2} = \frac{\alpha^2 a^3 \epsilon_f E^2 \theta_2}{8k}, \tag{4.11}$$

if the dielectric constants ϵ_f and ϵ_p have little change in the temperature range under the consideration. When $T > T_{c2}$, the ER system is a liquid. When $T_{c2} > T > T_{c1}$, the ER system is similar to nematic crystal state. When $T < T_{c1}$, the ER system is the solid bct lattice.

Acknowledgements

This research is supported by the Office of Naval Research grant N00014-90-J-4041, Illinois Department of Energy and Natural Resources grant SC-6, and grants from Materials Technology Center and Office of Research Development and Administration of Southern Illinois University at Carbondale.

REFERENCES

1. W. M. Winslow, J. Appl. Phys. **20**, 1137 (1949); U.S. Patent 2, 417, 850 (1947); H. Block and Kelly, US Patent 4,687,589 (1987); J.D. Carlson, US Patent no. 4,772,407 (1988); A.F. Sprecher, J.D. Carlson, and H. Conrad, Mater. Sci. Eng. **95**, 187 (1987);F. E. Filisko and W. E. Armstrong, US patent No. 4 744 914 (1988).
2. For a review also see, *Electrorheological Fluids*, edited by J. D. Carlson, A. F. Sprecher, and H. Conrad (Technomic Publishing, Lancaster, PA, 1990).
3. R. Tao, "Electric field induced solidification — theory of electro-rheology fluids", in *Condensed Matter Theories, V 6* 155-160 (Edited by S. Fantoni and S. Rosati, Plenum, 1991).
4. R. Tao, J. T. Woestman, and N. K. Jaggi, Appl. Phys. Lett.**55**, 1844 (1989).
5. R. Tao and J. M. Sun, Phys. Rev. Lett. **67**, 398 (1991).
6. T. C. Halsey and W. Toor, Phys. Rev. Lett. **65**, 2820 (1990)

7. R. Tao and J. M. Sun, Phys. Rev. A, **44**, R6181 (1991).

8. D. J. Klingenberg, F. van Swol, and C. F. Zukoski, J. Chem. Phys. **91**, 7888 (1989).

9. G. Bossis and J. F. Brady, J. Chem. Phys. **91**, 1866 (1989).

10. For example, see, P. W. Anderson, *Basic Notions of Condensed Matter Physics* (The Benjamin/Cummings Publishing Com., London, 1984), pp8-69.

11. T. J. Chen, R. N. Zitter, and R. Tao in the present volume (1992).

12. L. D. Landau and E. M. Lifshitz, *Electrodynamics of Continuous Media* (Pergamon Press, Oxford, 1983), pp36-91.

13. H Fröhlich, *Theory of Dielectrics* 2nd edition (Oxford, 1958), Chap. 2.

14. N. Metropolis, A. Rosenbluth, M. Rosenbluth, A. Teller, and E. Teller, J. Chem. Phys. **21** 1087 (1953).

15. S. Kirkpatrick, C. D. Gelatt, and M. P. Vecchi, Science **220**, 671 (1983); S. Kirkpatrick, J. of Stat. Phys. **34**, 975 (1984).

16. For example, see, *Monte Carlo Methods in Statistical Physics*, 2nd edition, ed. K. Binder (Springer-Verlag, New York, 1986).

17. For example, see, P. G. de Gennes, *The Physics of Liquid Crystals* (Clarendon Press, Oxford, 1974) pp7-56.

CRYSTALLINE STRUCTURE OF AN ELECTRORHEOLOGICAL FLUID DETERMINED BY LASER DIFFRACTION

TIAN-JIE CHEN*, R. N. ZITTER, and R. TAO
Department of Physics, Southern Illinois University
Carbondale, Illinois 62901, USA

ABSTRACT

The electric field-induced structures of small glass spheres in silicone oil produce laser diffraction patterns which verify, for the first time, the body-centered tetragonal (bct) lattice structure predicted by theory. By a mechanism totally unlike that of conventional X-ray scattering, the incident laser beam propagates through the lattice of close-packed spheres via stable optic modes along regular arrays of transparent spheres, then produces diffraction patterns after exiting from the lattice. The bct lattice is found to have three conventional vectors $\sqrt{6}R\hat{x}$, $\sqrt{6}R\hat{y}$ and $2R\hat{z}$ where R is the sphere radius and the field is along the z direction.

1. Introduction

An electrorheological (ER) fluid generally consists of a suspension of particles of high dielectric constant in a fluid of low dielectric constant.[1] Phenomenologically, when an applied electric field reaches a critical value, the fluid undergoes a transition to a gel-like state. A detailed description of this state is crucial to an understanding of ER phenomena, and considerable progress toward that end, both theoretical and experimental, has been made in recent years. As observed by Conrad *et al*, the particles first form chains along the field direction, which then coalesce into columns if the particle concentration is not too dilute.[2] Halsey and Toor discussed the thermodynamics of the process and estimated that the columns have a width $\sim R(L/R)^{2/3}$ where R is the particle radius and L is the distance between the field electrodes.[3] Tao and Sun investigated the structure within columns in more detail and predicted that for uniform spherical particles, the column would have a body-centered tetragonal (bct) lattice structure.[4]

The prediction of this structure as the many-body ground state of an ER fluid has recently been borne out by Monte Carlo simulation, which also shows that the lattice can be realized at room temperature.[5] There has, however, been no experimental confirmation, and since the electric field-induced solidification of an ER fluid is a relatively fast phase transition, it is not certain that a regular lattice would form in the process. In ordinary observations by optical microscope, the structure within the columns was not apparent.[2] Furthermore, most existing theoretical models and calculations of ER fluid properties assume particles in separated chains rather than in columns. Experimental evidence of the existence of lattice structure would be highly significant for future research and applications.

Here we report an investigation of column structure by laser beam diffraction. The results are entirely consistent with the bct lattice predicted by Tao and Sun. The diffraction

---

⋆ **On leave from Physics Department, Peking University, Beijing, China, 100871.**

mechanism, however, is totally unlike that of conventional X-ray scattering by crystals. For X-rays, a crystal is essentially transparent, with a 3-dimensional array of small scattering centers (nuclei and electrons) that produce far-field interference patterns. By contrast, in our experiments which use glass spheres much larger than the laser wavelength, a close-packed structure of spheres has few clear paths for light to travel. Here we describe a new process by which a close-packed lattice of transparent spheres produces diffraction patterns.

2. New Mechanism to Produce Diffraction Patterns

To understand this mechanism, it is important first to consider the proposed lattice in some detail. A unit cell of the body-centered tetragonal (bct) structure is illustrated in Fig. 1(a). The three conventional lattice vectors here are $\sqrt{6}R\hat{x}$, $\sqrt{6}R\hat{y}$ and $2R\hat{z}$ where the field is along the z direction. This close-packed lattice is "optically dense" to visible light of wavelength close to 1/10 of sphere diameter, in the sense that there are few if any paths for light to travel without encountering a number of spheres. In the [110] direction, for example, there are no direct paths, while in the [100] or equivalent [010] direction, the only clear paths between spheres have a cross-section size of the order of, or smaller than the optical wavelength. Accordingly, while one might expect mainly diffuse scattering of a laser beam due to multiple reflections and refractions by spheres, a mechanism for producing regular diffraction patterns is not immediately obvious.

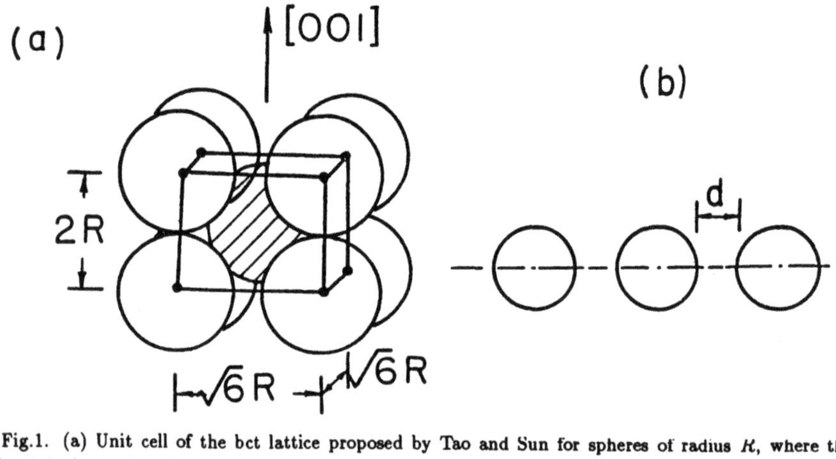

Fig.1. (a) Unit cell of the bct lattice proposed by Tao and Sun for spheres of radius R, where the [001] axis is the applied field direction. (b) A linear periodic array of glass spheres of radius R separated by distance d.

We propose the following process. Since each transparent sphere acts as a "thick" lens, a linear periodically spaced set of spheres, as shown in Fig. 1(b), constitutes a periodic array of focusing elements. An analysis of such arrays can be found in many texts, where it is shown that light rays can propagate stably and indefinitely along a properly spaced set of such elements.[6] In other words, there are stable ray modes that do not diverge from the axis or leave the array. The analysis is often employed in laser design, where an infinite array is used to represent the end-reflectors of an optical cavity.[6]

In the matrix formalism of geometric optics, a ray at a point P is described by a two-component vector (x,y) where x is the distance of P from the optic axis and y is the sine of the angle between the ray and the axis. At another point P' the same ray is represented by a vector (x',y') which is related to (x,y) by a linear transformation, or a 2×2 matrix. The mathematics is considerably simplified for "paraxial" rays that are close to and have only small angles with the axis. Reference tables can be found in the texts for basic matrices representing, e.g.,reflection and refraction at surfaces. Thus, for the array of spheres shown in Fig. 1(b) the matrix describing the transformation of a ray over one spatial period from the front surface of one sphere to the front surface of the next is given by

$$M = \begin{pmatrix} 1 & d \\ 0 & 1 \end{pmatrix} \begin{pmatrix} 1 & 0 \\ (1-n)/R & n \end{pmatrix} \begin{pmatrix} 1 & 2R \\ 0 & 1 \end{pmatrix} \begin{pmatrix} 1 & 0 \\ (1-n)/nR & 1/n \end{pmatrix}$$

$$= \begin{pmatrix} [(2-n)/n] + [2d(1-n)/nR] & [2R/n] + [d(2-n)/n] \\ 2(1-n)/nR & (2-n)/n \end{pmatrix} \tag{1}$$

where R and n are the radius and refractive index of the spheres, respectively. The matrices in the product, reading from right to left, represent respectively (i) refraction entering a sphere, (ii) traverse through a sphere, (iii) refraction leaving the sphere, and (iv) traverse of the space d to the next sphere. Passage along an array of N spheres is then given by M^N, and for N approaching infinity, analysis shows that stable, non-diverging rays propagate when the trace of M does not exceed 2.[6] Therefore, from Eq.(1) we obtain the stability condition

$$d < 2R/(n-1). \tag{2}$$

Our glass spheres have n $= 1.51$ and thus we require a spacing between spheres of $d < 3.9R$ for stable mode propagation. For the close-packed bct lattice shown in Fig. 1(a), all directions [m, n, l] with low indices (m, n, l= 0, ± 1) satisfy the above condition and, accordingly, stable paraxial ray modes may propagate along these directions. More extensive calculations for non-paraxial rays show that for the bct lattice, stable modes exist for rays that are an appreciable distance from the axis, up to about 3/4 of the sphere radius.

Thus, a beam of stable modes can travel along each linear array and emerge from the final sphere to a quasi-focus beyond the last lattice plane. The focal waist size here can be estimated as only a few micrometers in diameter, since it is small compared to sphere radius but is at least as large as the optical wavelength. The set of such focal "points" is essentially a 2-dimensional array (not 3-dimensional as in X-rays) that produces a diffraction pattern in the far field. Since far-field diffraction is just the spatial Fourier transform of the diffraction source, the result here has the form of the reciprocal lattice of the exit plane. More specifically, using a 2-dimensional version of the 3-dimensional analysis given in X-ray texts, diffraction maxima occur when

$$\vec{k} \cdot (l\vec{a} + m\vec{b}) = 2N\pi \quad (3)$$

where $\vec{k}$ is a diffracted wave vector, $\vec{a}$ and $\vec{b}$ are the lattice vectors in the exit plane, and l, m and N are integers. The solutions of $\vec{k}$ which satisfy Eq. (3) are the reciprocal lattice vectors of the exit plane.

In short, the optical output of the lattice of glass spheres is a 2-dimensional set of bright focal spots that produce a far-field diffraction pattern with the form of the reciprocal lattice of the exit plane.

3. Experimental Results

Details of the experiments are as follows. Our ER fluid consists of a low viscosity silicone oil containing size-selected glass micro-spheres of highly uniform diameter, either 10.0 ± 1.8 μm or 40.7 ± 1.7 μm (SPI Supplies, Westchester, PA, USA). The glass spheres were pre-treated to develop an absorbed H_2O film which greatly enhances their effective dielectric constant.[2,7] Only a small drop of the liquid suspension of silicone oil and glass spheres is needed to fill the space between brass electrodes in 3 mm apart on a microscope slide. An applied field, $\sim$ 1 kV/mm, induces the formation of columns averaging 0.6 mm in width. We note that this width is close to the theoretical estimation by Halsey and Toor mentioned earlier.[3] The average separation between columns is about 2 mm, and the column thickness perpendicular to the slide is about 0.15 mm. The oil eventually leaks out of the cell and if the field is maintained for several hours, the structures become "locked" in place and remain after removal of the field.

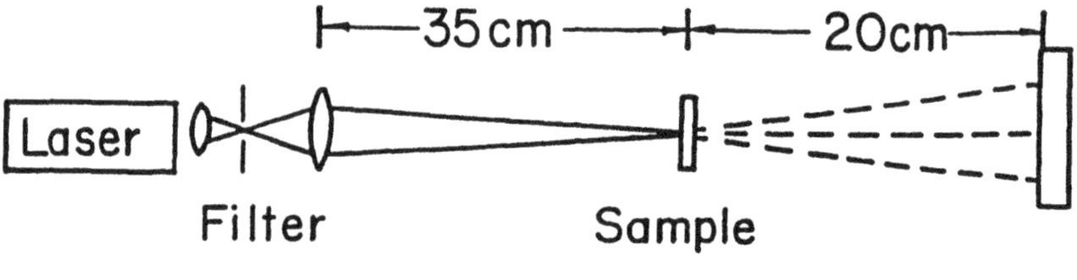

Fig.2 Layout of the diffraction experiment. The laser beam is perpendicular to the columnar direction in the sample.

The experimental optical diffraction arrangement is shown in Fig. 2. A 5 mw He-Ne laser beam is passed through a spatial pinhole filter to obtain a clean Gaussian intensity profile and then is focused at the sample. The focused beam has a Gaussian diameter about 0.1 mm or 1/6 of the average column width. The sample is on a micrometer adjustable mount for translation and angular rotation. We note a fundamental difference between our experiment and conventional X-ray diffraction by a crystal. In the latter case, the Bragg condition requires a specific orientation of the incident beam relative to crystal planes, and all orientations must be used to obtain a complete pattern. By contrast, our experiment is insensitive to the incident beam direction and an entire planar pattern for the lattice exit plane is obtained for a single orientation of the sample relative to the incident beam. This is because each sphere in the first layer encountered by the beam produces refracted rays in various directions, from which propagating array modes are selected by deeper-lying layers.

Fig. 3(a) is a microscopic view of typical columns. Fig. 3(b) shows a diffraction pattern produced by a sample with 20.0 μm diameter spheres. The geometry of this pattern is precisely the reciprocal lattice of the (110) plane in the bct structure proposed by Tao and Sun and the structure constants derived from the pattern are precisely those expected for this plane.[4] This implies that the transmission axis of stable ray modes is along axis [110]. Fig. 3(c) shows a pattern from a sample with 40.7 μm diameter spheres. Here the pattern is also that of a reciprocal lattice of the (110) plane, but twice as dense as in Fig. 3(b) because with spheres twice as large the reciprocal spacing is smaller by 1/2. In Fig. 3(d), the sphere diameter is 40.7 μm and the pattern is obviously more rectangular than the preceding cases; it is in fact, precisely the pattern expected for a (100) plane. In this case, the transmission axis of stable ray modes is [100].

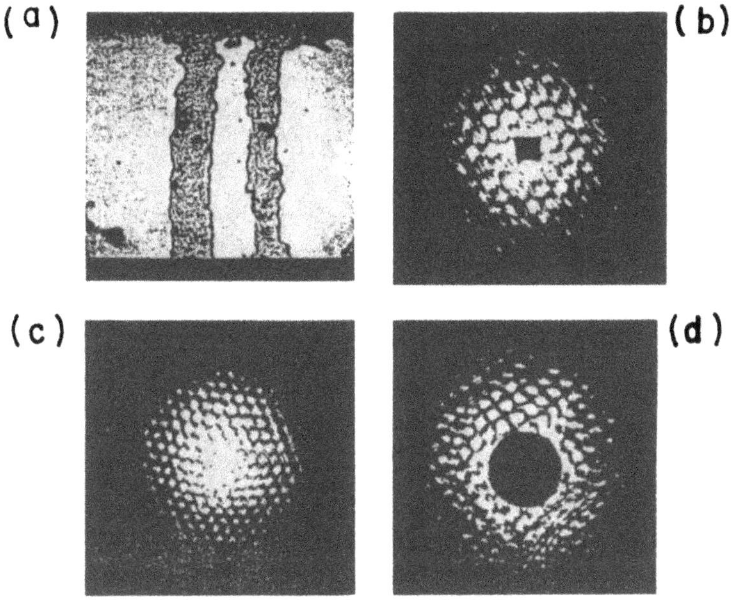

Fig.3. (a) View of typical columns. (b) Diffraction pattern of a (110) plane for 20.0 μm diameter spheres. (c) Pattern of a (110) plane for 40.7 μm diameter spheres. (d) Pattern of a (100) plane for 40.7 μm diameter spheres. In (b) and (d), the centers are masked to suppress overexposure.

Table I lists experimental and theoretical values of the structure constants. In all cases, there is agreement to within a few percent. As mentioned earlier, the diameter of our spheres are known to be non-uniform to this degree. Accordingly, the data are entirely consistent with the proposed bct lattice.

Table I. Structure constants

Lattice Plane	Sphere Diameter	Experiment	Theory
(110)	$20.0\mu m$	$a = 34.1\mu m$	$34.6\mu m$
		$b = 21.1\mu m$	$20.0\mu m$
(110)	$40.7\mu m$	$a = 69.1\mu m$	$70.5\mu m$
		$b = 38.9\mu m$	$40.7\mu m$
(100)	$40.7\mu m$	$a = 54.8\mu m$	$49.8\mu m$
		$b = 43.8\mu m$	$40.7\mu m$

With the small size of our laser beam focus, we can probe the structure at different points of a column, and the (100) or (110) patterns noted above are observed at different locations along a given column. Thus the overall column structure can be described as "polycrystalline", with the provision that the [001] applied-field axis is maintained.

Our sample geometry, in which columns are formed between brass electrodes on a glass slide, is well suited for microscopic observation of the columns as in Fig. 3(a) and for diffraction experiments with the laser beam perpendicular to the applied field direction (axis [001]). Clearly, the arrangement is not suitable for obtaining a (001) pattern, which would require a beam along the [001] direction, tangential to the slide and penetrating the electrodes.

Acknowledgements

This work is supported by Office of Naval Research Grant N00014-90-J-4041. We wish to thank Dr. Xuesong Zhang for valuable assistance.

References

1. For a recent overview see, *Electrorheological Fluids*, edited by J. D. Carlson, A. F. Sprecher and H. Conrad (Technomic Publishing, Lancaster, PA, 1990).
2. H. Conrad, M. Fisher and A. F. Sprecher, in Ref.1, p.63.
3. T. C. Halsey and W.Toor, Phys. Rev. Lett. **65**, 2820 (1990).
4. R. Tao and J. M. Sun, Phys. Rev. Lett. **67**, 398 (1991).
5. R. Tao and J. M. Sun, Phys. Rev. A, **44**, R6181 (1991).
6. See, for example, A. E. Siegman, *Lasers* (University Science Books, Mill Valley CA, 1986), Chap.15.
7. L. Hollan, *The Properties of Glass Surfaces* (John Wiley, New York, 1964).

OPTICAL PROBES OF STRUCTURE FORMATION IN ELECTRORHEOLOGICAL FLUIDS

J. M. GINDER and L. D. ELIE
Research Staff, Ford Motor Company
P.O. Box 2053, M.D. 3028
20,000 Rotunda Drive
Dearborn, MI 48121-2053 (U.S.A.)

ABSTRACT

The strong light scattering from the particles suspended in electrorheological (ER) fluids prohibits most optical studies of bulk samples. We have taken advantage of this strong scattering by measuring the diffuse transmittance of a commercial ER fluid. The sample-length dependence of the fluid transmittance is consistent with the predictions of photon diffusion theory for weakly absorbing random media. Accompanying the electric-field-induced structural change in ER fluids is an increase in the diffuse transmittance, providing a useful noninvasive probe of the time scale for structure formation in these materials.

1. Introduction

Electrorheological (ER) fluids, which generally consist of polarizable particles suspended in nonpolar fluids, have recently stimulated considerable interest within the scientific and engineering communities[1,2]. Open questions include the limitations of the polarization model for the electric-field-induced interparticle forces[3,4,5,6], as well as the nature of the field-induced particle chain structure[7,8]. Much effort is being devoted to the synthesis of new ER fluids possessing desirable rheological, electrical, chemical, and tribological properties. Receiving particular attention are those materials that exhibit large yield stresses, a characteristic of the field-induced ER solid that is particularly important from the device design perspective[9]. Another figure of merit, the response time of the field-induced structure change and concomitant enhancement of the rheological properties[10], has received less attention. Successful utilization of ER fluids in high-bandwidth active devices, however, will require detailed knowledge of the temporal response of ER fluids to changes in the applied electric field. It is thus important to understand the mechanism by which the ER structure is formed, the nature of the structural kinetics, and the role of structure in determining the mechanical properties of the ER solid.

The mismatch between the refractive indices of the suspended particles and suspending liquids results in high light scattering efficiency and the consequent diffuse white appearance of most ER fluids. Previous studies of the field-induced structure in ER fluids and their magnetorheological fluid analogs have thus been limited to optical microscopy or light-scattering probes of two-dimensional fluid layers, or dichroism, birefringence, and turbidity studies of nearly transparent three-dimensional model fluids.

Dipole-dipole aggregation of particles into chains and columns (also termed 'fibrilation' or 'fibrination') was first observed in optical microscopy of thin ferrofluid layers[11]. Electric-field-induced aggregation has been studied extensively in dilute suspensions of poly(styrene) spheres in water, so-called 'dielectric hole' systems. The low-field aggregation kinetics in thin layers of such suspensions was determined from measurements of the time-dependent distribution of chain lengths[12]. Classical light scattering from these suspensions was used to measure the field dependence of the interparticle separation for particles in chains[13], as well as to infer the diffusion coefficient describing particle dissociation upon removing the aggregating field[14]. Structure formation in nonpolar fluids more representative of typical ER materials has also been studied by optical microscopy. The interparticle potential between sterically stabilized poly(methyl methacrylate) spheres suspended in hexane was inferred by observing the onset of particle aggregation into pairs at low applied electric fields[15]. The time required to form a field-induced percolating chain has been measured in suspensions of silica in silicone oil[10]; this characteristic time was found to depend exponentially on the volume fraction of particles. A stereological analysis has been applied to the rather more complicated steady-state structures produced by high electric fields in dense suspensions[16].

Optical studies of field-induced structure formation in bulk fluids have previously been confined to those materials exhibiting weak light scattering, usually requiring the use of model systems in which particle and solvent have nearly identical refractive indices. The electric-field-induced birefringence and dichroism of concentrated suspensions of sterically-stabilized silica particles in cyclohexane was studied[17], allowing the measurement of the response time and the limiting structural anisotropy. The field-induced change in turbidity of suspensions of various anisotropic materials has also been reported[18,19]; it is not clear whether these measurements correspond to the weak-scattering regime. Two characteristic response times were observed[19]; the shorter time scale was assigned to the alignment of individual anisotropic particles along the field direction, while the longer time scale was assigned to that required for structure formation.

Understanding the optical properties of real ER fluids, or any colloidal suspension, requires knowledge of the scattering properties of the suspended particles[20]. In poorly index-matched suspensions -- i.e., most ER fluids -- photons entering the sample are scattered more than once[21]. This complicated scattering problem is simplified in the limit of many scattering events, in which case the photons can be assumed to travel diffusively through the sample[22,23]. From transport theory, the photon diffusion coefficient D is given by $D = c\ell^*/3$, where c is the velocity of light in the medium and ℓ^* is the transport (or momentum transfer) mean free path for photons. The diffuse transmittance is then related to the photon number density inside the sample, which is obtained by solving the diffusion equation for photons[20-23]. The diffuse transmittance of a slab of thickness L illuminated by a collimated beam is approximately[24]

$$T \approx \frac{5\ell^*}{3\ell_{abs}\sinh(L/\ell_{abs})} \tag{1}$$

where $\ell_{abs} = [\ell^*/(3\alpha)]^{1/2}$ is the effective absorption length for photons diffusing in a medium of absorption coefficient α, and where corrections for the nonzero internal reflectance of the sample boundaries[25] are omitted. For sample thicknesses such that $\ell_{abs} >> L > \ell^*$, the sample-length dependence of the diffuse transmittance simplifies to

$$T \approx \frac{5\ell^*}{3L} \tag{2}$$

while for $L \gtrsim \ell_{abs}$, the diffuse transmittance is given by

$$T \approx \frac{10\ell^*}{3\ell_{abs}} e^{-L/\ell_{abs}}. \tag{3}$$

The utility of diffuse transmittance lies in that, while direct structural information is lost due to the angular averaging that occurs in multiple scatterings, information about the size and spatial arrangement of the scatterers is contained in the measured transport mean free path ℓ^*[26,27]. The mean free path for a collection of isolated scatterers, ℓ_0^*, is calculated by taking an average over scattering solid angle of $1 - \cos\theta$ (reflecting that scattering at small angles θ barely effects the photon momentum) weighted by the particle scattering form factor F(q) obtained from Mie scattering theory[20], i.e.

$$\frac{1}{\ell_0^*}=n\sigma_m=n\int_0^{2k_i}F(q)(1-\cos\theta)q\,dq \tag{4}$$

where n is the number density of scatterers, σ_m is the momentum transfer cross-section, $|\mathbf{q}|=(2\pi/\lambda)\sin(\theta/2)$ is the magnitude of the scattering wavevector $\mathbf{q}$, λ is the wavelength of light in the medium, and $k_i = 2\pi/\lambda$ is the magnitude of the incident wavevector in the medium. When interparticle spatial correlations are not negligible -- i.e., in concentrated or interacting suspensions -- the modified scattering length ℓ^* is given by[27]:

$$\frac{\ell_0^*}{\ell^*}=\frac{\int_0^{2k_i}F(q)S(q)q^3dq}{\int_0^{2k_i}F(q)q^3dq} \tag{5}$$

where the structure factor S(q) of a collection of N particles is given by

$$S(q)=\frac{1}{N}\sum_i^N\sum_j^N e^{i\mathbf{q}\cdot(\mathbf{r}_i-\mathbf{r}_j)} \tag{6}$$

Diffuse optical transmittance can be applied to materials that are not amenable to optical microscopy or classical light-scattering probes, yielding information on the properties of the individual scatterers, the spatial distribution of scatterers, or both. Recent applications include the use of diffuse transmittance to monitor the kinetics of bubble growth in soap foams[28], as well as the observation of a correlation between the optical transmittance and the fluid permeability of porous sandstones[29].

We have measured the diffuse optical transmittance of a commercially-available ER fluid. The sample-length dependence of the transmittance is in accord with the predictions of photon diffusion theory. Time-resolved diffuse transmittance was used to study the time scale for structural changes induced by the application of large dc and ac electric fields. The results of these studies demonstrate that diffuse transmittance is a useful noninvasive probe of structure formation in ER fluids.

2. Experimental

The ER fluid utilized in these studies was obtained commercially[30]; the solid phase consisted of spheroidal particles of a lithium salt of poly(methacrylic acid) containing 12.5% by weight water. The average particle diameter was estimated by optical microscopy to be $2a \sim 0.7$ μm. The particles were suspended at volume fraction $\phi=0.35$ in a chlorinated hydrocarbon oil. The suspension was reported[30] to have a field-off viscosity $\nu = 0.35$ Pa-s at a temperature T=303 K. The suspending fluid was isolated by centrifuging the ER fluid mixture, allowing a determination of the base fluid refractive index n = 1.495 and absorption coefficient $\alpha = 0.11$ cm^{-1} at T=295 K and a wavelength $\lambda_0=0.488$ μm. Room-temperature dielectric permittivity measurements were performed with a Schlumberger impedance bridge. The measured permittivity of the ER fluid was strongly frequency-dependent[31]; its static conductivity $\sigma(0) \approx 4.1 \times 10^{-9}$ Ω^{-1}m^{-1}, while its 1 kHz dielectric constant ϵ(1 kHz) $\approx$ 9.1. For the base fluid, $\sigma_f(0) \approx 4.6 \times 10^{-9}$ Ω^{-1}m^{-1}, while ϵ(1 kHz) $\approx$ 8.1.

The apparatus used for the diffuse transmittance measurements is shown schematically in Fig. 1. An argon-ion laser ($\lambda_0=0.488$ μm, horizontally polarized) is beam-expanded by the lens pair L1, L2 to a $1/e^2$ diameter of 4 cm. An adjustable aperture A1 defines a collimated beam of diameter ~ 1.5 cm that illuminates the sample. The ER fluid is contained between two parallel semitransparent conducting glass electrodes that define the entrance and exit faces of the sample; the cell is sealed by a rubber membrane. The front electrode is mounted on an x-y translation stage to allow the application of shear and longitudinal stresses. An aperture A2 of diameter 0.15 cm limits the transmitted light detected by the photodiode PD (UDT PIN10, 1 cm diameter active area). High voltage is applied to the sample electrodes via a high-voltage amplifier (Trek 10/10) driven by a signal generator (Wavetek 75). The incident intensity, transmitted intensity, sample voltage and current are recorded with an 8-channel, 100 kHz analog-to-digital converter controlled by a personal computer.

The diffuse transmittance T was calculated as the ratio of the intensity transmitted by the sample to the intensity transmitted by the empty cell; corrections were made for specular reflection from the inner faces of the empty cell as well as for the dependence of the detector efficiency on the angular distribution of the diffusely transmitted light. Electric-field-induced changes in the diffuse transmittance $T_\parallel$ for light propagation parallel to the applied electric field E_0 were normalized to the zero-field transmittance by calculating $\Delta T_\parallel/T(0) \equiv (T_\parallel(t)-T(0))/T(0)$, a measure of the transmittance change that is independent of the throughput and angular res-

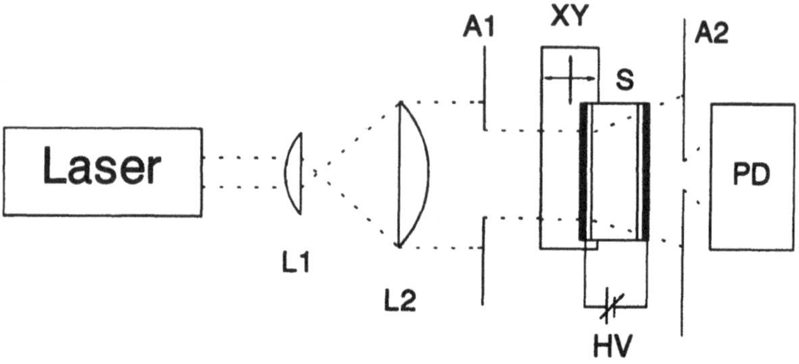

Figure 1. Schematic of the diffuse transmittance experiment. Laser, Argon ion laser; L1, L2, lenses; A1, A2, apertures; S, sample and transparent electrodes; XY, x-y translation stage; HV, high voltage amplifier; PD, silicon photodiode detector.

ponse of the apparatus so long as the angular distribution of the transmitted light is unchanged.

3. Results and Discussion

The length dependence of the diffuse transmittance of the ER fluid in zero applied electric field is shown in Fig. 2. For sample lengths $L \gtrsim 0.7$ mm, the diffuse transmittance decreases exponentially with length, yielding a decay length $\ell_{abs} = 2.0$ mm. For $0.45 \text{ mm} \lesssim L \lesssim 0.7$ mm, the transmittance varies inversely with L, consistent with the predictions of photon diffusion theory. Also supporting the role of multiple scattering is the observation that the transmitted light is completely depolarized.

The magnitude of the zero-field transmittance in the length regime in which $T \propto 1/L$ implies a transport mean free path $\ell^* \approx 160$ μm. Using the Mie scattering form factor[20] and the Percus-Yevick approximation for the hard-sphere structure factor[32], a particle refractive index $n_p \sim 1.44$ is estimated. An absorption coefficient $\alpha \approx 0.13$ cm^{-1} is obtained from the measured transport and absorption lengths, consistent with that measured for the suspending fluid. We conclude that the length dependence of the ER fluid transmittance is entirely consistent with that expected for absorbing, multiply-scattering media.

The diffuse transmittance changes $\Delta T_\parallel/T(0)$ of an ER fluid sample of path

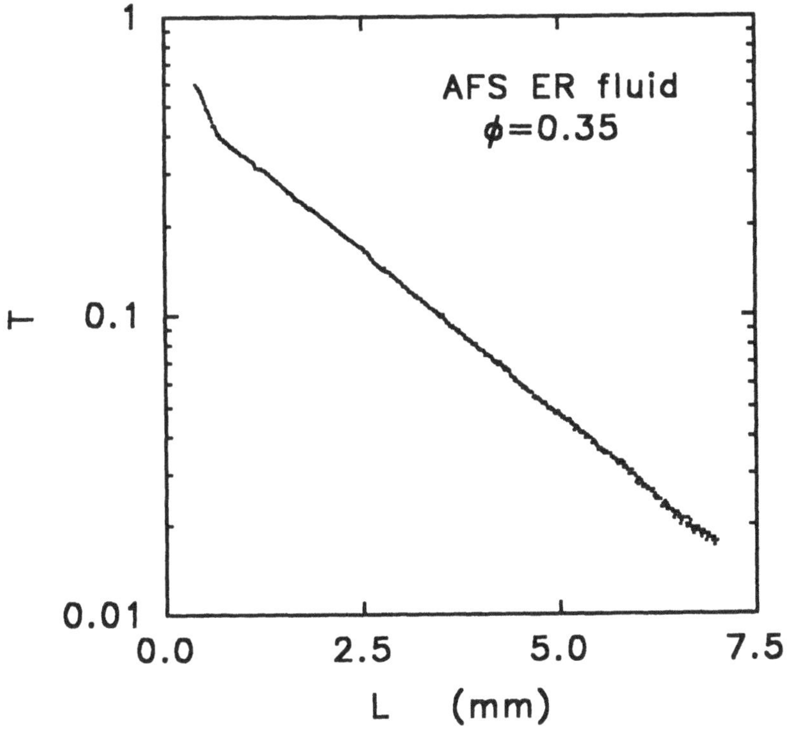

Figure 2. Sample-length dependence of the ER fluid diffuse transmittance, measured at T = 296 K with light of wavelength λ_0 = 488 nm.

length L = 2.00 mm induced by a dc electric field of magnitude E_0 = 1.02 kV/mm and by an ac sine wave field of frequency 300 Hz and rms amplitude 1.06 kV/mm are shown in Fig. 3. The transmittance is observed to increase upon the application of the field. The dc-field-induced $\Delta T_f/T(0)$ grows rapidly, reaching a steady-state value of order 0.1. By contrast, the ac-field-induced $\Delta T_f/T(0)$ is well-fit by a sum of fast and slow exponential growth processes[31], suggestive of the initial formation of chains and their eventual ripening into a dense columnar phase[7]. The dc-field-induced growth kinetics were characterized by the time required to reach 1- 1/e (63.2%) of the steady-state transmittance change, while the ac kinetics were characterized by the fast exponential time constant; the field dependence of these characteristic times τ

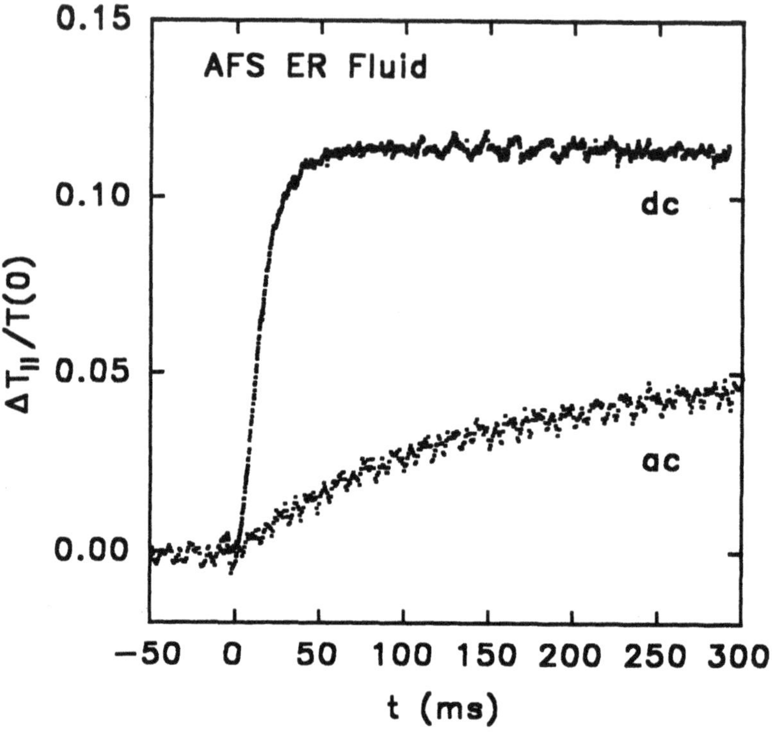

Figure 3. Time-resolved diffuse transmittance change induced by a 1.02 kV/mm static field ('dc'), and by a 1.06 kv/mm rms ac (sine wave) field at 300 Hz ('ac').

is shown in Fig. 4. The steady-state transmittance change does not decay appreciably for tens of minutes after the field is removed; however, the initial transmittance is recovered by mechanically straining the sample, apparently destroying the residual field-induced structure.

The observed field-induced transmittance increase agrees qualitatively with the intuitively expected 'shadowing' of particles in chains, and can be quantitatively related to the change in the transport mean free path $\Delta\ell_{\parallel}^{*}$. Differentiating the dif-

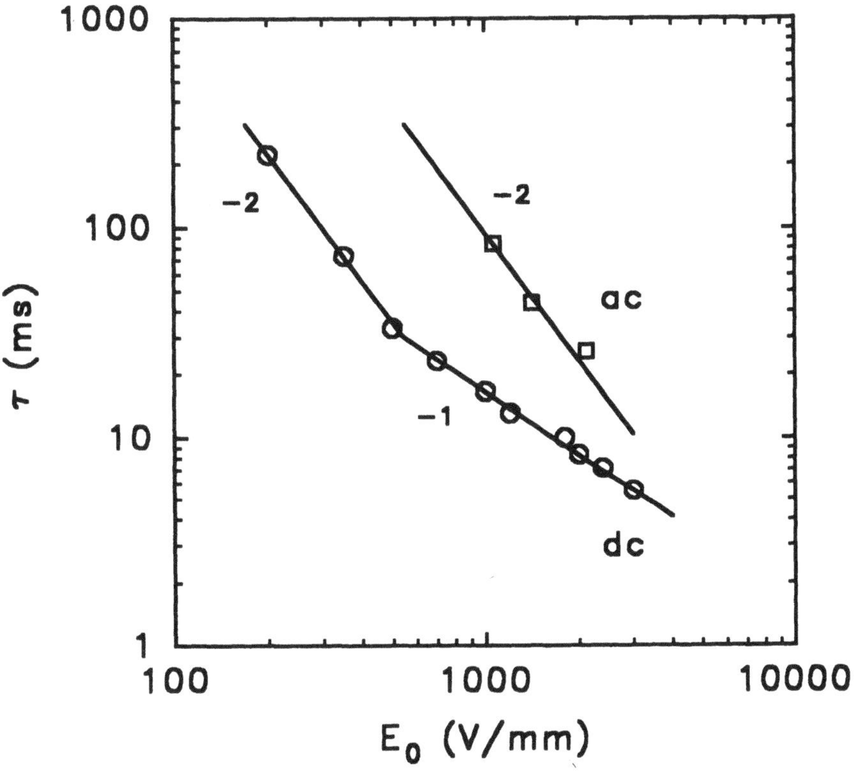

Figure 4. The dependence of the characteristic structure formation time on rms electric field, obtained from the time-resolved diffuse transmittance measurements. The power-law exponents describing the field dependences are indicated.

fuse transmittance in the absorptive regime (Eq. 3) with respect to ℓ^* yields

$$\frac{\Delta T_{\parallel}(t)}{T(0)} \approx \frac{1}{2}[1+\frac{L}{\ell_{abs}}]\frac{\Delta \ell_{\parallel}^*(t)}{\ell^*(0)}. \qquad (7)$$

Since L ≈ ℓ_{abs} for this sample, the observed steady-state transmittance increases imply increases in the transport mean free path of order 10%. The diffuse transmittance measures the field-induced structural change through the dependence of ℓ^* on the structure factor. Hayter and Pynn[33] have shown that the formation of particle chains parallel to the applied field is reflected in the appearance and growth of anisotropy in the structure factor S(q); in particular, S(q) is reduced at small $|q|$ for $q \perp E_0$. The expected increase in the calculated transport mean free path (see Eq. 5) is completely consistent with the observed $\Delta T_{\parallel}/T(0) > 0$. A reduction in the diffuse transmittance for directions perpendicular to E_0 is also predicted.

The characteristic time for structure formation inferred from the diffuse transmittance measurements (Fig. 4) can be compared with the theoretical pair-formation or flocculation time τ_f predicted for dilute suspensions of particles experiencing induced dipole and viscous drag forces,[34]

$$\tau_f = \frac{2\nu}{5\epsilon_0\epsilon_f\beta^2 E_0^2}\left(\frac{\pi}{6\phi}\right)^{5/3}, \tag{8}$$

where ν is the solvent viscosity, ϵ_0 is the permittivity of free space, ϵ_f is the real part of the solvent dielectric constant, and β is the relative polarizability[3] of the suspended particles. While the interparticle interactions in dense suspensions are complicated by many-body hydrodynamics and electrostatics, τ_f provides a reasonable first approximation for the aggregation time in concentrated ER fluids. As shown in Fig. 4, both the ac-field-induced and low-field dc-induced characteristic times exhibit an inverse-square dependence on the electric field, consistent with the role of induced dipole interactions. A relative polarizability $\beta^2(300\text{ Hz}) \approx 0.040$ is estimated from the measured ac characteristic times and the expression for the theoretical flocculation time; many-particle hydrodynamic effects have been approximated by substituting the zero-field ER suspension viscosity for the suspending fluid viscosity ν. This value of the relative polarizability can be compared to that inferred directly from the measured[29] complex permittivities of the suspending fluid, $\epsilon_f(300\text{ Hz}) \approx 8.1 + 0.9i$, and the ER fluid suspension, $\epsilon(300\text{ Hz}) \approx 9.4 + 2.2i$. Applying the Maxwell relation[35] $\epsilon \approx \epsilon_f(1 + 2\beta\phi)/(1 - \beta\phi)$, the squared magnitude of the complex relative polarizability $|\beta(300\text{ Hz})|^2 \approx 0.041$ is obtained, in excellent agreement with that estimated from the time-resolved diffuse transmittance measurements. The low-field dc characteristic times are consistent with a relative polarizability $\beta^2(0) \approx 0.43$; this value cannot be compared directly with a permittivity-based polarizability due to uncertainties in the zero-frequency permittivities.

The *linear* dependence of τ on E_0^{-1} observed for high dc fields (Fig. 4) contradicts the quadratic dependence expected for induced-dipole attractive forces. A possible origin of this linear dependence is that the suspended particles possess a nonzero surface charge, and thus they are subject to strong electrophoretic forces[36] in dc or low-frequency ac electric fields. The mean electrophoretic mobility $<\mu>$ of particles similar to those utilized in this study has previously been measured[37]; in dilute suspensions, $<\mu> = 6.8 \times 10^{-7}$ cm^{2}V^{-1}s^{-1}. Wholesale electrophoretic transport of particles toward one electrode cannot be responsible for the observed transmittance increase, however, since the estimated particle sweepout time $\tau_s \sim L/(2<\mu>E_0)$ $\sim$ 15 s for L = 0.2 cm and E_0 = 1 kV/mm, three orders of magnitude larger than the observed time scale for the transmittance change. On the other hand, a *distribution* of particle mobilities[38] would give rise to time-dependent changes in the relative spacings between particles when the field is applied; as the particles overtake one another, the structure factor -- and thus the diffuse transmittance -- would be altered. Assuming that the measured τ reflects the mobility variance required to traverse a typical interparticle separation $\Delta R = [(4\pi/3\phi)^{1/3} - 2]a = <\Delta\mu^2>^{1/2}E_0\tau$, a mobility variance $<\Delta\mu^2>^{1/2} \approx 0.4 \times 10^{-7}$ cm^{2}V^{-1}s^{-1} is inferred, a value not inconsistent with previous mobility measurements[37]. While these results suggest that electrophoretic effects cannot be ignored, other interparticle forces are also important: the ER fluid exhibits a > 1 kPa yield stress with dc applied fields, and the dc-field-induced $\Delta T_\parallel/T(0)$ persists after the field is removed. A better understanding of the effects of surface and suspension charge on the interparticle interactions, aggregation processes, and rheological and dielectric properties of ER fluids is clearly needed.

4. Conclusions

The diffuse transmittance of a commercial ER fluid has been studied. The length dependence of the transmittance is consistent with that expected for absorbing, multiply-scattering materials. Large dc or ac electric fields induce particle chaining in these materials, and a consequent increase in the diffuse transmittance measured parallel to the applied field. The temporal response of the diffuse transmittance provides a measure of the characteristic times for structure formation; the field dependence of the ac response time is consistent with the dipole-dipole interparticle interactions calculated from the measured complex permittivity, while the field dependence of the dc response suggests that particle electrophoresis plays a role in field-induced structure formation.

5. Acknowledgements

The authors are grateful to Jim Anderson, Jerry Chaundy, Craig Davis, Dale

Hartsock, Ken Hass, Bob Hetrick, Bob Novak, Irv Salmeen, and Mike Tamor for useful discussions and assistance.

6. References

1. *Proceedings of the First International Symposium on Electrorheological Fluids*, ed. by H. Conrad, A. F. Sprecher, and J. D. Carlson (North Carolina State University, Raleigh, NC, 1989).

2. *Proceedings of the Second International Conference on Electrorheological Fluids*, ed. by J. D. Carlson, A. F. Sprecher, and H. Conrad (Technomic, Lancaster, PA, 1990).

3. A. P. Gast and C. F. Zukoski, *Adv. Coll. Int. Sci.* **30**, (1989) 153.

4. H. Block and J. P. Kelly, *J. Phys. D: Appl. Phys.* **21** (1988) 1661.

5. T. C. Jordan and M. T. Shaw, *IEEE Trans. Electr. Insul.* **24** (1989) 849.

6. L. C. Davis, *Appl. Phys. Lett.* **60** (1992), in press.

7. T. C. Halsey and W. Toor, *Phys. Rev. Lett.* **65** (1990) 2820.

8. R. Tao and J. M. Sun, *Phys. Rev. Lett.* **67** (1991) 398.

9. D. L. Hartsock, R. F. Novak, and G. J. Chaundy, *J. Rheol.* **35** (1991) 1305.

10. J. C. Hill and T. H. Van Steenkiste, *J. Appl. Phys.* **70** (1991) 1207.

11. C. F. Hayes, *J. Coll. Int. Sci.* **52** (1975) 239; A. T. Skjeltorp, *Phys. Rev. Lett.* 51 (1983) 2306; G. Hegelson, A. T. Skjeltorp, P. M. Mors, R. Botet, and R. Jullien, *Phys. Rev. Lett.* 61 (1988) 1736.

12. S. Fraden, A. J. Hurd, and R. B. Meyer, *Phys. Rev. Lett.* **63** (1989) 2373.

13. F. Richetti, J. Prost, and N. A. Clark, in *Physics of Complex and Supramolecular Fluids*, ed. by S. A. Safran and N. A. Clark (Wiley, New York, 1987).

14. Z. Mimouni, G. Bossis, C. Mathis, A. Meunier, and C. Paparoditis, *Prog. Coll. Polym. Sci.* **81** (1990) 120.

15. P. M. Adriani and A. P. Gast, *Faraday Discuss. Chem. Soc.* **90** (1990) 1.

16. H. Conrad, M. Fisher, and A. F. Sprecher, in Ref. 2.

17. K. L. Smith and G. G. Fuller, in Ref. 1.

18. C. M. Cerda, R. T. Foister, and S. G. Mason, *J. Coll. Int. Sci.* **82** (1981) 577.

19. T. C. Jordan and M. T. Shaw, in Ref. 2.

20. See, e.g., H. C. van de Hulst, *Light Scattering by Small Particles* (Dover, New York, 1981).

21. A. Ishimaru, *Wave Propagation and Scattering in Random Media, vol. I* (Academic, New York, 1978).

22. *Scattering and Localization of Classical Waves in Random Media*, ed. by P. Sheng (World Scientific, Singapore, 1990).

23. P. M. Morse and H. Feshbach, *Methods of Theoretical Physics, parts I and II* (McGraw-Hill, New York, 1953); S. Glasstone and M. C. Edlund, *The Elements of Nuclear Reactor Theory* (Van Nostrand, Princeton, 1952).

24. See, e.g., A. Z. Genack in Ref. 22.

25. J. X. Zhu, D. J. Pine, and D. A. Weitz, *Phys. Rev. A* **44** (1991) 3948.

26. D. J. Pine, D. A. Weitz, G. Maret, P. E. Wolf, E. Herbolzheimer, and P. M. Chaikin, in Ref. 22.

27. S. Fraden and G. Maret, *Phys. Rev. Lett.* **65** (1990) 512.

28. D. J. Durian, D. A. Weitz, and D. J. Pine, *Science* **252** (1991) 686.

29. B. R. De, I. Hoffmann, D. Stroud, and M. A. Nelson, *Phys. Rev. B*, in press.

30. Advanced Fluid Systems, Ltd., London, England.

31. J. M. Ginder, *et al.*, to be published.

32. M. S. Wertheim, *Phys. Rev. Lett.* **10** (1963) 321.

33. J. B. Hayter and R. Pynn, *Phys. Rev. Lett.* **49** (1983) 1103.

34. M. Shapiro, A. L. Shalom, and I. J. Lin, *J. Appl. Phys.* **58** (1985) 1028.

35. See, e.g., G. Schwarz, *J. Phys. Chem.* **66** (1962) 2636.

36. See, e.g., W. B. Russel, D. A. Saville, and W. R. Schowalter, *Colloidal Dispersions* (Cambridge University Press, New York, 1989).

37. D. Brooks, J. Goodwin, C. Hjelm, L. Marshall, and C. Zukoski, *Coll. Surf.* **18** (1986) 293.

38. B. Robertson, *J. Chem. Phys.* **95** (1991) 3873.

THE STRUCTURE AND DYNAMICS OF ELECTRORHEOLOGICAL FLUIDS

THOMAS C. HALSEY

The James Franck Institute and Department of Physics
The University of Chicago
5640 South Ellis Avenue
Chicago, Illinois 60637

ABSTRACT

In a strong electric field, the behavior of electrorheological fluids is dominated by the dipolar interactions between the colloidal particles. When thermal effects can be ignored, the ground state of a system of dipolar particles is a body-centered tetragonal solid. However, such long-range ordered structures are rarely seen in experiment. When a high electric field is applied to a fluid, dielectric columns oriented parallel to the field rapidly form. The relaxation of these columns is a much slower process. I calculate the shape of the columns formed and the time scale for the relaxation of the columnar structure. I also discuss how this structure formation is altered in the presence of a shear flow.

1. Introduction

The other articles in this volume make clear the reasons for the strong practical, engineering interest in electrorheological (ER) fluids. But these fluids are also of great interest from a scientific point of view. The dominant force between particles in ER fluids is generally dipolar; thermal and hydrodynamic forces are also significant in some circumstances. The statistical mechanics of particles interacting via dipolar forces is quite subtle, because one cannot assume that the dominant influence over a particle is due to its local environment. With an interaction that scales with the distance r as r^{-3}, the interaction energy of a particle with its surroundings may be determined by the macroscopic shape of the sample. The dipolar force is also anisotropic, which leads to additional effects. Microscopic dipole forces tend to be rather weak; dipolar effects generally appear either at large length scales, as in the domain structure of a magnet, or at low temperatures, as in a system of magnetic spins in an insulator.[1,2] In ER fluids, the importance of the dipolar interactions arises from the large size of the particles.

In this contribution, I will examine the special role of dipolar interactions in a number of ER phenomena. In Section 2, I will discuss the ground states at high field of systems of dipoles; this section will partially duplicate a more complete discussion by R. Tao elsewhere in this volume. In Section 3, I will discuss the kinetics of structure formation in ER fluids, and in Section 4, the long-time relaxation of the structures formed. In Section 5, I will review some recent ideas concerning the shear-thinning viscosity of ER fluids. Finally, in Section 6 I will briefly discuss some outstanding theoretical problems in the field.

2. Dipolar Lattice Energies

Although in quiescent ER fluids one generally sees chains or columns of particles aligned with the field, spanning the distance between the electrodes, the true ground state of interacting dipoles in this geometry is not a dispersed phase, in contrast to analogous magnetic systems such as ferrofluids. Essentially this is because of the presence of image forces arising from the constant potential boundary condition at the electrodes; these image forces suppress the long-range dipolar repulsion in the directions perpendicular to the field axis.[3]

I will model an ER fluid as a suspension of spherical dielectric particles of radius r_d and dielectric constant ϵ_d in a non-conducting solvent with unit dielectric constant. It is certainly true that non-dipolar short-range interactions, particularly multipolar interactions, have a strong effect on the quantitative properties of an ER fluid.[4] Nevertheless, for the qualitative properties that I shall consider in this contribution, the additional short-range interactions are largely an irrelevant distraction.

The fluid is confined between conducting plates of an infinite parallel plate capacitor with plate separation L. When there is no electric field present, the particles are distributed randomly throughout the fluid. When a field is applied, the spheres acquire dipole moments $\vec{d} = \beta r_d^3 \vec{\mathcal{E}}_{loc}$, where $\vec{\mathcal{E}}_{loc}$ is the local electric field and $\beta = (\epsilon_d - 1)/(\epsilon_d + 2)$ is the polarizability of the spheres. If $\beta \ll 1$ then we can write $\vec{\mathcal{E}}_{loc} \approx \mathcal{E}\hat{z}$, where $\hat{z}$ is the normal to the plates. The interaction energy of two spheres separated by $\vec{r}$, with an angle θ between $\vec{r}$ and $\hat{z}$, is $u(r,\theta) = -(d^2/r^3)(3\cos^2\theta - 1)$. The relative strength of dipolar interactions and thermal effects can be characterized by the dimensionless parameter[5]

$$\lambda = \frac{r_d^3(\beta\mathcal{E})^2}{4k_BT} \tag{1}$$

Because of the requirement that the electrodes be held at constant potential, each dipole comes equipped with an infinite set of image dipoles, whose field must be included in computing the field of a single dipolar particle. In this way, we may easily compute the z-component of the field at a point in a chain of dipoles of dipole moment $\vec{d}$ parallel to the chain, and separation c (see Figure 1). This local field is given by[3]

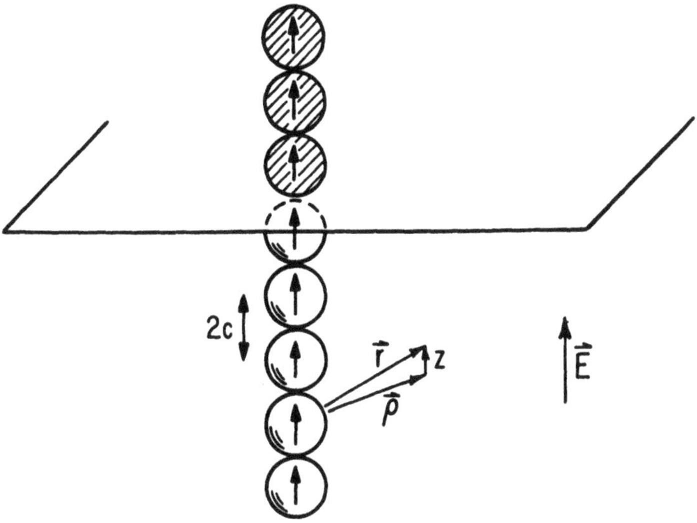

Figure 1. A chain of particles of spacing c. In calculating the field, one must include the effect of image dipoles (shaded) outside the electrodes. Polar coordinates $(\vec{\rho}, z)$ are also displayed.

$$\hat{z} \cdot \vec{\mathcal{E}} = 2d \sum_{n=1}^{\infty} \frac{2}{(nc)^3} = \frac{4d}{c^3}\zeta(3) = \frac{4.8060d}{c^3}, \tag{2}$$

where ζ is the Riemann zeta function. One can similarly compute the field at a position $(\vec{\rho}, z)$ with respect to such a chain. Here $\vec{\rho}$ is a two-dimensional vector in the plane perpendicular to the chain, and z is the direction parallel to the chain (or to the applied field). Because of the infinite extent of the "effective" chain, this field dies off exponentially as one moves away from the chain. A simple computation shows that the field in the z-direction of such a chain of dipoles is[6]

$$\hat{z} \cdot \vec{\mathcal{E}}(\vec{\rho}, z) = -\frac{2d}{c} \sum_{m=1}^{\infty} k_m^2 \cos(k_m z) K_0(k_m \rho). \tag{3}$$

where $k_m = 2\pi m/c$.

With these results it is straightforward to compute the field at the lattice sites of any Bravais lattice. For such a lattice, the dipole moments of the different particles

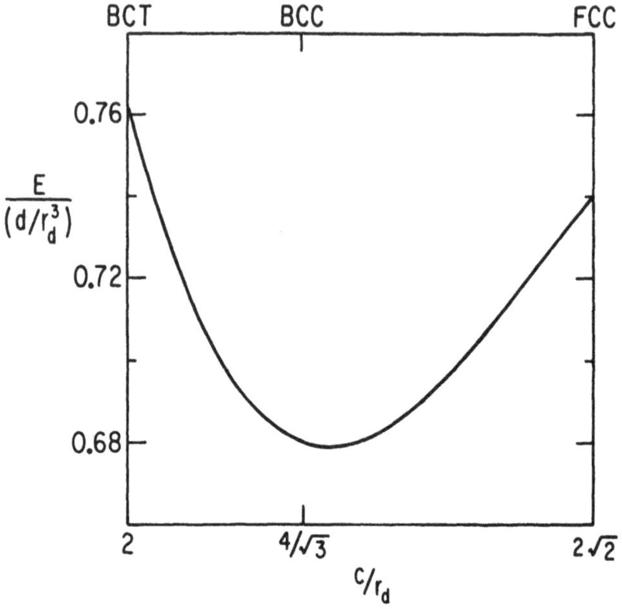

Figure 2. The local field E as a function of the vertical distance c between the particles for body-centered tetragonal lattices, given in units of d/r_d^3. The parameter c is measured in units of the particle radius r_d. The closest possible packing is chosen for each value of c. By varying c, we can go from the compressed BCT lattice of Tao and Sun to the FCC lattice via the BCC lattice.

are not a function of position, since all positions are equivalent. For Bravais lattices with cubic symmetry, such as FCC or BCC lattices, this computation is unnecessary, as the local field is given by the Clausius-Mossotti relation,[3,7]

$$\vec{\mathcal{E}}_{loc} = \frac{4\pi\vec{d}}{3v_c} \tag{4}$$

where v_c is the volume of a unit cell of the lattice. Given the local field, the electrostatic energy density is given simply by $\mathcal{F} = -(1/2v_c)\vec{d}\cdot\vec{\mathcal{E}}_{loc}$. We neglect the contribution of the applied field to $\vec{\mathcal{E}}_{loc}$, as this does not depend on the lattice structure.

For BCC and FCC lattices, Eq. (4) already gives a local field that is higher than that for a close-packed single chain, and thus a lower energy than the chain value.

This is because of the attractive interaction between nearest neighbor chains in such lattices. For the close-packed FCC lattice, $\mathcal{E}_{loc} = .7405d/r_d^3$. Tao and Sun have pointed out that one can obtain a higher local field by going to a non-close packed lattice: a body-centered tetragonal lattice that is collapsed along the z direction (parallel to $\vec{d}$).[8] Actually, the FCC lattice can also be viewed as a body-centered tetragonal lattice, but with a unit cell expanded along the z direction. By varying the value of c of a body-centered tetragonal lattice from the Tao and Sun value $c = 2r_d$ to the FCC value $c = 2\sqrt{2}r_d$ (the BCC lattice is at the intermediate value $c = 4r_d/\sqrt{3}$), one obtains the local field shown in Figure 2.[9] Note that the value of the local field for FCC lattices is within 3% of its value for the Tao and Sun structure. Thus entropic or multipolar effects could possibly change the preferred structure, although a recent experiment suggests that the Tao and Sun BCT lattice may persist in at least some physical systems.[10]

3. Column Formation

An equilibrium approach, based on an understanding of the thermodynamic ground states of dipolar systems, thus predicts that at very high fields one will see a condensation of the dipolar particles into a lattice structure, probably body-centered tetragonal. At lower fields, where thermal effects are more important, one might see solid phases of different structure, or perhaps a condensed (albeit anisotropic) liquid phase. Finally, for values of $\lambda \sim 1$, one would expect to see a gaseous phase corresponding to an ordinary suspension. Unfortunately, this picture has very little in common with what is actually seen in typical experiments.

When a strong electric field ($\lambda \gg 1$) is applied, chains and columns form over a time scale of milliseconds.[5] While there is some, mostly anecdotal, evidence that these chains or columns may densify on longer time scales, a bulk phase separation into a condensed phase is not seen. We shall now attempt to explain this.

Consider the situation immediately after an electric field has been applied to an electrorheological sample. Particles will begin to aggregate to one another, forming larger and larger condensed regions. Eventually, as these "droplets" approach the size L corresponding to the distance between the plates, they will begin to interact with their image droplets, and will be pulled into columnar structures spanning the plates.

We will derive a scaling of characteristic column width with system size by considering a simplified model of column growth.[3] The essential features of the model are:

1. The droplet is always in an equilibrium shape-i.e., the relaxation time for disturbances in the shape of the droplet, τ_r, is small compared to the time scale for aggregation onto the droplet, τ_a. We point out below that this assumption probably fails for large λ.

2. The interaction energy with other droplets is small compared to the interaction energy with the external field or to the surface energy of a droplet. Strictly speaking, this assumption is true only in the limit of vanishing dipole volume fraction ϕ. Nevertheless, we hope that even at moderate concentrations it makes a reasonable starting point.
3. The effect of the image dipoles is negligible until the length of the droplet approaches L.

Then the droplet shape is set by a competition between the field energy, which favors an elongated shape, and surface tension, which favors a spherical shape.

This surface tension is a somewhat subtle effect. We are explicitly assuming dipolar forces only, so this surface energy arises from the dipolar interactions, as does the bulk depolarization energy. It might therefore seem difficult to separate the two energies. However, we compute the bulk depolarization energy from the energy of a uniform dielectric; thus, while the dipole density per unit volume appears in this energy, the lattice constant of the dipolar lattice does not appear explicitly. The surface energy is specifically a lattice effect, and the lattice constant does appear explicitly in the surface energy.

We have computed the dipolar surface energies for a variety of lattices, and for a variety of directions within each lattice.[9] In the limit $\beta \ll 1$, the surface tension is of magnitude $\sigma = c(\hat{n})\beta^2\mathcal{E}^2 r_d/4\pi$, where $c(\hat{n}) \sim .1$ is dependent upon the normal direction $\hat{n}$ of the surface, and we assume close-packing so that the lattice constant is $\sim r_d$. The constant $c(\hat{n})$ actually becomes negative when the angle between $\hat{n}$ and $\vec{d}$ becomes small; however, the droplets we will consider below are quite elongated, and thus have relatively little surface area perpendicular to the field direction. We will therefore ignore this anisotropy, and assume that the surface contribution to the energy is $\mathcal{F}_\sigma \approx \tilde{s}(\epsilon - 1)\mathcal{E}^2 S/4\pi$, where S is the surface area of the droplet, $\tilde{s}$ is a constant, and the factor $\epsilon - 1$ has been included for convenience. ϵ is the macroscopic dielectric constant of the condensed phase.

Determining the field energy of a dielectric of arbitrary shape in a uniform external field is difficult. To simplify the calculation we assume that the droplet is a prolate spheroid, with symmetry axis along the z direction, radius $\tilde{b}$ in the directions $\perp \hat{z}$ and radius $\tilde{c}$ in the direction $\| \hat{z}$. The surface area of such a spheroid is

$$S = 2\pi\tilde{b}^2 + \pi\frac{\tilde{b}\tilde{c}}{E}\sin^{-1} E, \tag{5}$$

where $E = \sqrt{1 - \tilde{b}^2/\tilde{c}^2}$ is the eccentricity. To lowest order in the depolarization factor n_z, the field energy for a droplet of volume V is[11]

$$\mathcal{F}_{\mathcal{E}} = \frac{1}{8\pi}\Big(-(\epsilon - 1)\mathcal{E}^2 V + (\epsilon - 1)^2 \mathcal{E}^2 V n_z\Big). \tag{6}$$

The first term is independent of shape, and depends only on the total volume, which we shall take as a parameter. The depolarization factor, n_z, is[11]

$$n_z = \frac{1 - E^2}{2E^3}\Big[\log[(1+E)/(1-E)] - 2E\Big], \tag{7}$$

The total energy is $\mathcal{F} = \mathcal{F}_{\mathcal{E}} + \mathcal{F}_{\sigma}$. The equilibrium shape is independent of the electric field, and only depends upon the ratio of the droplet volume to the particle volume. We can solve $\partial\mathcal{F}/\partial E|_V = 0$ in the limits $V/r_d^3 \to 0$ and $V/r_d^3 \to \infty$. For small droplets, the preferred shape is approximately spherical, with $\tilde{c} \sim \tilde{b} \sim V^{1/3}$. But when $V \gg r_d^3$, we have instead that

$$\tilde{c} \sim r_d (V/r_d^3)^{3/7}, \tag{8}$$

and

$$\tilde{b} \sim r_d (V/r_d^3)^{2/7}. \tag{9}$$

Thus the droplet becomes more and more elongated as it grows larger. When $\tilde{c}$ approaches L, the droplet begins to feel the effect of the image forces, and is pulled into a cylinder stretching between the electrodes. Thus if the aggregation of particles proceeds slowly enough that the shape of the droplets formed has sufficient time to equilibrate, then we expect to see columns with a radius

$$\tilde{b} \sim r_d (L/r_d)^{2/3}. \tag{10}$$

Since the aggregation of new particles to the droplets will take place preferentially at the tips, this will serve as an upper bound on the column width if the droplets do not have time to equilibrate.

If β is larger, then our calculation of $c(\hat{n})$ is not valid. Nevertheless, both the surface tension energies and the depolarization energies will still be proportional to $\mathcal{E}^2$, thus the droplet shape will still scale according to Eqs. (8,9) for large V.

This has assumed that $\tau_r < \tau_a$, where τ_r is the time scale for relaxation of disturbances to the equilibrium droplet shape, and τ_a is the time scale for droplet aggregation. Typically, the relaxation time of a fluctuating droplet of size r is

$\sim \mu_d r/\sigma$, where μ_d is the viscosity of the droplet.[12] Estimation of the internal viscosity of the condensed droplet is difficult. If we take the solvent viscosity μ for μ_d, and $\sigma \sim \mathcal{E}^2 r_d$, then we obtain $\tau_r \sim (\mu/\mathcal{E}^2)(r/r_d)$. For large values of λ, one expects a much larger internal droplet viscosity, perhaps $\mu_d \propto \exp(\lambda)$. Thus this estimate should be considered a lower bound on τ_r.

Suppose that droplets of characteristic size r have formed. The dipolar force is $F_{\mathcal{E}} \sim \mathcal{E}r^3 \times \mathcal{E}r^3/l^4$, where l is the distance between droplets, and the viscous force is $F_{visc} \sim \mu r\dot{l}$. The characteristic separation l of droplets of size r is $l \sim r/\phi^{1/3}$. Solving for τ_a, we have $\tau_a \sim l/\dot{l} \sim \mu/\mathcal{E}^2\phi^{5/3}$, independent of the droplet size. Comparing to the result above for τ_r, we conclude that $\tau_a < \tau_r$.

Thus for large electric fields we do not expect to see equilibrium droplet shapes. However, since droplets will aggregate preferentially at their tips (where the dipole-dipole interaction is attractive), we can still make the bound $\tilde{b} \lesssim r_d^{1/3}\tilde{c}^{2/3}$. However, as we go towards values of $\lambda \sim 1$ our equilibrium assumption should become more valid. Presuming that the original condensation phase transition is first order, the droplet surface tension σ remains finite as one goes towards the transition, thus τ_r remains finite. However, the aggregation time τ_a goes to infinity at threshold. Thus there will be a finite range of λ near threshold where $\tau_r < \tau_a$, and the above calculation will be valid. Numerical results of W. Toor bear out this claim.[13]

4. Densification

Of course, the above still leads unanswered the question of whether there are any mechanisms that will lead to relaxation of the columnar structures into macroscopically phase-separated regions. Perfectly cylindrical columns will experience a negligible electrostatic interaction with one another, provided that they are separated by a distance large compared to the particle size r_d. However, there are other, thermally generated, interactions, which are stronger than the direct electrostatic interaction, even at large values of λ.

Being essentially a one dimensional structure, a dipole chain or column will experience strong Landau-Peierls fluctuations. Fluctuations with a wavelength comparable to the distance l between chains will interact strongly with similar fluctuations in other chains. We have calculated the free energy $\mathcal{F}(l)$ due to interactions between fluctuations in two chains or columns of radius $\tilde{b}$ with the result[6]

$$\mathcal{F}(l) = -177.76\frac{k_B T}{2\pi}\frac{L\tilde{b}^4}{l^5}. \tag{11}$$

For colloidal particles, the dominant Van der Waals' type interaction is the Keesom interaction, which arises from the coupling between thermal fluctuations in the dipole moments of the individual particles.[14] The interaction energy between

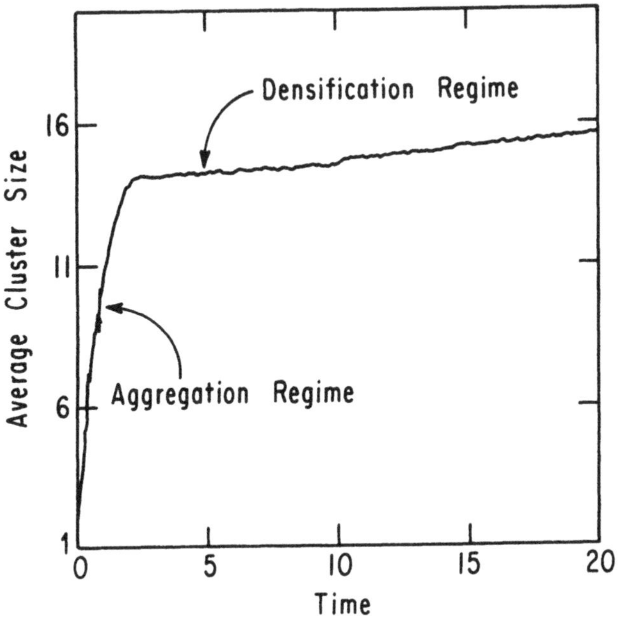

Figure 3. This shows the average size of dipole clusters versus time for 100 particles of radius $r_d = 0.5$ with $L = 10$ and $\lambda = 0.576$, averaged over 5 runs. The clusters grow rapidly by aggregation until they span the plates. Then they drift together more slowly. This long time densification is clearly visible. These results are from the molecular dynamics computation of W. Toor, and are reprinted by his permission.

lines is $\mathcal{F}_{Keesom} = -Aa^4L/l^5$, where A is the Hamaker constant. For ER fluids, A varies from about $0.2 - 2.5k_BT$, which is an order of magnitude smaller than the fluctuation coupling discussed above.[5] Thus, chain fluctuations increase the effective Hamaker constant by a large factor.

Although the fluctuations of chains or columns will generate this long-range interaction, we do not expect that these fluctuations will lead to collisions, and thus steric interactions. The distance wandered by a chain or column is $\propto (L/\lambda)^{1/2}$.[15] For $\lambda \sim 1$, the columns will have formed in equilibrium, and the distance between columns will be $l \sim L^{2/3} \gg L^{1/2}$. Thus the distance between columns is much larger than the typical distance that columns wander. At large λ, although the columns will be smaller and more closely spaced, the distance wandered is suppressed by the electric field. The distinctive feature of the fluctuation-induced and Keesom

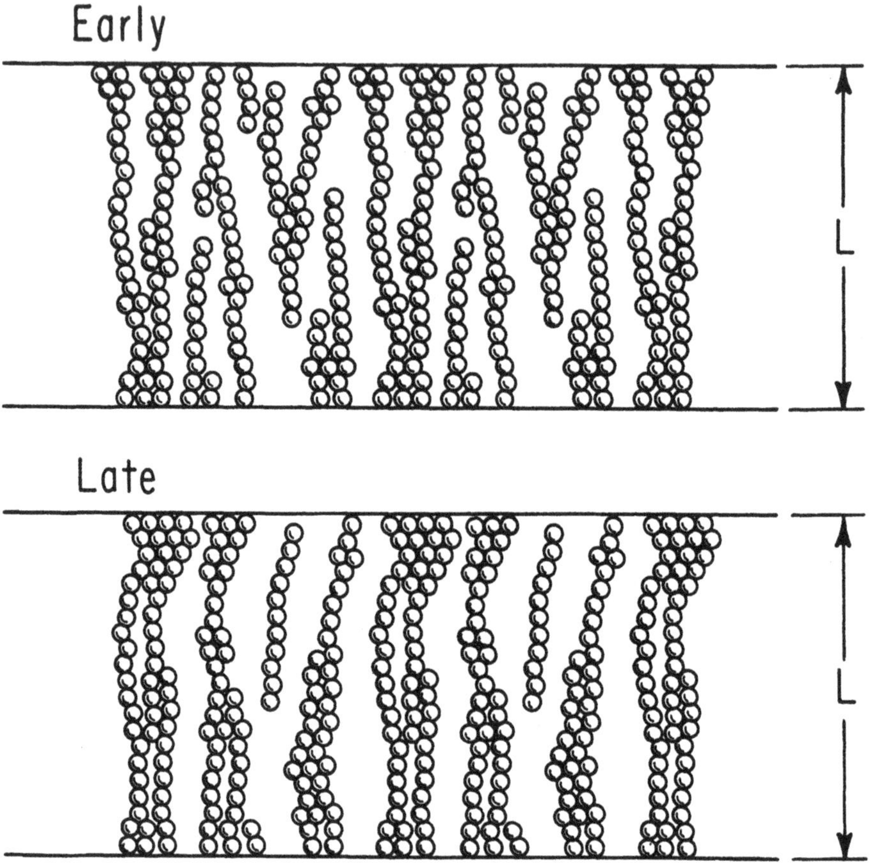

Figure 4. Configurations of particles for early and late times for the numerical computation of Figure 3 (in this case $L = 20$). The particle chains drift together slowly, so that the late time structure is noticeably coarser than the early time structure. These results are from the molecular dynamics computation of W. Toor, and are reprinted by his permission.

interactions discussed above is that their magnitude does not depend on the electric field, thus they survive at arbitrarily large electric fields.

If the viscosity of the fluid is μ then given an attractive force of the form shown in Eq. (11), the time scale τ_c over which columns separated by a distance l will

Table I. Densification time scale τ_C as extracted from results like those of Figure 3, for various values of λ. Within the accuracy of these results, there is no dependence of τ_C on λ. This is in agreement with the hypothesis that long-time densification is driven by thermal forces. The units of τ_C are arbitrary. These results are from the molecular dynamics computation of W. Toor, and are reprinted by his permission.

λ	τ_C
0.575	520 ± 90
1.0	600 ± 90
2.5	560 ± 60

aggregate into larger columns is $\tau_C \sim (\mu/k_B T \phi^{3/2}) l^3$. For typical ER fluids $\mu \sim 1$ barye-sec, so this becomes $\tau_C \sim 10^2$ sec $(l/\mu m)^3$. This sets the time scale on which one expects to see macroscopic phase separation in the fluid. Again, I emphasize that this time scale does not depend on the electric field. Thus even if λ is quite large, one still expects thermal relaxation to be significant in the long-time regime.

These predictions are in agreement with numerical results from a two dimensional molecular dynamics calculation performed by W. Toor.[13] Unlike many other studies, this computation explicitly included both thermal effects and image forces. Figure 3 shows the number of particles in a typical columnar structure as a function of time after the electric field was applied. A column is here defined simply as a set of particles in contact with one another. The separation of the column growth into an early aggregation regime and a late densification regime is clearly evident. Figure 4 shows pictures of the particle structure at early and late times in the densification regime. It is clear that the smaller columns have moved together.

From the slope of the growth curve in the densification regime, one can extract the time scale τ_C. Table I shows this time scale at three different values of λ (in arbitrary units). Within the accuracy of these simulations, this time scale does not vary with λ, in agreement with the hypothesis that thermal forces are responsible for densification.

5. Structure Formation in Shear Flows

ER fluids show a shear-thinning viscosity, which has been modelled as Bingham plastic behavior. Klingenberg and Zukoski suggested that this shear-thinning behavior was due to the formation of condensed boundary layers near the electrodes, so that the velocity gradients appeared only over a portion of the sample.[16] In this section, I will present an alternative view, in which one obtains a shear-thinning viscosity not as a boundary effect, but due to the bulk properties of the fluid.[17] Of course, which scenario is more valid is a question for experimentalists to answer; it is quite possible that different systems may behave in the Klingenberg and Zukoski manner, may follow my scenario, or may follow some combination of the two pictures.

As in the discussion of column formation in Section 3 above, I will assume an independent droplet picture. This is not necessarily an extremely restrictive assumption. If droplets are arranged in a random fashion, then nearby droplets will not strongly disturb the behavior of a single droplet, but will only change the average overall field by a factor that can be obtained from the Clausius-Mossotti relation. At any rate, both the calculation of this section and that of Section 3 will certainly be valid in the limit of low particle concentration.

The dimensionless group relevant to this problem is the Mason number Mn, which gives the ratio of polarization forces to hydrodynamic forces;

$$\mathrm{Mn} = \frac{24\pi\mu\dot{\gamma}}{(\beta\mathcal{E})^2}. \tag{12}$$

where $\dot{\gamma}$ is the strain rate. On dimensional grounds, we expect the total suspension viscosity $\tilde{\mu}$ to behave as

$$\tilde{\mu} \sim \mu\,\mathrm{Mn}^{-\Delta}, \tag{13}$$

defining the shear-thinning exponent Δ. I should add that in the experiments of Martin and Adolf, the data could not be collapsed onto a universal function of Mason number, which suggests that some other parameter may be important in determining the interparticle forces in these experiments.[17] Nevertheless, I shall continue to use the dipolar model for the inter-particle forces.

Consider a droplet of the condensed phase, of roughly ellipsoidal shape. A shear flow will rotate an ellipsoidal droplet, so that its long axis is no longer parallel to the field. As we shall see, the larger the droplet, the greater the rotation in a shear flow. This rotation reduces the depolarization energy of a droplet, raising the possibility that a balance between depolarization energy and surface energy will determine a characteristic droplet size. I will compute this size, which leads immediately to a shear-rate dependent fluid viscosity.

To compute the torque exerted by a shear flow upon the droplet, I assume that the internal viscosity of the electrorheological droplet is much larger than the viscosity of the solvent. This should be valid at large values of λ (see Section 3

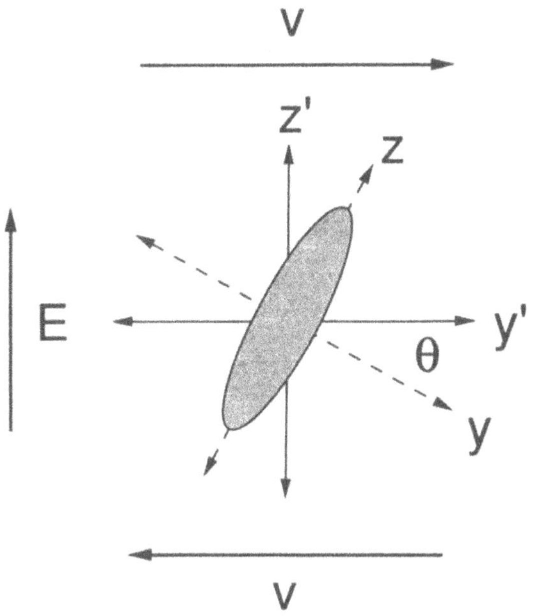

Figure 5. An ellipsoidal droplet will be rotated by a shear flow. The axes fixed in the body of the ellipsoid are the $y - z$ axes, while those fixed in space are the $y' - z'$ axes. The electric field is in the z'-direction, and the flow is in the y'-direction. The droplet is rotated an angle θ by the flow.

above). In this case, the effect of circulation in the interior of the droplet may be neglected, so that I can use the result for the torque on a rigid ellipsoid.

At this point it is useful to define some coordinates. The axes fixed in the body of the ellipsoid are $\{x, y, z\}$, and the spatial axes are $\{x', y', z'\}$. The electric field is oriented along the z' direction, and the undisturbed flow $\vec{v}$ is along the y' direction, $\vec{v} = (0, \dot{\gamma} z', 0)$. The $y - z$ axes are rotated with respect to the $y' - z'$ axes by a clockwise rotation of an angle θ (see Figure 5). The x direction ellipsoidal axis is of length $2\tilde{a}$, the y direction axis is of length $2\tilde{b}$, and the z direction axis is of length $2\tilde{c}$, so that the ellipsoidal surface is defined by $(x^2/\tilde{a}^2) + (y^2/\tilde{b}^2) + (z^2/\tilde{c}^2) = 1$. To lowest order in the Mason number we can take $\tilde{a} = \tilde{b}$, so that the ellipsoid is a spheroid.

The hydrodynamic torque about the y and z axes vanishes. The torque K_h about the x (or x') axis is given for low Reynolds numbers $\mathrm{Re} = \dot{\gamma}\tilde{c}^2\rho/\mu$ (ρ is the solvent density) by[18]

$$K_h = \frac{2\mu V\dot{\gamma}}{(\tilde{b}^2 n_y + \tilde{c}^2 n_z)}\left\{\tilde{c}^2\cos^2\theta + \tilde{b}^2\sin^2\theta\right\}, \tag{14}$$

where V is the spheroid volume and $n_{x,y,z}$ are the depolarization factors of the spheroid.

If the droplet is in equilibrium, then this torque must balance the electrical torque $K_{\mathcal{E}}$. Since the spheroid is prolate, we write $n_z \equiv n$, $n_y = (1-n)/2$. n is given by Eq. (7).

The torque $K_{\mathcal{E}}$ on a spheroid of dielectric constant ϵ inclined at an angle θ to the field is[11]

$$\begin{aligned} K_{\mathcal{E}} &= \frac{(\epsilon-1)^2[1-3n]\mathcal{E}^2 V\sin 2\theta}{8\pi[n\epsilon+1-n][(1-n)\epsilon+1+n]} \\ &= \frac{(\epsilon-1)^2}{8\pi(\epsilon+1)}\mathcal{E}^2 V\sin 2\theta + O(n) \end{aligned}, \tag{15}$$

where $\vec{\mathcal{E}}$ is the undisturbed electric field.

If we assume that θ is small, then balancing these two terms we obtain

$$\theta \approx \frac{16\pi(\epsilon+1)\mu\dot{\gamma}}{(\epsilon-1)^2\mathcal{E}^2}\frac{g^{-1}}{1-(1/2)\log(g/g^*)} \propto \frac{\mathrm{Mn}\, g^{-1}}{1-(1/2)\log(g/g^*)}. \tag{16}$$

where $g = \tilde{b}^2/\tilde{c}^2$, and $g^* = 4/e^2$.

Having determined the tilt angle θ as a function of g, we can now minimize the energy of a droplet over all possible sizes and ratios $\tilde{b}/\tilde{c}$. This is done by balancing the depolarization energy (which will be a function both of n and of θ – both small parameters) against the surface tension of a droplet.

Keeping only the lowest order terms in θ and n separately, the depolarization energy of a spheroid is[11]

$$\mathcal{F}_{\mathcal{E}} \approx -\frac{V(\epsilon-1)\mathcal{E}^2}{8\pi}\{1-(\frac{\epsilon-1}{\epsilon+1})\theta^2 - (\epsilon-1)n\} \tag{17}$$

The surface energy $\mathcal{F}_s$ is,

$$\mathcal{F}_s = \frac{\tilde{s}(\epsilon-1)\mathcal{E}^2}{4\pi}S \tag{18}$$

as in Section 3 above, where S is the surface area of the spheroid, and we expect $\tilde{s} \sim r_d$.

If we ignore the logarithmic factors, then by minimizing the total energy we obtain $g \propto \mathrm{Mn}^{2/3}$, $\tilde{b} \propto \mathrm{Mn}^{-2/3}$, and $\tilde{c} \propto \mathrm{Mn}^{-1}$. Since $\mathrm{Mn}/g \propto \mathrm{Mn}^{1/3}$, $\theta \ll 1$ for small Mn. Also, $\tilde{b} \propto \tilde{c}^{2/3}$, as in the static theory.

We then expect on dimensional grounds alone that the observed fluid viscosity $\tilde{\mu}$ will be given by $\tilde{\mu}\dot{\gamma} \sim (\phi/V)|K_h|$ where ϕ is the particle volume fraction. This immediately gives the power law

$$\tilde{\mu} \sim \mu g^{-1} \propto \mu \dot{\gamma}^{-2/3} \tag{19}$$

for the shear-thinning viscosity, so that we predict $\Delta = 2/3$.

The experiments of Martin and Adolf see values of Δ in the range $0.68 < \Delta < .91$ as the electric field increases, which suggests that refinements of this theory might be able to account quantitatively for the data. In particular, when the droplets become large enough to interact with their images, one might expect them to become roughly cylindrical, rather than ellipsoidal. A simple calculation based on the above considerations then gives $\Delta = 0.8$, so that one does anticipate, at the very least, that Δ will increase with field. However, a full treatment must also include the effect of collisions between droplets.

These experiments also failed to see any yield stress. This suggests that, at least under some circumstances, one has slip boundary conditions for the particles at the plates. These boundary conditions will necessarily invalidate the picture of Klingenberg and Zukoski.

6. Discussion

It is clear from the discussion of Sections 3 and 4 is that thermal effects have been very imperfectly explored in ER fluids up to now. Some thermal effects, such as densification, may persist at high fields. Furthermore, the threshold regime, while perhaps of little practical importance, should be of great interest to statistical physicists.

The formation of structure at high fields is likely to be dominated by aggregation effects. Here one has a type of cluster-cluster ballistic aggregation; however, unlike standard ballistic aggrgation, the force under which the particles move is anisotropic. There has been only preliminary work on this problem.[19]

Finally, the model that I have presented for the viscosity is undeniably somewhat crude. Droplet-droplet interactions, image forces, and the role of pinning at the boundaries remain largely unexamined. It is clear that the theory of ER fluids is a field that has barely even been mapped out.

Acknowledgements

The work discussed in Sections 2, 3 and 4 was performed in collaboration with W. Toor. The work of Section 5 was stimulated by the experiments of J.E. Martin and D. Adolf. I would like to acknowledge stimulating conversations with R. Rosensweig, B. Shaw, T.A. Witten, and C.F. Zukoski. I am grateful to the National Science Foundation for the support of this research through the Presidential Young Investigators program, grant DMR-9057156. Acknowledgement is made to the Donors of The Petroleum Research Fund, administered by the American Chemical Society, for the partial support of this research.

References

1. R.E. Rosensweig, *Ferrohydrodynamics* (Cambridge University Press, New York, 1985).
2. D.H. Reich, T.F. Rosenbaum, G. Aeppli, and H.J. Guggenheim, Phys. Rev. **B34** (1986) 4956 .
3. T.C. Halsey and W. Toor, Phys. Rev. Letts. **65** (1990) 2820.
4. D.J. Klingenberg, Masters Thesis, University of Illinois, Urbana (1988).
5. A.P. Gast and C.F. Zukoski, Adv. Colloid Interface Sci. **30** (1989) 153.
6. T.C. Halsey and W.R. Toor, J. Stat. Phys. **61** (1990) 1257.
7. H. Fröhlich, *Theory of Dielectrics, 2nd ed.* (Oxford University Press, New York, 1986) Chapter 2.
8. R. Tao and J.M. Sun, Phys. Rev. Letts. **67** (1991) 398.
9. W.R. Toor and T.C. Halsey, unpublished.
10. T.-j. Chen, R.N. Zitter, and R. Tao, unpublished, and in this volume.
11. L.D. Landau, E.M. Lifshitz, and L.P. Pitaevski, *Electrodynamics of Continuous Media, 2nd ed.* (Pergamon Press, New York, 1984) Chapter 2.
12. J.-C. Bacri and D. Salin, J. Phys. (Paris) Letts. **43** (1982) L-649.
13. W.R. Toor, submitted to J. Coll. Interface Sci.
14. W.H. Keesom, Phys. Z. **22** (1921) 129; J. Mahanty and B.W. Ninham, *Dispersion Forces* (Academic Press, San Francisco, 1976) 1-33.
15. E.M. Lifshitz and L.P. Pitaevski, *Statistical Physics, 3rd ed., Vol. 1* (Pergamon Press, New York, 1980) 432ff.
16. D.J. Klingenberg and C.F. Zukoski, Langmuir **6** (1990) 15.
17. T.C. Halsey, J.E. Martin, and D. Adolf, submitted to Phys. Rev. Letts.
18. G.B. Jeffery, Proc. Roy. Soc. (London) **A102** (1922) 161.
19. H. See and M. Doi, J. Phys. Soc. Japan **60**, 2778 (1991).

COMPUTER SIMULATION OF MODEL ELECTRORHEOLOGICAL FLUIDS

JUN-ICHI TAKIMOTO
Scientific Research Laboratory, Nissan Motor Co., Ltd.
1 Natsushima-cho, Yokosuka, 237 Japan

ABSTRACT

Dispersed particles in electrorheological fluids form chain-like clusters under electric field. Our theory and computer simulations show that, although the clusters become shorter as the shear rate $\dot{\gamma}$ increases, the induced stress τ_e due to the clusters remains roughly independent of $\dot{\gamma}$, as experimentally observed. We also show that τ_e is mostly determined by the strength of the electrostatic interaction between particles in contact with each other, and the distance dependence of the interaction has little effect on τ_e.

1. Introduction

Electrorheological fluid is a dispersion system consisting of polarizable particles (typically of 1 μm size) and insulating liquid.[1] Under the applied electric field E, it behaves as a so called Bingham fluid; the electrically induced shear stress (or yield stress) τ_e is roughly independent of the shear rate $\dot{\gamma}$.

It is well known that the dispersed particles form chain-like clusters under the electric field. Ascribing the induced stress τ_e to these clusters is an attractive idea. To confirm this idea, we have studied the shear stress due to these clusters by an intuitive theory and computer simulations. It is found that, although the clusters become shorter in proportion to $\dot{\gamma}^{-1/2}$, the induced stress τ_e is roughly independent of $\dot{\gamma}$, in agreement with experiments. It is also found that the magnitude of τ_e is mostly determined by the strength of the electrostatic interaction between particles in contact with each other, and the distance dependence of the interaction has little effect on τ_e.

2. Theory

2.1 Cluster Length

We first estimate the maximum length of the clusters which can exist under given shear rate $\dot{\gamma}$. Let us consider a cluster of length $n=2m+1$ in simple shear flow $u_x = \dot{\gamma}y$ and electric field E (Fig.1). There are two kinds of forces acting on each particle; hydrodynamic resistance and electrostatic force. In a stationary state, these two forces should be balanced with each other.

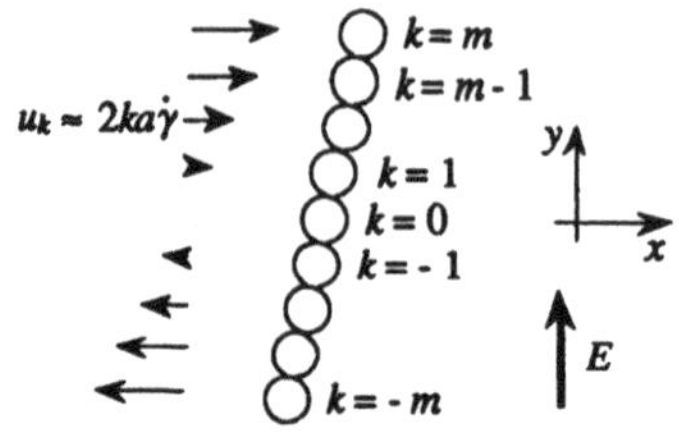

Fig.1 Cluster of length n=2m+1 in simple shear flow.

In a coordinate system moving with the cluster, the average flow velocity around the k-th particle (k=-m,-m+1,. . .,m) is roughly given by $u_k \sim 2ka\dot{\gamma}$, and the hydrodynamic resistance f_k on the k-th particle is approximately given by the Stokes resistance

$$f_k \approx 6\pi\eta_0 a u_k \approx 12\pi\eta_0\dot{\gamma}a^2k. \tag{1}$$

Here η_0 is the viscosity of the dispersing fluid, and a is the radius of the particles.

As the electrostatic force, we only consider the nearest neighbor interaction, and let F_k be the x-component of the electrostatic attractive force between particles k-1 and k. Balance of forces on the particle k is then expressed as

$$f_k - F_k + F_{k+1} = 0. \tag{2}$$

Solving this equation in terms of F_k, we get

$$F_k = 6\pi\eta_0\dot{\gamma}a^2(m+1-k)(m+k). \tag{3}$$

There should be this amount of attractive force between particles k-1 and k in order for the cluster to be stable. F_k is largest at the center (k=0) of the cluster. As the cluster becomes longer, F_0 increases as the square of the cluster length n. Let us define a parameter F_c as the upper limit of the x-component of the electrostatic attractive force between nearest neighbor particles. Then the upper limit n_c of the cluster length is determined from the inequality $F_0 < F_c$ as

$$n < n_c \equiv \left(\frac{2}{3\pi}\frac{F_c}{\eta_0\dot{\gamma}a^2}\right)^{1/2}. \tag{4}$$

Thus, as $\dot{\gamma}$ increases, the length of the clusters decreases in proportion to $\dot{\gamma}^{-1/2}$.

2.2 Induced Stress

Now let us estimate the electrically induced stress τ_e. Assuming that the clusters are the origin of τ_e, it can be estimated by

$$\tau_e = \Phi \langle F_k \rangle. \tag{5}$$

Here, Φ is the number of clusters crossing the xz-plane per unit area; using the volume fraction ϕ of the dispersed particles, it can be expressed as

$$\Phi = \frac{3\phi}{2\pi a^2}. \tag{6}$$

$\langle \overline{F_k} \rangle$ is the average of the x-component of the electrostatic forces between nearest neighbor particles; $\overline{}$ indicates the average within a cluster, and $\langle \cdots \rangle$ means the average over different clusters. Using Eq.3 the average within a cluster can be easily carried out:

$$\langle \overline{F_k} \rangle \approx \pi \eta_0 \dot{\gamma} a^2 \langle n^2 \rangle. \tag{7}$$

As a rough estimate, $\langle n^2 \rangle$ can be approximated by $n_c{}^2$. Then $\langle \overline{F_k} \rangle$ is given as

$$\langle \overline{F_k} \rangle \approx \frac{2}{3} F_c, \tag{8}$$

and the induced stress τ_e is estimated as

$$\tau_e \approx \frac{1}{\pi} \phi \cdot \frac{F_c}{a^2}. \tag{9}$$

Thus, the induced stress τ_e is independent of the shear rate $\dot{\gamma}$, as experimentally observed.

Repeating the derivation of Eq.9 for the 2-dimensional model is straightforward (only the expression for F need be modified), and we get

$$\tau_e \approx \frac{4}{3\pi} \phi_2 \cdot \frac{F_c}{a}. \quad \text{(2-dimension)} \tag{10}$$

Here, ϕ_2 is the area fraction of the dispersed particles. This expression is compared with the result of simulations in the next section.

3. Computer Simulation

The simulation method is similar to that used by Klingenberg et.al.[2] Hydrodynamic resistance on each particle is approximated by the Stokes resistance (hydrodynamic interaction between particles is neglected). The velocity of the particle is so determined that this hydrodynamic resistance is exactly balanced with the electrostatic force on the particle.

As the electrostatic interaction, we first use the simplest dipole-dipole interaction $\mathbf{F}_d$:

$$\mathbf{F}_d = \mathbf{D}_d(\theta) F_d(r) \quad \text{with} \quad \mathbf{D}_d(\theta) \equiv \frac{3}{16}\left[(5\cos^2\theta - 1)\hat{\mathbf{r}} - 2\cos\theta\hat{\mathbf{e}}\right]$$

$$F_d(r) \equiv G \cdot \left(\frac{2a}{r}\right)^4. \tag{11}$$

Here, $\mathbf{r}$ is the radial vector between interacting particles, $\hat{\mathbf{e}}$ and $\hat{\mathbf{r}}$ are the unit vectors in the direction of $\mathbf{E}$ and $\mathbf{r}$, respectively, and θ is the angle between $\hat{\mathbf{e}}$ and $\hat{\mathbf{r}}$. If the origin of the

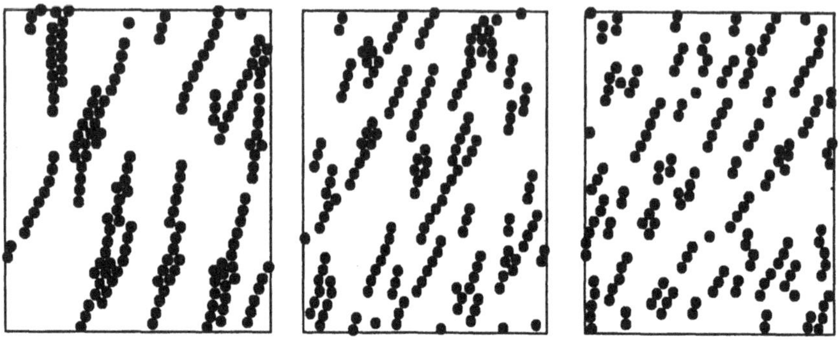

(a) $\dot{\gamma} = 0.01$ (b) $\dot{\gamma} = 0.04$ (c) $\dot{\gamma} = 0.10$

Fig.2 Typical configurations of particles under three different shear rates. ϕ_2=0.196. Shear rate $\dot{\gamma}$ is measured in units of $G/6\pi\eta_0$.

dipole on each particle is the difference of the dielectric constant of the dispersing fluid ε_0 and the particle ε_1, the parameter G can be written as $G=\alpha^2\varepsilon_0 a^2 E^2$ with $\alpha=(\varepsilon_1-\varepsilon_0)/(\varepsilon_1+2\varepsilon_0)$. But the actual interaction may be much stronger than $\mathbf{F}_d$ due to mutual enhancement of the polarization, especially for short distance. We will return to this problem later in this paper, but for the moment let us use $\mathbf{F}_d$ as the electrostatic interaction.

Simulations are carried out in 2-dimensional model (center of each particle is restricted to be on a single plane), and sheared periodic boundary condition is used. The shear rate $\dot{\gamma}$ is measured in units of $G/6\pi\eta_0$, and 2-dimensional stress (force per unit length) is measured in units of G/a.

Fig.2 shows typical configurations of particles under three different shear rates. As expected, the clusters become shorter as the shear rate increases. The average size of clusters $\langle n \rangle$ is shown in Fig.3(a) as a function of $\gamma^{-1/2}$. The dashed line in the figure is the theoretical estimate of the upper limit of the cluster length (Eq.4). The size of a cluster, which is calculated in the present simulation, is defined as the total number of particles in the cluster; since some clusters have branchings as in Fig.2, the "size" of clusters is somewhat larger than the "length" of clusters. Considering this, the results of simulation and theory are in reasonable agreement.

The induced shear stress τ_e is calculated by[3]

$$\tau_e = \frac{1}{S} \sum_{<jk>} \left\langle F^x_{jk}(y_j - y_k) \right\rangle . \qquad (12)$$

Here, F^x_{jk} is the x-component of the force between particle j and k, and S is the total area of the system. Fig.3(b) shows the calculated induced stress. As far as the shear rate is not so large that the average cluster size is greater than ~2, the induced stress is roughly independent of $\dot{\gamma}$, in agreement with the theory of the previous section. If, however, the

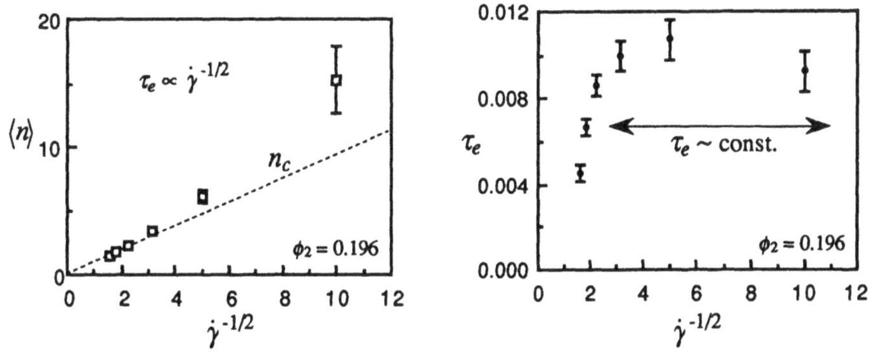

(a) cluster size (b) induced stress

Fig.3 Average cluster size and the induced stress τ_e as functions of $\dot{\gamma}^{-1/2}$. τ_e is measured in units of G/a.

shear rate is so large that the clusters are almost completely broken (n~1), the induced stress starts to decrease. But, our method of simulation (hydrodynamic interaction between particles is neglected) is not so reliable under such a high shear rate. So, the correct behavior in this high shear rate region may be different from the results of the present simulation. This point is left for future studies.

Fig.4 shows the induced stress as a function of the area fraction ϕ_2. For small and intermediate ϕ_2, τ_e is proportional to ϕ_2, in agreement with Eq.10. Since the hydrodynamic interactions are important at high density, whether τ_e saturates at large ϕ_2 or not is difficult to predict from the current simulation.

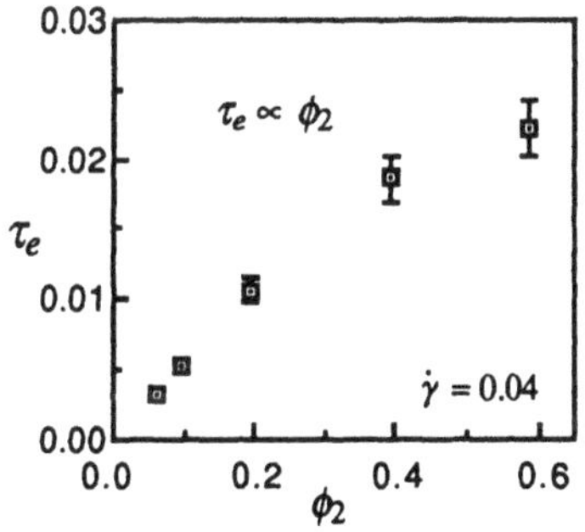

Fig.4 Area fraction dependence of the induces stress.

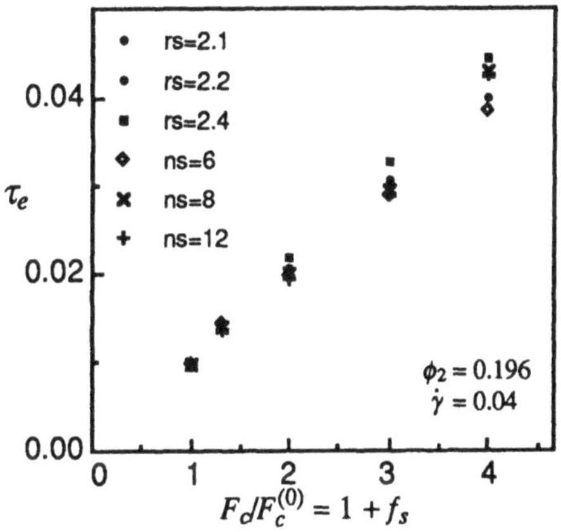

Fig.5 The induced stress for six model interactions.

We now proceed to the study of the relation $\tau_e \propto F_C$. To do that, we should return to the discussion of the electrostatic interactions.

As has been mentioned before, the actual electrostatic interaction may be much stronger than simple dipole-dipole interaction $\mathbf{F}_d$ (Eq.11), especially for short distances. So we write the total interaction $\mathbf{F}$ as a sum of $\mathbf{F}_d$ and a short range interaction $\mathbf{F}_s$

$$\mathbf{F} = \mathbf{F}_d + \mathbf{F}_s. \tag{13}$$

Exact form of $\mathbf{F}_s$ is very difficult to know (it may depend on the surface condition of the particles, etc.). So, instead of deriving it form the first principle, we use the following model interactions: The angular dependence of $\mathbf{F}_s$ is taken to be same as that of $\mathbf{F}_d$;

$$\mathbf{F}_s = \mathbf{D}_d(\theta) F_s(r). \tag{14}$$

For the distance dependence $F_s(r)$, we use one of the following two types of functions:

(i) power law: $$F_s(r) = G \cdot f_s \cdot \left(\frac{2a}{r}\right)^{n_s} \tag{15}$$

(ii) step function: $$F_s(r) = G \cdot f_s \quad (r < r_s)$$
$$= 0 \quad (r > r_s) \tag{16}$$

The dimensionless parameter f_s specifies the relative strength of $\mathbf{F}_s$ and $\mathbf{F}_d$ at r=2a. Using this f_s, the parameter F_C, the maximum of the x-component $\mathbf{F}$, is given as

$$F_C = \frac{G}{\sqrt{15}}(1 + f_s) \equiv F_C^{(0)} \cdot (1 + f_s). \tag{17}$$

Fig.5 shows the induced stress as a function of $F_C/F_C^{(0)} = 1+f_s$ for six model interactions: power law functions with $n_s = 6, 8, 12$, and step-functions with $r_s/a = 2.1$, 2.2, 2.4. This figure clearly shows that the induced stress is mostly determined by F_C.

The results of our simulations can be summarized as

$$\tau_e \approx 0.2 \cdot \phi_2 \cdot \frac{F_C}{a}, \tag{18}$$

which is about factor 2 smaller than the prediction of the theory (Eq.10). Considering the crudeness of the theory, the agreement is fairly good.

Eq.18 (or Eq.9 for 3-dimension) indicates that all the material dependence of τ_e is contained in the parameter F_C. To compare the results of simulations with experiments, simulations in 3-dimension and estimate of F_C for various materials are necessary. These studies are now under way.

4. References

1. See for example, H. Block and J. P. Kelly, *J. Phys. D* **21** (1988) 1661.
2. D. J. Klingenberg, F. van Swol and C. F. Zukoski, *J. Chem. Phys.* **94** (1991) 6160.
3. M. Doi and S. F. Edwards, *The Theory of Polymer Dynamics* (Oxford University Press, Oxford, 1986) p.72.

ELECTRICALLY INDUCED STRESSES IN ER FLUIDS: THE ROLE OF PARTICLE CHAIN STRUCTURE

Andrew M. Kraynik
Computational Fluid Dynamics, Division 1511
Sandia National Laboratories, Albuquerque, NM 87185

Roger T. Bonnecaze† and John F. Brady
Department of Chemical Engineering
California Institute of Technology, Pasadena, CA 91125

ABSTRACT

The technical and commercial feasibility of ER technology depends on the formulation of ER fluids with large yield stresses. We have investigated the connection between suspension structure and electrically induced stress transfer by analyzing the nonlinear elastic behavior of model ER fluids composed of spherical particles arranged in perfectly ordered single chains or thicker double chains. We conjecture that the single–chain structure maximizes stress transfer stemming from electrostatic polarization forces. If true, predictions of maximum stress transfer for single chains provide an upper bound on the ER effect. Our results suggest that pair–wise electrostatic interactions with particles off the chain axis, whether they are intrachain or interchain, reduce stress transfer as measured by the shear stress.

Introduction

Electrorheological (ER) fluid technology centers around suspensions that exhibit enhanced stress transfer when subjected to an external electric field. The technical and commercial feasibility of applications based on the ER effect depends, in large part, on the formulation of "stronger" ER fluids. This strengh is often reported as a *yield stress*, which has become the primary measure of electrically induced stress transfer in ER fluids. The desire to understand the connection between yield stresses and elastic behavior and suspension structure motivates this investigation.

† Current address: Department of Applied Mathematics and Theoretical Physics, University of Cambridge, Cambridge ENGLAND CB3 9EW

This work performed at Sandia National Laboratories and supported by the U.S. Department of Energy under contract #DE-AC04-76DP00789.

The term "yield stress" is used to describe several related but very different phenomena, as suggested by the title of our presentation at the International Conference on Electrorheological Fluids at the Southern Illinois University at Carbondale: "The Three Yield Stresses of ER Fluids: Role of Suspension Structure." The definitions of yield stress are so different that even the existence of a finite yield stress depends on the exact physical situation under consideration. To minimize confusion and ambiguity, consider the following. Viscometric flow data for ER fluids are often fit to an empirical equation of the form

$$\mu = \frac{\sigma}{\dot{\gamma}} = \frac{\sigma_y}{\dot{\gamma}} + \mu_p(\dot{\gamma}) \,, \tag{1}$$

where μ is the viscosity, σ is the shear stress, $\dot{\gamma}$ is the shear rate, σ_y is called the (*dynamic* or *viscometric*) yield stress, and $\mu_p(\dot{\gamma})$ is the plastic viscosity function. If one only intends to use Eq. (1) to represent experimental data over the finite shear rate range in which they were taken, then the yield stress σ_y is *just* an empirical fitting parameter even though it may be very useful. Because the shear rate range of measurements is finite, and, strictly speaking, never includes zero, the usual concern surrounding data extrapolation arises when Eq. (1) is assumed to be valid in the limit $\dot{\gamma} \to 0$. This assumption implies the much stronger constitutive relation for a viscoplastic fluid given by

$$\begin{aligned} \sigma &= \sigma_y + \mu_p(\dot{\gamma})\,\dot{\gamma} \,, && \sigma > \sigma_y \,, \\ \dot{\gamma} &= 0 \,, && \sigma \leq \sigma_y \,. \end{aligned} \tag{2}$$

When the plastic viscosity does not depend on $\dot{\gamma}$, Eq. (2) corresponds to the ubiquitous Bingham fluid model. In these equations, σ_y has also become the absolute threshold in stress below which $\dot{\gamma} = 0$. Even though the material cannot sustain steady flow below σ_y, it can support elastic–plastic deformations and exhibit other solid–like characteristics. As discussed by Barnes & Walters[1], it is impossible to prove the inequalities in Eq. (2) through experiments. Because ER fluids, like many highly structured fluids, do not appear to flow – or more precisely, *do not flow enough to matter* – below some threshold in stress, it is useful to model them as viscoplastic fluids.

Microrheological studies[2–7] of model ER fluids composed of rigid, non–Brownian, dielectric spheres suspended in a viscous fluid with lower dielectric constant, are consistent with the stricter conditions on the yield stress required by Eq. (2). In particular, these studies indicate that the chain–like particulate structures observed to form in ER fluids, can support elastic–plastic deformations – at least in the absence of Brownian motion.

This elastic–plastic behavior can be quite complex and exhibit three different yield phenomena[8]. These phenomena can be assigned three distinct yield stresses: 1) the elastic limit, 2) the static or initial yield stress, and 3) the dynamic or viscometric yield stress. The elastic limit is associated with the loss of reversible elastic behavior and the onset of hysteresis in stress–strain response. The static yield stress refers to a distinct maximum in stress with increasing strain, and the dynamic yield stress is the asymptotic value of stress for large strains. These distinctions can be important because controlled–stress methods determine the static yield stress, while controlled–rate techniques provide the dynamic yield stress.

Recent experimental and theoretical investigations support the original proposal of Winslow[9] that electrostatic polarization is the primary mechanism responsible for the ER effect. This mechanism suggests that the dimensionless stress tensor for an ER fluid under static conditions can be expressed as $\sigma(\alpha, \phi)$, which has been scaled by $\epsilon(\beta E)^2$, where ϕ is the volume fraction of particles, E is the magnitude of the electric field, $\beta = (\alpha - 1)/(\alpha + 2)$, and $\alpha = \epsilon_p/\epsilon$ is the ratio of dielectric constants for the particles and fluid. We assert that all yield stresses obey the same scaling and are bounded from above by the maximum stress that can be supported under static deformations.

Klingenberg & Zukoski[2] first considered a model ER suspension with idealized structure, which we will refer to as perfectly ordered single chains (see Figure 1). Using multipole expansions to calculate the electrostatic force between two spheres, assuming pair–wise additivity of forces, and considering only nearest–neighbor interactions, they calculated the elastic behavior under simple shearing deformations. With these results, they developed a model for the dynamic yield stress, which showed favorable agreement with the data of Marshall, Goodwin and Zukoski[10]. Klingenberg[4] extended this strategy to calculate the dynamic yield stress for more realistic chain structures, which were obtained from dynamic simulations. He developed a clever but *ad hoc* "perturbation approach" to account for many–body and multipole electrostatic interactions.

Bonnecaze & Brady[5] have developed a molecular dynamics–like simulation method to accurately and efficiently compute the electrostatic interactions in suspensions with arbitrary particle configuration. Their method avoids solving for the complete potential field, which is neither necessary nor efficient; includes both many–body far–field and near–field interactions; and properly accounts for the long–range interactions common to electrostatics. This method in combination with the Stokesian dynamics method developed by Brady and coworkers[11], provides a tractable theoretical foundation for simulating transport in ER suspensions stemming from hydrodynamic interactions, electrostatic polarization, and Brownian motion.

Bonnecaze & Brady[6,7] also used their simulation method to determine the elastic behavior of ER suspensions composed of perfectly ordered single chains and to develop a model for the dynamic yield stress. When α is not too large, the yield stress saturates with increasing particle concentration, indicating that this effect can be caused by interchain electrostatic interactions. The formation of thick chains with increasing particle concentration is not necessary to observe this saturation phenomenon, as previously suggested[4].

Recognizing that the dramatic rheological properties of ER fluids stem from the

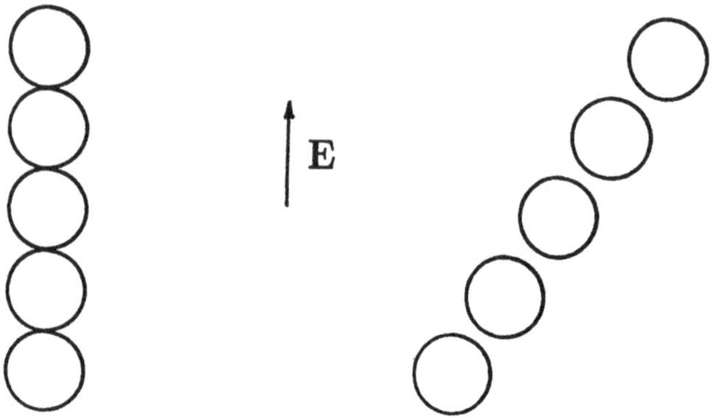

Figure 1 Schematic of perfectly ordered single chain structure. The undeformed chain on the left is aligned with the electric field **E** and the particles touch when $d = 0$, as shown. No particles touch in the deformed chain on the right.

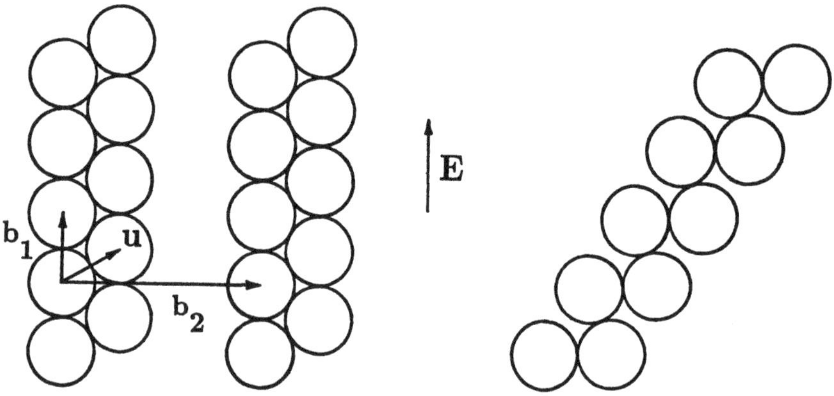

Figure 2 The structure of the undeformed double chains on the left is specified through the basis vectors $\mathbf{b}_i$ and the position vector **u**. Chain connectivity is preserved under finite deformations for the deformed chain on the right.

development of chain-like structures and their response to macroscopic deformation and flow, we employ the above simulation method to extend previous studies on nonlinear elasticity of perfectly ordered single chains to an idealized double chain structure.

Chain Structure

To investigate the effect of chain structure on electrostatically induced stress transfer in ER suspensions, we consider two idealized chain configurations, which will be referred to as single chains and double chains. A perfectly ordered single chain is shown in Figure 1. In the initial, unstrained configuration, identical touching spheres form a chain that is aligned with the electric field. We assume that under homogeneous simple shearing deformations of the ER suspension, the displacement of individual particles is completely determined by the imposed shear. This so-called affine displacement maintains perfect order within the chain, however, connectivity between neighboring particles is lost when the shear strain $\gamma > 0$, as shown in Figure 1.

The double chain structure, shown in Figure 2, can be viewed as two single chains in contact with each other. Even for finite shear strains, each particle in one chain touches two particles in the other chain, preserving the connectivity of the entire structure. Obviously, this connectivity cannot be maintained if the separation between adjacent particles in the component single chains exceeds one particle diameter, which corresponds to $\gamma > \sqrt{3}$. While this threshold is purely geometric, we will show that the particles can separate at much smaller values of γ for a different reason.

We model the ER fluid as spatially periodic with particles confined to a monolayer in the x–y plane. This condition retains essential physics of ER suspensions. The unit cell is specified through the two basis vectors shown in Figure 2 and given by

$$\mathbf{b_1} = \left(\frac{\pi N}{(2+d)\phi_a}, 0 \right) , \quad \mathbf{b_2} = (2+d)\ (\gamma, 1) , \tag{3}$$

where N is the number of particles in the unit cell (equal to one for single chains and two for double chains), ϕ_a is the area fraction of particles, d is a "small" distance that allows for the possibility of a minimum separation between particles (*e.g.* due to attached polymer chains or adsorbed surfactants), and all lengths have been scaled by the particle radius a. We fix the origin of a Cartesian coordinate system at the center of an arbitrary particle. For single chains, all particle centers occupy lattice sites that move with the basis vectors. For double chains, the location of particles on the "other side" of the chain is specified by the vector $\mathbf{u}$, which satisfies the arbitrary but physically reasonable connectivity requirement: $|\mathbf{u}| = |\mathbf{u} - \mathbf{b_2}| = 2 + d$. Even though the displacement of $\mathbf{u}$ is not affine, the highly symmetric double chain structure that we choose to analyze is determined by purely geometric considerations.

Isolated Point Dipole Limit

We base this microrheological study on the electrostatic polarization mechanism for ER stress transfer. This widely accepted mechanism captures many salient features of the ER effect and provides a tractable theoretical foundation for incorporating other relevant colloidal and electrical transport phenomena such as Brownian motion, finite electrical conductivity, *etc.* We consider a model ER fluid composed of spherical particles of radius a and dielectric constant ϵ_p suspended in a fluid with smaller dielectric constant ϵ. When the distance between particles is very large or when the ratio of dielectric constants $\alpha = \epsilon_p/\epsilon$ is near unity, the electrostatic force $\mathbf{F}_{ij}^E$ induced on sphere i due to the presence of sphere j is well approximated by the interaction between isolated point dipoles. In dimensionless form,

$$\mathbf{F}_{ij}^E = r_{ij}^{-4} \left[(3\cos^2\theta_{ij} - 1)\, \mathbf{e}_r^{ij} + \sin 2\theta_{ij}\, \mathbf{e}_\theta^{ij}\right] , \tag{4}$$

where the vector $\mathbf{x}_{ij} = r_{ij}\mathbf{e}_r^{ij}$ connects the center of sphere i to sphere j, θ_{ij} is the angle between the electric field vector $\mathbf{E}$ and $\mathbf{x}_{ij}$, and $\mathbf{e}_r^{ij}$ and $\mathbf{e}_\theta^{ij}$ are unit vectors associated with $\mathbf{x}_{ij}$ and θ_{ij}, respectively. All lengths have been scaled by a and the force has been scaled by $12\pi\epsilon(a\beta E)^2$, where $\beta = (\alpha-1)/(\alpha+2)$. The net electrostatic force $\mathbf{F}_i^{\Sigma E}$ on sphere i is given by

$$\mathbf{F}_i^{\Sigma E} = \sum_j^{\infty} \mathbf{F}_{ij}^E , \tag{5}$$

where the sum is taken over all spheres in the unbounded suspension. From the symmetry of Eq. (4), $\mathbf{F}_i^{\Sigma E}$ vanishes for particles in single chains because the force contributed by particles on opposite sides cancel. However, this is not true for double chains because each particle on the "other side of the chain" contributes to the sum in Eq. (4). We balance the net electrostatic force on each particle in a double chain by invoking contact forces $\mathbf{F}_{ik}^C$ between both particles that it touches. These contact forces of unspecified origin could stem from, for example, small elastic deformations of the particles or their attached stearic groups, or from the singular lubrication forces associated with viscous flow between nearly touching particles. For simplicity, we make the not unreasonable assumption that this contact force acts only along the line of particle centers and is given by

$$\mathbf{F}_{ik}^C = -F_{ik}^C\, \mathbf{e}_r^{ik} , \tag{6}$$

where the unit vector $\mathbf{e}_r^{ik}$ is directed along the line of centers of the ik pair. The force balance on particle i gives

$$\mathbf{F}_{ik_1}^C + \mathbf{F}_{ik_2}^C + \mathbf{F}_i^{\Sigma E} = \mathbf{0} . \tag{7}$$

Equations (5), (6), and (7) determine the magnitudes of the induced contact forces F_{ik}^C. The total stress tensor $\boldsymbol{\sigma}$ for an ER suspension is given by[12,13]

$$\boldsymbol{\sigma} = \frac{1}{V}\sum_{i}^{N}\sum_{j}^{\infty} \mathbf{x}_{ij}\mathbf{F}_{ij}^{E} + \frac{1}{V}\sum_{i\neq k}^{N_C} \mathbf{x}_{ik}\mathbf{F}_{ik}^{C} , \tag{8}$$

where N_C in the total number of particle contacts within a unit cell of volume V. The first sum, which is taken over all pair–wise electrostatic interactions in the unbounded suspension for all particles in the unit cell, represents the "ER stress" stemming from induced polarization forces. Neglecting interchain interactions, which is correct in the limit where chains are far apart or $\phi_a \ll 1$, the double sum can be expressed as

$$\sum_{i}^{N}\sum_{j}^{\infty} \mathbf{x}_{ij}\mathbf{F}_{ij}^{E} = 2N\sum_{k=1}^{\infty} k\,\mathbf{b_2}\mathbf{F}_{k\mathbf{b_2}}^{E} + 2(N-1)\sum_{k=-\infty}^{\infty} (\mathbf{u}+k\mathbf{b_2})\mathbf{F}_{\mathbf{u}+k\mathbf{b_2}}^{E} , \tag{9}$$

where, for example, $\mathbf{F}_{k\mathbf{b_2}}^{E}$ is the polarization force acting between two particles with relative position $k\mathbf{b_2}$, and we have used symmetry to simplify the result. The first sum on the RHS represents the single–chain contribution to the ER stress. The other term represents pair–wise interactions between particles on different sides of a double chain. The contact stress term in Eq. (8) is given by

$$\sum_{i\neq k}^{N_C} \mathbf{x}_{ik}\mathbf{F}_{ik}^{C} = -F_{\mathbf{u}}^{C}\mathbf{u}\mathbf{u} - F_{\mathbf{u}-\mathbf{b_2}}^{C}(\mathbf{u}-\mathbf{b_2})(\mathbf{u}-\mathbf{b_2}) . \tag{10}$$

The stress components for a single chain are plotted as a function of strain in Figure 3. The shear stress σ_{yx} is the primary quantity of engineering interest. We have shown all stress components to illustrate two complex rheological features of ER fluids. First, under an imposed electric field, the stress is anisotropic ($\sigma_{xx} \neq \sigma_{yy}$) even when $\gamma = 0$ and the chains are unstrained. This occurs because the interparticle forces are strongly attractive along the electric field vector $\mathbf{E}$ and much less so or even repulsive perpendicular to it. Second, since $\sigma_{xy} \neq \sigma_{yx}$, the stress tensor is asymmetric. This asymmetry stems from the external couple acting on the strained chains, which behave like a single dipole wanting to align with the electric field. Interestingly, there is no couple on the individual particles because of their spherical symmetry. Two very useful measures of stress transfer are the maximum ER stress σ_{yx}^{max}, and the shear modulus G, given by the slope of σ_{yx} for small γ. The quantity σ_{yx}^{max} is the maximum stress that can be sustained by the ER fluid under simple shearing deformations.

The function $\sigma_{yx}(\gamma)$ for both single and double chains is shown in Figure 4. These curves have been normalized so that they reflect the contribution of each particle to the stress. The double–chain curve terminates when $\gamma \approx 0.341$ because the magnitude of a contact force has changed sign, indicating that an "attractive" contact force

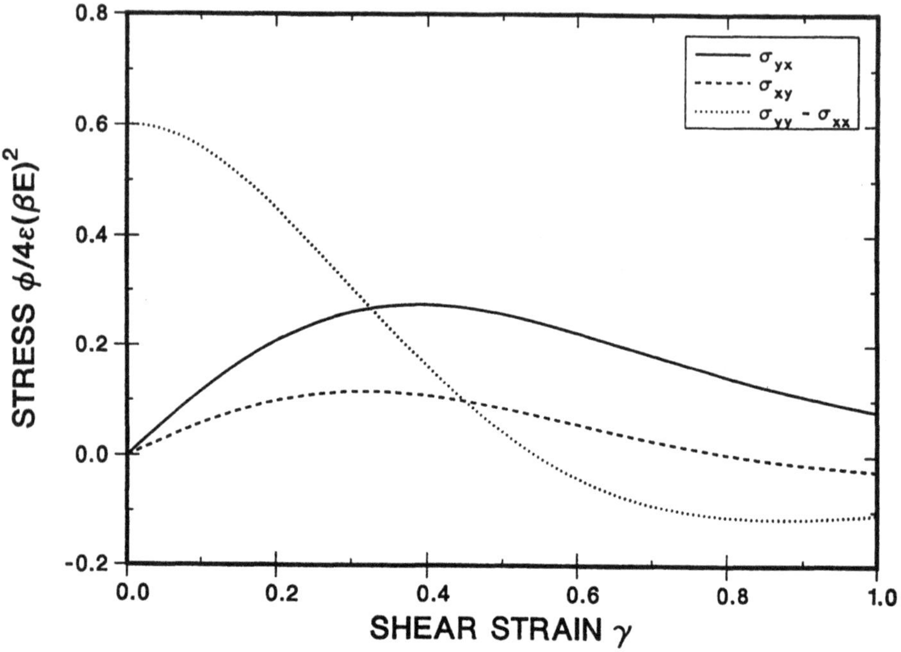

Figure 3 Dependence of stress components on shear strain γ for large elastic deformations of dilute single chains with induced polarization forces given by point dipoles. The solid curve corresponds to the shear stress σ_{yx}, which is of engineering interest.

would be required to maintain connectivity of the simple double chain structure under consideration.

It is likely that connectivity could be maintained by allowing less constrained unfolding of the double chain; *e.g.* by including more particles in the unit cell and pinning fewer particles to the lattice. This might delay the loss of connectivity until $\sigma_{yx}(\gamma)$ achieves a maximum, however, we do not feel that it would change the conclusion that the net ER stress transfer capacity is reduced for double chains.

Figure 4 also shows the magnitude of the terms that contribute to $\sigma_{yx}(\gamma)$ for double chains. The contact stress actually enhances σ_{yx}, but not enough to overcome the detrimental intrachain contributions to the $\mathbf{x}_{ij}\mathbf{F}^E_{ij}$ stress, which stem from certain pair-wise electrostatic interactions between particles on different sides of the chain. Because $\mathbf{F}^E_{ij}$ goes as r_{ij}^{-4}, the detrimental intrachain effect can be understood by considering the contributions of the two nearest neighbors in the other chain to the yx-component of $\mathbf{x}_{ij}\mathbf{F}^E_{ij}$. The "upper" and "lower" nearest intrachain neighbors are located at $\mathbf{u}$ and $\mathbf{u} - \mathbf{b_2}$, respectively, as shown in Figure 2. When $\gamma = 0$ the

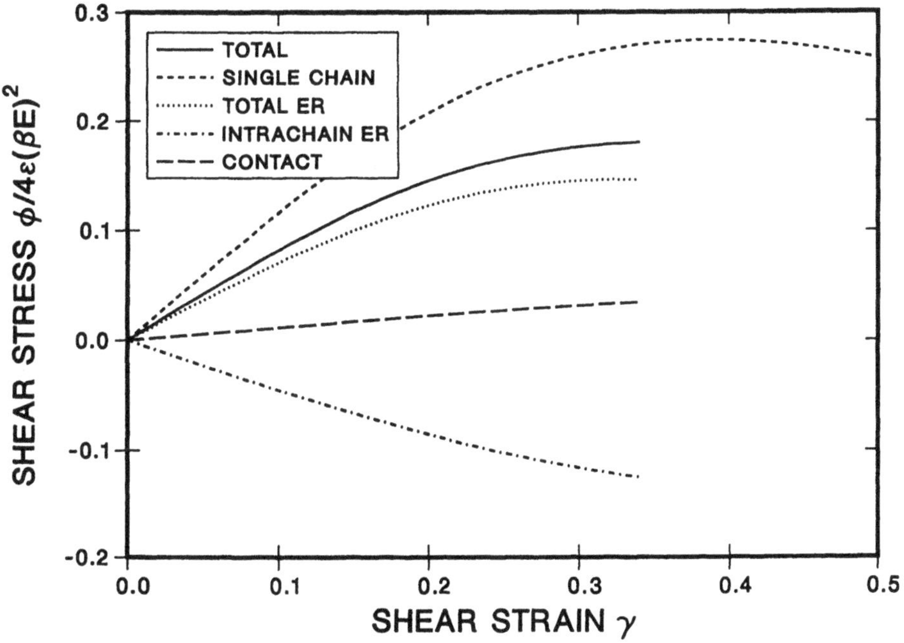

Figure 4 Contributions to the total shear stress σ_{yx} for dilute double chains under large elastic deformations. The "intrachain ER" term is responsible for the reduced shear stress for double chains and comes from pair-wise interactions between particles on different sides of the chain. The "single chain" and "intrachain ER" terms combine to give the "total ER" contribution. The curves corresponding to double chains terminate when $\gamma \approx 0.341$ because an attractive contact force would be required to maintain chain connectivity for larger strains.

x-component of $\mathbf{F}_{ij}^E$ for both particles is the same and the y-component of the position vectors is equal in magnitude but opposite in sign; therefore, the upper particle stress is positive and the lower particle stress is equal in magnitude but negative. As the shear strain increases, both y-components of position decrease, the upper particle force decreases and eventually goes negative, and the lower particle force increases. Starting from the undeformed condition, the contribution of both particles to the shear stress decreases and is responsible for lowering $\sigma_{yx}(\gamma)$ for double chains.

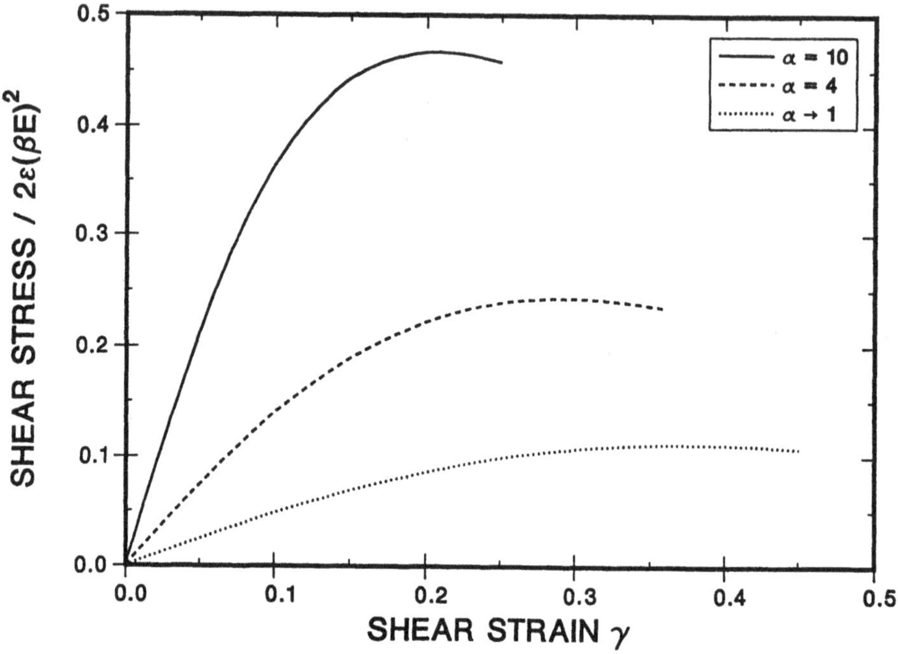

Figure 5 Stress–strain curves $\sigma_{yx}(\gamma)$ for single chains with $\phi_a = 0.30$. The polarization forces include far–field multi–body as well as near–field electrostatic interactions. The lower curve corresponds to point dipoles.

Many–Body and Near–Field Electrostatic Interactions

While simple models of ER fluids that provide important insights can be contructed from the point dipole approximation, they severely underestimate the magnitude of the electrically induced interparticle forces and hence the stress of the suspension. First, the point dipole approximation neglects the enhanced electric field experienced by each particle due to the many–body reflections. Second, when α is large and the particles are nearly in contact, their is a singular concentation of positive and negative charge at opposite ends of each particle which greatly enhances the interparticle attractive force. An infinite sum of higher multipoles is necessary to capture this near–field interaction, which is responsible for large ER stresses.

We use the method developed by Bonnecaze & Brady to determine $\mathbf{F}_{ij}^E(\alpha)$. This method accurately and efficiently accounts for many–body far–field and near–field

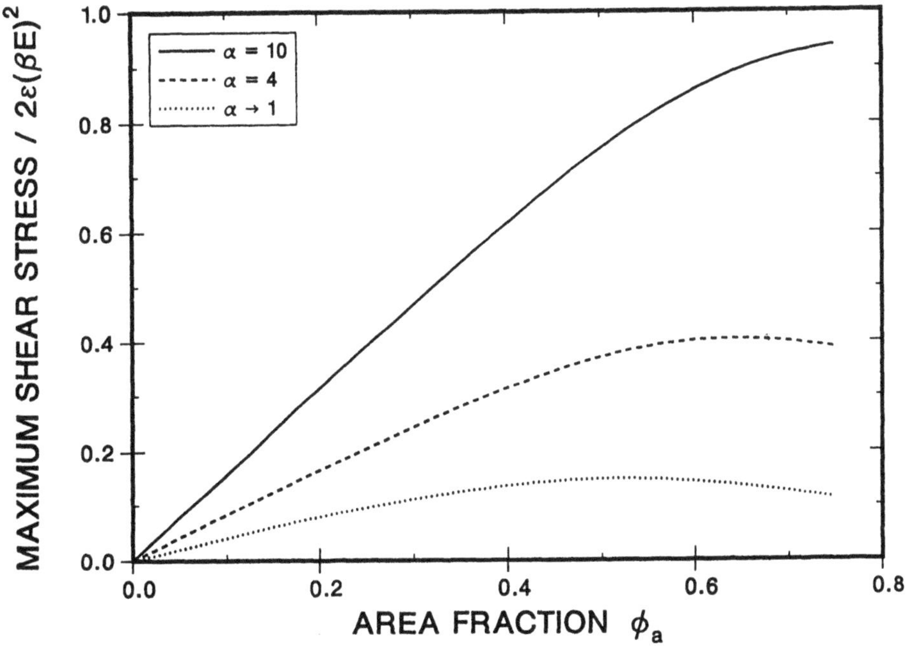

Figure 6 Maximum shear stress σ_{yx} as a function of area fraction ϕ_a for single chains.

interactions. The considerable details can be found elsewhere[5–7]. We will not explain the method here, but focus on results that explore the effect on $\sigma_{yx}(\gamma)$ of chain structure and particle concentration when α is large.

The effect of large α (or similarly finite β) is shown in Figure 5, which contains the stress–strain curves for single chains with $\phi_a = 0.30$. The curve for $\alpha \to 1$ corresponds to the point dipole limit. Because $\sigma_{yx}(\gamma)$ has been scaled by $2\epsilon(\beta E)^2$, any variation from the point dipole curve represents an enhancement from the many–body and near–field electrostatic effects. The curves in Figure 5 show that stress transfer – as measured by $\sigma_{yx}(\gamma)$, σ_{yx}^{max}, and G – increases with α. Moreover, the value of strain γ^{max}, which corresponds to σ_{yx}^{max}, decreases with increasing α. This indicates that as α increases, the maximum stress transfer occurs for smaller departures of the chain axis from the electric field vector $\mathbf{E}$. The ratio σ_{yx}^{max}/G and γ^{max} show similar dependence on α and ϕ_a. The spacing between chains $|\mathbf{b}_1|$ decreases with increasing area fraction ϕ_a, introducing the possibility of strong interchain interactions. Figure 6 shows the primary effect of increasing particle concentration. The maximum shear stress σ_{yx}^{max} saturates – or even goes through a maximum – as ϕ_a increases. In the

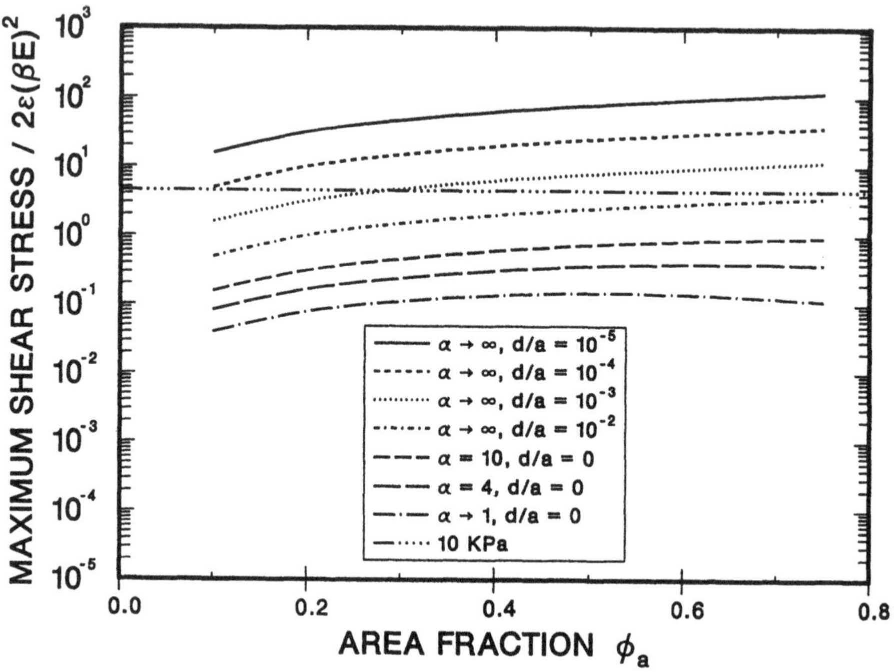

Figure 7 Maximum shear stress σ_{yx} as a function of area fraction ϕ_a for perfectly ordered single chains. The horizontal line corresponds to a shear stress of 10 KPa for perfectly conducting spheres, $E = 5$ KV/mm, and $\epsilon = 5\epsilon_0$.

point dipole limit, this effect can be attributed entirely to pair-wise electrostatic interactions between particles in different chains. In general the saturation is due to the increased importance of interchain interactions relative to the intrachain interactions. The pronounced maximum exhibited by $\sigma_{yx}^{max}(\phi_a)$ for point dipoles, which occurs long before neighboring chains touch (when $\phi_a = \pi/4 \approx 0.785$), indicates strong interchain interactions, especially when compared to the curve for $\alpha = 10$, which is monotone increasing for $\phi_a \leq 0.75$. A linear relation between σ_{yx}^{max} and ϕ_a indicates negligible interchain interactions – this linearity is observed when α is very large and the particles are perfect conductors, and all interactions are between nearest neighbors. When $\alpha \to \infty$ the induced force $\mathbf{F}_{ij}^E$ diverges as two particles aligned with the electric field approach contact. For this reason, the stress–strain curves are calculated under the assumption that the particles are initially separated by a small distance d, which is accounted for in Eq. (3). For given $d \leq 10^{-4}$, $\sigma_{yx}(\gamma)/\phi_a$, collapses to a single curve, and σ_{yx}^{max}/ϕ_a is essentially constant. Under these conditions, γ^{max} is less than 0.01, which means that the chains are essentially oriented with the electric field at σ_{yx}^{max}.

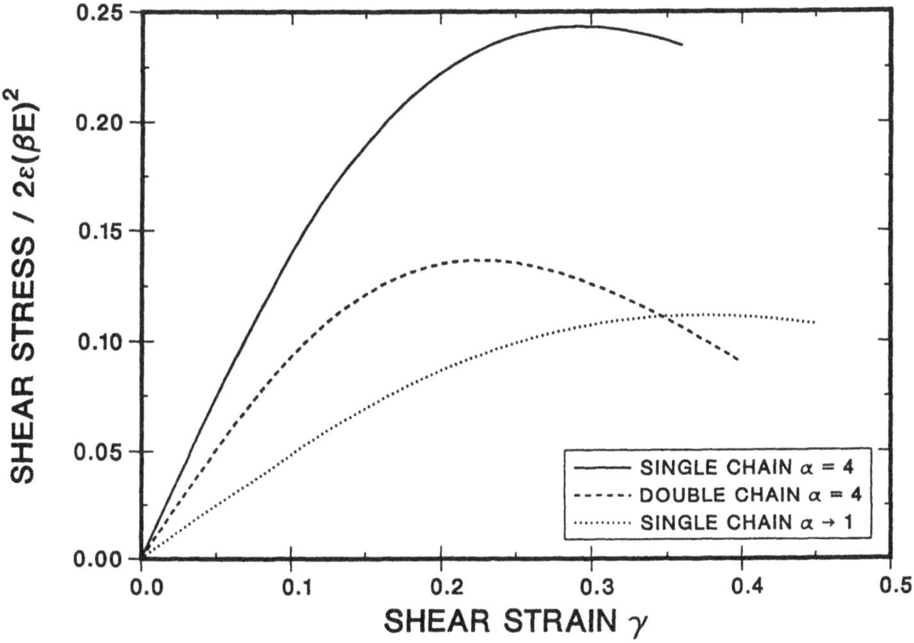

Figure 8 Stress–strain curves $\sigma_{yx}(\gamma)$ for single and double chains with $\alpha = 4.0$ and $\phi_a = 0.30$. The polarization forces include far–field multi–body and near–field electrostatic interactions. The lower curve corresponds to single chains and point dipoles.

A summary of $\sigma_{yx}^{max}(\phi_a)$ for single chains with α and d as parameters, is shown in Figure 7. Clearly, stress transfer predictions incorporating many–body and near–field electrostatic effects can exceed, by over three orders of magnitude, those based on the point dipole approximation. The stress transfer for conducting particles can be made arbitrarily large by making d very small, however, there are physical limitations on particle size and stabilizing layer thickness. The upper curve for $\alpha \to \infty$ and $d = 10^{-5}$ would correspond, for example, to perfectly conducting particles, 200 μm in diameter, separated by a stabilization distance of 1 nm by surfactants or the like. Many other phenomena influence the upper bound on stress transfer in ER fluids; *e.g.* the local electric field in the gap separating conducting particles would be highly intensified and can be estimated as $2E/d$. The fluid would be required to withstand 2 MV/mm without breaking down, for an external electric field of 1 KV/mm and $d = 10^{-3}$.

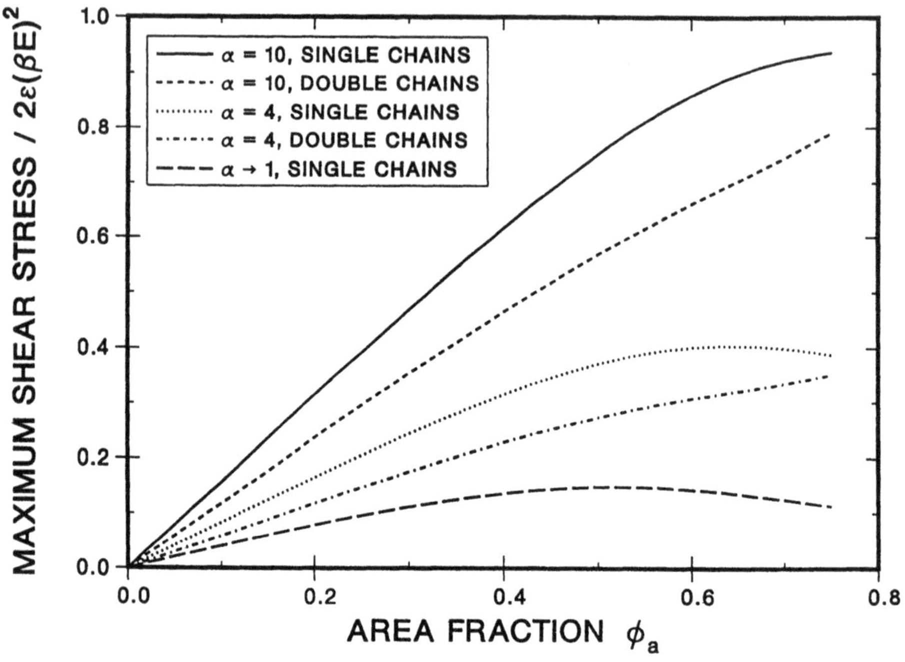

Figure 9 Comparison of the maximum shear stress σ_{yx}^{max} as a function of area fraction ϕ_a for single and double chains with $\alpha = 4.0$ and 10.0.

The stress–strain curves contained in Figure 8, for $\alpha = 4$ and $\phi_a = 0.30$, show that stress transfer also decreases for double chains when finite α effects are included. As for single chains, when α is large, the maximum shear stress is achieved at smaller values of γ. Therefore, when α is large, there is a distinct maximum in $\sigma_{yx}(\gamma)$ before a contact force becomes attractive and the double chain loses connectivity, which was not the case in the point dipole limit.

Our calculations for double chains and large α are summarized in Figure 9, which indicates that σ_{yx}^{max} is lower than for single chains at the same particle concentration – the same is true for $\sigma_{yx}(\gamma)$ and therefore the shear modulus G.

Discussion and Conclusions

By adopting the point dipole approximation, one can develop relatively simple analytical expressions for electrically induced stress transfer in ER fluids composed of single chains or double chains. For double–chain structure, we can conclude that pairwise electrostatic interactions that are not directed parallel to the chain axis reduce stress transfer as measured by the maximum shear stress σ_{yx}^{max}, the stress–strain relation σ_{yx}, and the shear modulus G. Results for α equal to 4 and 10 suggest, but do not prove, that off–axis electrostatic interactions, which may be either intrachain or interchain, also reduce stress transfer when many–body and near–field interactions are important.

We have extended the point dipole analysis to triple chains and thicker ones and have been able to infer the behavior of an "infinitely thick" chain. We have also begun analyzing branched chain structures. Those results support the conclusions reached in this paper but will be reported elsewhere.

These microrheological studies will benefit the development of ER technology in many ways. We believe that single–chain structure maximizes stress transfer that stems from induced electrostatic polarization. If this conjecture is true, predictions of σ_{yx}^{max}, such as these nonlinear elasticity calculations contained in Figure 7, would provide upper bounds on the ER effect. The quantity σ_{yx}^{max} can be identified as a static yield stress, which is an upper bound on the dynamic yield stress[6–8]. This information enables applications engineers to develop improved estimates of capabilities that could be supported by ER technology. The horizontal line in Figure 7 corresponds to a shear stress of 10 KPa for an ER fluid in an electric field of 5 KV/mm. This hypothetical ER suspension is composed of perfectly conducting spheres suspended in a dielectric fluid with $\epsilon = 5\epsilon_0$, where ϵ_0 is the permittivity in vacuum. This example indicates that when large α effects are included, the electrostatic polarization mechanism can account for the largest yield stresses measured. If our results do represent upper bounds on the ER effect, then potential applications of ER technology that demand shear stresses orders of magnitude larger than 10 KPa, may not be feasible. This conclusion does neglect the possibility of other physical mechanisms leading to stronger ER fluids.

The shear stress would be even greater if the particles in single chains (oriented away from **E**) remained in contact (to within d) because $\mathbf{F}_{ij}^{E}$ decreases with r_{ij}. However, perfectly ordered single chains cannot maintain connectivity under the affine shearing kinematics considered here. We are currently investigating the higher bounds on the shear stress associated with touching spheres.

The current study suggests that the ER effect could be enhanced through judicious control of suspension structure, *e.g.* by minimizing physical processes that coarsen chain structure. Halsey & Toor[14] have proposed a mechanism that would lead to relatively slow formation of thick chain structures following application of an electric field. Simulations of ER fluids under shearing flow[3–5,7], which follow the complete evolution of suspension structure, also predict chain thickening with time. This aggregation of particles into fewer, thicker chains with flow, provides a mechanism for thixotropy, the decrease in viscosity with time. Both of these mechanisms would affect rheology through structure changes.

From this study it should be clear that the feasibilty and development of ER technology will benefit from continuing investigation of microstructural iteractions in these suspensions.

References

1. H. A. Barnes and K. Walters, *Rheologica Acta,* **24** (1985) 323.
2. D. J. Klingenberg and C. F. Zukoski, *Langmuir,* **6** (1990) 15.
3. D. J. Klingenberg, F. van Swol and C. F. Zukoski, *J. Chem. Phys.* **91** (1989) 7888.
4. D. J. Klingenberg, PhD Thesis, Dept. Chem. Eng., University of Illinois, Urbana (1990).
5. R. T. Bonnecaze and J. F. Brady, *J. Chem. Phys.* (1992) Submitted.
6. R. T. Bonnecaze and J. F. Brady, *J. Rheology* (1992) Submitted.
7. R. T. Bonnecaze, PhD Thesis, Dept. Chem. Eng., California Institute of Technology, Pasadena (1991).
8. A. M. Kraynik, in *Proceedings of the Second International Conference on ER Fluids,* ed. J. D. Carlson, A. F. Sprecher and H. Conrad (Technomic, Lancaster, PA, 1990) 445.
9. W. M. Winslow, *J. Appl. Phys.* **20** (1949) 1137.
10. L. Marshall, J. W. Goodwin and C. F. Zukoski, *J. Chem. Soc. Faraday I,* **85** (1989) 2785.
11. J. F. Brady and G. Bossis, *Ann. Rev. Fluid Mech.* **20** (1988) 111.
12. G. K. Batchelor, *J. Fluid Mech.* **83** (1977) 97.
13. Y. M. Bashir and J. D. Goddard, *J. Rheology,* **35** (1991) 849.
14. T. C. Halsey and W. Toor, *Phys. Rev. Letters,* **65** (1990) 2820.

YIELD STRESS AND FIELD INDUCED STRUCTURE IN ELECTRO AND MAGNETORHEOLOGICAL SUSPENSIONS

G. BOSSIS, Y. GRASSELLI, E. LEMAIRE
Laboratoire de Physique de la Matière Condensée - URA 190
Parc Valrose - 06108 NICE Cedex 2 - France

and

L. PETIT
Laboratoire de Physique, Ecole Normale Supérieure de Lyon
46, Allée d'Italie, 69 LYON Cedex 7 - France

and

J. PERSELLO
Rhône Poulenc Recherches
52, rue de la Haie Coq
93302 AUBERVILLIERS

ABSTRACT

Static yield stresses obtained in magnetorheological fluids are compared with the predictions of the single chain model. This model is shown to overpredict the experimental results. The structure formed by the application of a magnetic field is studied by optical microscopy and image analysis. The structuring field and the average size of the aggregates measured for different volume fractions well compare with the predictions of a model based on free energy minimization. The dynamic structure obtained for a silica based electrorheological fluid in an oscillating shear flow shows the formation of stripes aligned in the direction of the flow.

1. Introduction

Magnetorheological (M.R.) as well as electrorheological (E.R.) suspensions exhibit a fundamental change of structure when submitted respectively to a magnetic or an electric field. In both cases this change of structure comes from the tendency for two polarized particles to align themselves in the direction of the applied field. This dipolar-type interaction is responsible for the formation of aggregates which are stretched in the direction of the field. These aggregates easily span the interval between two plates which are a few millimeters apart and act as solid links between them; in this state the suspension does not flow and can be considered as a gel. The consequence is that we need to apply a finite stress, the static yield stress, τ_s, in order to break this gel and to obtain a flow. Almost any type of suspension, when submitted to a high enough electric field, shows this kind of behavior. Actually the only essential prerequisites are the presence of an insulating suspending fluid which will allow to apply a high electric field and the existence of a difference of permittivity between the particles and the suspending fluid. Numerous observations of static yield stresses have been done on different kinds of suspensions in the presence of an electric field which aim at the discovering of the best ER fluid[1] (characterized by the higher ratio of the yield stress to the square of the electric field). In comparison,magnetorheological suspensions have been less studied, principally because small enough magnetic particles (a few microns) are less easy to obtain and furthermore their high density does not facilitate their dispersion and their study. A few years ago, Rhône Poulenc synthesized some polystyrene particles containing up to 60% in volume of magnetite. These particles have a reasonable density ($\rho \approx 2$) and do not sediment too quickly. We have shown that, as expected, they present a magneto-rheological effect and more interestingly, that **the yield stress we can measure depends fundamentally on the interactions between the particles and the walls.** If the disks of the cell are made of a magnetizable material like iron with a roughness much larger than the diameter of the particles ,we believe that the attraction between the particles and the wall surface is stronger than the interparticle interactions. In such a case we can hope to measure an intrinsic static yield stress and we have found that the experimental results were about 60% lower than the theoretical yield stress predicted by the single chain model.

In the next section we shall discuss the application of this model to the prediction of the static yield stress both for dielectric and magnetic particles. Then we shall try to extract some experimental information on the average shape of the aggregates and to understand how this can be used to predict the static yield stress or more generally the rheological behaviour of these fluids. Finally we shall briefly describe how the static structure obtained for a silica suspension can change in the presence of an oscillating flow and exhibit some quite unexpected features.

2. Single chain model - a comparison with experiment

The static yield stress can be predicted if we assume a simple ideal structure where all the particles are spherical, monodisperse and belong to single chains which span the interval between the two disks of the measurement cell. If N_c/S is the number of chains per unit surface of the cell and F_r the rupture force needed to separate two spheres, then we have :

$$\tau_s = \frac{N_c}{S} F_r = \frac{\varphi}{\frac{2}{3}\pi a^2} F_r \qquad (1)$$

where φ is the volume fraction of the solid particles and a their radius.

The rupture force is the maximum restoring force calculated by Klingenberg for two dielectric spheres[3,4].

$$F_r = 3\,\varepsilon\, a^2 E^2 \beta^2 f \qquad (2)$$

In this expression ε is the permittivity of the suspending fluid , ε_p the permittivity of the solid particle,and $\beta = (\varepsilon_p - \varepsilon) / (\varepsilon_p + 2\varepsilon)$. The quantity f is actually a complicated function of the separation between two spheres, of the orientation of the doublet relatively to the imposed field, and of the ratio $\alpha = \varepsilon_p/\varepsilon$. In the case of conducting spheres ($\varepsilon_p \to \infty$) the quantity f diverges at contact as the inverse of the separation between the spheres[5]. The relations (1) and (2) reflect two different levels of approximation. The first one is related to the organization of the particles in ideal chains of equally spaced particles, the second one to the kind of interaction between the particles which is assumed to come uniquely from electronic polarization. An ionic polarization can also communicate to the particle an apparently high permittivity but the forces can no longer be calculated by solving Maxwell's equations in the presence of two dielectric spheres, and Eq(2) is not applicable.

The application of Eq. (2) to magnetic particles is valid as long as the magnetization M of the sample remains proportional to the magnetic field :

$$M = \chi H \qquad (3)$$

In this case we only need to replace E by H, the average magnetic field inside the suspension and the permittivity ε by the permeability μ. We nevertheless want to stress two things :

1. The Maxwell field inside the suspension is just given by $E = \Delta V/l$ where ΔV is the

difference of potential between the two electrodes separated by a distance l, whereas $H = H_0 - 4\pi M$ with H_0 being the external field in the absence of the sample. The demagnetizing field lowers the field inside the sample by the quantity $4\pi M$. This is not the case for the electric field since we are not operating with constant charges on the plates but with a constant potential.

2. The permittivity ε or the permeability μ in the expression of F_r should, in a mean field approach, characterize the average medium surrounding the two test spheres. Consequently we have to take for ε (or μ) the average permittivity (permeability) of the suspension instead of that of the solvent. For E.R. fluids this can explain why the static yield stress experimentally does not increase as the volume fraction (Eq. 1) but more slowly; this effect was demonstrated for M.R. fluids[2].

The condition of linearity, Eq. (3), is only satisfied at low fields (typically H < 100 Oersted). For higher fields it is difficult to estimate the error, due to the saturation of the magnetization, on the derivation of Eq. (2). We have verified the applicability of Eq. (2) with the help of a macroscopic experiment on centimetric spheres placed in a magnetic field. Our previous results[6] were obtained for a chain of iron spheres and we found a rupture force about three times larger than the theoretical one calculated with the help of Eq. (2) and of the measured permeability of iron. We have repeated this experiment with steel spheres used in ball bearings and this time the measured rupture force is within a few percent of the predicted one. It turns out that the higher value obtained with iron is due to the larger surface roughness of these spheres:the presence of sharp bumps enhances the local magnetic field and so increases the attraction between the spheres.

Despite this good prediction of the rupture force by Eq.(2) the static yield stress predicted from Eqs. (1) and (2) is about 60% higher than the experimental one. It remains to question Eq. (1) i.e. the single chain model. First of all we must note that our suspension of magnetic particles was polydisperse and so could not fit with the model of ideal chains of spheres. Furthermore we know that the structure developed under the action of an external field is constituted of linear aggregates whose cross section can be many times that of a single particle. In order to get some information on the size of these aggregates, and also on the field necessary to form them, we have devised an experiment to analyze the structure with an optical microscope coupled to an image analysis.

3. Field induced structure in M.R. fluids

The suspension of magnetic particles was sealed between two glass plates separated by parafilm sheets and placed under the microscope objective. The magnetic field is perpendicular to the glass plates and is applied progressively with the help of a current generator which is monitored by a computer. At the same time we record the intensity of the light which is transmitted through the sample. A typical record is shown in fig. 1.

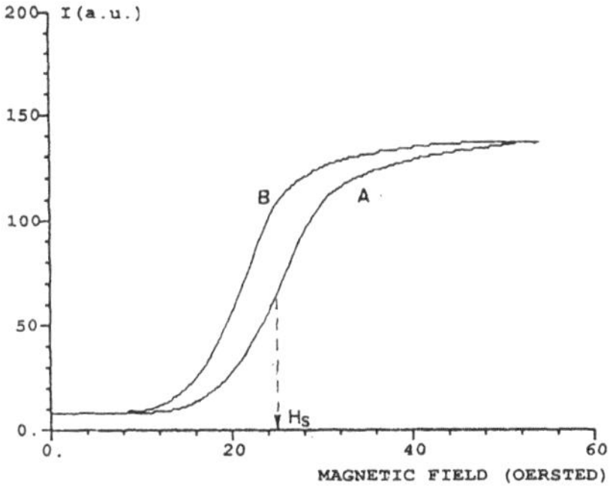

Fig. 1. Intensity of the transmitted light versus the magnetic field for a M.R. suspension containing a volume fraction $\Phi = 0.1$ of magnetic particles
A : rising time 1 Oersted/mn B : falling time 5 Oersted/mn.

The intensity of the transmitted light remains constant or slightly decreases up to a value of 15 Oersted for the magnetic field. Then it increases quickly between 20 and 35 Oersted and more slowly for higher fields. The increase of the transmitted light is due to the formation of paths inside the suspension which are free of particles an do not absorb the light. With this experiment we can define a structuring field, H_S, by the value of the field corresponding to the inflexion point on the rising curve A. This field can characterize the completion of the aggregation process only if the structure is in equilibrium with the applied field. This will not be the case if we do not take care to increase the field slowly enough. The curve B which corresponds to a falling time which is five times the rising time is clearly out of equilibrium : the aggregates formed at a higher field need some time to disaggregate by Brownian motion. For this suspension we need a rate of order 1 Oersted/mn in order to attain equilibrium in which case we do not observe any hysteresis.

If we look at an image obtained for a field higher than the structuring field we see the structure shown in Fig. 2. The black parts are the projections of the sections of the aggregates. The magnetic field is perpendicular to the plane of the picture. We can see for this suspension whose volume fraction is 10%, that the cross section of the aggregates become irregular with the formation of lateral fingers. This phenomenon is more and more pronounced as the volume fraction increases. On the contrary ,at low volume fractions , ($\Phi < 5\%$) the cross section is quite circular. This irregular and somewhat elongated cross section is reminiscent of the labyrinthine instability observed in ferrofluids[7]. Note that we have observed the same shapes for the cross section of aggregates formed in an E.R. fluid[8].

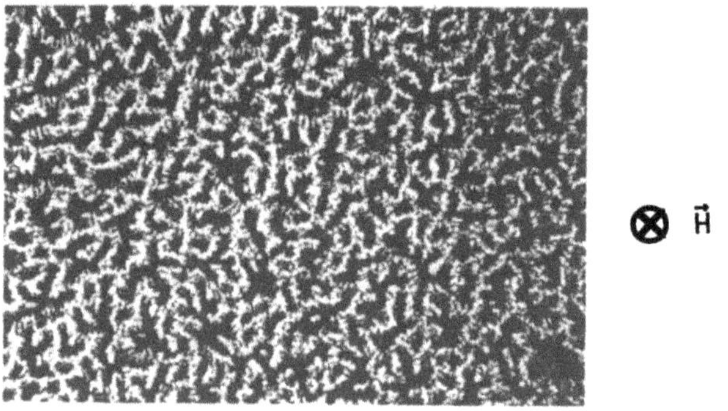

Fig. 2. Structure of a M.R. suspension with a volume fraction of 10%. The magnetic field is perpendicular to the plane of the picture; lengthscale of the picture : 250μ.

From these images we have determined by thresholding and image analysis the center of gravity of each aggregate. Then we can calculate a pair correlation function between the positions of these centers of gravity and deduce the average distance $<d>$ between them.

We have found that the ratio $<d>/l$ where l is the thickness of the sample is constant within the uncertainty of our results in the range $100 < l < 300$ μm. The ratio $<d>/l$ has a value of 0.1 at the lower densities and slightly increases with the volume fraction, to attain 0.15 for a volume fraction of 10%. The corresponding average radius, b, of the aggregate is easily calculated if we assume that the volume fraction inside the aggregates is that of a randomly packed suspension : $\Phi = 0.64$. It is interesting to note that in order to obtain for the radius of the aggregates a value close to the radius of a particle, we need a separation between the plates of a few tens of microns; in yield stress measurements with a plate-plate configuration the gap is often larger than 100 μm so we do not have any chance to get single chains of spheres between the plates.

Concerning the structuring field ,H_S ,we find that it increases linearly with the volume fraction : it is easier to obtain a structuration of the suspension at low volume fraction than at high volume fraction. These results - $\langle d\rangle/l \approx 0,1$ and the linear behaviour of the structuring field - can be theoretically obtained from a minimization of the magnetic energy of an assembly of ellipsoids with a mean field approach[8]. Furthermore this model allows to relate the slope of the curve $H_S = f(\varphi)$ to the internal permeability of the particles, as has been verified experimentally. These results do not concern flowing suspensions but they surely could be useful to build a model based on thick fibers of particles instead of single chains, and then contribute to improve the theoretical prediction of the static yield stress.

4. Structure of E.R. fluids in an oscillating shear flow

In order to understand the rheology of E.R. or M.R. fluids we need to visualize the change of structure induced by the interplay between shear and electric forces. We begun this study with an E.R. fluid, composed of silica spheres in a silicone oil ,placed between two glass plates covered with semi-transparent electrodes. One glass plate is fixed and the other is mounted on a vibrating arm; this apparatus allows us to observe the change of structure with a microscope and at the same time to measure the rigidity modulus. The electric field is parallel to the optical axis of the microscope and when the suspension is at rest we obtain a structure which looks like the one represented in Fig. 2. In the presence of an oscillating flow we can follow with a stroboscoped light the motion of each individual fiber, and we have first established that there was no slipping on the plates - at least when the strain was less than unity. For a given field and a given frequency of the oscillating plate (typically E = 40 KV/cm and ν = 60 Hz) we observe different types of structure depending of the amplitude of the strain. At low strain amplitude ($\gamma < 0.1$) the fibers remain more or less cylindrical. When the amplitude increases ($0.1 < \gamma < 1$)they become anisotropic and take the form of sheets whose plane is perpendicular to the velocity lines. These sheets progressively arrange themselves in a well-ordered structure with alignments which are perpendicular to the velocity lines (cf. Fig. 3).

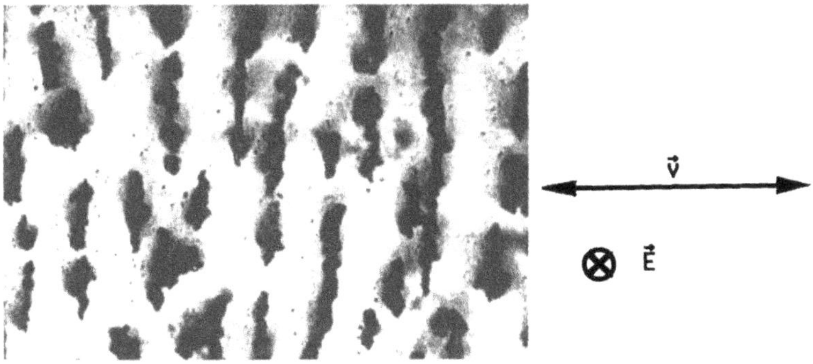

Fig. 3. Structure of an E.R. suspension in an oscillating shear flow at a frequency $\nu = 63$ Hz and a strain amplitude $\gamma = 0.2$. The electric field E = 40 KV/cm is perpendicular to the plane of the picture. The velocity lines are perpendicular to the alignments.

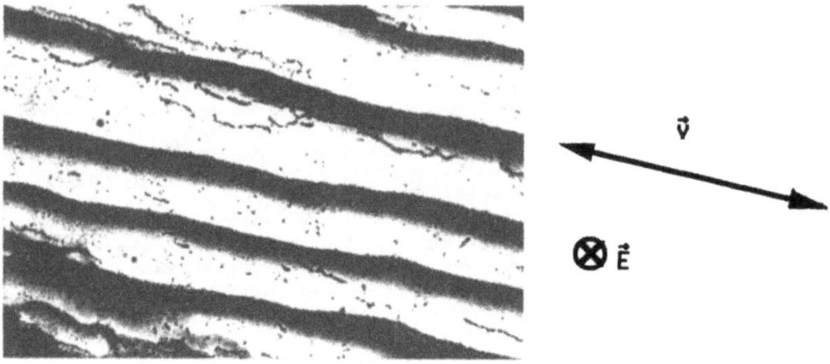

Fig. 4. Same conditions as in figure 3 but with a higher strain amplitude $\gamma = 1$. We observe the formation of stripes which are aligned on the velocity lines

If we increase the strain amplitude to about $\gamma = 1$ the picture 3 transforms into picture 4 where we now observe continuous stripes of particles which follow the direction of the velocity lines. These structures are quite puzzling and their explanation involves on one hand the minimization of the electric energy which can favour the formation of sheets of particles since it lowers the depolarization field and on the other hand the kinetic effect of the flow which can arrange the particles in lines or planes even in the absence of an electric field. This is a rich domain which asks a systematic investigation in order to quantify these structural changes as a function of the ratio of electric and shear forces.

We have also found that the shear modulus was very sensitive to the kind of structure which was developed and it seems clear that all the rheological behaviour will also be sensitive to these changes in the spatial organization. We believe that these results underline the need for the development of rheooptical studies to be able to interpret the rheological behaviour of E.R. and M.R. fluids.

5. Acknowledgments

This work was realized with the financial support of Rhône-Poulenc. We are also very grateful to A. Audoly for his technical support and to Dr. Cagan who helped us to measure the magnetization of our samples.

6. References

1. H. Bloch and J.P. Kelly, *J. of Physics D : Appl. Phys.*, **21** (1988) 1661.
2. E. Lemaire and G. Bossis, *J. Phys. D : Appl. Phys.*, **24** (1991) 1473.
3. D. J. Klingenberg, Master Thesis University of Illinois, (1989).
4. D. J. Klingenberg and C.F. Zukoski, Langmuir, **6** (1990) 15.
5. P.A. Arp and S.G. Mason, *Colloid and Polymer Science*, **255** (1977) 566.
6. G. Bossis and E. Lemaire, *J. of Rheology*, **35**(1991).
7. R.E. Rosensweig, *Ferrohydrodynamics*, Cambridge University Press (New York, 1985).
8. E. Lemaire, Y. Grasselli, G. Bossis, *J. de Physique II*, 1992, in press.
9. G. Bossis, E. Lemaire, J. Persello, L. Petit., *Prog. in Coll. and Polym. Sci.*, 1992, in press.

EFFECTS OF FINITE CONDUCTIVITY IN ELECTRORHEOLOGICAL FLUIDS

R. A. ANDERSON
Division 1812, Sandia National Laboratories
Albuquerque, NM 87185, USA

ABSTRACT

Electrorheological (ER) effects arise from electrostatic forces between particles suspended in a carrier fluid, induced when an electric field is applied. These forces are a consequence of field distortion in the neighborhood of each particle. Although it is known that finite conductivity can influence ER effects, the standard assumption has been that the contrast between the dielectric constants of particles and surrounding fluid is responsible for the field distortion. This is likely to be seriously in error with a low-frequency or dc applied field if measurable conductivity is present in either the particles or fluid. Below a crossover frequency, f_c, field distortion becomes dominated by conductivities, and a variety of other complications may arise. Conductive particles could acquire a net charge and fail to participate in chain formation, and a relatively high fluid conductivity may lead to competing processes, including the possibility of a significant distortion of the field distribution in an ER cell. The latter phenomenon is not directly related to f_c and might occur at frequencies above it. Any of these effects could lead to field dependent or frequency dependent behavior of experimental results.

1. Introduction

Within the approximation of isolated point dipoles, the force between a pair of particles in an electrorheological (ER) suspension is proportional to the square of the applied electric field, the dielectric constant of the liquid phase, and the square of a quantity β which is a measure of the dipolar field distortion around a particle. The standard assumption is that β depends on the contrast between the dielectric constants of a particle and the surrounding fluid. For a spherical particle,

$$\beta = (\kappa_p - \kappa_f) / (\kappa_p + 2\kappa_f) \tag{1}$$

where the subscripts of the relative dielectric constants, κ, denote particle or fluid (the solid or liquid phases of the ER fluid). This simple expression, which incorporates only the capacitive part of the complex permittivity, is likely to be seriously in error with a low-frequency or dc applied field if a measurable conductivity is present in either the particles or fluid. Forces calculated from this basis do not reliably predict the presence or absence of ER effects at very low frequencies, nor do they account for the behavior of some commercial fluids where ER effects are confined to frequencies approaching dc.

The fact that finite conductivity can influence ER effects is well known, and recent review articles point out a number of conductivity related effects[1,2]. However, it may not be widely recognized how significant such effects might be. It is the intent of this paper to demonstrate, with the aid of a few examples, that the effects of finite conductivity may need to be taken into consideration.

The most straightforward effect of finite conductivity is to alter particle polarization at low-frequency or dc applied fields. Fig. 1(a) depicts the electric field lines in the neighborhood of a particle having a larger permittivity than the surrounding fluid, and the value of β is near unity. With finite conductivity, the polarization of the particle under dc conditions will eventually depend on conductivities, σ, rather than dielectric constants. For simplicity the conductivities are presumed to be uniform, and the β parameter becomes

$$\beta = (\sigma_p - \sigma_f) / (\sigma_p + 2\sigma_f). \tag{2}$$

The analogy between this expression and Eq. 1 is easily seen. The quantities within the parentheses in these two equations simply reflect the two frequency limits of the complex permittivity, $\kappa\varepsilon_0 - i\sigma/\omega$.

The behavior of a particle with a conductivity lower than the fluid helps to illustrate the transition between Eqs. 1 and 2. Charges traveling along the field lines in Fig. 1(a) will accumulate on the surface of the particle until the electric field becomes distorted, as in Fig. 1(b), so that current is diverted around the particle. Here the value of β is approximately -1/2.

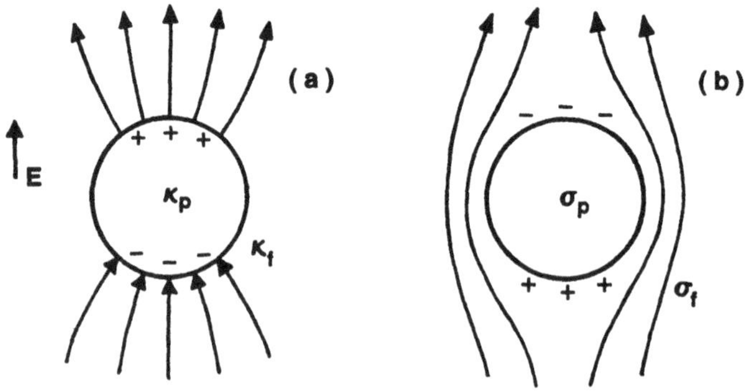

Figure 1. Electric field distortion near an isolated spherical particle in a surrounding fluid. (a) Capacitance dominated regime: As shown, $\kappa_p > \kappa_f$ and $\beta \approx 1$. (b) Conductance dominated regime: As shown, $\sigma_f > \sigma_p$ and $\beta \approx -1/2$.

The transition from capacitance to conductance dominated regimes at low frequencies is governed by a resistance-capacitance time constant,

$$\tau = (\kappa_p + 2 \kappa_f) \varepsilon_0 / (\sigma_p + 2 \sigma_f) \tag{3}$$

where ε_0 is the permittivity of free space. The crossover frequency, inversely proportional to τ, is given by

$$f_c = (2 \pi \tau)^{-1} . \tag{4}$$

As an example, $f_c \approx 1$ Hz if $\kappa_p = 300$ and $\sigma_f = 10^{-10}$S/cm.

It should be pointed out that Eq. 3 was obtained under the assumption of an isolated spherical particle. Nearby particles in ER suspensions clearly will have an effect on τ. Shifts in the crossover frequency of less than a factor of two, at particle concentrations of 30% by volume, were deduced from treating the surrounding fluid as an effective medium[3]. (These calculations were performed with several different parameter choices, including the values in the above example.) A larger perturbation might be expected if the neighboring particles were in chains or columns rather than distributed uniformly according to an effective medium treatment. Nevertheless, Eq. 3 provides a useful, order-of-magnitude criterion for the onset of conductance domination.

The above example shows how readily the conductance dominated regime can be entered; the value of fluid conductivity that was used is not unrealistically high. Changes in particle polarization, however, are not the only complications that can arise with finite conductivity, and that is the focus of the remainder of this paper.

2. Particle Conductivity

With conductive particles in a nonconductive fluid, β approaches unity at frequencies smaller than f_c. Thus conductive particles might be expected to mimic the behavior of high-permittivity particles. However, chains of highly conductive particles would surely short out a test cell or cause catastrophic electrical breakdowns. In fact, it is far from obvious what happens with even moderately low levels of conductivity.

Consider a pair of conductive spheres, which have become polarized as depicted in Fig. 2(a). These particles are drawn together by their induced moments. If the spheres make electrical contact when they touch, a redistribution of charge as in Fig. 2(b) is required to make the pair an equipotential. The new charge arrangement results in a net charge on each sphere, giving rise to a force in the applied field which tends to pull the spheres apart. If the charge distribution were fully equilibrated, this disruptive force would be several times stronger than the attraction between the opposite charges on the two spheres. Clearly, the pair will eventually come apart after making electrical contact. The consequence of electrical contact between a conductive particle and an electrode is similar.

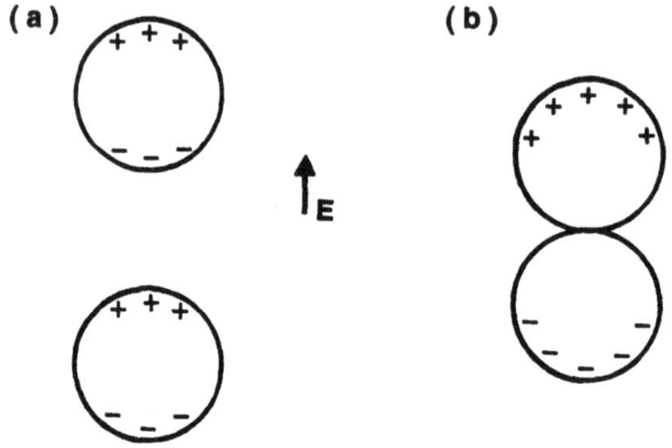

Figure 2. Interaction of two conductive particles. (a) Initial polarization results in attraction. (b) Charge distribution is altered after electrical contact, and the particles will be pulled apart in the applied field.

How can these considerations be reconciled with the observation of strong low-frequency or dc ER effects in some suspensions of weakly conductive particles? Perhaps the stabilizing agent provides a thin layer of insulation on the surface of each particle that prevents intimate electrical contact. With weak interparticle electrical contacts, one might expect frequency dependences arising from the slow redistribution of charge. Since the pressure between particles is expected to vary as the square of the applied field, and the electrical contact might be pressure dependent (for example, from elastic deformation of thin insulating layers on the particles), field dependences or even field-dependent frequency dependences might be expected.

Another explanation of strong dc ER effects with conductive particles can be proposed. Chains or columns may be able to bridge the electrodes before being disrupted by the growth of net charge on individual particles. Contact at both electrodes would allow a steady current to percolate through a column, as indicated by the arrows in the enlarged portion of Fig. 3, and the high electric field near each weak electrical contact would provide the electrostatic force that binds the structure together. Ragged edges such as the darkened particle near the center of the column, however, might tend to acquire a sizeable net charge and be torn away. With this mechanism of chain stabilization, one could expect to observe several maverick particles shuttling back and forth between the electrodes and not participating in stable aggregations.

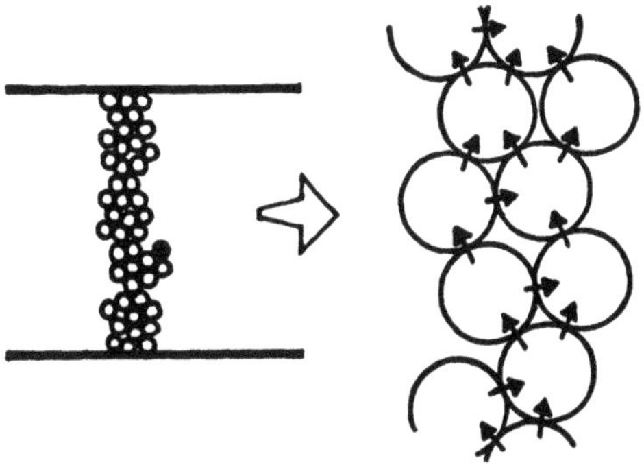

Figure 3. Column of conducting particles bridging the electrodes. A steady current can percolate along the chain (small arrows in enlargement). High fields near the electrical contacts stabilize the structure. Protruding particle (shown darkened) may acquire a net charge and be torn away.

An obvious question is prompted by this steady-current hypothesis. Is it reasonable to assume that chains have time to form before being disrupted by the growth of net charge on the particles? A quantitative evaluation of the time scale for net charge growth is impracticable since it would depend on unknown geometric details at the interparticle contact. Nevertheless, it is immediately clear that the localized electrical resistance at a contact region is likely to be much larger than the effective resistance between the two halves of a single particle. Growth of net charge would therefore require a much longer time than individual-particle polarization, characterized by τ in Eq. 3. Since chain formation will start as soon as the individual particles polarize, chains can bridge the electrodes before being disrupted if the chain formation is rapid compared to net charge growth.

3. Fluid Conductivity

3.1 General Considerations

A variety of physical phenomena can give rise to measurable conductivity in a dielectric liquid[4]. Electronic carriers in some fluids may account for the conductivity. With polar molecules, the equilibrium ionization of the fluid or of an impurity may be responsible. Electrode processes also could be important, electronic carrier injection for example, and electrochemistry could provide a source or sink for ions in solution. The presence of significant amounts of impurity ions might be expected in an ER fluid due to the large surface area of the solid phase. It seems likely that the conductivity of many ER fluids is ionic in nature.

Ionic conduction immediately raises the complication of diffuse-layer phenomena. The Debye screening length for a singly charged ion is given by[5]

$$L_D = (\kappa_f \, \varepsilon_0 \, k_B T / 2 \, n \, e^2)^{1/2} \tag{5}$$

where $k_B T$ is the Boltzmann thermal energy, n is the ion density, and e is the electronic charge. This length becomes comparable to the particle size or spacing (of the order of μm) at an ion density of $10^{13}/cm^3$. However, the potential drop over this length at typical electric fields in ER studies is of the order of a volt, which is much larger than $k_B T/e$ at room temperature ($\approx$ 0.025 V), and the approximation that L_D is a simple exponential scale length no longer applies[5]. Under this circumstance nearly all of the ions in the diffuse screening layer would be packed tightly against the surface of a particle, as implicitly assumed in Eq. 2. To illustrate another aspect of this point, we can estimate the distance over which diffusion affects ionic transport in an ER fluid. In other words, at what length scale is the diffusion length, $2(Dt)^{1/2}$, comparable to the drift distance, μEt, in the electric field? The diffusivity, D, tracks the mobility, μ, according to the Einstein relation

$$D = \mu \, k_B T / e \tag{6}$$

and we find that the length of diffusive influence is given by

$$\xi = 4 \, k_B T / e \, E \, . \tag{7}$$

At room temperature ξ is $\approx$ 0.03 μm at a field of 3 kV/mm, and is considerably smaller than typical particle dimensions. From this discussion it would appear that ionic charge carriers in ER fluids, to a good approximation, simply flow along electric field lines until stopped by a surface barrier where charge accumulates.

A further simplification is available if there is a constant density of charge carriers in the liquid phase, for example from an equilibrium ionization process. This may approximate the usual case with conductive ER fluids. A constant density of carriers leads to a conductivity that is uniform and independent of time, and Eqs. 2 thru 4 provide a good description of the behavior of isolated particles. Alternatively, if the ion supply is fixed so that ions swept to the electrodes are lost, then ion-depletion layers will form which may significantly alter the electric field in the fluid. Such a possibility will be discussed in section 3.3.

3.2 Ion Mobilities and Densities

Thermal considerations impose an upper bound on the conductivity of a practical ER fluid. A reasonable value might be 10^{-7} S/cm, at which the Joule heating reaches 90 W/cm^3 with a field of 3 kV/mm. This value can be used to deduce the maximum ion density in a practical ER fluid, which will be a useful quantity in section 3.3. In order to accomplish this, we require an estimate for the mobility of an impurity ion.

Ionic mobility is strongly dependent on the ion species[6], but an order-of-magnitude value for an impurity ion in a low-viscosity (≈ 1 cp) fluid is 10^{-3} cm²/Vs. With a fluid with higher viscosity, for example silicone oil (≈ 30 cp), proportionately smaller ion mobilities could be expected. For the purposes of this discussion we will adopt the order-of-magnitude value 3×10^{-5} cm²/Vs for a 30-cp fluid. In the interest of simplicity, ions of either sign are assumed to be singly charged and to have the same mobility. The ion densities at the Joule heating limit are found to be about 3×10^{14}/cm³ or 10^{16}/cm³, at 1 or 30-cp viscosity, respectively. As will be seen in section 3.3, ionic space charge effects can occur at ion concentrations orders of magnitude smaller than these values.

It is informative to determine the interelectrode drift time for ions having the above mobilities. At $\mu = 10^{-3}$cm²/Vs or 3×10^{-5}cm²/Vs, and an electric field of 3 kV/mm, an ion traverses the gap in about 3 or 100 ms, respectively. Accordingly, frequency dependences related to ionic drift in a 1-mm-thick ER cell might be expected to occur in the range from about 10 Hz to a few kHz, depending on the fluid viscosity. The low end of this range would correspond to the higher viscosity.

3.3 Ion-Depletion Layers

Fig. 4 depicts ion drift in a dc electric field when the supply of ions is limited. The positive and negative populations drift in opposite directions, causing depletion layers to appear at both electrodes. These layers of space charge eventually meet near the center of the interelectrode gap (if the mobilities of positive and negative ions are comparable), then shrink back as the remaining ions are swept away. As discussed in section 3.2, the time scale of this process falls in the range of a few ms to about 100 ms, depending on fluid viscosity, in a 1-mm-thick ER cell.

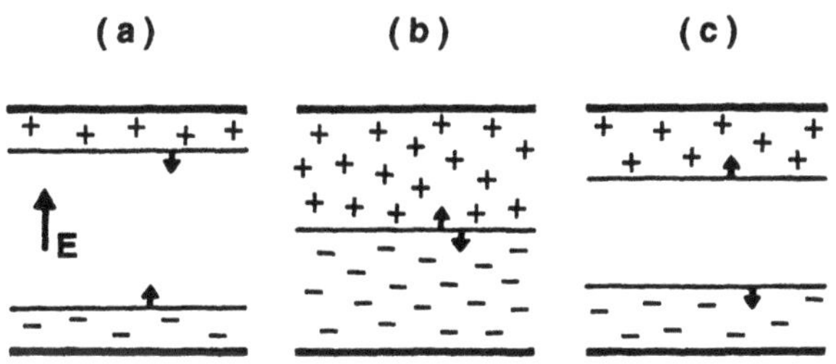

Figure 4. Ion drift in an electric field with a fixed supply of ions. Ion depletion causes space charge layers to form adjacent to the electrodes; (a), (b), and (c) indicate the time evolution of the space charge.

The consequences of these space charge layers clearly depend on the original density of ions in the fluid. At ion concentrations well below some critical density, n_c, the ions are simply swept away as depicted in Fig. 4 with very little distortion of the applied electric field. In this case the fluid conductivity would decay rapidly to near zero as ions are lost. By contrast, at ion concentrations near or above n_c, the electric field would become strongly enhanced near each electrode, but reduced over much of the interelectrode volume.

The density n_c can be estimated from the situation shown in Fig. 4(b), by requiring the field at gap center to be zero. Under the approximation the space charge corresponds to singly charged ions, at the original ion concentration (which ignores the effect of field distortion in the ion drift), the critical density is

$$n_c = 4\, \kappa_f\, \varepsilon_0\, E / e\, d \tag{8}$$

where d is the electrode spacing. Evaluating Eq. 8 with a relative dielectric constant of 3, a field of 3 kV/mm, and a gap of 1 mm yields a density of $2 \times 10^{12}/\mathrm{cm}^3$. This is two or more orders of magnitude below the upper bounds for ion concentrations from section 3.2 and corresponds to a rather innocuous conductivity of 2×10^{-11}S/cm in a 30-cp-viscosity fluid.

A limited supply of ions could also affect particle polarization at frequencies below f_c. A rough estimate of the amount of ionic charge immobilized at the surface of particles can be made under the assumptions leading to Eq. 2 (isolated spherical particles in the dc limit). If the particles are nonconductive, the number of singly charged ions normalized to the ER fluid volume is given by

$$n_p = 1.125\, c\, \kappa_p\, \varepsilon_0\, E / e\, a \tag{9}$$

where c is the volume fraction of the particles in the ER fluid and a is the particle radius. From Eqs. 8 and 9, the ratio n_p/n_c is about $0.1(\kappa_p/\kappa_f)(d/a)$ at a particle concentration of 30% by volume. This ratio is likely to be at least 10^4 with high-κ particles and electrode spacings of the order of 1 mm. Clearly, with a fixed amount of ions, a much larger initial ion density is required for the conductivity dependent polarization of the particles than for a significant field distortion from ion-depletion space charge layers. At ion densities near n_c we would expect to observe effects mainly related to the distortion of the applied field, although the particles would mitigate this effect by attempting to immobilize ions of either sign. Particle polarization and space charge effects may occur together at higher ion densities.

4. Combined Conductivity

As a final example, we consider an interesting case in which both particles and fluid have appreciable conductivity. Each particle, however, is presumed to be covered with a very thin dielectric layer, which could represent a barrier for ion transport, an effect of the stabilizing agent, or some intentionally applied surface treatment[7]. Fig. 5 depicts the polarization charges that develop upon application of an electric field, with an isolated spherical particle of this type. Notice that the internal and external polarizations are in contention. The internal conductivity is attempting to collapse the potential difference between the top and bottom of the particle while the external charges seek to increase the potential difference in order to divert the current, as in the case of a nonconductive particle. The consequence of this would be a very high electric field in the thin dielectric layer, with an enhancement factor of approximately 1.5 a/δ relative to the macroscopic applied field, where a/δ is the ratio of particle radius and dielectric layer thickness. The enhanced field could approach typical dielectric strengths if δ is of the order of tens of nanometers. Clearly, the effective β will depend on the high-field electrical properties of the dielectric layer.

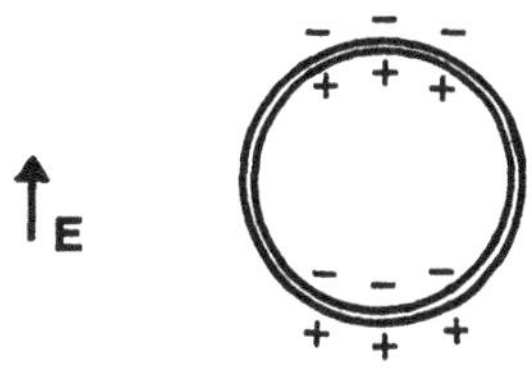

Figure 5. Polarization of a conductive particle having a thin dielectric coating, surrounded by a conductive fluid. Opposing internal and external polarizations greatly enhance the electric field in the thin dielectric layer.

5. Summary

Interparticle forces in ER fluids can be described with comparative simplicity if the electric field is applied at an ac frequency well above a crossover frequency, f_c. Under this condition the field distortion around the particles is governed by the contrast between the dielectric constants of the particles and the surrounding fluid, according to the familiar expression for the β parameter. It is a further simplification if the conductivity of the liquid phase is low, less than roughly 10^{-11} S/cm with a viscosity of 30 cp, so that any field modification from ion-depletion space charge layers is insignificant.

If either of the above conditions fails to hold, a variety of complications may arise. At frequencies below f_c the β parameter becomes dominated by the contrast between particle and fluid conductivities, rather than dielectric constants. Furthermore, conductive particles may acquire a net charge and fail to participate in chain formation, and a relatively high fluid conductivity may lead to competing processes, including the possibility of a significant distortion of the field distribution in an ER cell if the supply of mobile ions in the liquid phase is fixed. The latter phenomenon is not directly related to f_c and might occur at frequencies above it. If both particles and fluid are conductive, particle polarization may becomes sensitive to the high-field electrical properties of the particle-fluid interface. Effects such as these could lead to field, frequency, and time dependent behavior that must be accounted for when experimental results are interpreted.

6. Acknowledgements

The author acknowledges helpful technical discussions with several colleagues at Sandia National Laboratories, especially D. B. Adolf, A. M. Kraynik, and J. E. Martin. This work was performed at Sandia National Laboratories, supported by the U. S. Department of Energy under contract number DE-AC04-76DP00789.

7. References

1. H. Block and J. P. Kelly, *J. Phys. D: Appl. Phys.* **21** (1988) 1661.
2. A. P. Gast and C. F. Zukoski, *Adv. Colloid and Interface Sci.* **30** (1989) 153.
3. B. E. Springett, *Phys. Rev. Lett.* **31** (1973) 1463.
4. T. J. Gallagher, *Simple Dielectric Liquids: Mobility, Conduction, and Breakdown* (Clarendon Press, Oxford, 1975) chapter 2.
5. C. A. Barlow, Jr., in *Physical Chemistry: An Advanced Treatise, Vol. IXA/Electrochemistry,* ed. H. Eyring (Academic Press, New York, 1970) chapter 2.
6. Ref. 4, chapter 1.
7. A. Inoue, in *Electrorheological Fluids: Proceedings of the Second International Conference on ER Fluids,* ed. J. D. Carlson, A. F. Sprecher, and H. Conrad (Technomic Publ. Co., Lancaster, PA, 1990) p. 176.

II. MATERIALS TECHNOLOGY OF ER FLUIDS

Semi-Conducting Polymers as ER Fluid Substrates.

by

H Block, P Rattray and T Watson

Centre for Molecular Electronics
School of Industrial and Manufacturing Science
Cranfield Institute of Technology
Cranfield, Bedford MK43 OAL, UK

ABSTRACT

Selected semi-conducting polymers provide effective substrates for ER fluids for which role they are particularly useful in that promoting additives such as water are not needed. Their mode of action, and indeed that of water promoted fluids, is believed to be related to the polarisation of such fluids under fields. For this reason, investigations of the mechanism of ER should involve the study of both their rheology and their polarisation under electric fields.

Below we present such results, involving both conventional dielectric measurements on ER fluids and the extension of these to situations where both high dc fields and flow fields are acting. Also given are rheological data relating to the Bingham behaviour of these fluids under dc fields and their responses under flow to a train of pulsed fields ($\leq$ 500 Hz). The ER fluids studied were mostly dispersed polymeric semiconductors, which were not promoted, with some data on water promoted polyelectrolyte based fluids. The results are discussed in terms of the ER outcome of polarisation and its time dependence and consequential suggestions made as to the mechanism of the ER effect.

Introduction

It is becoming accepted that the ER phenomenon starts with the polarisation of the active ER promoting particles dispersed in the electrically insulating base fluid. The method by which this is most effectively achieved often involves some form of interfacial polarisation caused by carriers drifting under the field until they are trapped at the particle fluid boundaries. It seems on the basis of experience[1-4] that neither the nature of the carrier (ionic, electronic) nor the region within the particles where carriers are moving (bulk, surface or even within a double layer adjacent to the surface) appear to limit the choice of ER substrate. What is important is that the carriers are mobile. In this connection it is now becoming accepted that the role of the classical

ER promoters such as water act because they reduce ion association thereby increasing their mobility, as well as providing some of the ionic carriers. Such classes of fluids, often designated "hydrated", depend upon the intrinsic presence of ions or their extraneous provision from the promoters, surfactants or impurities: they are ineffective when unpromoted. More recently, ER fluids have been developed which are free of promoting additives[3-8]. They also act by interfacial polarisation but do so because the carriers are mobile in the "anhydrous" state either because of rather specific structural conditions (for example ionic conduction in zeolites) or because the carriers are electronic.

There are of course mechanisms other than carrier drift to an interface which can cause polarisation (eg dipolar orientation) but the magnitudes of interfacial polarisations can be much larger, because the distances of charge separation are not limited to the order of bond dimensions but rather to the much larger interface separations. Thus it does not require many carriers per particle to provide a substantial induced dipole and hence bulk polarisation.

The rheological effects which are consequent upon polarisation involve induced dipole - induced dipole interactions which, if given the opportunity, result in the well known structuring and fibrillation of ER fluids[9,10]. It is this structuring when allowed to fully form which leads to the solidified state, and it is this state which is disrupted when a stress exceeding the static yield stress is applied. Then, initially and at low flows, the solid appears to fracture leaving considerable structuring nearer one electrode[10]. It is unlikely that such separation into a fluid and a solid region is retained at higher shear rates and very probable that further degradation occurs to give rather looser clusters or broken chains throughout the body of the fluid. However, there does not appear to be direct evidence about the extent of such clustering. The dipole-dipole interactions which are involved in the particle associations on forming or degrading clusters depend not only upon the ability of the field to induce significant dipoles but also upon their fluctuation in both direction and magnitude under the rotational tendencies of shear flows. In consequence a time element is introduced, even under dc fields.

For these reasons, a full understanding of the ER phenomenon and the consequent design of effective ER fluids involves both their dielectric and the rheological properties, and the dynamic interplay between them. Studying the dielectric response of ER fluids under flow (known as flow modified permittivity or FMP[11,12]) is an important adjunct to this, as are studies of the rheology of ER fluids under fields, both dc and time varying. Furthermore, such investigations should extend to cover conditions of high shear rates when the ER effect may be significantly curtailed or even disappear.

Polarisations and Interactions in Flowing Dispersions.

The significance of the dynamics on the ER mechanism is best introduced by considering how an individual polarisable particle in rotational flow might react to the imposition of an electric field. However, it must be stressed at the outset that this simple situation which ignores particle-particle interactions, is not what provides a strong ER effect, rather we start our considerations with this simple case to draw some initial conclusions which are then taken further to the more realistic situation of particle-particle interactions.

In flow and before any field is applied, the particle is in rotational motion and its angular velocity together with its rate of polarisation determines how effectively a dc field induces a dipole in the particle. Extremes can readily be identified. With rapid polarisation, the induced dipole is quickly established and aligned in the field direction and remains essentially co-linear with the field and of constant magnitude during particle rotation. On the other hand, particles whose polarisation is sluggish do not polarise well in flow in a dc field because the direction of the field in the particle frame is rotating too fast to give sufficient time to establish the dipole. The relative rates of the polarisation to the shear rate establish whether either of these conditions appertain, but in either event, there is no effect on the rheology even under large fields. The intermediate situation is different. A dipole will develop but its magnitude will be shear rate dependent because of the rotation of the particle. Also, because the decay of the induced dipole is also time dependent, the direction of the dipole is not co-linear with the field. There is a resistance to rotation of the particle because of the consequent torque between field and dipole, which implies the limiting viscosity number increases. The situation for spherical particles has recently been analysed[13] with the interesting prediction that particle rotation can be halted. This occurs when the energy dissipated in constantly rearranging the charge distribution within the particle exceeds that which would be dissipated in changing the flow field to accommodate the now non-rotating particle. The rotational behaviour of the particle depends upon the shear stress and field strength as well as its permittivity and conductivity and the viscosity of the dispersing liquid. It is predicted that the change between a state of flow involving locked spheres to rotating ones can occur discontinuously. This switching occurs at different stress levels depending on whether the locked state is being disrupted (increasing shear) or re-established (decreasing shear) with the implication that the shear stress - shear rate dependence then exhibits hysteresis. This is a very interesting result since the hysteresis observed in some concentrated ER fluids[3,4] may originate in an analagous manner. At the present time these recent theoretical prediction have not been put to the test. To do so would require shear stress - shear rate data to be measured at high dilutions.

The situation in concentrated dispersions is much more complex since particle - particle interactions play a major role whenever induced dipoles are generated. For sluggish polarisers the end result is similar to the case of single particles, since the rotational shear field attenuates or destroys the polarisation. If the fluid has been quiescent or flowing at such low shear rates that dipoles are induced and structuring is occurring, then the effect of increasing the flow rate in such fluids is to wipe out the underlying polarisation, with the result that the fluid will revert in its flow characteristics to that appertaining in the absence of field. If the field is switched on at higher shear-rates then there is no polarisation, interaction or ER effect. In either case the fluid would essentially fail in its ER role.

With ER fluids based on particles that polarise rapidly the field would induce dipoles even under high shear rates but these would not then result in chaining or clustering for the following reason. Any incipient chaining in the field direction would cause such a structure to be rotated into the flow lines, the polarisation directions would simultaneously relax so as to keep the induced dipoles aligned in the field direction which has the consequence that as the rotation proceeds, the dipole interactions become increasingly repulsive. The incipient chain would have no permanence, rather there would be a complex precessional motion between interacting particles. This may have an effect on the viscosity but it would not be as great as would be the case in an intermediate situation where some permanence of attractive dipole interactions survive, because of the time lag in their relaxation. In still conditions, fibrillation would still occur but the ease of dipole adjustment to any shear could imply that such fibrils are weak. Also, such rapidly interfacially polarising materials have a level of conductance which can cause problems if chains of particles short the electrodes. Even if the dc source can still cope with the current demand, the effective field influencing particles between chains declines as the fibrils form so that the driving force for fibrillation is adversely effected once fibrillation sets in. Whatever the causes, practical experience to date suggests that rapidly polarisable particles, which in the context of interfacial polarisation implies too conductive materials for them, make poor ER substrates.

The intermediate situation, which in our experience corresponds to conductivities for the material of the particles to be within the range 10^{-2} to 10^{-4} Sm^{-1}, leads to the best ER fluids, both in terms of high static yield stresses and the retention of excess field induced stresses in flow. Particle-particle interactions begin to survive rotation leading to some permanence in the clustering. Extra energy dissipation in rotational flow occurs because of the extra size of the clusters, the constant rearrangement of the charge distributions within them and the torque between the induced dipoles and the field. Furthermore, as structures are building up, the total dipole of these could reach levels where they cease to rotate against the electric field, with the flow changing to accommodate this circumstance. This process has similarities to the single particle case discussed above but could well lead to a much more dramatic outcome. Initially formed non-rotating clusters with their larger non-fluctuating dipoles would interact strongly and could then readily grow further into the fibrils which form across the shear gap to give the structures of the solidified Bingham state. Reversing the process with an electrically set Bingham body proceeds initially by disrupting this structure to give broken fibrils which are thereby weakened and degrade further with increasing shear and time until rotation of clusters sets in. The flow field would then change character. If the switching between the fluid flows involving non-rotating and rotating clusters occurs at different stress levels when clusters are forming or disrupting, then hysteresis similar to that predicted for isolated particles would occur. However, in the present situation the non-rotating state would correspond to the Bingham solid. As illustrated below, hysteresis is observed with a return from flow following, in its latter stages, a shear stress - shear rate dependence which leads to a lower stress level for the Bingham state, which is not fully structured. The hysteresis could also originate by a somewhat different mechanism involving the following cycle. After rupture of the Bingham solid, there will initially be a period of accelerating flow until dynamic equilibrium is achieved, following which there is a steady climb in the shear stress - shear rate curve. On gradually decreasing the stress, the flow declines without immediately reverting to the same level of structuring as is the case in initiating the flow. This then leads to a lower intercept for the quiescent state, which is less structured and which requires ageing to revert to its initial state.

There is much experimental evidence in support of the above mechanism with possibly the most important feature being the apparent window of dielectric relaxation frequencies which characterise ER fluids. We have previously stressed[3] the importance of dispersions required for an ER role to have dielectric relaxations in the range 10^2 to 10^5 Hz and that dispersions whose interfacial polarisation processes lie outside that range do not provide strong ER responses. The reasons for such requirements are clear in the light of the above model, especially under conditions of flow, but in practice even seem to apply to the level of static yield stress. For some time now our criteria for designing ER fluids has been based on these considerations which also led to the development of "anhydrous" ER fluids[5].

Experimental

Prior to presenting some of the experimental results which provide the basis for the mechanism outlined above, it is relevant to discuss the experimental procedures which were used. The perceived objectives were to study the dielectric and rheological behaviour of ER fluids under conditions of shear and high fields as a function of shear rate and field frequency, and to do so with a variety of ER fluids. Certain experimental criteria, conditions and constraints must be considered in designing and implementing such studies *viz.*

i. The only cell geometry able to give both a uniform electric field and a constant shear rate across a narrow gap is the concentric cylinder cell.

ii. In order to implement dielectric measurements in flow the free fluid surface must be totally screened out from these measurements, else meniscus shape changes are sensed by the bridge.

iii. Measurements of complex permittivity under high dc fields can be made by applying a small ac sensing voltage to the cell, which then has to have in series blocking capacitors to protect the bridge. The alternative of using high voltage ac sources requires the construction of special bridges and involves analysis of the non-linear responses which the ER fluids show with field. Such large ac field measurements were not undertaken.

iv. Since Bingham bodies are involved, constant stress rather than constant strain rheometry is preferred.

v. The rheological response of the fluids should be studied under square wave pulsed fields, with the ability to vary the on-off duty cycle, and not only under steady dc fields. The rise and decay of the flow would then provide information on the response times for the ER effect.

vi. Since the programmes are to investigate a variety of fluids most of which have to be synthesised, the volumes required for tests should be kept low.

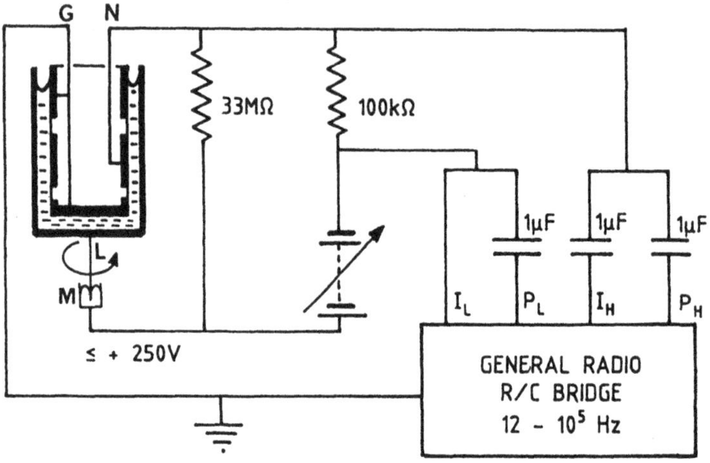

Figure 1 A schematic of the FMP cell and circuits for applying a dc voltage. For description see text.

These factors are germane to the cell and peripheral equipment designs which are shown schematically in Figures 1 and 2. The construction of the dielectric cell (Figure 1) was based on one described previously[10,11] with the simplification that it carried no torque beam for viscosity measurements, the latter being undertaken with the equipment shown in Figure 2. The dielectric cell electrodes were constructed of stainless steel, with the neutral electrode N and guard electrodes G mounted as surface rings on PTFE and the live electrode L acting as the Couette cup. This was motor driven via a belt and noise free electrical contact was made to it via a mercury cup (M). The electrode/shear gap was 0.25 mm and the stator diameter was 44 mm. A GenRad 1689 RLC Digibridge capable of operating in the frequency range 12 to 10^5 Hz was used for RC measurments. In order to isolate the dc field from the bridge, capacitors and fail safe current limiting resistors were interposed between the cell and bridge as shown. The impedance of these were such that they made an insignificant addition to the cell impedances although had this not been the case their contribution could easily have been corrected for. Since all the electronic operated power sources tried exhibited significant and complicated frequency varying impedances, a stack of 12 V sealed lead-acid rechargable batteries were used to supply the dc field. This proved satisfactory in that test RC circuits gave correct measured values to 2% at all fields for frequencies $\geq 10^2$ Hz. Larger errors were observed at lower frequencies and in consequence no dielectric data under fields was recorded below 10^2 Hz.

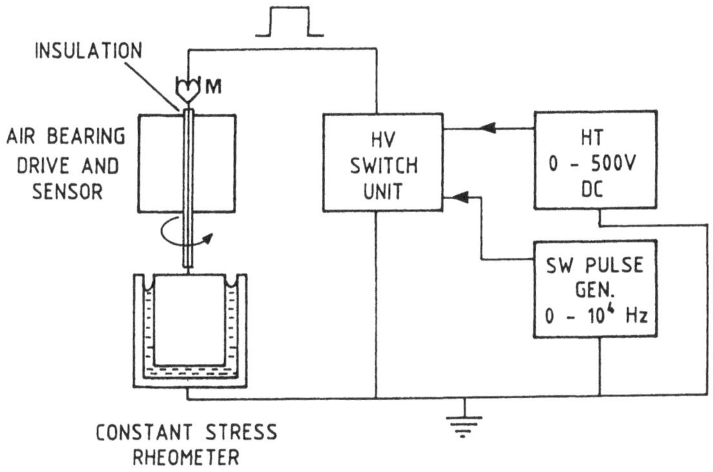

Figure 2 A schematic of the controlled-stress electrorheometer and the peripherals for supplying the high field pulse train. For description see text.

A modified Carri-Med controlled-stress rheometer fitted with a Couette cell (gap 0.2 mm, bob diameter 39.6 mm, length 25mm) was used for the rheometry measurements (Figure 2). The modification involved insulating the drive shaft from the air bearing by covering it with heat shrink sleeving and mounting a noise free contact in the form of a mercury cup M on top. Fields were applied as shown using a specially constructed high voltage switch unit designed and constructed at The Wolfson Consultancy Service, University of

Southampton, which was energised from a Wallis Hivolt 0 to 500 V continuously adjustable power supply. The switch unit was designed to overcome the problem of rapidly charging the Couette cell which acted as a lossy capacitor, particularly when filled with fluid. Its response is capable of delivering square wave outputs under these loads with a rise and fall time $\leq 10^{-5}$s and pulse trains with on-times down to 10^{-4}s and repeat rates up to 10^4 Hz. The square wave pulse train controlling the switch was generated from a Thandar TG501 function generator operating in square wave mode up to 10^4 Hz with a variable on-off duty cycle limited by the bounds of this frequency. Further details of the rheometry system are to be published.

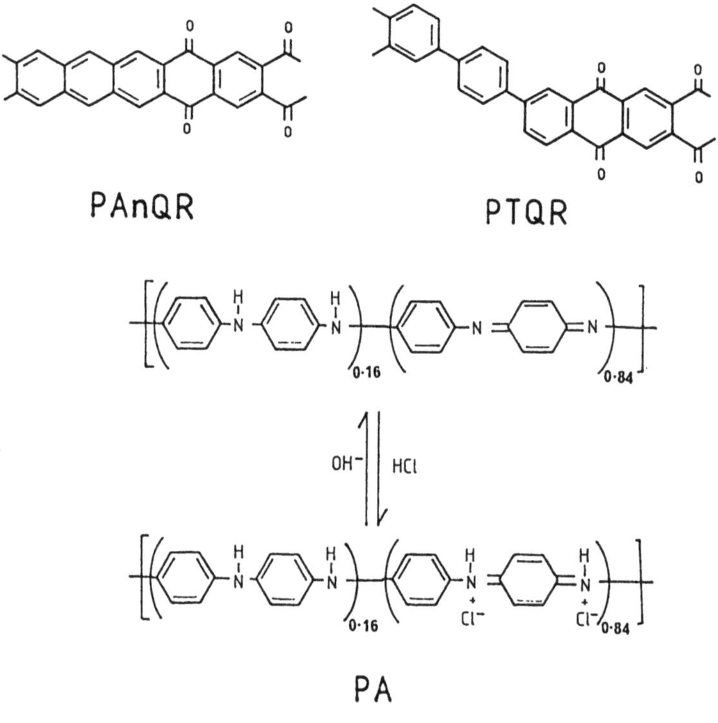

Figure 3 The semi-conducting polymers referred to and their codings. PAnQR, poly(anthracenequinone radical); PTQR, poly(p-terphenylquinone radical); PA, "emeraldine" form of poly(aniline) as prepared by the method of ref. 15 and its interconversion between acid and base forms.

Some static yield stress measurements were taken using the static yield stress meter previously described[3]. A number of dielectric measurements were made on still fluids and some of these were done using a General Radio D321 micrometer cell with a thin layer of ER fluid placed in the gap between the upper live electrode and the lower compound neutral - guard electrodes.

With the exception of the hydrated fluids provided by Advanced Fluid Systems Ltd., all active substrates were based on well dried semi-conducting polymers synthesised as previously described[3,7,14,15]. The structures, names and codings are shown in Figure 3. They were dispersed as finely ground powders (particle diameters $\leq$ 30 μm) in 100 cS silicone fluid previously dried over silica gel.

ER Fluid Behaviour under Field and Flow

There is a considerable amount of experimental evidence which is consistent with the mechanism for the ER effect given above. Here some of this evidence is presented and discussed.

Dielectric spectra of ER fluids in the absence of shear provide information, in terms of the relative permittivity (ϵ') and dielectric loss (ϵ''), on the achievable polarisation and the rate of its establishment. The difference between the low and high frequency values of relative permittivity ($\Delta\epsilon'$) is an indicator of the former and the relaxation frequency (f_R), corresponding to the position of the loss maximum in ϵ'' or point of inflection in ϵ', relates to the latter. A striking example of the influence of dielectric relaxation rate on ER effectiveness has been shown[4] by comparing a series of water promoted poly(methacrylate) salts dispersed at 0.35 volume fraction in various base fluids. Figure 4 illustrates the dielectric spectra of the series taken in the absence of either flow or dc fields. This particular set of fluids have an approximately common low frequency limit of $\approx 25 \pm 3$ for ϵ', but much more divergence in the positions of their relaxation frequencies. Figure 5 shows the static yield stresses of these fluids as a function of the square of the field strength, to which the static yield stress is proportional at low fields. The ER effectiveness on that

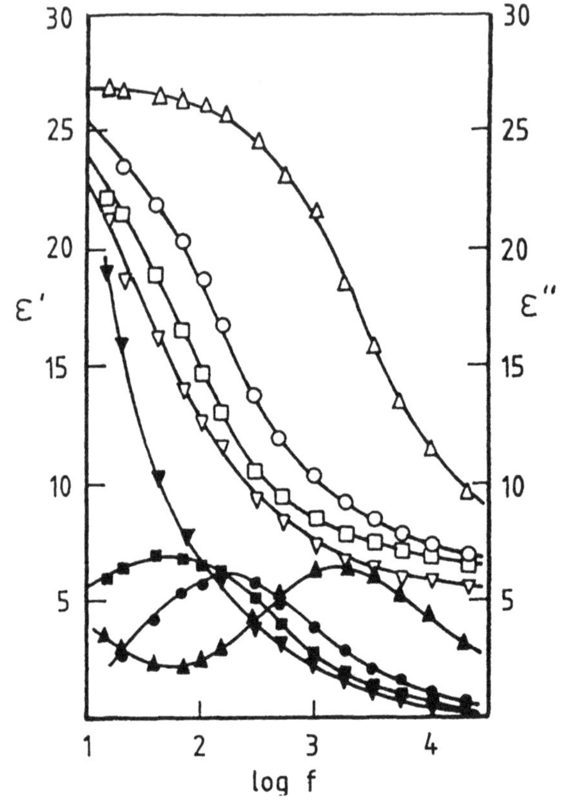

Figure 4 The dielectric spectra of still dispersions of various polymethacrylate salts containing 15% water and dispersed at a volume fraction of 35% in various base oils.o, lithium salt in pyralane 3010; ▲, potassium salt in pyrelene 3010; □, lithium salt in pyrelene 3010 and ▽, lithium salt in dibromo-diphenylmethane. Taken from ref. 5.

basis follows the trend of f_R rather than ε' and indeed, the most effective fluid is not the most polarisable. Previously reported work with fluids based on a series of poly(acenequinones) with different conductances and in consequence f_R, show the same correlations with f_R^3. Yet another similar situation occurs when one examines the effectiveness of poly(aniline) (PA) as an ER substrate. It is well known that this material can have its conductivity adjusted in the range 10^{-8} to 10^2 Sm^{-1} by treatment with bases or buffers[15]. After

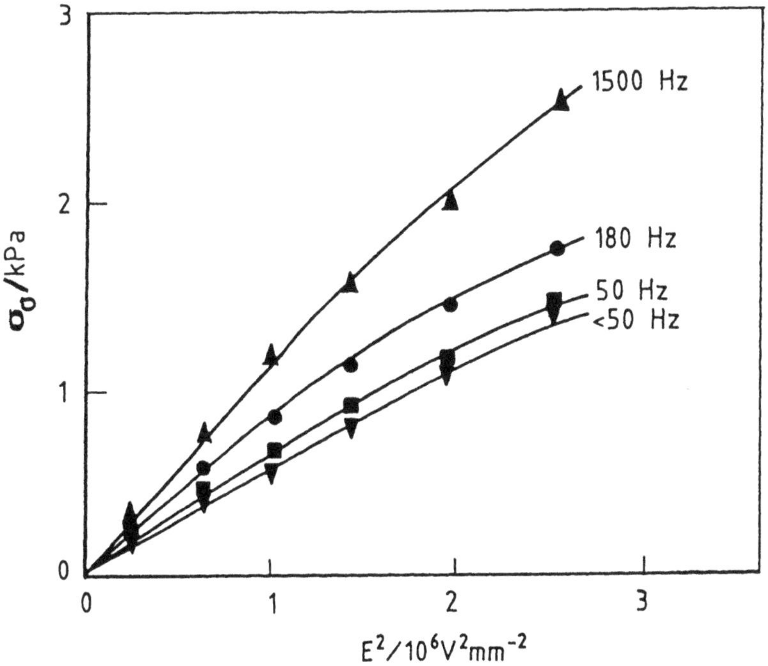

Figure 5 The field (E) dependence of the static yield stress (σ_o) of the poly(methacrylate) salt based ER fluids described in the legend to Figure 4. The f_R values are indicated on the left of the curves. Taken from ref. 5.

drying and dispersion in silicone fluid to 30% solids, the dielectric spectra reflect the level and rate of interfacial polarisation as is illustrated by the examples shown in Figure 6. The conductive acid "emeraldine" form of PA has a very large polarisation and a relaxation frequency above 10^5 Hz, probably in the microwave region. Its conductance was so high that it drew too much current from the dc source, preventing static yield stress measurements. Complete conversion to the free base produces a material with a $f_R \approx 30$ Hz and a static yield stress of 2.1 kPa under a dc field of 4 $kVmm^{-1}$, whilst neutralization to pH 10 provides a fluid relaxing at ≈ 300 Hz and showing an optimum static yield stress of 2.7 kPa under the same field. All these examples are consistent with the model in that they show an optimum in effectiveness involving both polarisation rate and magnitude.

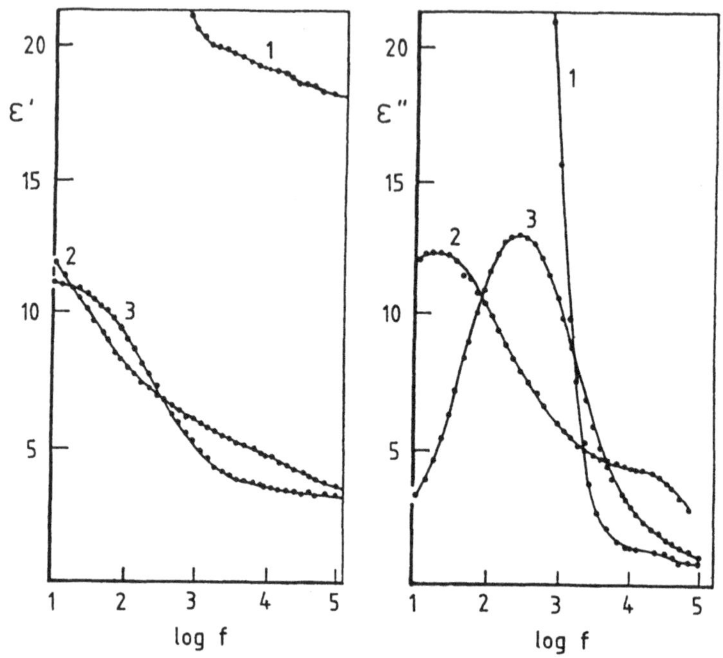

Figure 6 The dielectric spectra of various PAs dispersed at a volume fraction of 0.3 in silicone fluid. Type 1 - acid form, type 2 - fully neutralised and type 3 - buffered to pH 10. Taken from ref. 5.

Examples of the shear stress - shear rate behaviour under dc fields of selected anhydrous ER fluids based upon semi-conducting polymers are shown in Figure 7. All these fluids show hysteresis when run through a cycle of flows from increasing to decreasing shear stresses and thus they all show aspects of the behaviour discussed above. The static yield stress levels measured in the static yield test meter are higher than the enhanced shear stresses shown under flow, even at the lowest shear rates, and these enhancements decrease further before dynamic equilibrium is established. For PAnQR and for the neutralised form of PA this enhancement survives and remains approximately constant over the remaining experimental shear rate range. The situation with PTQR is different. It is not as an effective an ER substrate, not only because fluids based upon it show lower static yield stresses, but because there is no field enhancement of yield stress once dynamic equilibrium is established. This is a

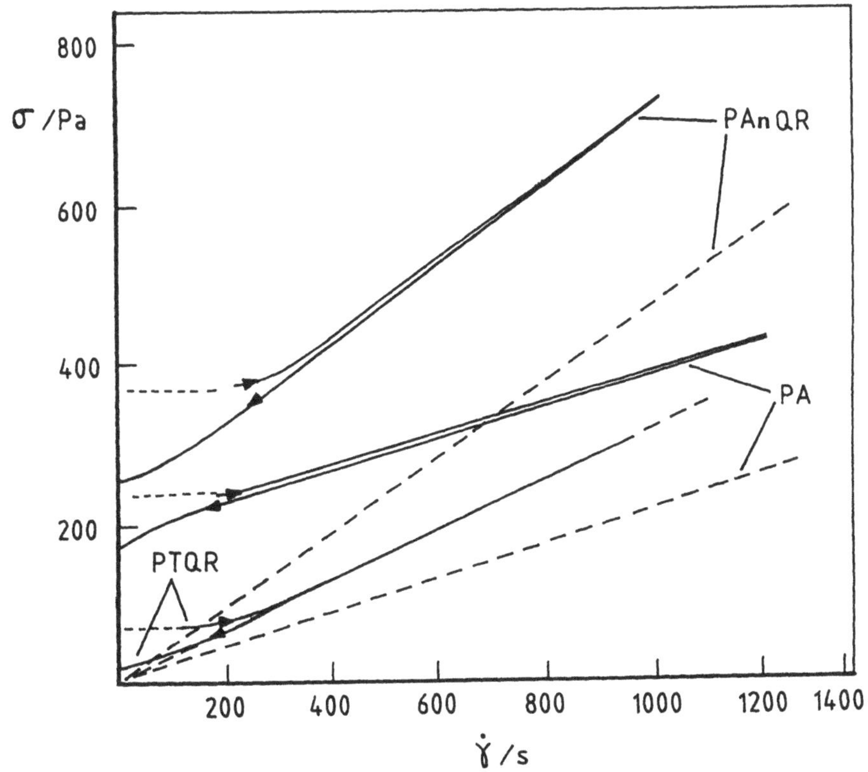

Figure 7 The shear stress (σ) - shear rate ($\dot{\gamma}$) behaviour of the semiconducting polymer based ER fluids all at 0.2 volume fraction in 100 cS silicone fluid. Solid lines indicate the response at 2 kVmm^{-1} and broken lines with no field.

consequence of the sluggish polarisation of PTQR preventing an effective ER response in flow. Fluids of this type would not respond to controlling fields in such flows and fail in their ER role.

PTQR particles do however structure in still dispersions as is demonstrated by the polarisation of these fluids under dc field. Figure 8 shows the dielectric spectra of PTQR in silicone under a field of 2 kVmm^{-1} (curve 1, with curve 2 showing the dielectric spectra under zero dc field for comparison). There is a dramatic enhancement of the low frequency ϵ' which is due to the fibrillation of particles in the field direction leading to enhanced interfacial

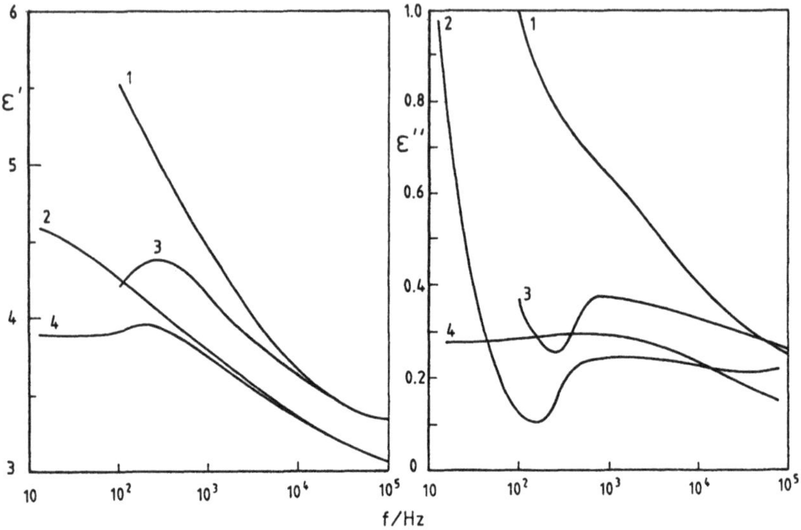

Figure 8 The dielectric behaviour of a PTQR (0.10 volume fraction) in 100 cS silicone fluid. Conditions of measurement: curve 1 - 1 $kVmm^{-1}$ and zero flow; curve 2 - zero dc field and no flow; curve 3 - 1 $kVmm^{-1}$ plus flow at a shear rate of 3000 s^{-1}; curve 4 - zero field plus flow at a shear rate of 3000 s^{-1}.

polarisation[16]. This structuring is destroyed under shear (curve 3) to approach the FMP behaviour in the absence of a dc field (curve 4). The FMP resonance positions which lie between 10^2 and 10^3 Hz are possibly shifted slightly to higher frequency but this is only just at the level of accuracy and the genuineness of the shift still requires confirmation. We have examined the enhancement of the low frequency ϵ' in still conditions and under varying levels of dc fields by analysing the dielectric data using Cole-Cole plots (Figure 9). This enabled us to estimate both $\Delta\epsilon'$ due to the interfacial polarisation and f_R. Figure 10 shows that the polarisation for this fluid, as reflected in $\Delta\epsilon'$, is dependent upon the square of the field over the range examined. Such a dependence is also often seen in the rheology of ER, particularly at low levels of field, and stems from the pairwise interactions of induced dipoles[17].

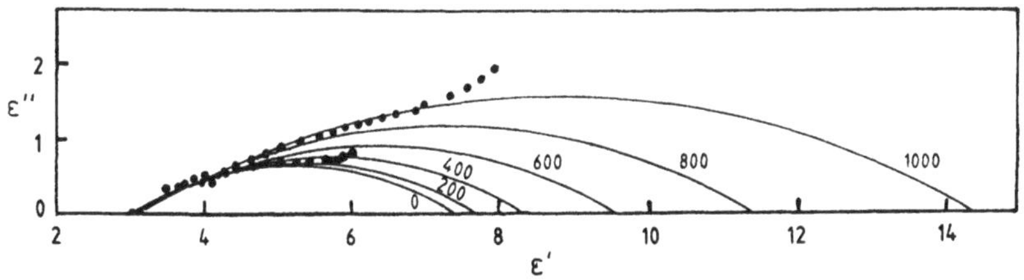

Figure 9 Cole-Cole plots for dispersions of PTQR at 0.2 volume fraction in 100 cS silicone fluid under dc fields. Field level in Vmm^{-1} indicated on best fit arcs with only data points for 0 and 1000 Vmm^{-1} shown for the sake of clarity.

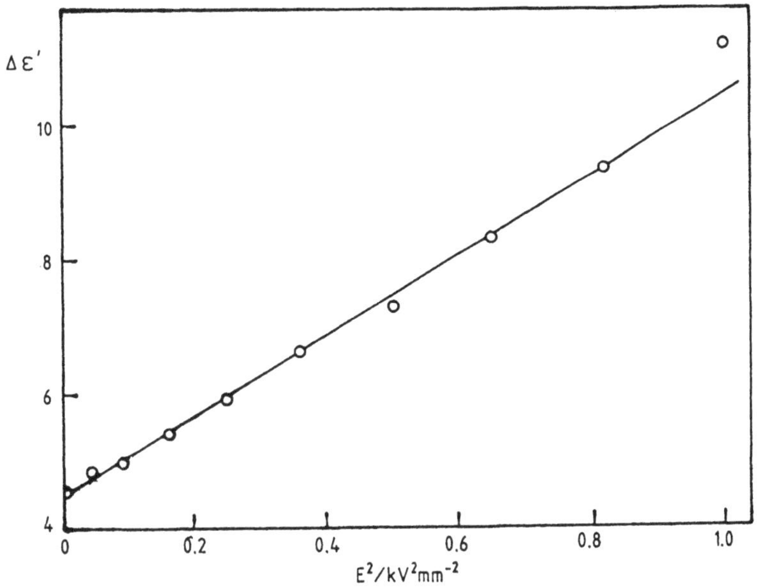

Figure 10 The field (E) dependence of the dielectric increment ($\Delta\epsilon'$) for the PTQR based fluid of Figure 9. $\Delta\epsilon'$ values corresponds to the arc lengths in that figure.

Figure 11 shows how f_R is influenced by the structuring, with a slowing of the polarisation as that increases. It can also be seen that f_R under zero fields is concentration dependent, with the polarisation taking longer at higher concentrations. Presumably carrier drift within the particles is coupled between neighbours and of an extent depending upon their separations. Figure 11 also shows that this

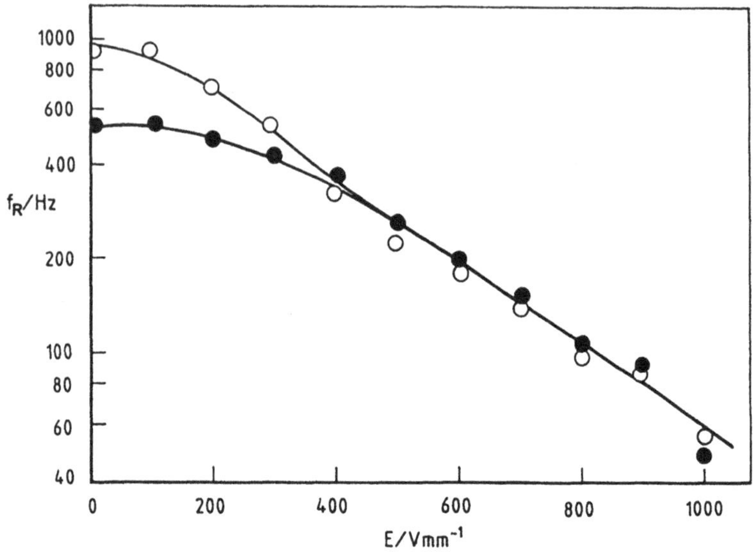

Figure 11 The field (E) dependence of the relaxation frequency (f_R) for PTQR based fluids in 100 cS silicone oil. The open points refer to 0.1 and the solid points to 0.2 volume fraction fluids.

concentration effect tends to disappear as the field increases. These observations are consistent with clustering under field to give closer contacting particles within the fibrils and increased separations between them, so that interchain coupling becomes negligible at the concentrations used. Similar dielectric effects have been found to occur with PAnQR and hydrated poly(lithium methacrylate) dispersions[18].

Some preliminary results on the time dependence of the rheological response of PAnQR dispersed in a silicone fluid are illustrated in figures 12 to 14. In each case the fluid was sheared at a fixed stress of 805 Nm^{-2} and pulsed fields of 1.5 kVmm^{-1} applied, and the response monitored in terms of the shear rate, which was recorded with time at intervals of 0.6s. At the low pulse rate of 0.05 Hz the response follows the field without any detectable lag in flow when judged on the basis of a 20 s pulse interval. Increasing the

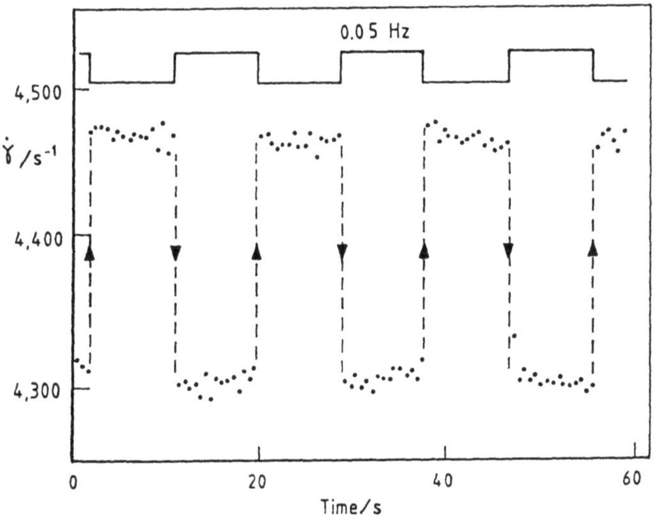

Figure 12 The response of a PAnQR based fluid (0.15 volume fraction in 100 cS silicone oil) to a 1.5 $kVmm^{-1}$, 0.05Hz pulsed field under a constant stress of 805 Nm^{-2}. The on-off states of the pulsed field are shown together with the recorded shear rate ($\dot{\gamma}$) and the direction of switching.

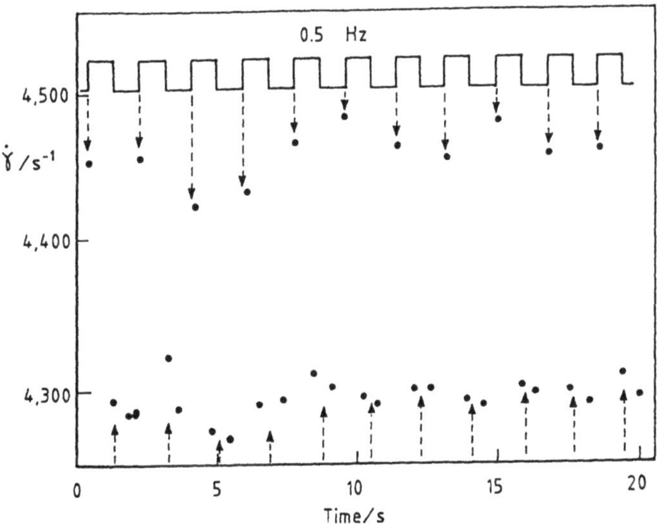

Figure 13 Similar data to that illustrated in Figure 12 using the identical fluid and field strength but with the pulse rate set to 0.5 Hz.

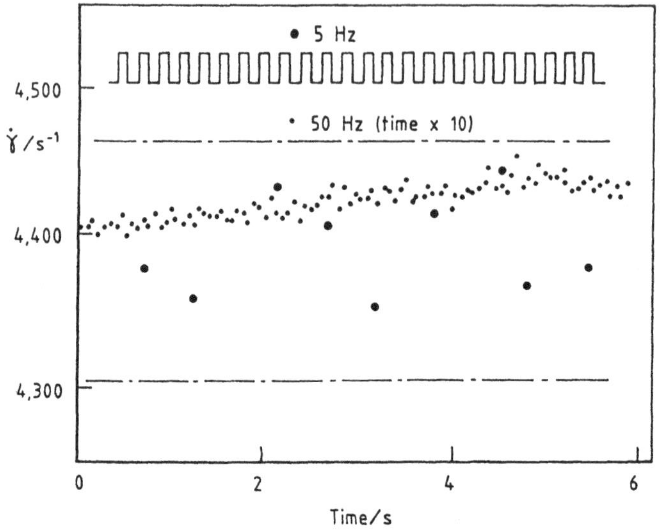

Figure 14 Similar data to that illustrated in Figure 12 using the identical fluid and field strength but with a 5 Hz (●) or 50 Hz (•) pulsed field (5 Hz pulse train only shown). The position for the dc and zero field levels are shown by hatched lines.

frequency to 0.5 Hz (Figure 13) firstly highlights the difficulty of our current equipment in following the changing flows with sufficient resolution. The fluid switches within this time scale but the data acquisition rate is too slow to see accurately the pulse shape of the responding flow. There is however indication that the response to the field switching on is more attenuated than the off step. At 5 Hz (Figure 14) the situation is worse, with the data acquisition rate being far too slow to see the shape of response, but response there is, although it may be that full switching between the fielded and non-fielded flows does not occur. With a pulse rate of 50 Hz (Figure 14) any switching between fielded and non-fielded flow levels does not occur, rather the flow rate settles to a constant level which is quite close to the non-fielded value. The attenuation in the levels between which switching occurs at 5 Hz is at least in part due to the inertia of the system, whilst the poor efficiency of the field at 50 Hz in even effecting the level of steady flow is probably the first sign of the polarisation relaxing out. At 500 Hz the PAnQR based fluid completely fails to react to a 1.5 $kVmm^{-1}$ pulsed field, indicating a lack of polarisation.

Conclusions

We have shown that the ER phenomenon appears to be controled by the polarisation of the ER fluids which, for maximum effectiveness, should be interfacial. The rate of polarisation also plays a role in ER effectiveness. In consequence the selection of good ER active substrates must take both factors into account and such considerations were the basis of the development of "anhydrous" ER fluids. Experimental charcterisation of ER fluids and analysis of the response requires both dielectric and rheological measurements to be carried out in parallel, and this approach is the one most likely to lead to further understanding of the ER phenomenon and to improvements in ER fluids.

Acknowledgements

The authors wish to thank the SERC and MOD for the support given to our ER research. We also wish to thank Advanced Fluid Systems Ltd for providing samples of the hydrated fluids.

References

1. H. Block and J. P. Kelly, *J. Phys. D Appl. Phys.* **21** (1988) 1661.
2. H. Block and J P Kelly, *First Internat. Symp. on ER Fluids*, ed. H Conrad, J D Carlson and A F Sprecher, (North Carolina State Univ. Engin. Pub., Raleigh, NC,1989) p. 1.
3. H. Block, J. P. Kelly, A. Qin and T. Watson *Langmuir* **6** (1990) 6.
4. H. Block, J P Kelly and T Watson *High Value Polymers*, ed. A. H. Fawcett (RSC Special Pub. No 87, 1991) p. 151.
5. H. Block and J P Kelly, *Proc. IEE Colloq.* **14** (1985) 1; *US Patent* (1987) 4687589.
6. J. D. Carlson *Eur. Pat. App.* (1988) 88311167.6; Y. Ishino, S. Endo, S. Tomita, Y. Fukayama and T. Saito *Eur. Pat. App.* (1989) 89115820.6; M. Kambara and H. Tomizawa *Eur. Pat. App.* (1989) 88309910.1; F. E. Filisko and W. E. Armstrong

US Patent (1988) 4744914; F. E. Filisko and L. H. Radzilowski *J. Rheol.* **34** (1990) 539.
7. H. Block, J. Chapples and T. Watson, *UK Pat. App.* (1990) 8908825.6.
8. C. J. Gow and C. F. Zukoski, *J. Coll. Interface Sci.* **136** (1990) 175.
9. W. M. Winslow, *US Patent* (1947) 2417850; *J. App. Phys.* **20** (1949) 1137; N. Sugimoto, *Bull JSME* **20** (1977) 1476; J. E. Stangroom, *Phys. Technol.* **14** (1983) 290; M. Whittle, *Phys. World* **2** (1989) 39; N. Webb *Chem. Brit.* (1990) 338.
10. A. P. Gast and C. F. Zukoski, *Adv. Coll. Interface Sci.* **30** (1989) 153; D. J. Klingenberg and C. F Zukoski *Langmuir* **6** (1990) 15.
11. H. Block *Polymers in Solution* ed. W. C. Forsman (Plenum Press, New York and London, 1986) p. 111; H. Block, W. D. Ions, G. Powell, R. ,P. Singh and S. M. Walker *Proc. Royal Soc. ser. A* **352** (1976) 153; H. Block, K. M. Goodwin, E. M. Gregson and S. M. Walker, *Nature* **275** (1978) 632; H. Block, W. D. Ions and S. M. Walker *J. Polymer Sci. Polymer Phys. Edn* **16** (1978) 989; H. Block, E. M. Gregson, A. Ritchie and S. M. Walker *Polymer* **24** (1983) 859; H. Block, E. Kluk, J. McConnell and B. K. P. Scaiffe *J. Coll. Interface Sci.* **101** (1984) 320.
12. H. Block, E. M. Gregson, A. Qin, G. Tsangaris and S. M. Walker *J. Phys.E Sci. Inst.* **16** (1983) 896.
13. J. Hemp *Proc. Royal Soc. ser. A* **434** (1991) 297.
14. H. A. Pohl and E. H. Engelhart *J. Phys. Chem* **66** (1962) 2085; H. A. Pohl and D. A. Opp *J. Phys. Chem* **66** (1962) 2121; R. Rosen and H. A. Pohl *J. Polym. Sci. Polym. Chem. Edn* **4** (1966) 1135.
15. A. G. MacDiarmid, J. C. Chiang, A. F. Richter and A. J. Epstein *Synthetic Metals* **18** (1987) 105.
16. R. W. Sillars *J. Inst. Elect. Eng.* **80** (1937) 378.
17. D. L. Klass and T. W. Martinek *J. App. Phys.* **38** (1967) 67; G. G. Petrzhik, A. A. Trapeznikov, E. V. Makarova, V. P. Nikiforov and R. A. Davletshin *Russian J. Phys. Chem.* **47** (1973) 1522; G. G. Petrzhik, O. A. Chertkova and A. A. Trapeznikov *Dokl. Akad. Nauk. SSSR* **253** (1980) 173;

Yu. F. Deinega *Inzh.-Fiz. Zh.* **18** (1970) 994; P. M. Adriani and A. P. Gast *Phys. Fluids* **31** (1988) 2757; A. P. Gast and C. F. Zukoski *Adv. Coll. Interface Sci.* **30** (1989) 153; L. Marshall, C. F Zukoski and J. W. Goodwin *J. Chem. Soc. Faraday Trans. 1* **85** (1989) 2785.
18. H. Block and P. Rattray *To be published*.

RHEOLOGICAL PROPERTIES AND MODELS OF DRY ER MATERIALS

FRANK E. FILISKO
Materials Science & Engineering, The University of Michigan
H.H.Dow Building
Ann Arbor, Michigan 48109

ABSTRACT

Within the last few years activity in the field of electrorheology has been reawakened by the discovery of particulate systems which produce ER active suspensions without the need for the large amounts of adsorbed water, i.e. dry systems. In fact it can be argued that such systems operate by means of mechanisms intrinsic to the chemistry of the materials. Such systems include the crystalline alumino-silicates, various polyelectrolytes, and polymeric semiconductors including phthalocyanine and poly(acene-quinone radicals). Although some question that they are completely cry, there is no question that the amount of water remaining is many times lower than the minimum amounts required for activity of the water based systems and further that the water remaining is not loosely adsorbed but very strongly bound. The most important aspect of these discoveries however concern the molecular mechanisms associated with electrorheological activity. Various categories of dry or intrinsic ER active systems are discussed in terms of orientational and interfacial polarization mechanisms as associated with their ER activity.

Description

Electro-rheological materials, first reported in the 1940's (Winslow 1949), are materials whose rheological properties, i.e. flow and deformation properties under stress, are a strong function of the electric field strength imposed on them. These materials are typically fluids (in the absence of an electric field) whose viscosities under steady shear increase markedly with applied electric fields, but under constant shear stress, at high enough fields, the materials can solidify. In their solid state, various properties of the solid such as shear and tensile strengths are as well a function of electric field. Further, all changes induced by the applied electric field are almost instantaneously reversible upon removal of the field, i.e. such a material can be almost instantaneously solidified and liquified merely by application and removal of an electric field. In the liquid state (post yield) under constant rate of shear, these materials can have viscosities which range from a few poise under zero fields but approaching infinity at very low shear rates under the influence of fields on the order of 1000 volts/mm. In the solid state (pre-yield) , these materials can have shear strengths on the order of 5,000-10,000 Nt/M^2 for fields of around 5000 volts/mm.

Basic Composition

ER materials are typically suspensions or dispersions of fine hygroscopic particles in a hydrophobic non-electrically conducting dispersion media. Particle sizes in the range of 0.1 - 10 μm are most common although particles much larger have demonstrated ER effectiveness, and there have been some reports that macromolecules in solution may exhibit the effect. The vast majority of systems require additionally that significant amounts of water (10%-30%) be adsorbed onto the particulate phase. It is this requirement which posed severe limitations on the potential use of these materials.

Recent discoveries however have been made of particulate phases which produce ER active materials without the need for water or other activator and produce similar properties levels(Block and Kelly 1987, Filisko and Armstrong 1988, Inoue 1990). Most importantly, this discovery implies that the mechanism can be an inherent part of the molecular structure or chemistry of the particulate phase and therefore, as soon as they are understood more clearly, materials will be able to be synthesized to enhance desireable properties of the materials.

Intrinsic vs. Extrinsic ER Activity

Before attempting to discuss water-free systems, it is necessary to establish criteria by which particulate systems may be classified as water-free ER active. The most obvious definition might simply be to define them in terms of the presence or absence of water. However this can lead to some ambiguities since it is impossible to eliminate all traces of water from materials specifically for various polymeric materials, which may degrade readily at temperatures necessary for drying. It must then be based upon the functionality of the water present in regard to ER activity, its necessity in order for a system to be ER active, various characteristics of the ER fluids associated with the presence of water, and the relative amounts of water present.

In the following discussion, water mediated ER activity will be termed "extrinsic" while activity associated with the chemistry and physics of a material will be called "intrinsic'. The term extrinsic here implies more than just water, since other materials may be adsorbed onto the powders such as glycerol, silanols, etc. and produce some ER activity. However our concern here is just for water. Thus intrinsic systems, when wet, may demonstrate both intrinsic and extrinsic activity. However, a water dependent system will demonstrate only extrinsic activity, i.e. upon drying, it will become inactive electrorheologically.

In defining intrinsic systems, it is more appropriate to discuss characteristics of water mediated or extrinsic activity and contrast the behavior. Extrinsic activity is ER activity which requires the presence of significant amounts (>5%) of adsorbed water and has the following characteristics:

1. Extrinsic activity decreases as the amount of adsorbed water decreases and goes to zero around 5%. A corollary of this is that the water is easily removed by heating to temperatures as low as 50 °C .

2. The bulk currents are relatively high but decrease rapidly upon drying as does the ER activity.

3. Activity increases upon heating (which is true for all systems) but begins decreasing at temperatures as low as 60 °C (Fig.1) . A corollary of this is the irreversibility of extrinsic activity following heating.

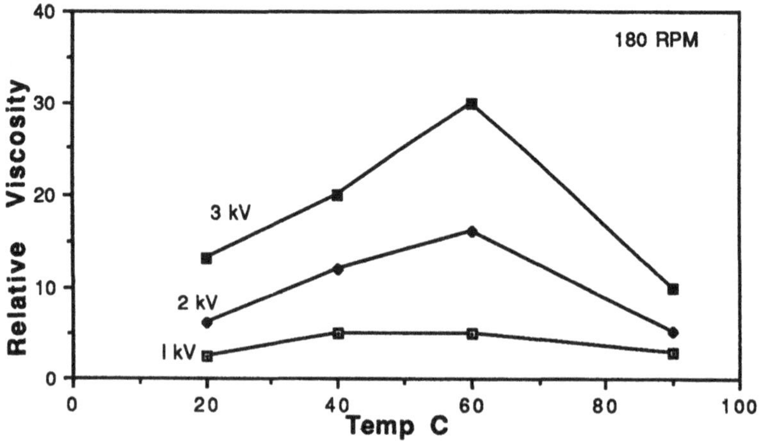

Fig.1: Electrorheological activity of a hydrous or extrinsic ER fluid incorporating poly(lithium methacrylate) particulates.

4. Activity diminishes to zero below 0 °C and irreversibility above approximately 100 °C. These temperatures can be modified by various additives but the temperature limitations are a consequence of those transitions associated with water.

In contrast dry or intrinsic systems may operationally be defined as those which:

1. demonstrate ER activity despite severest attempts at drying and with only trace amounts of water, well below the minimum amounts required for extrinsic systems.

2. Those which contain no adsorbed water but only that which is structural or tightly bound.

3. Activity continues well below 0 °C and well above 100 °C (or upper temperature consistent with stability of the fluid components) and the properties are reversible with temperature.

The above criteria are the more practical, but fundamentally it would be more desireable to define intrinsic and/or extrinsic activity in terms of basic chemical and physical parameters which govern ER activity. However at present these are not that clearly known. What has been established however are strong correlations between certain system parameters, although apparent exceptions to these still remain unexplained.

An excellent discussion is this regard of the current status of this research was recently presented by H. Block et. al.(1990) . In the most general terms, it is expected that the ER phenomenon involves some sort of polarization since by definition polarization is an induced response of a dielectric to an electric field. The most relevant polarization mechanisms to consider in ER activity include interfacial and/or orientational since electronic, ionic(if present), and atomic are of relatively minor magnitude but more importantly are associated with characteristic frequencies greater than about 10^{10}/sec. Further these are resonant adsorptions, i.e. they do not demonstrate a temporal activity which is characteristically associated with both interfacial and orientational polarization mechanisms. On the other hand ER activity seems to be associated with the presence of relaxational dielectric adsorptions with characteristic frequencies lower than about 10^5/sec. Orientational mechanisms are a consequence of charge assymetries associated with the basic molecular or crystal structure and which are able to orient or rotate about fixed positions in response to the applied electric field. Interfacial polarization however is associated with heterogenous systems and is a consequence of the presence of locally mobile charge carriers associated with the dispersed phase which are able to move predominantly within the confines of the dispersed particles in response to the electric field and accumulate to various degrees at the interface of the phases. In the theoretical treatment of this phenomenon by Maxwell(1892), Wagner(1914), and Sillars(1937), they were able to show that relatively large losses observed dielectrically could be attributed to the presence of small amounts of conducting or semiconducting inclusions in a nonconducting matrix. Sillars went on to show that the power loss in such materials was proportional to the conductivity and the square of the field strength within the dispersed phase.

In order to assess the relative importance of each of these in regard to ER activity, it is worthwhile considering what function adsorbed water may serve in producing extrinsic activity. The first is that the water, by virtue of it relatively high permittivity, may impart a high polarizeability to the particles thus increasing the electrostatic force of attraction between the particles which is proportional to $(k_p-k_f)^2$ in a point dipole approximation (Gast and Zukoski,1989). The second is that the adsorbed water creates a conductive layer on the surfacre of the particles by which ions in the water may drift across the surface in response to an electric field. In a nonconductive hydrophobic environment, the water layer serves as a mechanism by which local conduction of the charge carriers (ions) can occur, thus creating the possibility of interfacial polarization.

It would seem that the relative effectiveness of the orientational mechanism can be assessed rather readily by compounding materials containing particles of high dielectric constant and it would seem straightforward to produce fluids of exceptional strengths. Such approaches have apparently met with disappointing results. The use of metallic inclusions would seem to represent the ultimate particles, but they have not met with success possibly due to difficulties in imposing an electric field over them. Other better examples include titanium dioxide (TiO_2) which has a dielectric constant of around 200, yet produces an active ER fluid only when wet (Fig. 2).

When dried , it completely looses it's ER activity even though it still has a high bulk dielectric constant and a corresponding dipolar mediated attractive interaction between the particles. This dipolar related polarization, although it must result in an attractive interaction between the particles and trivially produces chain structures under electric fields in static situations, does not seem to be able to develop the relatively large electric field mediated viscous effects characteristic of ER activity.

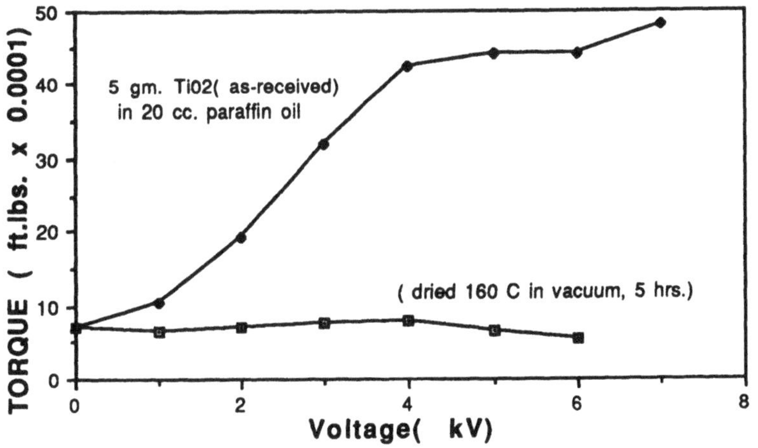

Fig. 2: Electrorheological activity of an ER fluid containing titanium dioxide particulates with adsorbed water and dried.

The interfacial polarization mechanism, being related to charge carriers which are able to migrate on but not between particles, would also seem to be optimized by employing metallic particulates. Unfortuneately, metallic inclusions, even if adequately insulated so as to not cause direct shorts in these concentrated suspensions, have not resulted in highly active ER materials.

Interfacial polarization is associated with the presence of charge carriers (ions or electrons) on the particles which are able to migrate primarily on the particles under the influence of the electric field and which accumulate at the interface region between the particles. The phenomenon first proposed by Maxwell and subsequently refined by Wagner and Sillars is commonly referred to as Maxwell Wagner Sillars (MWS) polarization. Its occurrence is well established and related to mixtures of materials with different dielectric constants and conductivities, the dispersed phase being of higher dielectric constrant constant and conductivity. The theories however are for dilute suspensions because of obvious difficulties in treating concentrated systems of which ER materials are an example. The phenomenon is most commonly related to the occurrence of large dielectric dispersions in relatively low frequency regions ($<10^5$). Thus the association of such dielectric dispersions with ER activity are indirect but nonetheless highly probable. Studies have been made of dielectric spectrum for ER materials under flow (Block, et. al., 1990) but without a voltage applied, and under flow conditions with biasing voltages applied(Deinega et. al. 1978). Under all conditions ER active materials generally show the presence of dielectric dispersions. The latter were however obtained on water based systems.

The most direct association of mobile charge carriers with ER activity is afforded by materials which are photo-electrorheologically active. In this case ER materials are compounded using photoconductors as the dispersed phase. This involves construction of an appropriate testing device in which the walls of the cup in a couette device are made of a conductive but transparent coating on the inside surface. With this apparatus under shear and with a d.c.

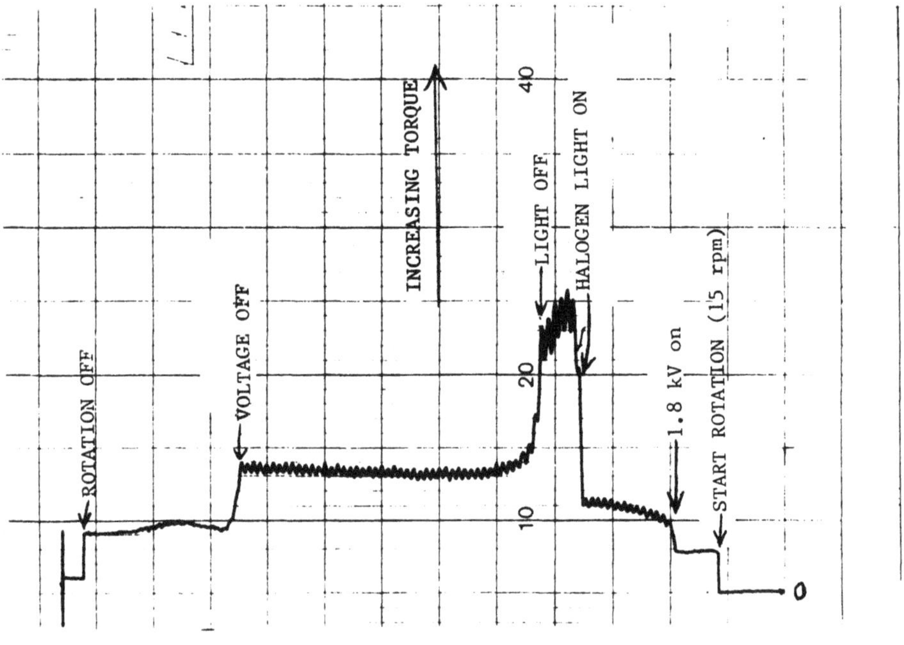

Fig. 3: Activity of a photoelectrorheological material (PHERM) consisting of dried phenothiazine dispersed into paraffin oil.

field applied, exposing the materials to a light of appropriate wavelength, an increase in viscosity is observed co-incident with the application of light(Fig.3). The light activates donor electrons into the conduction band allowing them to migrate under the influence of the imposed field. The phenomenon as well as a number of active systems were reported by Carreira and Mihajlov (1971).

In summary, the two primary polarization mechanisms which seem to be circumstantially related to intrinsic ER activity of dry systems involve orientational as well as interfacial polarization. Both mechanisms should be operational and the relative importance of each may vary with system. However, although both would trivially give a chaining effect observed in static situations, the true indication of ER activity is performance under flow. And as such, the mechanisms by which these two polarizations result in the large viscosity increases associated with ER fluids has not yet been established.

Intrinsic (or dry) Particulate ER Systems

The various materials which have been reported as producing water-free or intrinsically active ER suspensions can be classified into 4 different categories.
1. Ionic conductors, 2. Metallic conductors, 3. Semiconductors,
4. Polyelectrolytes

Ionic Conductors

The alumino-silicates, which are the most studied of these materials, contain metallic cations not as an impurity but as an essential part of their chemistry. These cations are required to satisfy charge stoichiometry as a consequence of the aluminum atoms replacing silicon atoms within the SiO_2 tetrahedra making up the alumino-silicate structure. The cations however cannot fit into the lattice so they reside on the surfaces of the numerous cavities and channels characteristic of these highly porous materials. As a consequence, the cations are held Coulombically onto the surface but are fairly free to move laterally, thus giving rise to the potential for mobile charge carriers which are implicated in their ER activity.

Because these materials are very stable thermally, they can be extensively dried as high as 900 ^oC for short times without structural changes. However these materials have an enormous affinity for water and special precautions must be followed if they are to be kept moisture free. One such precaution is that after drying, they must be cooled either in a sealed container of under vacuum which must be liquid nitrogen trapped.

In terms of polarization mechanisms, since the reported dielectric constants are about 1.6-1.8 (Breck 1974) , the interaction between particles due to orientational polarization are not substantial. On the other hand, as mentioned, these materials in their dry state, contain locally mobile cations capable of migration under an electric field. All materials clearly show 'he presence of MWS dielectric dispersions (Fig.4).

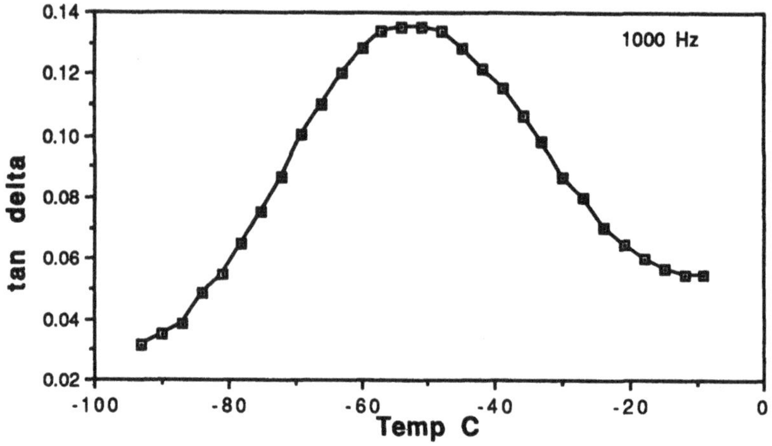

Fig.4: MWS interfacial dielectric dispersion for an extrinsic ER fluid containing alumino-silicate (Zeolite A) particulates.

We feel therefore that these function intrinsically as ER materials predominantly by MWS interfacial polarization mechanisms.

To our knowledge the alumino-silicates are the only ionic conductors which have been tested for ER activity in their dry state. Other materials such as β-alumina however become conductive only at elevated temperatures.

Metallic Conductors

These materials have been looked at to a much greater degree then reported but have been disappointing in their performance. Further they represent substantial difficulties in making fluids since high fields cannot be imposed over them due to the high conductivity of the particles. In one study(Sasada, et. al.1974) , aluminum particles were coated with $Al(OH)_3$ by boiling the particles in water prior to dispersing them in oil. A significant ER activity was demonstrated but the shear rates of the test were extremely low.

Additionally there was no discussion of drying procedures, therefore it is possible that these particles contained water. In a similar study by Inoue(1990) , the insulated aluminum particles were dried at 400 ^{o}C for 1 hr. under nitrogen before testing. He reports that they contain less than 0.5% water and they demonstrate significant ER activity. However , if the water is confined to the skin layer, it would be substantial and probably involved in the ER activity.

Semiconductors

Much of the research being performed on semiconductor based ER fluids is by H. Block and co-workers (1985, 1988,1990). Although there are many different types of materials which produce active fluids when presumeably dry, and many that don't, the most notable work is on various poly(acenequinones). These materials show substantial ER activity without substantial amounts of water and presumeably no adsorbed water. The drying procedures are not severe due to the thermal instability of the materials and involve drying at 70 ^{o}C under vacuum for three days. The water remaining is structural and tightly bound . The intrinsic mechanism by which these materials presumeably function electrorheologically involves interfacial polarization as a result of electronic charge carriers characteristic of these materials, i.e. being semiconductors. A characteristic of these materials however is the occurrence of substantial bulk currents upon application of the electric field, presumeably because the interfacial electrons can easily jump or tunnel between particles.

Another group of reportedly ER active semiconductors include the photoelectrorheological materials. These materials are simply dispersions of photoconducting materials which become ER active upon application of an appropriate light source, i.e. of correct wavelength to activate electrons into the conduction band. In our work, phenothiazene was used and dried at 120 ^{o}C under vacuum for 1 day before being dispersed into paraffin oil. An example of the behavior is shown in Fig. 3. Many other materials are summarized by Carreira and Mihajlov(1971) , however they do not discuss whether the materials are dry.

Polyelectrolytes

Although various polyelectrolytes have been commonly used in ER fluids, only one study was reported involving dry materials

(Treasurer et. al. 1991). In this study it was shown that the activity does not depend strongly on the cation type or on the type of macroion. However there is a strong correlation between the ionic dissociation energies. An interesting finding was that many such materials did not produce active fluids at room temperature, but the same fluids were very active at 100 °C.

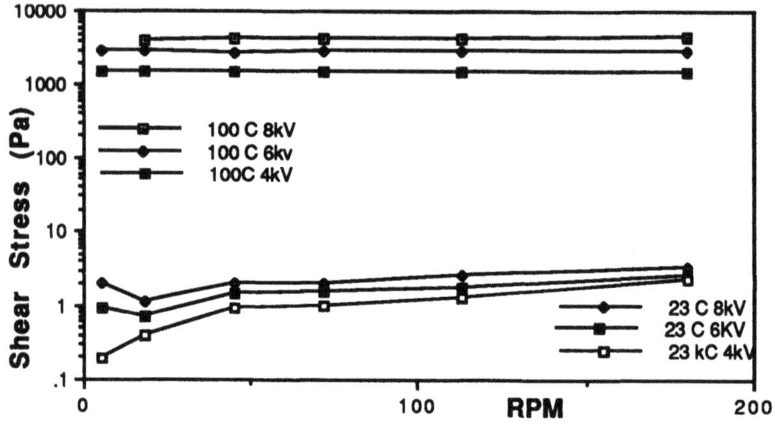

Fig. 5: Activity of an intrinsic ER fluid containing dried polyelectrolyte particulates (Sephadex QAE) at room temperature and 100 °C.

This behavior we feel is associated with the presence of the MWS dispersion which is around 50 °C (Fig. 6).

The mechanism by which these materials function is unclear although the presence of clear MWS dispersion implicates an interfacial polarization mechanism. However the source of the mobile charge carriers is uncertain. It is speculated that the cations have a certain degree of mobility within the macromolecular coils despite the absence of an electrolyte as suggested by Oosawa(1971). These cations can move around within the coil in response to an imposed electric field but cannot migrate outside of the coil, as if in the presence of an electrolyte.

Although these materials were dried at 120 °C under vacuum for one day, they all showed amounts of residual water ranging from 0.1 - 2%. No correlation however existed between the amounts of water present and the ER activity. Furthermore, this water is clearly tightly bound and not simply adsorbed.

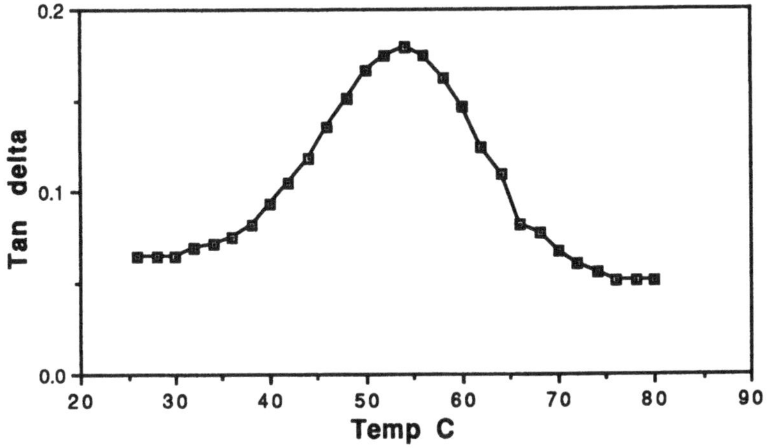

Fig 6: MWS interfacial dispersion of an ER fluid containing dried polyelectrolyte particulates (Sephadex QAE).

Acknowledgements

The authors would like to acknowledge the efforts of many in these studies including Mr. Robert Ervin, Dr. Zheng Lou, Dr. Urvee Treasurer, Dan Gamota, Amy Wang, Andrea Smith, and the support of the University of Michigan Industrial Affiliates Program in Electrorheology, which includes the following sponsors: Aeroquip Corporation, Chrysler Corporation, Ethyl corporation, Ford Motor Company, Monroe Automotive Equipment Co., Naval Underwater Systems Center, Nissan Motors, and Tremec Trading Company.

References

Block, H. and J.P. Kelly (1985). Proc. IEE Colloq. 14:1.

Block, H. and J.P.Kelly(1987),"Electrorheological fluids" (U.S.Pat) 18 Aug., No. 4,687,589.

Block, H. and J.P. Kelly(1988). "Electrorheology", J. Phys. D: Appl. Phys. 21:1661-1677.

Block, H., J.P. Kelly, A. Qin, and T. Watson(1990). " Materials and mechanisms in electrorheology", Langmuir 6(1):6-14.

Breck, S.W. (1974). Zeolite Molecular Sieves, John Wiley & Sons, Inc., New York.

Carriera,L. and V.S Mihajlov(1971),"Recording apparatus and method employing photoelectroviscous ink", (US Pat), 5 Jan., No. 3,553,708.

Deinega,Y.F., K.K.Popko and N.Y. Kovgfanich(1978)," Temperature dependence of the electroviscous effect and dielectric parameters of suspensions of hydrated substances in hydrocarbons", Heat Transfer-Soviet Research 10(1), 50(1978).

Filisko, F.E. and W.E. Armstrong (1988). "Electric field dependent fluids", (U.S.Pat.), 17 May 1988, No. 4744914.

Filisko, F.E. and L.H. Radzilowski (1990). " Intrinsic mechanism for activity of alumino-silicate based electrorheological fluids", J. Rheol. 34(4): 539-552

Gast, A.P. and C.F. Zukoski (1989), "Electrorheological fluids as colloidal suspensions", Adv. Colloid Int. Sci.,30, 153.

Inoue, A.(1990), " Study of a new electrorheological fluid", Proc. 2nd. Conf. ER Fluids, Raleigh NC, Aug 7-9(1989), Technomic Pub. Co.

Maxwell, J.C. (1892). Electricity and Magnetism, Vol.1. 452, Claredon Press, Oxford .

Oosawa, F. (1971). Polyelectolytes, Marcel Dekker, Inc., New York.

Sasada,T, T. Kishi, and K. Kamijo(1974), " On the electroviscous effect in colloidal liquids", Proc. Jpn. Congr. Mater. Res., 17, 228-31.

Sillars,R.W.(1937),"Properties of a dielectric containing semi-conducting particles of various shapes" J. Inst. Elec. Engrs.(London) 80, 378.

Treasurer, U., L. H. Radzilowski and F.E. Filisko (1991). "Polyelectrolytes as inclusions in electrorheologically active materials: effect of chemical characteristics on ER activity" J. Rheol. 35(6), 1051-68.

Wagner, K.W.(1914), Arch. Elektrotech.2, 371(1914).

Winslow,W.M. (1949) "Induced fibration of suspensions", J. Appl. Phys. 20:1137-1140.

DESIGN OF HIGH PERFORMANCE DRY ELECTRORHEOLOGICAL FLUID

Yuan-Ze Xu, Rei-feng Liang, Tian Hao,
Qi-Zhong Zhang, Zheng-Miao Xu, Yi-Hong Chen
Polymer Physics Laboratory, Institute of chemistry,
Academia Sinica, Beijing 100080, China

ABSTRACT

A new kind of ER fluids created based on semi-conducting polymer with partially delocalized conjugated electron structure is introduced which shows high performances: high yield stress, low current density, and higher strength at elevated temperatures. The working mechanism and the principles of designing are discussed based on comprehensive rheological arguments.

1. Introduction - a new approach.

In spite of almost half a century's history of electrorheology, considerable scintific and industrial interests are still shown currently in this field. [1-5] With this level of interest, now seems to be the time to convert scientific curiosity to the realisation of its industrial potential. The most critical challenge in ER techniques still lies in the increase of comprehensive performances of ER fluids.

Early developed ER fluids are mostly wet systems, where water plays a specific role in the ER effect. There are a number of theories to explain the function of water. In most theories water increases the interfacial polarization of the particles by allowing or enhancing ionic migration and the formation of polarizable double layer. Both theory and practice indicate that moist fluids are essentially limited to a narrow temperature range and show undesirable levels of conductance arising from mobile ions. The side effect seems inevitably accompany ER effect for they arise from the same origin.

A common point of all theories emphasized is that the ER effect has its origin in Coulombic forces that are generated between polarized particles under electric field. Some authors further indicated that the attracted particles aggregate into fibrous structures spanning between electrodes, which causes enormous large stress against deformation, which makes the ER fluid become solid-like. Latest theoretical calculation and molecular dynamics-like simulations as well as experimental work support this 'polarization dominant' physical scheme and present it in clearer and more quantitative way.[6-10]

A new way to design ER fluid according to this model would be to find highly polarized particle while not relying on water content and other ion migration as in the case of ion-conducting particles. This could not be realized with the traditional ER material, e.g. silica, cellulose, starch, clay or polyelectrolytes.

Partially delocalized Π-conjugated electron structure in organic semiconductors provide a new approach to realize large polarization while very low tunnel current through the particles. Based on this idea and also considering the neccessary mechanical strength and density we created a new semiconducting polymer (NC- series) based ER suspension. The semiconducting NC polymers were made by well controlled oxidation reactions of polyacrylonitrile.

Different levels of semiconductivity were obtained by controlling the reaction conditions in certain stages. The process is similar to that involved in the manufacture of high strength carbon fibers from polyacrylonitrile, where successive steps are involved: cyclization of the PAN precursor, oxidation, carbonization, and graphitization stages. Graphitized carbon fibers with good conductance do not induce any measurable ER effects.

The present work provides a summary of recent studies of NC- ER fluid in our laboratory. [10, 11] We will discuss the structural model mainly based on the rheological arguments in the second part. The knowledge of the relations of ER properties to the structural and compositional variables comprises the main basis of ER fluid design, which will be discussed in the third part.

2. Physical model of NC- ER fluid - Rheological arguments

2.1. Morphological observations.

The formation of particle structure under an electric field was observed with a long focus microscope. The gap between stainless steel electrodes was set at 1 mm. Typical picture is shown in Figure 1.

All our samples exhibit percolated fibrous structures. More dilute suspensions show more perfect spanned fibers or columns, while in more concentrated systems the particle agglomarates form a porous, fibrous network structure without much movement and rearrangement when the field is applied.

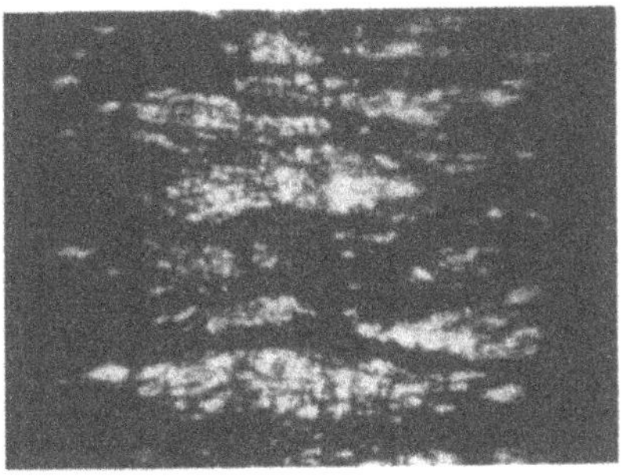

Fig.1. Fibrous structure of NC- ER suspension in silicone oil under electric field. The electrode distance is 1 mm.

The observation of the structure change during deformation is available when we make one electrode shiftable. It is noticed that the strands and network adhere well with electrode walls even during shearing. Also, the structure inclines by shearing and breaks down in the central zone.

2.2. Rheological evidences to physical model.

We would not describe the rheological properties comprehensively here.[10] But summarize the structure informations from the rheological measurement would be interesting.

(1) *Bingham type model with yield stress term dominant.* The flow curve of stress versus shear rate obeys Bingham law:

$$\tau(E,\gamma) = \tau_y(E) + \eta_\infty \dot{\gamma} \qquad (1)$$

and runs almost parallel to the rate axis in broad shear rate range. This means the hydrodynamic dissipation force plays a role only at high shear rate. The viscosity term seems to be near the zero field value.

The physical meaning implied is: all the drag in flow is contributed by the electric attraction between polarized particles in spanned structure, the break and recover of which could be so fast as to keep the stress almost unchanged during flow.

(2) *Polarization force dominates various rheological quantities.*

For high performance ER fluids the static and dynamic yield stresses show very little difference. The yield stress increases quadratically with field strength as predicted by all theories of electric polarization.[4,6,8] A Mason number:

$$Mn = \eta_{\infty}\dot{\gamma}/kE^2 \tag{2}$$

can be generated, where k represents dielectric constant, and the exact expression is not very clear yet.

The dynamic moduli G' and G" show very small linear range and decrease with strain amplitude:

$$G'', G', \eta^* \propto \gamma^{-a} \tag{3}$$

Considering that a is a constant nearly 1, the dynamic stress being the product of modulus and strain becomes a constant, which is near the yield value. We found that the dynamic moduli also obey a quadratic relation with E for some high performance NC- ER fluids. Therefore, the dynamic properties are also dominated by the polarization force like in the steady case.

At this stage, it seems possible to correlate all the rheological behaviors within a simple physical scheme where the driving force is the polarization attraction of the spanned fibrous structure between the electrodes. Some concrete feature of this structure will be seen below.

(3) *Transient responses reveal the inclined spanned structure.*

An impressive structural evidence was found by the stress recovery experiment when flow ceases. Usual structured media exhibit a monotonic stress decay after flow stop. But for ER fluid under high E field, after a short stress decay as in usual media, the stress gradually recovers in some extent depending on the field strength as shown in Figure 2.

A possible explanation for this could be based on the assumption that the inclined structure formed by shearing would remain inclined and even improved as the flow stops. It is not clear yet, this 'long memory' of the inclined fibrous strucutre is generally originating from the short distance interaction of polarized particles or is related with the non-spherical particle form of our suspensions.

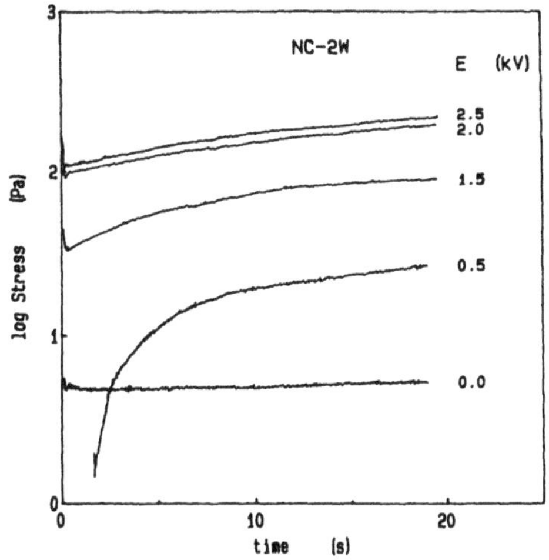

Fig. 2. Stress recovery after a short flow step ($\dot{\gamma}$= 10 s^{-1}, t=0.3 s) for NC-2W fluid. Measured in parallel pate rheometer at 300 K.[10]

With this idea we can also explain why thixotropic loop exists while there is no overshoot in step flow. In the case of decreasing shear rate, the weaker inclined strands at higher rate need some time to be improved, while at start of flow the force increases accompaning with the orientation of strands from vertical to inclined direction. This is a kind of *orientation related thixotropy*.

Searching the current density during oscillation of shear, we observed the double frequency of the current signal and the maximum current at low strain, which corresponds to the compression rather than the stretch of the strands.

(4) *Characteristic critical concentration.*

We found that it is possible to correlate the relation between the rheological propertier e.g.dynamic modudus G' and volume concentration at different electric field strength by means of reduced coordinate. The curves of relative modulus G'/G'_c versus relative concentration ϕ/ϕ_c of NC-D suspensions overlap in spite of the field E changing from 0 up to 2.5 KV/mm as shown in Figure 3. The values of critical concentration and critical modulus are listed in Table 1.

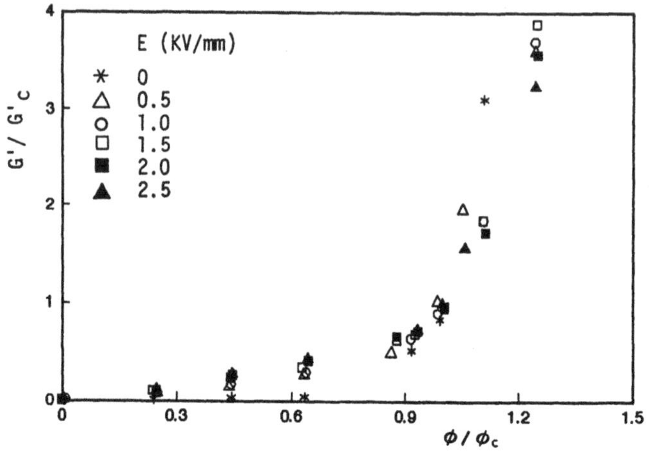

Fig.3. Concentration dependence of modulus for NC-3D suspension at 300 K. Plotted in reduced coordinate.

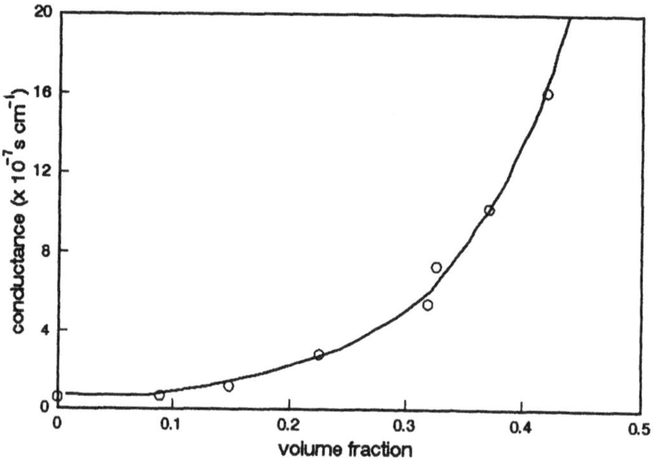

Fig.4. Concentration dependence of conductance of NC-3D suspension.

Table 1. Critical concentration and critical modulus for NC-3D ER suspension at various electric field strength.

Field strength (KV/mm)	0	0.5	1.0	1.5	2.0	2.5
critical concentration φ_c	.371	.370	.377	.375	.374	.374
critical modulus G'_c (Pa)	3.64	73.6	300.	600.	967.	1350

Similar values of critical concentration exists for various rheological properties: G', G", η as well as for electric conductivity σ as shown in Figure 4. This concentration is independent of field strength. This point may correspond a percolation transition of the fractal aggromerates in ER suspension. Furthermore, the electric field seems not change this fractal network structure. Indeed not much particle migration was observed in concentrated suspensions as electric field is given. Of course, the network is mechanically very anisotropic in electric field and behaves like fibres, strands or columns. We should expect some porosity in the structure, which may influence the viscosity both in high E or zero field. Further work should be done in this subject.

3. On fluid design

3.1. *Requirements and variables.*

Depending on the application direction, the suggested ER specifications seems scattering. The requirements for ER devices have some limits above which it brings industrial significance.[12-14] An estimated main requirements of ER fluid according to recent device research could be summarized as follow:

zero field viscosity	1 Pa.s
current density	<10 $\mu A/cm^2$ at 3.5 KV mm^{-1}
yield stress	4-15 Pa
working temperature	240-420 K
non-toxic and stability	

There are many structural, compositional, and processing variables, which influence the performances of the end products, as shown in Table 2.

Table 2. Some variables influencing ER performances of dry ER fluids.

Structural	particle and medium conductivity
	dielectric constants
	interface condition
	particle size and geometry
	mismatch of density
Compositional	volume concentration
	particle size distribution
	additives and contaminants
Processing	reaction parameters in semiconducting material preparation
	control parameters in particle preparation
	control parameters in suspension preparation

We are not yet at the stage to discuss systematically all the relations between these variables and their influence to the performances. Some important factors and results will be described to convince this promising approach.

3.2. *Concentration.*

As we see from Figures 3 & 4, zero field viscosity and conductivity increase drastically above the critical concentration of nearly 37 vol.%. This critical value depends on the porosity or fractal dimension of the suspension, accordingly, on the interface energy of the particle and medium. The maximum concentration is generally limited by zero field viscosity considering the undesired power comsumption and the heating effect. The high concentration also leads unacceptable current density for systems with higher conducting particles. However, we did not find the saturation of the yield stress in our working range. Figure 5 is an example for NC-3D ER fluids.

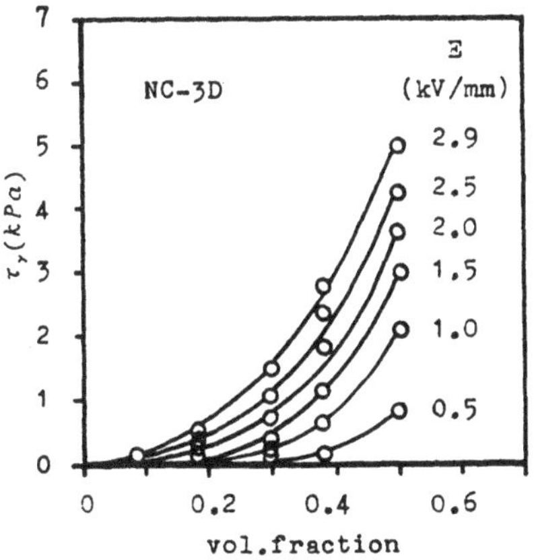

Fig.5. Yield stress as a function of volume concentration for NC-3D fluids. T= 300 K.

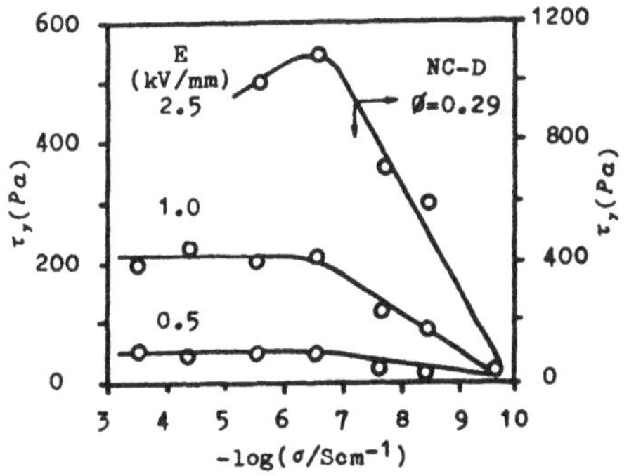

Fig.6. Yield stress as a function of particle conductance for NC-D fluids. T= 300 K.

3.3. *Particle conductance.*

To estimate the polarization ability of the particle, it is not only easier to measure the conductivity of the particle than its dielectric constants, but also a relative direct way of measuring polarization, if the conductivity is mainly caused by the tunnel current of the conjugated electron clouds rather than by the charge carriers of contaminants. Figure 6 shows the effect of the particle conductivity on the yield stress. A maximum of yield stress appears at certain conductance range under higher electric field. That means too high conductivity also reduces the particle interaction in addtion to the problems caused by high current density.

Therefore, to design a useful ER fluid we should make optimizations and concentrate our efforts in the stripped region of the schematic plot in Figure 7, where curve a means the necessary yield stress , curve b means the limit of allowed maximum current and curve c is the limit of allowed zero field viscosity.

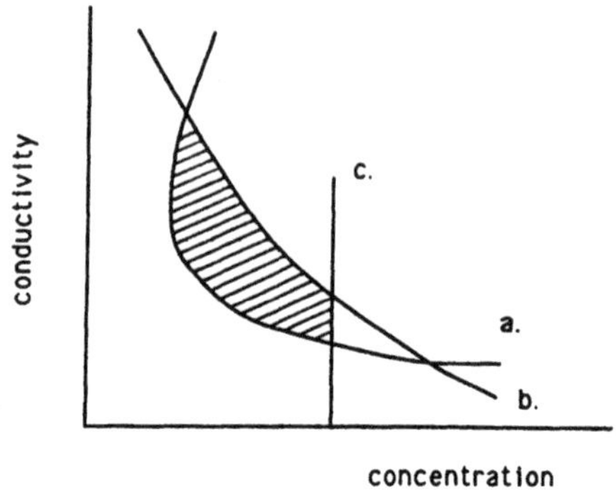

Fig.7. Schematic view of the working area according to the requirements.

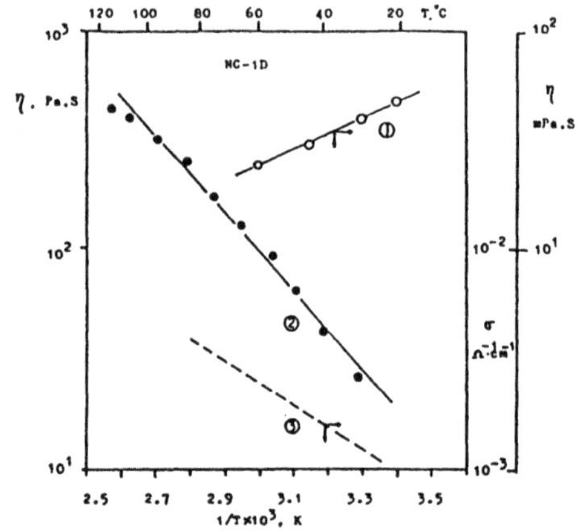

Fig.8. The temperature dependence of viscosity and particle conductance for NC-1D system.

3.4. *Temperature dependence*

The dry ER fluids show prevail advantages over other systems in its favorable temperature dependence. As shown in Figure 8, the yield stress increases with temperature exponentially. An Arrhenius type equation of Eq.4 is adequate to describe the temperature dependences of suspension as well as particle conductivity and the viscosity of the medium- silicone oil.

$$\eta, \sigma = Ki \exp(Ei/RT) \quad (4)$$

However, the energy barriers E_i are very different. E_1= 14.0 KJ/mol, the usual activation energy of viscous flow for silicone oil. E_2= -35.6 KJ/mol for NC-1D ER suspension, and E_3= -18.3 KJ/mol for the conductance of semi-conducting particles exhibit typical temperature dependences of semiconducting systems. This phenomenon provides a promising application possibility if we can achieve a favorable compromise between the yield stress and the conductance of ER suspensions at high temperatures. The mechanism of the conductance depending on electric field and temperatute will be discussed in a coming paper.

At last, the influences of water content should be discussed.[10,11] It is clear that water is not necessary at all for our system - a *true* dry ER suspension. But water content will enhance the ER effect below boiling temperature. NC-W systems means the suspension containing NC particles balanced at atmospheric humidity with water content ca. 5% in particle before being mixed with silicone oil. The comparison of humidified NC-5W fluid with corresponding dry fluid is shown in Figure 9, where the yield stress and current desity show a change for wet system below boiling point. Although water enhances the ER effect at room temperature, it causes unacceptable current density increase with temperature until water being exhausted at elevated temperature.

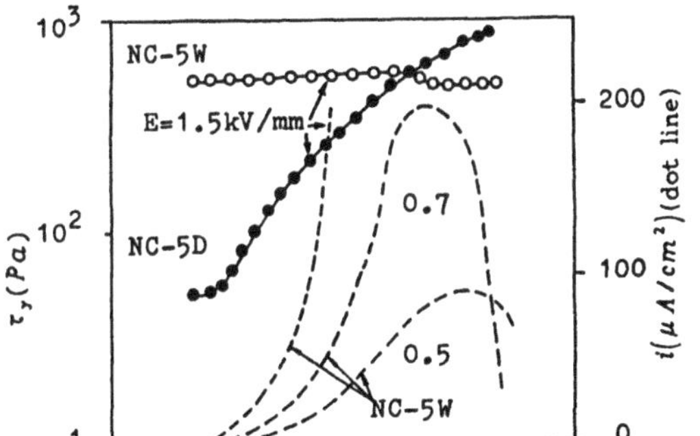

Fig.9. The yield stress and current density as functions of temperature depending on water content for NC-5 ER suspensions.

Along the above described mechanism and approach we created a series of NC-D ER fluids. Typical properties of NC- ER suspensions up to now are summarised in Table 3. Further improvement of this system is being done and seems not exhausted.

Typical Properties of NC- ER Suspensions

sample	volume fraction (%)	water in particle wt.%	test temp. °C	zero-field viscosity Pa.s	field strength KV/mm	yield stress KPa	current density (μA/cm^2)
NC-1D	13.1	0	25	0.3	2.5	.22	<1
NC-1W	10.8	7.4	25	0.36	2.5	.42	<1
NC-2D	13.1	0	25	0.4	2.5	.18	<1
NC-2W	14.4	6.5	25	1.3	2.5	.49	<1
NC-2W*	18.8	6.5	25	3.0	2.5	1.25	<1
NC-3D	11.2	0	25	0.3	2.5	.07	<1
NC-3D*	50.	0	20	7.3	1.5	3.0	15.
					2.5	4.2	80.
NC-4D	50.	0	25	2.2	1.5	1.28	.25
					2.9	3.0	3.
NC-4W	39.	7.3	25	3.0	1.5	1.62	4.5
					2.9	3.3	25.
NC-4W*	47.	7.3	20	7.4	1.5	4.36	5.
					2.0	6.58	12.5
NC-5D	22.	0.	25	.36	1.5	.05	<1
			120	2.0	1.5	.86	1.
NC-5W	22.	7.4	25	1.78	1.5	.52	5.
			120	2.6	1.5	.49	>200.
NC-6D	45.	0.	25	5.	2.25	2.8	<5. *
				1.	2.9	1.	20.
NC-7D	53.	0.	25	20.	2.5	11.	30.

* measured by Rheometrics Co.

5. Acknowledgement

This work is supported by Chinese National Science Foundation.

6. References

1. Y.F. Deinega and G.V. Vinogradov, Rheol. Acta 23, 636 (1984)
2. H. Block and J.P. Kelly, J. Phys. D 21, 1661 (1988)
3. T.C. Jordan and M.T. Shaw, IEEE Trans. Elect. Insul. 24, 849 (1989)
4. A.P. Gast and C.F. Zukoski, Adv. Coll. Int. Sci. 30, 153 (1989)

5. T.G. Duclos, Autom. Eng. 96, 46 (1988)
6. L. Marshall, J.W. Goodwin, and C.F. Zukoski, J. Chem. Soc. Farad. Trans. I 85, 2785 (1989)
7. D.J. Klingenberg, D. Dirking, and C.F. Zukoski, J. Chem. Soc. Faraday Trans. 87, 425 (1991)
8. D.J. Klingenberg, F.van Swol, and C.F. Zukoski, J. Chem. Phys. 91, 7888 (1989)
9. D.J. Klingenberg, F. van Swol, and C.F. Zukoski, J. Chem. Phys. 94, 6160, 6170 (1991)
10. Y.Z. Xu, and R.F. Liang, J. Rheol. 35(7) 1355 (1991)
11. R.F. Liang, Q.Z. Zhang, and Y.Z. Xu, in *Proceeding of China-Japan Intern. Congress on Rheology*, Beijing, Peking University Press (1991) 296
12. D.L. Hartsock, R.F. Novak, and G.J. Chaundy. J. Rheol. 35(7) 1305 (1991)
13. N. Webb, Chemistry in Britain, April 338 (1990)
14. J.L. Sproston and R. Stanway, Bulletin of British Soc. Rheol. vol.34, No.2, 39 (1991)

THE EFFECTS OF VARIOUS DIELECTRIC OILS AND VOLUME FRACTION OF SOLIDS ON THE MECHANICAL PROPERTIES OF HYDROUS BASED ELECTRORHEOLOGICAL FLUIDS

A. F. Sprecher, Jr.
Materials Science and Engineering Department
Box 7907
North Carolina State University
Raleigh, NC, 27695-7907 USA

and

Y. Chen, Y. Choi and H. Conrad
Materials Science and Engineering Department
Box 7907
North Carolina State University
Raleigh, NC, 27695-7907 USA

ABSTRACT

Electrorheological (ER) fluids consist of fine polarizable particles suspended in a dielectric liquid. It is the dielectric mismatch between these phases that give rise to the ER effect. This paper presents data on ER fluids with different dielectric oils employing a constant solid phase consisting of zeolite with adsorbed water. The effect of solid concentration was also studied, in order to develop a more complete understanding of the mechanisms involved. Results are explained in terms of the dielectric properties of the phases present using an electrostatic model developed by the authors and an additional structural feature termed as chain bundling. The electrostatic model takes onto account the characteristic chain-like structure which is commonly observed in these fluids, while the chain bundling concept incorporates multiple chain interactions.

Introduction

It is now generally believed that the electrostatic attraction of the suspended phase is responsible for the electrorheological (ER) effect in most fluid systems. When an external electric field is applied to these materials the solid particles align into a chain-like structure, which requires an appreciable amount of energy to disrupt. This structural observation has been reported by other researchers throughout the years[1,2], and has been characterized using quantitative stereology for a model ER fluid system[3,4].

Even though this electrostatic attraction stems from the dielectric mismatch between the phases present, calculations of ER fluid strengths based on simple point dipole interactions and number of chains present yield low estimations[5,6]. Even more sophisticated approaches which take into account the particle geometry effects on the electrostatics, generally yield values for the strengths that are appreciably lower than those measured[6-9].

The purpose of the present work was to examine the effects of different dielectric oils that have been commonly used in ER fluids. Their wide range of dielectric constants (2.19 to 5.35) made it possible to study how dielectric mismatch controls ER fluid strength. The common solid phase that was suspended in all these oils was a hydrated zeolite.

In order to obtain a clearer understanding of the mechanisms which control ER fluid strength, it is necessary to understand the role of solid concentration. To that end, a separate set of experiments was performed on one oil system where only the volume fraction of zeolite was varied.

All tests were conducted at a relatively low strain rate and low values of strain so that the particle structural aspect could be more easily controlled in terms of a relatively well defined chain-like arrangement.

Experimental

1. Materials investigated

Several commonly used dielectric oils were selected in order to examine their effect on the mechanical properties of ER fluids. These included silicone oil, corn oil and several varieties of hydrocarbon oils. They were chosen because they have been used by other investigators in ER fluid research and that they have a relatively wide range in their dielectric constants (2.19 to 5.35). The specific oils are presented in Table I along with their viscosities and dielectric properties. All the ER fluids used in the present investigation used zeolite 3125 from Sigma Chemical Co. (with 8 wt.% water adsorbed) as the suspended solid phase. The zeolite was heated at 160°C for six hours in order that it and the water would come to an equilibrium with the desired moisture content. This procedure is described in more detail in reference [10].

Two series of tests were performed. The first series was designed to examine the effect of solid particle phase concentration. These experiments used only mineral oil and varying amounts (ϕ = 5 - 25 vol %) of zeolite particles. The second set of tests used the aforementioned dielectric oils with a fixed concentration of zeolite of about 20 vol%.

All data were obtained at room temperature and employed an electric field range of 0.5 to 4 kV/mm.

Table I. Various dielectric oils used to make the ER fluids for the present study

ER Fluid*	Source	K_f	K_p	τ (Pa) E=2kV/mm	Viscosity (mPa s)
Mineral oil	Ward Transformer Co.	2.19	14.4	161	35
Silicone oil	Dow Corning	2.61	23.7	163	52.8
Corn oil	CPC International Inc.	3.28	29.9	291	80
Caster oil	High Energy Corp.	4.24	31.2	298	980
DOP oil	High Energy Corp.	5.35	48.3	297	71.2

* All ER fluid samples were made with one of the above oils and Zeolite 3125 with 8wt% water. Source: Sigma Chemical Co.
K_f is the dielectric constant of the as received oil and K_p is the dielectric constant of the zeolite in the respective oil.

2. *Rheological characterization*

Rheological data consisted of shear stress versus shear strain and was obtained on two measuring systems. Measurements for the series of experiments which studied the concentration of solids effects were obtained using a Physica rheometer which was specially designed for electrorheological work. A constant strain rate mode ($\dot{\gamma} = 2s^{-1}$) was employed.

For the series using different dielectric oils, a lateral shear test similar to that described in [6,10,11] was employed using a strain rate of 0.085 s^{-1}. This technique, which was developed by the authors, utilizes a modified tensile test machine so that the stress-strain behavior can be studied at relatively low strains and strain rates.

To understand the mechanisms involved in these fluids, it is necessary to define the their strength. The definition of "yield stress" is not universal in rheological research. Differences in test procedures give stress values at different strains and strain rates as well as strain history. This fact along with certain fluid body assumptions that are sometimes made, makes comparisons of stress values from different sources difficult. In both sets of data, the "yield stress" was taken at a strain of 5 while being sheared at the aforementioned constant strain rate.

3. *Electrical characterization*

The dielectric constant for the oils (K_{oil}) was obtained by measuring the capacitance of a standard coaxial cell with and without fluid. The dielectric constant of the suspended zeolite particles (K_p) was obtained in a similar fashion and determined using the following rule of mixtures.

$$K_{erf} = K_{oil}\,(1-\phi) + K_p\,(\phi) \tag{1}$$

Where K_{erf} is the dielectric constant of the ER fluid. The capacitance meter uses a low signal level of the order of 1 volt in its measurement. This low level is desirable so that particle chaining can be avoided. Also, the sample was thoroughly stirred just prior to the measurement. The dielectric constant of the ER fluid that is measured can be higher if the particles are in the chained structure [6]. These conditions were maintained in order to minimize these effects and obtain a more accurate value for K_p.

Results

Table I presents a summary of the dielectric and viscosity data obtained for the as-received oils as well as the dielectric properties of the suspended zeolite in the oils. Also included is the strength of each fluid at E=2kV/mm. All the fluids tested exhibited a squared dependence of the strength with respect to the applied electric field. This is reasonable providing that the polarization on the suspended particles is linearly induced and the dielectric integrity of the fluid is maintained. All ER fluid systems may not exhibit this behavior if the polarization saturates at some lower value of electric field.

Upon examination of the dielectric properties, it can be seen that the situation is somewhat complicated. Even though the zeolite had the same initial amount of water in all oil systems, its dielectric constant K_p, varies with the different systems. K_p ranges from about 14 in the silicone oil to about 48 in the DOP oil system. Apparently there is a reaction between the oil, zeolite and water present in each fluid.

In Table I it can be seen that generally as the dielectric constant of the oil increased, the strength of the ER fluid increased. This observation, although true, is somewhat misleading since K_p was different for each system as well. Figure 1 presents the normalized yield stress ($\tau/E^2\phi$) versus the ratio K_f/K_p. As is seen from this Figure, for a given K_p, the strength of the fluid actually decreases as K_f increases over the range of K_f/K_p that data was obtained.

The effect of concentration, ϕ, of zeolite in the ER fluid can be seen in Figure 2, where the (τ/E^2K_f) versus ϕ is given. Although a linear relation exists, the function does not pass through the origin and the slope is substantially greater than unity. Both of these facts suggests a more complicated situation than that which can be predicted by simply accounting for the number of chains in the ER fluid. This has been reported by others as well [10,14].

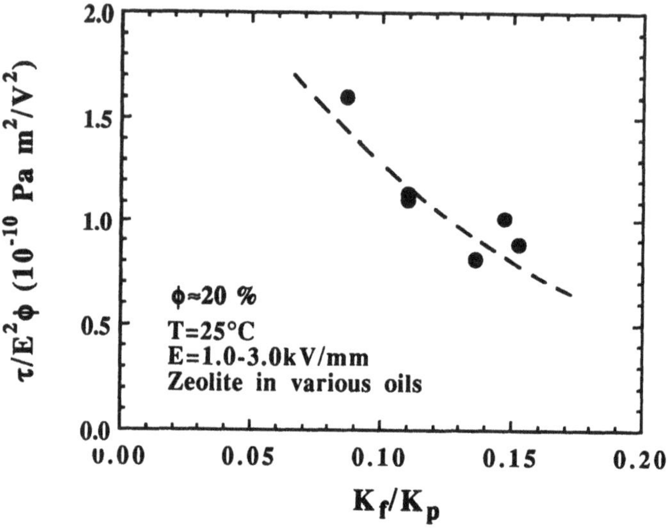

Fig. 1. Normalized yield stress, $\tau/E^2\phi$ versus the ratio of dielectric constant of the various oils to that of the zeolite particles, K_f/K_p. E ranges from 1.0 to 3 kV/mm and the concentration of zeolite is from 18 to 20 vol%.

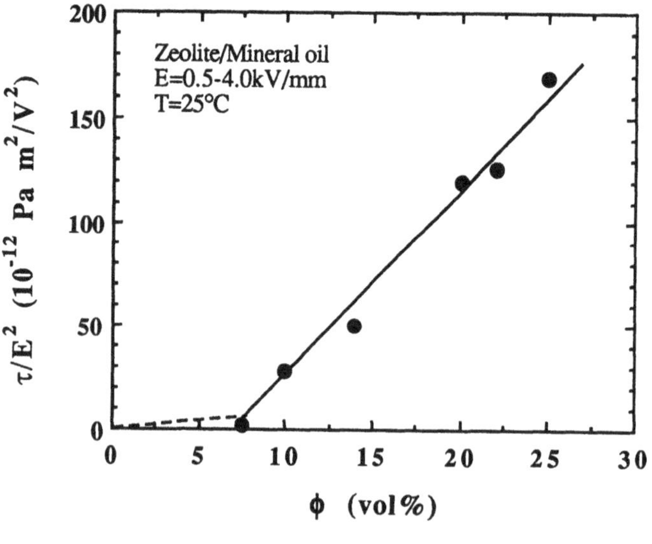

Fig. 2. Normalized yield stress, τ/E^2 versus volume fraction of zeolite, ϕ, in mineral oil.

Discussion of Results

To interpret the data shown in Figures 1 and 2 it is necessary to introduce a model for the strength of an ER fluid. In a fashion similar to that introduced in [6], the expression for the strength consists of a structure term and an electrostatic term.

$$\tau = (N_a) (F A_s) \mathrm{Sin}\theta \tag{2}$$

The $\sin\theta$ resolves the normal electrostatic forces that hold a chain together to the shearing geometry for which the data was obtained. This model is based on the most generic feature associated with these fluids, namely, particle chaining, and assumes a well-defined arrangement of particles in the chains. This most likely will hold true for low strain rates (where the present data was obtained).

The structure term, N_a, is simply the number of completed chains per unit area in the system and is given by

$$N_a = (3\phi/2\pi a^3)\, X_c \tag{3}$$

where ϕ is the volume fraction of solid phase, a is the particle radius and X_c is the fraction of chains that completely bridge the electrode gap. For very small concentrations, X_c may be less than unity, since the particles will be greatly scattered through out the oil, but for higher concentrations it will most likely be near one.

The electrostatic term, $F A_s$ consists of the strength of a single *isolated* chain of particles F, and a single chain enhancement factor, A_s which must take into account the interaction of neighboring chains. A_s is defined as the ratio of strength of an isolated chain to that of a chain in a bundle or cluster of chains. The product, $F A_s$, then represents the strength of a single chain of particles while in a bundle of surrounding chains.

The data presented in Figures 1and 2 can be analyzed using the above model provided the strength of a single isolated chain F, is known. This parameter has been theoretically calculated [9] and measured [7]. Figure 3 shows the results of these calculations, where the normalized force F/E^2K_f, is plotted against β^2 [9]. $\beta = (K_p-K_f)/(K_p+2K_f)$, is the dielectric mismatch term pertaining to the solid and liquid phases present in the system.

Briefly, the calculation in [9] is a numerical solution to an expansion of the electric potential between particles in a single isolated infinite chain of ER fluid particles. It satisfies Laplace's equation along with appropriate boundary conditions. The simulation indicated that the electric field distribution around the particles was highly perturbed and enhanced by an

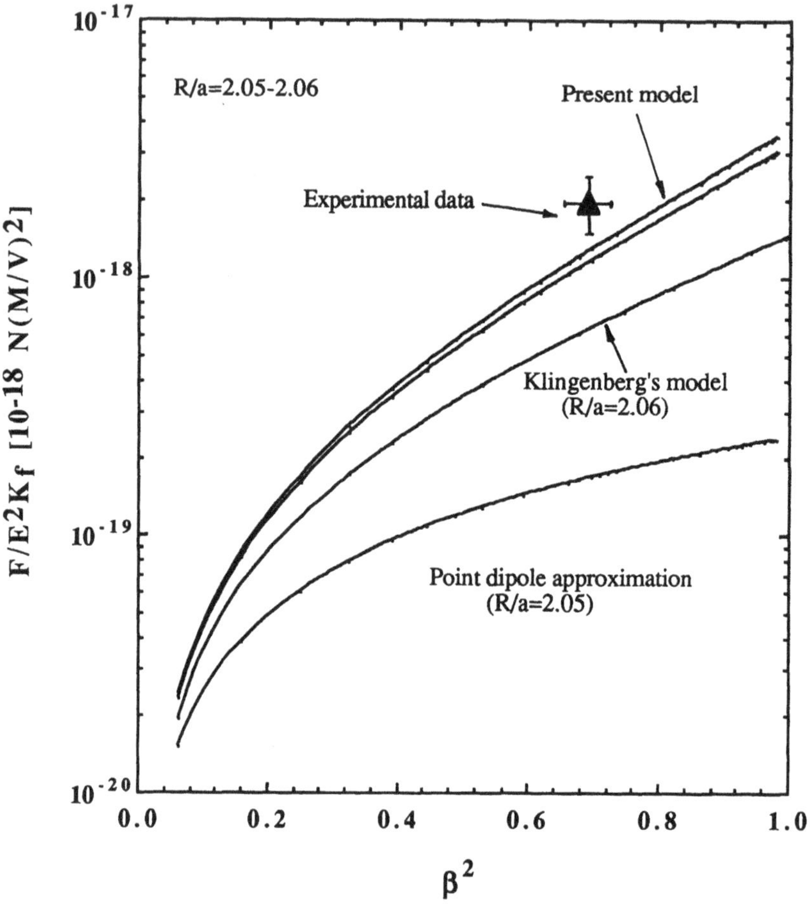

Fig. 3. Normalized force of interaction, F/E^2K_f, for a single, isolated chain of ER fluid particles versus the dielectric mismatch term squared, β^2, Chen et. al.[9].

order of magnitude compared to that of the applied electric field. This consequently leads to higher forces of interaction than that of the simple dipole approximation. It also predicts larger forces of interaction than does the calculations from Klingenberg [8], which is similar to [9] but employs only two ER fluid particles in an electric field. As can be seen, the experimental data point taken from Sprecher et al [7] for such a chain of model ER fluid particles is in reasonable agreement with the calculation.

For a given K_p, the decrease in stress with K_f which is shown in Figure 1, can be explained in terms of F. This is the term that is controlled by the dielectric properties of the two phases present in the ER fluid. Figure 4 presents F versus K_f from the calculations in [9]. Here it is seen that two regions can be defined. In region I, the force increases with increasing K_f, while region II shows just the opposite behavior. For a constant K_p, there exist a maximum value of force at a unique value of K_f. The calculations show that there is an optimum value of $K_p/K_f = 13.5$ where F is maximum. This variation in force with K_f is reasonable since F is approximately proportional to the product , $\beta^2 K_f$, [9]. In region I (low K_f), β is near unity and F increase in approximate proportion to K_f. However in region II (high K_f), β is less than one so that the overall product $\beta^2 K_f$ and consequently F decreases. The fact that the local electric field intensity near the particle contact area is much higher than that of the applied electric field, accentuates this K_f dependence. In the present set of experiments, the maximum K_p/K_f ratio is 11.7, so that the data obtained in the present set of experiments lies in region II, where increasing K_f results in a decrease ER fluid strength.

The data in Figure 1 can be shown in a more useful form ($\tau/E^2\phi\beta^2$ versus K_f) and is presented in Figure 5. In accordance with the model presented, by using the secant slope of this data and the results from [9] obtained for F, the single chain enhancement factor, A_s was determined to be in the range of 12 to 14. This is in keeping with values calculated in preliminary computer simulations on ER fluid particles in closed packed arrays by Shih et al [15].

Figure 6 illustrates schematically why the chains might be stronger in a bundle compared to when they are isolated. The figure shows a closed packed three chain bundle before and after shearing. Note the particle contact points *a* and *b*, where the *a* points are along the primary chain axis while the *b* points are those where adjacent chains contact. Provided that the interaction forces at points *a* and *b* are the same for a bundled configuration compared to the axial forces in an isolated chain, the bundled arrangement can be stronger since there are more particle contact points per chain. Also to be noted in the figure is that when the system is deformed the *b* particle points will stay closer together for larger shear strains. This in turn will also result in a stronger structure.

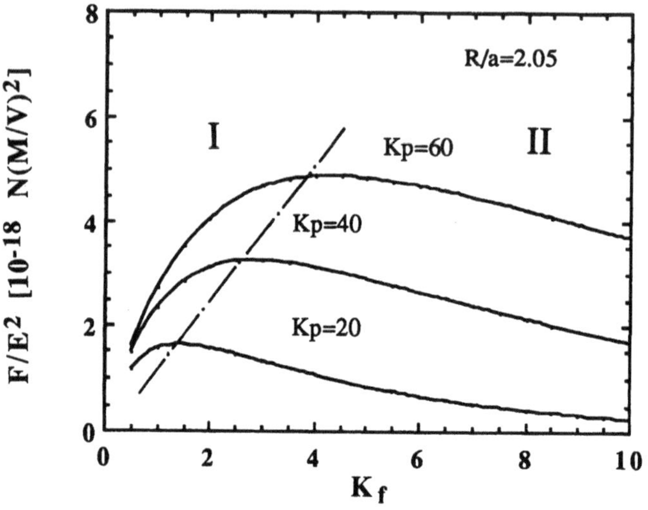

Fig. 4. Calculated normalized force of interaction, F/E^2, for a single, isolated chain of ER fluid particles versus the oil dielectric constant K_f for various solid phase dielectric constants.

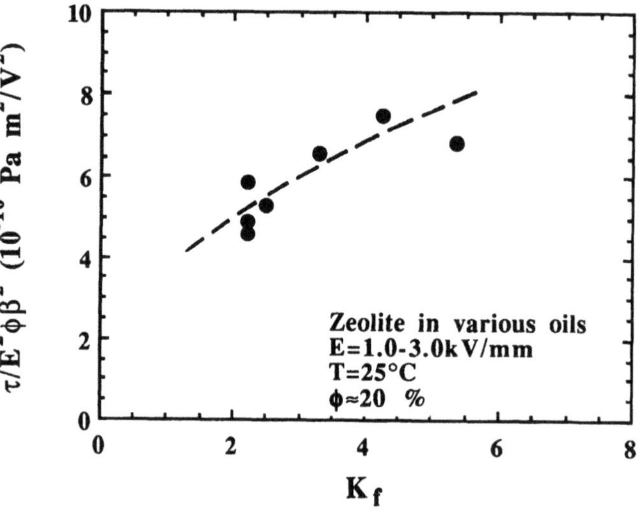

Fig. 5. $\tau/E^2\beta^2\phi$ versus K_f for the data presented in Figure 1.

Since A_s is dependent on neighboring chains to form bundles, it would stand to reason that it is concentration dependent. At very low ϕ, the chains will either be incomplete or isolated, so that A_s will be unity. At very high ϕ, complete bundle formation will exist (all possible particle nearest neighbors will be present), so that A_s would tend to saturate at some maximum value of ϕ. These effects can be determined from the experimental data found in Figure 2. A_s can be determined by taking the measured stress and dividing it by that predicted based solely on isolated chain properties. This data is presented in Figure 7 where A_s versus ϕ in shown. The figure does exhibit a minimum ϕ for which $A_s=1$, then increases in a linear fashion and tends to saturate at values of ϕ =20% and greater.

Summary and Conclusions

Two sets of experiments were performed to investigate the effects of different dielectric oils and volume fraction of solids on the mechanical properties of ER fluids. All fluids used as their solid phase zeolite loaded with 8 wt% water. In the first set of tests several ER fluids with different dielectric oils were tested mechanically and electrically at low strains and strain rates. The second set of experiments involved a single ER fluid system of zeolite and mineral oil and only the solids fraction of the zeolite was varied.

The strength of the various systems increased with increasing dielectric constant of the particular oil used. When the changing dielectric properties of the suspended zeolite was taken into account, the opposite behavior was observed. This fact is in keeping with calculations of the strength for a single chain of ER fluid particles. For high ER fluid strength, increases in the dielectric constant of the oil must be accompanied by appropriate increases the dielectric constant of the suspended particles.

The strength of a zeolite/mineral oil ER fluid was found to increase with increasing zeolite concentration. The functional dependence of the data did not suggest a simple accounting of chains present in the system. But rather was explained in terms of number of chains and an enhancement effect of chains grouped in bundles. This enhancement factor ranged from unity at low concentrations and increased in a linear fashion above concentrations of about 5 vol%. It subsequently tended to saturate at a value of about 12 to 14 for concentration levels of 20 vol% and greater. This concentration dependence is reasonable when the chain bundle density is considered.

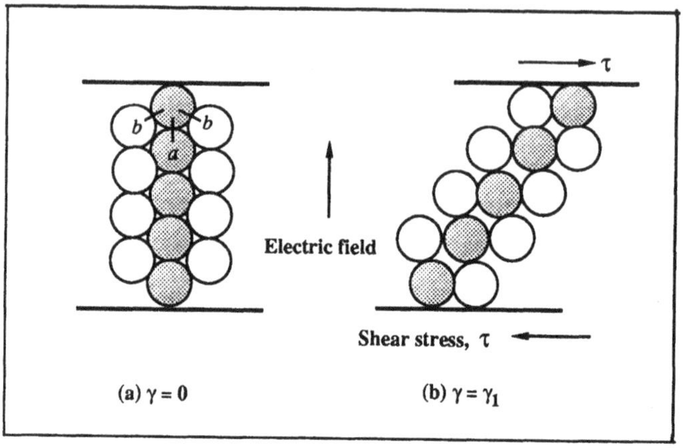

Fig. 6. Schematic of closed packed, three chain bundle configuration: (a) before shear straining, (b) after shear straining. "*a*" particle contact points are along the primary chain axis, while the "*b*" contact points are those from adjacent chains.

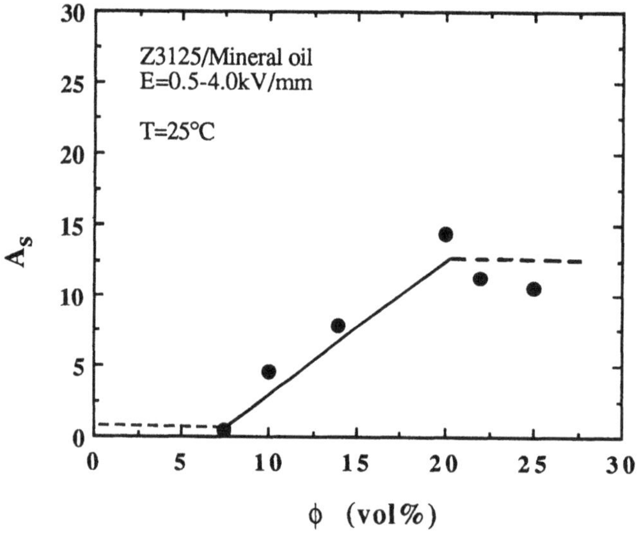

Fig. 7. Single chain enhancement factor, A_s, versus zeolite volume concentration, ϕ. A_s is defined as the ratio of strength of a ER particle chain in a bundle to that of the same chain isolated from neighboring chains.

Acknowledgement

The authors gratefully acknowledge support of their work jointly by the National Science Foundation under NSF award #CBT-8714515, the Ford Motor Company and the NCSU University/Industry Consortium on ER fluids.

References

1. W. M. Winslow, *J. Appl. Phys.*, **20** (1949) 1137.
2. D. L. Klass and T. W. Martinek, *J. Appl. Phys.*, **38** (1976) 67.
3. H. Conrad, M. Fisher and A. F. Sprecher, Jr., *Proc. 2nd Int. Conf. Electrorheological Fluids*, J. D. Carlson, A. F. Sprecher and H. Conrad, eds., Technomic Publishing Co. Lancaster-Basel (1990) 63.
4. M. D. Fisher, A. F. Sprecher, Jr., and H. Conrad, *Symposium on Metallography: 75 Years Later,* ASTM, Atlantic City, NJ, May (1991).
5. R. T. Bonnecaze and J. F. Brady, *Proc. 2nd Int. Conf. Electrorheological Fluids*, J. D. Carlson, A. F. Sprecher and H. Conrad, eds., Technomic Publishing Co. Lancaster-Basel (1990) 27.
6. H. Conrad, A. F. Sprecher, Jr., Y. Choi and Y. Chen, *J. of Rheolgy,* Annual Meeting of the Society of Rheology, Santa Fe, NM Oct. (1990).
7. A. F. Sprecher, Jr., Y. Chen, and H. Conrad, *Proc. 2nd Int. Conf. Electrorheological Fluids*, J. D. Carlson, A. F. Sprecher and H. Conrad, eds., Technomic Publishing Co. Lancaster-Basel (1990) 252.
8. D. J. Klingenberg, and C. F. Zukoski, *Langmuir*,**6**, (1990) 15.
9. Y. Chen, A. F. Sprecher, Jr., and H. Conrad, J. of Appl. Phys., accepted for publication Dec. (1991).
10. H. Conrad, Y. Chen, and A. F. Sprecher, *Proc. 2nd Int. Conf. Electrorheological Fluids*, J. D. Carlson, A. F. Sprecher and H. Conrad, eds., Technomic Publishing Co. Lancaster-Basel (1990) 252.
11. A. F. Sprecher Jr., J. D. Carlson and H. Conrad, *Mat. Sci. and Eng.*, **95** (197) 187-197.
12. R. D. Miller and T. B. Jones, *J. Phys. D: Appl. Phys.*, **21** (1988) 527.
13. V. N. Makatum and D. N. Vorobev, *Proc. 2nd Int. Conf. Electrorheological Fluids*, J. D. Carlson, A. F. Sprecher and H. Conrad, eds., Technomic Publishing Co. Lancaster-Basel (1990) 184.
14. L. Marshall, C. F. Zukoski and J. Goodwin, *J. Chem. Soc., Faraday Trans.1*, **85** (1989) 2958.
15. Y. Shih, A. F. Sprecher, Jr., and H. Conrad, unpublished research at North Carolina State University, Raleigh, NC (1991).

BASE LIQUIDS FOR ER FLUIDS

LINDA F. EVANS, IAN HARNESS, PETER R. KERMODE

and

JAMES E. STANGROOM
ER Fluid Developments Ltd, Brough, Sheffield S30 2HG, U.K.

ABSTRACT

Two predictions of the dielectric mechanism proposed for ER fluids were tested by examining 17 fluids made up using the same solid but different base liquids. The dependence of yield stress on field in all these fluids is well represented by both the linear and the square-law models, but detailed examination shows the former to be marginally better. However, the dielectric model fails completely to account for the effect of dielectric constant of the liquid and solid, and it is concluded that this model has only limited value, at least when applied to "Strong" ER fluids. On the other hand, the activity of a series of ER fluids is shown to correlate well with the energy of an interface between water and the base liquid, as required by the water bridge model; the latter also suggests a linear dependence on electric field.

1. Introduction

Whatever the differences in detail, it is probably true to say that the majority of workers attribute the fundamental behaviour of ER fluids to the electrostatic consequences of induced dipoles. An electric field applied to an ER fluid would be expected to induce dipoles in the particles, causing the latter to be attracted to each other, chiefly along the line of application of the field. This, in essence, is the "dielectric mechanism" proposed for ER fluids. There are variations on this basic outline. For example, the Mason Number model [1] compares the dielectric with viscous forces at various flow rates; Block and Kelly's approach [2] concentrates on the lag of the induced dipoles when the particles are spinning due to local differences in flow velocity. Regardless of the details, arguing from the empirical observation that ER fluids exhibit a yield stress under conditions where induced dipoles would be expected to arise to the assertion that these dipoles are ultimately responsible for the yield stress involves an assumption that is open to question. Some workers, such as Whittle [3], have expressed doubts that the dielectric mechanism by itself is sufficient to account for the strength of some ER fluids. However, quantitative arguments such as this are difficult to apply when the details of the supposed link between induced dipoles and yield stress are not understood. For example, it could plausibly be argued that some geometrical arrangements may greatly exaggerate the macroscopic effects of relatively weak electrostatic forces. This paper seeks to shed new light on the basic question by examining the influence of the base liquid on the response of ER fluids.

A field-dependent yield stress is the most basic feature of ER fluids, and this would be expected to be fairly directly related to the force between neighbouring pairs of particles. The electrostatic force between two spheres, radius a, dielectric constant K_p, separated by a distance R in a medium of dielectric constant K_f when exposed to an electric field E is given by the equation:-

$$F = 24 \frac{a^6 \varepsilon_o E^2}{R^4} \frac{K_f(K_p - K_f)^2}{(K_p + 2K_f)^2} \tag{1}$$

This leads to two predictions:-

a) That for a given ER fluid, the yield stress should be proportional to the square of the electric field strength.

b) If all other parameters apart from K_f (or K_p) are kept constant, the yield stress should vary according to Eq. 1: in particular, it should change with the difference in the dielectric constants, becoming zero when they are equal.

2. Experimental Approach

In spite of wide variation in chemical composition, there is broad correspondence in the behaviour of all the ER fluids that have been described, so it is likely that there is a common mechanism. However, there are many differences in detail, and it is essential to examine a series of ER fluids in order to distinguish features common to all ER fluids from those found in only a few. Examining a range of ER fluids has the further advantage of providing enough data to allow the effects of experimental error to be reduced by normal statistical techniques.

Only "Active" ER fluids, i.e. those which showed sufficient response to be worth considering in practical applications, were investigated, reflecting our primary interest in developing ER fluids for practical use. It could be argued that "Strong" ER fluids are a separate group, and results based on these alone cannot be extended to the whole range of ER fluids. While this is valid, the "Strong" ER fluids are the most difficult to understand in terms of the dielectric mechanism and therefore deserve close study. From the purely practical point of view, the stronger the ER fluid, the easier, more reliable and relatively more accurate are the measurements. In fact, given the range of ER fluids available, it seemed to us positively perverse to attempt to study the ER phenomenon in materials in which it is barely apparent.

The first prediction, the square-law dependence of yield stress on field, could have been tested on data selected at random from the many thousands of test results we have on file - indeed, this was done for dynamic results. However, to investigate the effect of dielectric constant, we needed a series of ER fluids in which the dielectric constant of one component varied while that of the other was held constant. It is well known that the properties of the solids used in ER fluids vary from batch to batch, and can be changed by subsequent treatment. To keep the properties of the solid constant throughout the series of experiments, all solid samples were taken from a single container. At the start of the series, and periodically thereafter, an ER fluid was made up with this solid and a "reference" base liquid and tested in the normal way. Within experimental error, the behaviour of these check fluids was unchanged, so we can assume that the solid was the same in all the tests. Base liquids, on the other hand, can be readily purified by distillation or other methods, and their properties, including dielectric constant, can be checked against literature values. Table 1 shows the base liquids finally used.

Table 1 - Liquids and Dielectric Constants

liquid	Dielectric Constant
p-Xylene	2.27
Benzene	2.28
m-Xylene	2.37
Toluene	2.38
o-Xylene	2.57
Methyl Naphthalene	2.67
Trichlorobenzene	3.87
o-Iodotoluene	4.02
Bromodiphenyl Ether	4.10
o-Fluorotoluene	4.18
o-Bromotoluene	4.45
Bromonaphthalene	4.65
Fluorobenzene	4.84
m-Dichlorobenzene	4.90
Bromobenzene	5.18
Chlorobenzene	5.50
o-Dichlorobenzene	9.73
Acetophenone	17.39

3. Experimental Details

In order to distinguish changes in the interparticle forces from the effects of geometrical arrangement, we used "Static" measurements. In this method, described in detail elsewhere [4], the ER fluid is placed in the cup of a Couette viscometer from which the motor drive has been removed. The field is turned on and a gradually increasing torque is applied to the bob. Shearing takes place very sharply; the torque at which this occurs, translated into shear stress at the surface of the bob, is recorded.

All the ER fluids were made up to 30% (V/V) with the standard solid (lithium polymethacrylate, cross-linked with methylene bis-acrylamide [4]) and tested at 30° C. For each fluid, the voltage applied across 0.5mm was increased in steps of 200V, two replicate observations being made at each voltage, until dielectric breakdown occurred.

Although it does not form part of this study, it is worth mentioning that the current/field curves of all the fluids tested were virtually the same, in spite of the differences in chemistry and activity of the fluids. This strongly suggests that the current is normally determined by the solid alone.

4. Results and Discussion

4.1 Field Dependence

A linear law [5] and a power law [6] have been suggested as alternatives to the square law dependence of yield stress on field. However, the power law implies a difference in the basic mechanism of ER fluids and so was not further considered in this study. To compare the linear and the square relationships, regression coefficients were calculated for each over the seventeen sets of raw data. It was necessary to include the "Threshold" in the linear model; this was found by a least-squares technique. These regression coefficients are presented in Table 2.

Table 2 -Regression Coefficients

Liquid	$(E - E_o)$	E^2
p-Xylene	0.9988	0.9696
Benzene	0.9952	0.9874
m-Xylene	0.9976	0.9796
Toluene	0.9931	0.9720
o-Xylene	0.9962	0.9879
Methyl Naphthalene	0.9986	0.9860
Trichlorobenzene	0.9966	0.9735
o-Iodotoluene	0.9956	0.9854
Bromodiphenyl Ether	0.9949	0.9627
o-Fluorotoluene	0.9986	0.9735
o-Bromotoluene	0.9951	0.9812
Bromonaphthalene	0.9939	0.9947
Fluorobenzene	0.9916	0.9824
m-Dichlorobenzene	0.9981	0.9837
Bromobenzene	0.9944	0.9944
Chlorobenzene	0.9925	0.9939
o-Dichlorobenzene	0.9933	0.9966
Acetophenone	0.9898	0.9825

The first point to note in Table 2 is that all the regression coefficients are high, i.e. both the linear and the square law models give excellent representations of the experimental data. On closer inspection, it is apparent that most values of the regression coefficient for the linear model are marginally higher than the corresponding values for the square law. An Analysis of Variance of the two sets of coefficients reveals that there is indeed a difference between them which is significant at the 0.1% level, i.e. the probability of this difference appearing purely by chance is less than one in a thousand.

Statistical arguments must be treated with caution, but in many circumstances, results as significant as these would be regarded as virtually incontrovertible evidence in favour of the linear model. Such an extreme interpretation is not reasonable in the present situation, but, nevertheless, it is clear that the linear model must be seriously considered, and experiments must be designed with this model in mind.

One feature of the linear model which has important consequences for the design of experiments is the "Threshold" field. If there is in fact a sharp change in the behaviour of ER fluids at some value of the field, it follows that performance at high fields cannot be inferred from low field behaviour. Indeed, it appears likely that extrapolations in other directions may be dangerous as well. Attempts were made to extend this study to "Dynamic" data, (i.e. data collected in the normal way in a Couette viscometer) by comparing the linear and square-law models applied to data taken at random from our files. As in the static case, the linear model gave a marginally better representation of the data than the square law, but the difference between the two was much less significant. This may be due to the fact that technical problems limited the voltage range over which the dynamic data was collected. On the other hand, it may also represent a real difference between static and dynamic behaviour. For example, the frequency of interparticle collisions would be expected to play a significant role in determining the dynamic yield stress, and this could well have a square law dependence on the applied field.

4.2 Base liquid Dielectric Constant

Marshall, Zukoski and Goodwin [1] estimated the dielectric constant of hydrated lithium polymethacrylate in ER fluids provided by Laser Engineering Developments Ltd. They measured the dielectric constants of the ER Fluid and the pure base liquid in a Wayne-Kerr bridge - the frequency was not specified, but the standard bridge works at about 1kHz. From these results, they obtained a value of 23 for the dielectric constant of the solid; this agreed well with calculations based on the known dielectric constants of methacrylate polymers and water. Combining the figures of Marshall, Zukoski and Goodwin with an estimate of the water content from conductivity data on the various ER fluids [7], we obtained a value of 8 for the dielectric constant of the lithium polymethacrylate used in our experiments, reflecting a slightly lower water content.

Eq.1 predicts a linear relationship between the quotient (Yield Stress/Square of Field Strength) and a function of the dielectric constants of the solid and liquid. This is plotted out in Fig 1, using a value of 8 for the dielectric constant of the solid.

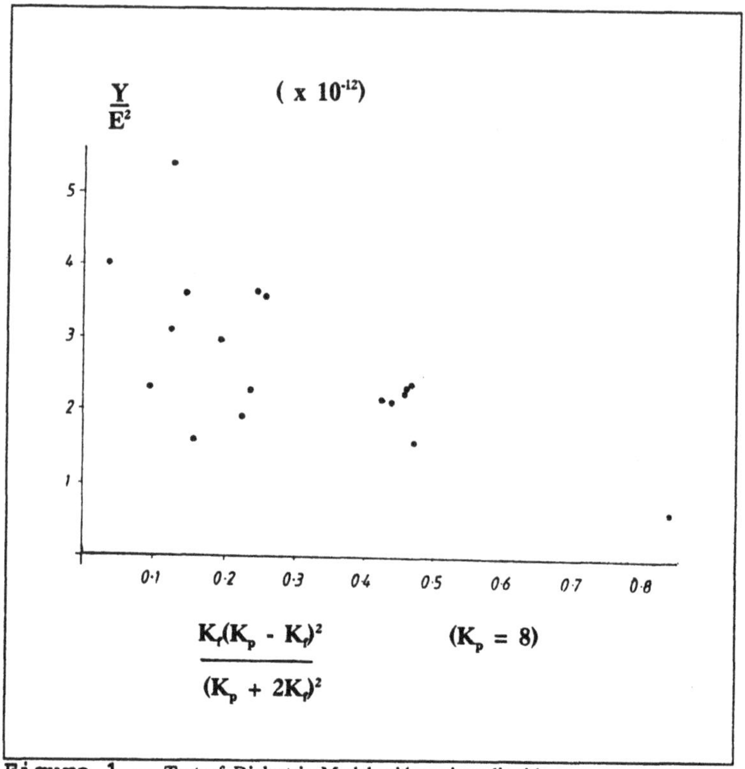

Figure 1 – Test of Dielectric Model with various liquids

By no stretch of the imagination can this be described as a straight line. Since we were by no means certain of the dielectric constant of the solid, we constructed a range of such plots, using different values of this parameter. Figs 2 and 3 show two such attempts, for solid dielectric constants of 5 and 11 respectively. These are no more convincing than Fig 1, and it is fair to say that we have failed to find any value which gives a reasonable line.

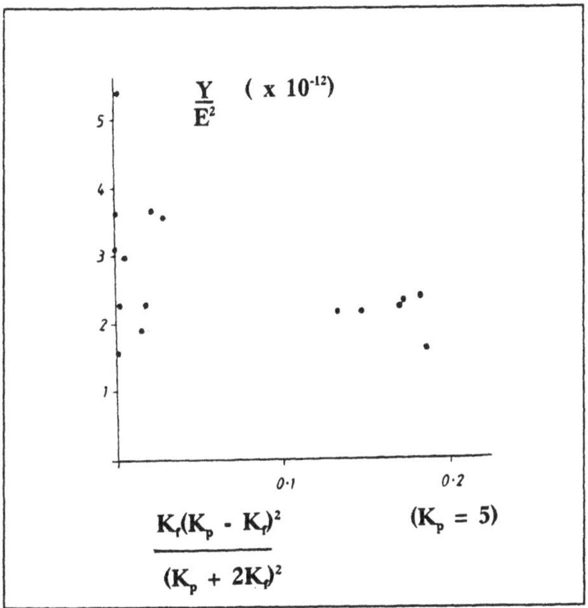

Figure 2 - Test of Dielectric Model with various liquids

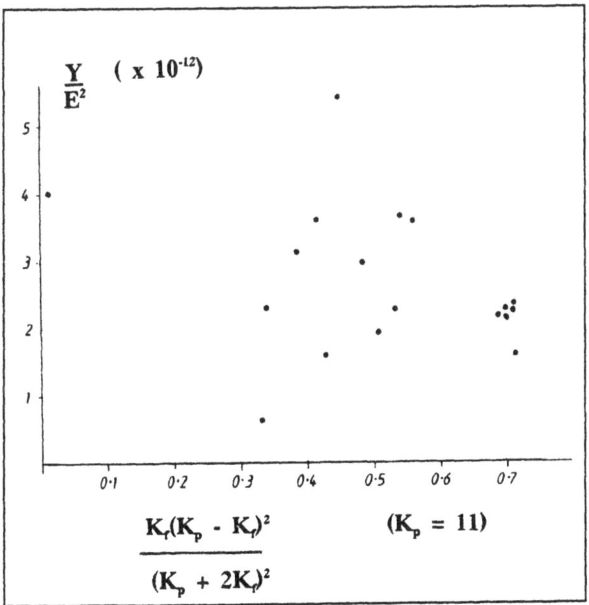

Figure 3 -Test of Dielectric Model with various liquids

Eq.1 also predicts zero response when the dielectric constant of the base oil is equal to that of the solid. Our estimated value of solid dielectric constant falls within the range covered by the base liquids, so we would expect to see some evidence of such a minimum. In Fig 4, the quotient (Yield Stress/Square of Field Strength) is plotted against the liquid dielectric constant. There does not appear to be any trace of a minimum. An equivalent plot can be obtained by using a single base liquid and varying the dielectric constant of the solid by adjusting the water content. Fig 5 shows the result. Water contents were estimated from fluid conductivity data as before, and the solid dielectric constant calculated in accordance with the figures of Marshall, Zukoski and Goodwin. Although the absolute values obtained by this procedure may be suspect, at the very least, it would be expected that the shape of the final graph would bear some relation to that predicted by Eq. 1, that is a minimum. However, the actual plot obtained, shown in Fig 5, shows a maximum.

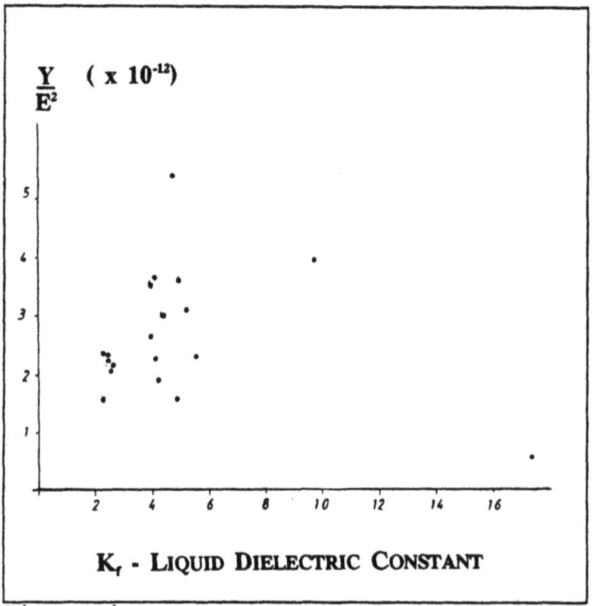

Figure 4 – ER Response and Dielectric Constant of Liquid

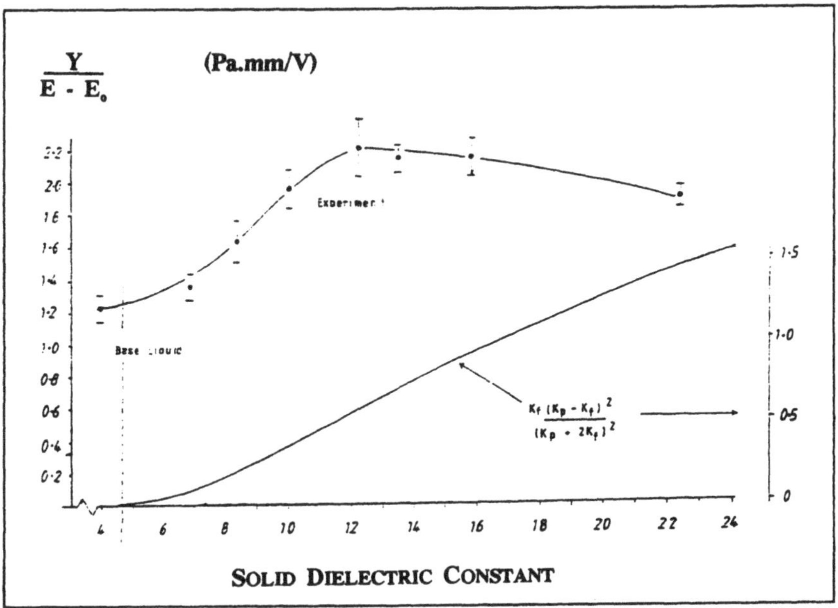

Figure 5 - ER response and Dielectric Constant of Solid

Figs 1 - 5 strongly suggest that the dielectric model embodied in Eq.1 cannot account for the effects of variations of dielectric constant; at best, it is therefore incomplete. Whittle [3], reached the same conclusion on the basis of very different evidence. He remarks:-"These arguments suggest that the dielectric mechanism alone is insufficient to account for the magnitude of the ER effect in "strong" fluids."

There is another aspect of the dielectric model which is worth mentioning in relation to these experimental results. The dielectric constants on which the model is based reflect the charge separations brought about electrostatically by the applied field. It is well known that ER fluids conduct small but measurable currents, but the implications of this for the dielectric model do not appear to have been explored. Dielectric constants can be measured independently of conductivity by using a.c. at moderate frequency and isolating the capacitative current, but there seems little justification for ignoring leakage in d.c. measurements on ER fluids. In these, the capacitative currents which arise when the field is first switched on will fall to zero within a few seconds at most; thereafter, the potential distribution will be determined purely by the differences in the conductivities of the particles and the base liquid.

Dielectric constants may still be involved, for example in the electrostatic forces resulting from local potential gradients, but the system as a whole will be very different from the basic dielectric model. Admittedly, these objections arise primarily with static d.c. measurements, but these would be expected to be the simplest of all. If a theory cannot deal with these, it is unlikely to be successful when time-dependent terms are introduced, as in measurements on flowing ER fluids or a.c. fields.

5. Alternatives to the Dielectric Model

Stangroom [8] has proposed an alternative to the dielectric model. He suggests that the applied field drives water normally adsorbed inside the particles to the outside by a process of electro-osmosis. The water then forms "water bridges" with neighbouring particles; these resist deformation due to the energy of the interface between the water and the base liquid.

Detailed mathematical analysis of the "Water Bridge" approach is not yet available, but some semi-quantitative predictions can be made. Stangroom suggests that the water acquires a charge with respect to the particle matrix from mobile hydrated ions. Movement of the water when an electrical field is applied (electro-osmosis) is therefore a form of dielectric polarisation in which the charge carriers are hydrated ions, and the movement of water will be linearly related to the field. However, the water holding two neighbouring particles together will be derived from only one of them, so the strength of the bridge would be expected to be linearly related to the field. In passing, it is worth noting that similar arguments from the water bridge approach predict that the current density in a static ER fluid should be proportional to the square of the field strength: this is observed in practice.

According to the water bridge model, the strength of an ER fluid should be proportional to the energy at a base liquid/water interface. The latter was measured for a series of liquids by a pendant drop method [9]:- the technique initially checked with standard liquids. ER fluids were then made up using a standard solid in the measured liquids; these were tested as previously described.

Fig 6 shows the slopes of the graphs of Static Yield Stress against Field above the "threshold" value plotted against the energy at the interface of the corresponding base liquid with water. A very good straight line is observed - the regression coefficient (0.96) is remarkably high. At first sight, it is surprising that the line does not go through zero, implying that a certain minimal interfacial energy is required before any ER activity becomes apparent. However, low interfacial energy implies chemical similarity, which will also favour mutual solubility. For example, diethyl ether, which is normally regarded as immiscible, has a low interfacial energy with water (10mN/m) and in fact is "fairly soluble". If water was only marginally soluble in the base liquid, it would be rapidly leached out, and there would be no possibility of forming water bridges.

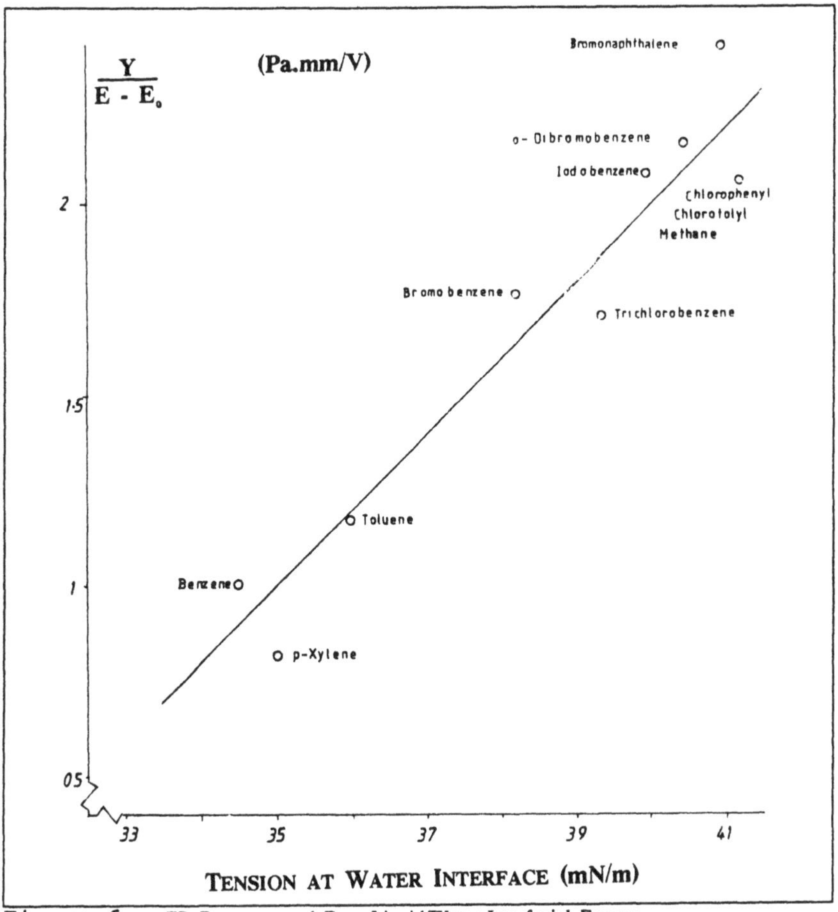

Figure 6 – ER Response and Base Liquid/Water Interfacial Energy

While this evidence lends strong support to the idea of water bridges, it is not conclusive. At the very least, many more results, covering a much wider range of solids and liquids are required. There is also the possibility that what is involved is not interfacial energy per se, but some other, more fundamental parameter of which interfacial energy is a reflection.

6. Conclusions

The results presented in this paper suggest that the simple dielectric model has very limited value as a description of the ER fluids examined. Experimental data is, at best, equivocal regarding one of its predictions, the dependence of yield stress on the square of the field strength. Predictions of the effects of variations of the dielectric constants of the liquid and solid components bear little or no relationship with experimental observations. In addition, it is by no means clear how a model based purely on dielectric effects can be applied to a d.c. system in which there are measurable leakage currents.

Although data is limited, the situation regarding the water bridge hypothesis seems to be much more encouraging. Semi-quantitative arguments suggest a linear relationship between yield stress and applied field; the experimental evidence supports this view. The second prediction, that the activity of ER fluids should be directly related to the energy at a water interface with the base liquid, has been verified experimentally. This must be regarded as firm evidence in favour of the basic hypothesis.

The data on which these conclusions were based were all obtained from ER fluids which have been described by some workers as "Strong". However, the validity of the distinction between these and the other ER fluids has yet to be established.

7. References

1. L.Marshall, C.F.Zukoski and J.W.Goodwin, *J. Chem Soc.* **85** (1989) 2785.

2. H.Block and J.P.Kelly, *J. Phys. D: Appl Phys.* **21** (1988) 1661.

3. M.Whittle, *J. Non-Newtonian Fluid Mech.* **37** (1990) 233.

4. J.E.Stangroom, U.K. Patent No. 1501635 (1974)

5. J.E.Stangroom, U.K. Patent No. 1570234 (1977)

6. H.Conrad and A.F.Sprecher, "Characteristics and Mechanisms of Electrorheological Fluids". Paper presented at the NATO workshop on Non-Linear Fluids, Los Alamos, 1990.

7. J.E.Stangroom, U.K. Patent No. 2100740B (1981)

8. J.E.Stangroom, *Phys. Technol.* **240** (1983) 290.

9. S.Fordham, *Proc. Roy. Soc.* **A149** (1948) 1.

THE EFFECT OF SOLVENT AND PARTICLE DIELECTRIC CONSTANTS ON THE ELECTRORHEOLOGICAL PROPERTIES OF WATER-FREE ER FLUIDS

TERRY GARINO, DOUG ADOLF AND BRAD HANCE
Sandia National Laboratories,
Albuquerque, NM 87185

ABSTRACT

The properties of water-free electrorheological (ER) fluids containing electrically insulating ceramic particles were studied. Results on fluids containing particles with different dielectric constants, ε_p, in dodecane (ε_s=2) indicate that strong ER effects can occur due to dielectric constant mismatch. For example, in a fluid containing 40 vol% $BaTiO_3$, the yield stress was >1.2 x 10^4 dyn/cm^2 at 700 V/mm. The yield stress of these fluids depends on ε_p much more strongly than predicted by the point dipole approximation, especially at high particle contents. The yield stress was also independent of frequency between 10 and 4000 Hz. However, when the fluid contained 2-methylcyclohexanol (ε_s=13), a much higher conductivity solvent, the yield stress increases strongly with frequency. This occurs because the dielectric mismatch mechanism only operates above a critical frequency that depends on the electrical conductivity of the liquid.

1. Introduction

Nearly all ER fluids described in the literature with high values of yield stress contain particles that are either inherently semiconducting or are made effectively semiconducting due to the presence of water.[1-3] However, an ER effect due to the dipole moments induced in insulating particles because of the difference between their dielectric constant, ε_p, and that of the solvent, ε_s, should occur in water-free systems.[4] On the other hand, for the dielectric mismatch mechanism to operate, the frequency of the field must be above a critical value, ν_{cr}, where capacitance effects dominate conductive effects. Below ν_{cr}, the distortion of the electric field will depend on the mismatch between the conductivities of the particles and the solvent, not on the dielectric constant mismatch.[5] Since the value of ν_{cr} increases with the solvent conductivity, an additional, and possibly more stringent, requirement on the solvent conductivity is imposed, besides that due to power loss, since ν_{cr} must be low enough for practical applications.

When dipoles are induced in particles by an electric field due to dielectric mismatch, the particles will form chains connecting the electrodes. The force between adjacent particles in a chain can be determined by adding the point dipole forces between adjacent particles, the point dipole forces between non-adjacent particles and the near-field forces between adjacent particles. For all of these forces, the magnitude of the force is proportional to the square of the magnitude of the electric field, E, and, approximately, to ε_s, since the field energy is given by one half $\underline{E}\cdot\underline{D} \sim \varepsilon_s E^2$. For the point dipole interactions, the force is also proportional to the square of a function of the dielectric constants, β, equal to $(\varepsilon_p-\varepsilon_s)/(\varepsilon_p+2\varepsilon_s)$, which is a measure of the dipolar field distorsion around a particle. Therefore, for point dipole interactions the force between particles is proportional to $\varepsilon_s(\beta E)^2$. This predicts that for a given ε_s, the force between particles, and thus the yield strength of the ER fluid, increases towards a limit as ε_p is increased, reaching 90% of that limit when $\varepsilon_p/\varepsilon_s \sim 55$. On the other hand, when ε_p is constant, the force reaches a maximum when $\varepsilon_p/\varepsilon_s \sim 5.4$. These trends are shown in Figure 1.

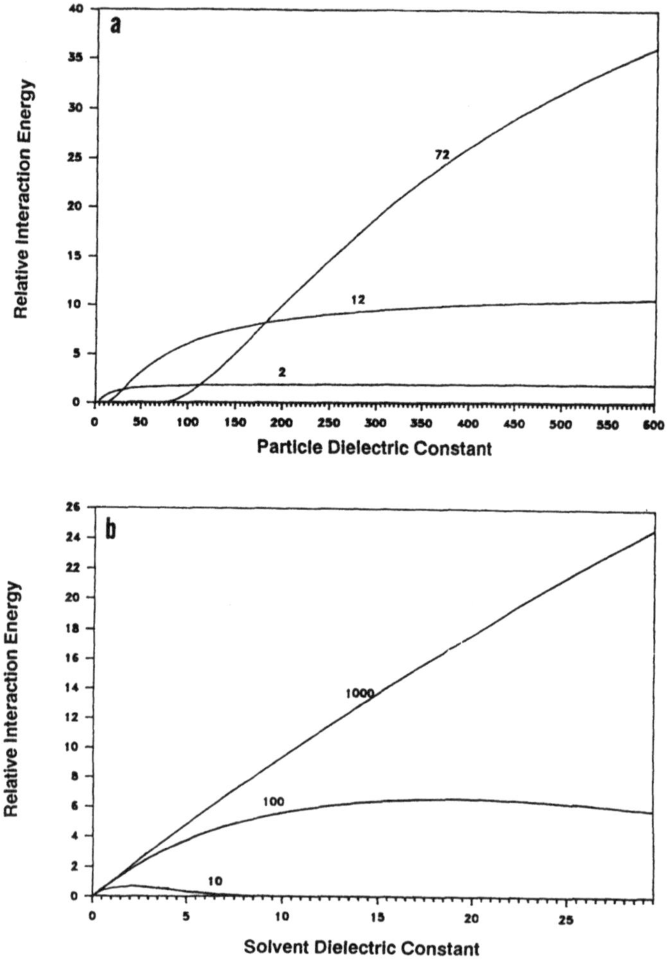

Figure 1. Interaction energy between adjacent spheres due to point dipole interactions as ε_p increases for three different values of ε_s (a), and as ε_s increases for three different values of ε_p (b).

The primary goals of this study were to determine whether ER effects occur due to dielectric constant mismatch in water-free fluids containing non-conducting particles and to determine if a critical frequency exists in such systems. This was done by studying fluids containing ceramic particles with dielectric constants covering a broad range and by studying the behavior as a function of frequency of fluids with high ε_p particles in solvents with different values of ε_s.

2. Experimental Procedure

The ER fluids studied were prepared from submicron ceramic powders with dielectric constants that vary over a wide range. The powders used were aluminum oxide (ε_p = 10, AKP-HP, Sumitomo Chemical Co.), titanium oxide (ε_p = 70, predominantly rutile, Aldrich Chemical Co.), strontium titanate (ε_p = 270, Johnson Matthey Electronics) and barium titanate (ε_p = 2000, grade 219-3, Ferro Corp.). The solvents used were dodecane (ε_s = 2, purified grade, Fisher Chemical Co.) and 2-methylcyclohexanol (Eastman Kodak Co., ε_s = 13). In dodecane, a poly-isobutene amide (OLOA 1200, Chevron) was used as a dispersant at a concentration of ~0.5 wt%. The fluids were prepared with 10 vol% solids and were sonicated in an ultrasonic bath and shaken until the powder was well dispersed. To prepare more concentrated fluids, some of the supernatant was removed from the 10% fluid after the powder had settled due to the mismatch in density between the powders and the solvents. All fluids were vigorously shaken immediately prior to ER measurements.

A Carri-Med controlled stress rheometer with parallel annuli (OD=60mm, ID=57mm) was used to measure the rheological properties of the fluids with and without an applied voltage. The maximum shear stress that the rheometer could apply was 1.2 x 10^4 dyn/cm^2. Runs were performed by increasing the stress at a constant rate to a maximum in 90 s. Apparent yield stress was determined by fitting the data to the Bingham plastic equation. A Trek Model 10110 power supply was used to apply a square wave with a frequency between 10 and 4,000 Hz.

Fluid resistivity was determined using a DC power supply to apply a voltage to the fluid between two parallel platinum plates which were 2 cm^2 in area and separated by ~0.4 mm. The current was then measured with a Keithley multimeter (model 619).

3. Experimental Results and Discussion

Figure 2 shows the apparent yield stresses at 1 kV/mm and 400 Hz for fluids containing Al_2O_3, TiO_2, $SrTiO_3$ or $BaTiO_3$ at three different concentrations in dodecane as a function of ε_p. The fact that the yield stress depends strongly on ε_p indicates that dielectric mismatch is responsible for the ER effect in these fluids. However, contrary to the prediction of the point dipole equation (see Figure 1), the ER effect does not saturate when $\varepsilon_p/\varepsilon_s$~55, but continues to increase with ε_p, especially at the higher solids concentrations. This indicates that near field forces and point dipole forces between non-adjacent particles are important and depend on ε_p more strongly than does β, as predicted by model calculations.[6]

Figure 3 shows the same data as Figure 2 only plotted as a function of concentration for three of the ceramic materials. The degree of nonlinearity of yield stress with concentration greatly increases with ε_p. Behavior similar to this has been observed in other systems, such as those containing semiconducting polymers.[1]

To understand the implications of this data, some discussion of the structure of ER fluids is necessary. When an electric field is applied to an ER fluid, the particles form chains, some of which completely span the gap between the electrodes. Various model calculations and simulations suggest that the measured yield stress is proportional both to the average strength of the chains that span the gap and to the

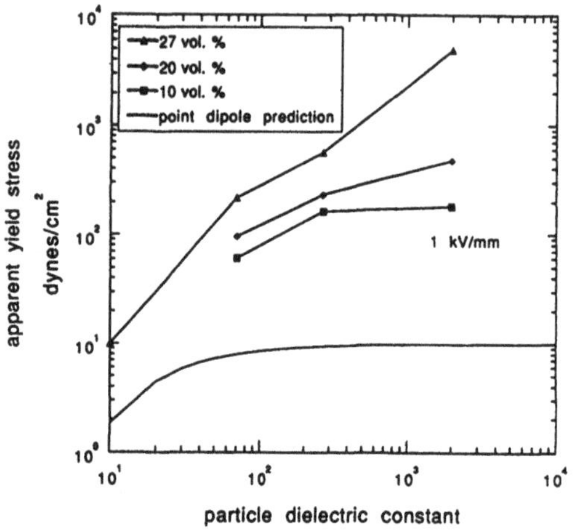

Figure 2. The dependence of the apparent yield stress at 1 kV/mm, 400 Hz, of ER fluids containing ceramic particles in dodecane as a function of the dielectric constant of the particles for three different solids concentrations. The prediction of the point dipole equation is shown for comparison of shape, not magnitude.

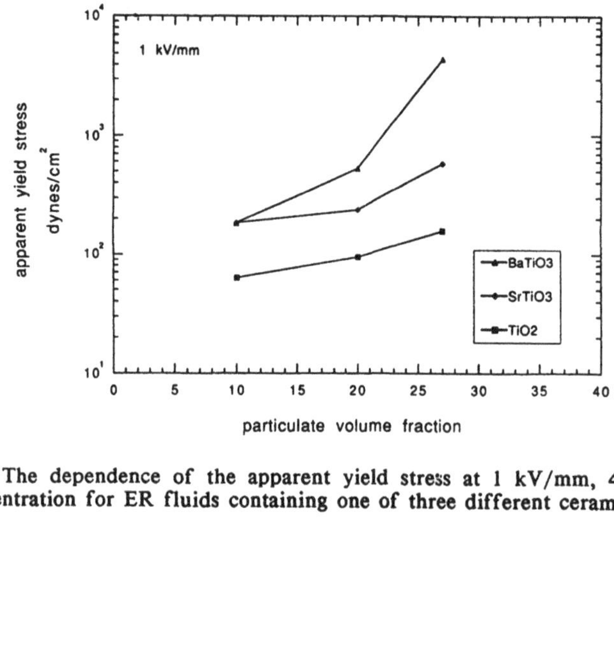

Figure 3. The dependence of the apparent yield stress at 1 kV/mm, 400 Hz on the solids concentration for ER fluids containing one of three different ceramic materials in dodecane.

number of such chains. The chains will have thicknesses distributed around some average value. The specific strength of a chain can be a function of its thickness; calculations differ as to whether thicker or thinner chains are stronger.[6,7] Intuitively, one would expect that the number of chains that span the electrodes to increase and the average chain thickness to either stay the same or, more likely, to increase as the solids concentration of the fluid increases.

Thus there are two possibilities for the behavior shown in Figure 3. The first is that the near field forces and point dipole forces between non-adjacent particles become increasingly important as the particle concentration increases, due to changes in structure, and depend more strongly on ϵ_p than does β.[6] This assumes that the chain structure formed in a fluid at a given concentration of solids is independent of ϵ_p. Thus, the $BaTiO_3$ and $SrTiO_3$ fluids have structures similar to each other at all concentrations, but the fluids with the higher ϵ_p $BaTiO_3$ particles have higher yield stresses only at the higher concentrations because near field forces and point dipole forces between non-adjacent particles are much more important in the structures formed at higher concentrations. Alternatively, it may be that due to these forces being stronger in a fluid with larger ϵ_p particles, a different structure is formed whose inherent strength increases with ϵ_p. For example, the larger magnitude of these forces in the $BaTiO_3$ fluid may lead to either more chains completely spanning the electrodes or to chains with a different average thickness at high particle concentrations than in the $SrTiO_3$ fluid. To determine which of these possibilities is actually occurring will require determination of the structures of real fluids by techniques such as freeze fracture microscopy.

Figure 3 also indicates that the magnitude of the ER effect in the $BaTiO_3$ fluids is very large: a yield stress of ~4000 dyn/cm^2 at 1 kV/mm for the 27 vol% fluid. The yield stress of a 40 vol% $BaTiO_3$ fluid was greater than 1.2 x 10^4 dyn/cm^2, the limit of our experimental apparatus, at 0.7 kV/mm. These values compare favorably with some of the highest yield stress values reported.[1]

The apparent yield stress of the fluids studied is shown in Figure 4 as a function of electric field. As expected, for all fluids the yield stress was proportional

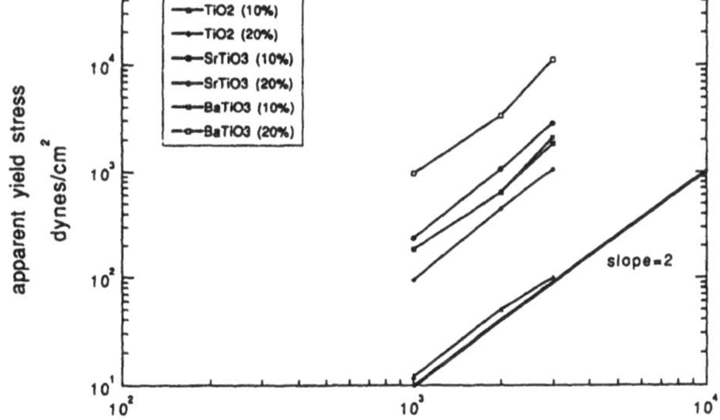

Figure 4. The dependence of the apparent yield stress of ER fluids consisting of various ceramic materials in dodecane on the magnitude of the applied field, at 400 Hz.

to the square of the field. Using the relationship to extrapolate the data for the 40 vol% $BaTiO_3$ fluid from 0.7 kV/mm to 5 kV/mm gives a yield stress of ~6.0×10^5 dyn/cm^2.

The DC resistivity of a 40 vol% $BaTiO_3$ in dodecane fluid as a function of electric field is shown in Figure 5. At high fields, the resistivity approaches 10^9 Ω-cm. Resistivity measurements on the fluid's supernatant after the particles settled out gave much higher values, indicating that the $BaTiO_3$ particles carry most of the current in the fluid. Since the bulk resistivity of ceramic $BaTiO_3$ is ~10^{12} Ω-cm, surface conduction on the $BaTiO_3$ particles may be occurring in the fluid.

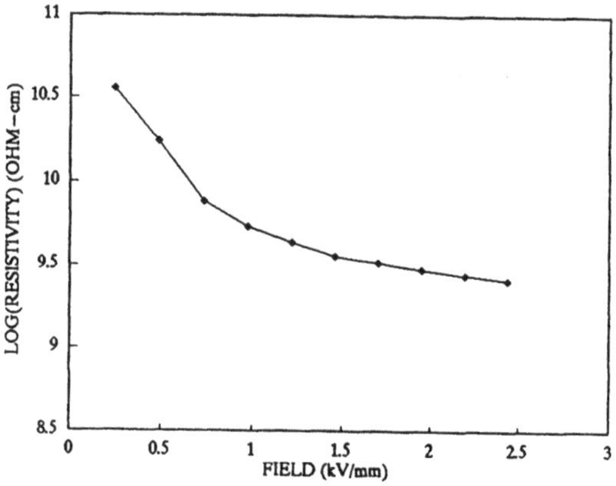

Figure 5. The DC resistivity as a function of applied field for a fluid consisting of 40 vol% $BaTiO_3$ in dodecane.

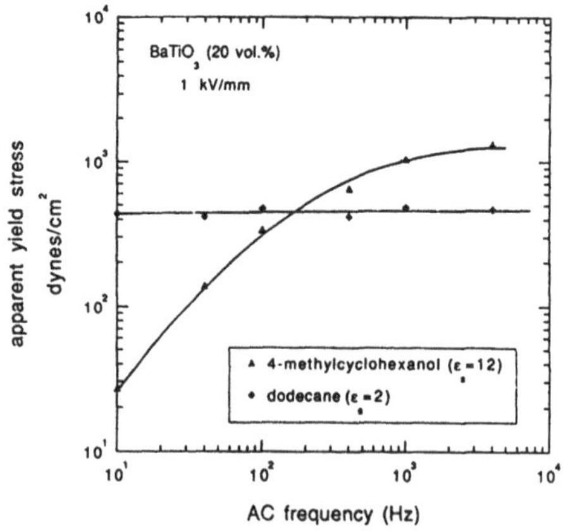

Figure 6. The apparent yield stress of two 20 vol% $BaTiO_3$ fluids, one with dodecane as the solvent and the other with 2-methylcyclohexanol, as a function of the frequency.

The apparent yield stress at 1 kV/mm as a function of frequency of two 20 vol% $BaTiO_3$ fluids, one with dodecane and the other with 2-methylcyclohexanol (2-mch) are shown in Figure 6. The yield stress of the dodecane containing fluid was independent of frequency, whereas that of the 2-mch containing fluid increased strongly with frequency. However, the yield stress of the dodecane fluid was very low when a DC voltage was applied. These results are consistent with the critical frequency concept: the critical frequency is <10 Hz when the solvent is dodecane since its resistivity is very high ($>10^{14}$ Ω-cm) whereas it is several hundred Hz when the solvent is the much more conductive 2-mch ($\rho \sim 10^8$ Ω-cm).

The value of yield stress approached at high frequency for the 2-mch fluid is about 3 times greater than that of the dodecane fluid. As discussed above, the point dipole equation predicts that since $\beta \sim 1$ in both fluids (since the dielectric constant of $BaTiO_3$ is ~2000) the yield stress should be directly proportional to the ϵ_s of each fluid, giving a ratio of the two yield stresses of 6.5. However, we found above that the point dipole equation is not consistent with our results of yield stress as a function of ϵ_p, so that we should replace β in the point dipole equation with a different function of ϵ_s and ϵ_p. The small difference between the experimental value of the yield stress and that predicted by the point dipole equation in the dodecane and 2-mch fluids could be due either to the fact that the correct function that replaces β does not become independent of ϵ_s until $\epsilon_p/\epsilon_s > 150$ or to differences in fluid structure caused by differing degrees of particle aggregation in the two suspensions.

4. Conclusions

Dielectric constant mismatch has been shown to be capable of producing large ER effects in water-free fluids containing insulating particles. The dependence of ER properties on the dielectric constant of the solid phase is much stronger than that predicted by the point dipole equation, especially at high solids concentrations. The increasing importance of near-field forces and dipole forces between non-adjacent particles as ϵ_p increases has been suggested as the cause of this behavior. The concept of a critical frequency has been demonstrated in two solvents with widely differing conductivities. These fluids also exhibited approximate proportionality between the yield stress and ϵ_s at frequencies sufficiently above the critical frequency.

5. Acknowledgements

The authors thank R. A. Anderson for critically reviewing this manuscript and M. S. Harrington for performing some of the experimental work. This work was performed at Sandia National Laboratories, supported by the U. S. Department of Energy under contract number DE-AC0476DP00789.

6. References

1. H. Block, J. P. Kelly, A. Qin and T. Watson, *Langmuir* **6** (1990) 6-14.
2. H. Block and J. P. Kelly, *J. Phys. D: Appl. Phys.* **21** (1988) 1661-1677.
3. J. E. Stangroom, *Phys. Technol.* **14** (1983) 290-296.

4. L. Marshall, C. F. Zukoski IV and J. W. Goodwin, *J. Chem. Soc., Faraday Trans. 1* **85** (1989) 2785-2794.
5. R. Anderson, this volume.
6. A. Kraynik, this volume.
7. A. F. Sprecher, Y. Chen H. Conrad, this volume.

A NUMERICAL INVESTIGATION OF ELECTRORHEOLOGICAL MATERIAL BEHAVIOR

MICHAEL J. CHRZAN

Lord Corporation
Thomas Lord Research Center
Cary, North Carolina 27511

and

JOHN P. COULTER

Department of Mechanical Engineering and Mechanics
Lehigh University
Bethlehem, Pennsylvania 18015

ABSTRACT

The inter-particle behavior that occurs within electrorheological (ER) materials was investigated numerically. In this initial phase of the study, a two-dimensional material assumption was adopted, and static situations were investigated. A unit cell geometry was selected, with the modeled region consisting of two particles separated by a fluidic carrier medium. The numerical approach selected is based on a modified finite-difference scheme encompassing numerically generated boundary-fitted coordinate grid systems. Hybrid cylindrical-numerically generated grid systems were utilized to represent the unit cell type selected. With appropriate boundary conditions applied, electric potential distributions throughout unit cells were determined. From these results, electric field and current distributions and resultant mechanical stresses were then calculated. Results are presented in the form of electric potential, field, and current distributions, as well as net inter-particle forces. It is shown that high electric field levels occur in the "cap region" between particles and that the bulk of the resultant force holding particle chains together can be attributed to this. The electric field distribution was found to be highly dependent upon the particle spacing, relative size, and particle/fluid electrical properties.

NOMENCLATURE

a, b, c, d	mesh concentration constants (carrier fluid region)	r^*, θ^*	dimensionless cylindrical spatial coordinates
A	surface area	R	particle radius
B_r, B_θ	mesh concentration constants (particle region)	x, y	Cartesian spatial coordinates (physical domain)
e_n	unit normal vector	x^*, y^*	dimensionless coordinates
E	electric field	θ_o	maximum angular coordinate value
F	interparticle force		
i, j	nodal point identification indices	α, β, γ	geometric mesh functions
I	electrical current density	δ_{ij}	Kronecker delta function
J	geometrical transformation Jacobian	ε	dielectric constant
		ε_o	universal dielectric constant
k_p, k_q	mesh concentration constants (carrier fluid region)	ξ, η	computational domain coordinates
		$\xi_i, \xi_j, \eta_i, \eta_j$	mesh concentration constants (carrier fluid region)
M, N	mesh size parameters (particle region)	ρ_{vu}	unpaired volume charge density
M_f, N_f	mesh size parameters (carrier fluid region)	ρ_{su}	unpaired surface charge density
		σ	electrical conductivity
n	normal spatial direction	τ	shear stress
P, Q	mesh concentration functions (carrier fluid region)	ϕ	electric potential
		ϕ^*	dimensionless electric potential
r, θ	cylindrical spatial coordinates	ϕ_t, ϕ_b	electric potential boundary values

1. Introduction

The electrorheological response has been described as that phenomenon which gives rise to an increase in suspension stress transfer upon the application of an external field. The underlying mechanisms contributing to this bulk effect have been identified as colloidal and electric field induced interparticle interactions and have been discussed in the literature[1-3]. Critical to any prediction of the overall suspension characteristics is a treatment of the interparticle electrostatics. Most "bulk property" models assume a point dipole approximation for the interaction force between two spheres. This is a convenient and generally accurate approximation for bulk suspension behavior. In general, however, the point dipole approximation is not valid, particularly at small particle separations and at high particle-to-fluid dielectric constant ratios. A more exact solution is necessary for a description of the electric field intensities around and between closely spaced particles.

In general, the electrostatic problem for two spheres in an external homogeneous field cannot be solved in a closed form, since more exact descriptions require successively higher order terms in the multipole expansion of the Laplacian solution. Although the problem has received extensive investigation, most solutions to the two-sphere problem involve cases where the spheres are perfectly conducting only[4-6]. Results for dielectric cases have been approximated by Keller[7], and Batchelor & O'Brien[8] for highly conductive spheres using slight deviations from the perfectly conductive case. Klingenburg[9], using a truncated multipole expansion, has calculated the forces for two equal spheres as a function of the dielectric constant ratio, however no one has done so for two particles of arbitrary or lossy dielectric constant. Even for the perfectly conducting case, results are conflicting depending on the method of solution used. Numerical results from Davis[5] who uses a bispherical coordinates method suggest that the electrostatic force between two charged spheres becomes infinite when the spheres touch while results from Smith & Baraket[10] indicate a finite force for touching spheres. Jeffery[6], who uses a singular perturbation expansion, claims both solutions are correct and attributes the disagreement to the different ways the authors formulate the problem. He concludes, however, that all the methods used are numerically inconvenient when the spheres are very close together.

Difficulties in the theoretical evaluation of these problems has made computational numerical methods an attractive alternative. The finite element method (FEM) and finite difference method (FDM) have long been employed to solve complex solid and fluid mechanical as well as thermodynamic problems (e.g., stress analysis, heat conduction). Only recently, however, have FEM and FDM been used to solve problems of electric and magnetic phenomena. Just as with their mechanical analogs, electrodynamic systems involving irregular domains, material inhomogeneities, and complex boundary conditions can be handled using either FE or FD methods.

In an effort to avoid some of the analytical difficulties of the theoretical solutions mentioned previously as well as provide a model which will allow greater geometric and material generalization, a computational model using the numerical method of finite differences has been developed. This model calculates the electrostatic potential between two particles in chains aligned parallel to an applied electric field. As it is an electrostatic solution, it solves either for insulating dielectric particles in an insulating dielectric fluid, or for conductive particles in a conductive fluid. The results from the latter case will be presented here. From the potential, the electric field, current density, and electrostatic force is calculated. As will be seen, composite particles or particles of irregularly shaped geometries are also solvable with this method. In addition, the model is expandable to include nonlinear behavior of governing equations or boundary conditions allowing future solutions to transient conditions or non-ohmic constitutive equations. Although, the model as presented solves only the two-dimensional case (i.e., infinitely long chained cylinders), it can be extended to solve the spherical case. The method of finite differences has been shown to provide a better numerical approximation to the operator ∇^2 than finite elements[11] and is hence the preferred alternative to the solution of the governing equation in this model.

2. Problem Formulation

A schematic representation of the electrostatic electrorheological material configuration investigated during the present study is shown in Figure 1. A two-dimensional configuration was assumed, representative of a suspension of rod-like rather than spherical particles. As illustrated in Figure 1a, the suspension was assumed to be monodisperse with particles configured in a repeating arrangement of single particle chains aligned with an applied electric field. The close examination of the configuration selected reveals the existence of a repeating unit cell consisting of two adjacent semi-circular particle elements and the irregularly shaped carrier fluid region surrounding them. A typical unit cell is expanded in Figure 1b, and configurations similar to this were focused on during the present study.

Concentrating on the unit cell depicted in Figure 1b and assuming piecewise isotropic and homogeneous dielectric or conductivity constants, either the dielectric insulating case or the steady-state conduction case can be solved with the same governing equations and boundary conditions. Theoretical formulations for these two cases follow.

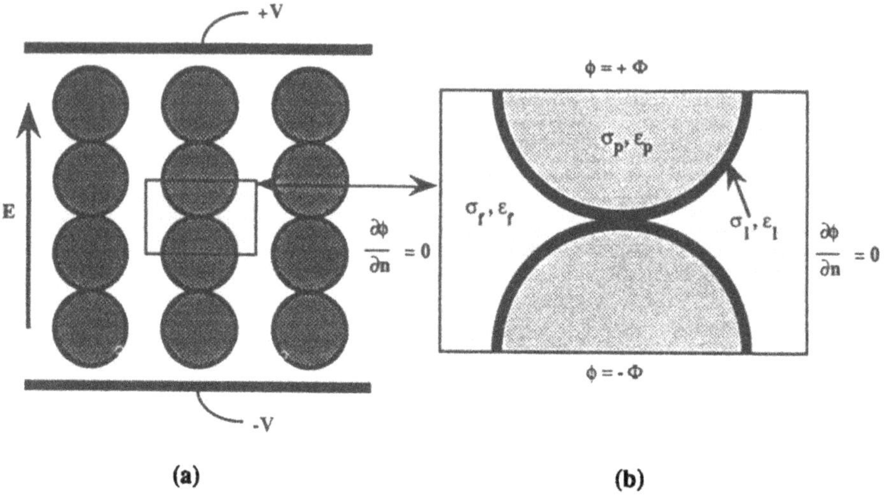

Figure 1: Geometric configuration of the chained particle problem.

2.1. Dielectric Insulating Case

The insulating, dielectric equation is derived from Gauss' law for linear, isotropic dielectrics, which is

$$\nabla \cdot \varepsilon \mathbf{E} = \rho_{vu} . \tag{1}$$

In Eq. 1, ε is the dielectric constant of the medium, $\mathbf{E}$ is the electric field vector, and ρ_{vu} is the unpaired or free volume charge density in the dielectric. In addition, the electrostatic form of Faraday's law requires that

$$\nabla \times \mathbf{E} = 0 , \tag{2}$$

which in turn implies that the electric field can be related to the scalar electric potential ϕ as follows:

$$\mathbf{E} = - \nabla \phi . \tag{3}$$

Hence the electric field may be found once the scalar electric potential is known. In regions where the permittivity is uniform and where there is no unpaired charge, Eqs. 1 and 3 combine to give

$$\nabla^2 \phi = 0, \tag{4}$$

which is Laplace's equation for the electric potential.

Because of the second order elliptic nature of the Laplace equation, the complete specification of unit cell boundary and carrier fluid/particle interface conditions is required. Across an interface separating media a and b having different permittivities, the continuity conditions implied by Eqs. 1 and 2 are in general

$$\hat{\mathbf{e}}_n \cdot (\varepsilon_a \mathbf{E}_a - \varepsilon_b \mathbf{E}_b) = \rho_{su} \tag{5}$$

where the jump in the normal component of the electric displacement is equal to the unpaired surface charge density across an interface ρ_{su}, and

$$\hat{\mathbf{e}}_n \times (\mathbf{E}_a - \mathbf{E}_b) = 0, \tag{6}$$

the tangential component of the electric field is continuous across an interface. In Eqs. 5 and 6, $\hat{\mathbf{e}}_n$ represents a unit vector normal to the fluid/particle interface. Assuming no unpaired charge density, the continuity equations for the electric potential become,

$$\varepsilon_a \frac{\partial \phi_a}{\partial n} = \varepsilon_b \frac{\partial \phi_b}{\partial n} \tag{7}$$

and,

$$\phi_a = \phi_b. \tag{8}$$

Along the upper and lower boundaries of the selected unit cell, constant electric potential Dirichlet type boundary conditions are appropriate, as shown in Figure 1b. Lastly, the boundary condition required for the sides of the unit cell is of Neumann type insuring parallel field lines:

$$\frac{\partial \phi}{\partial n} = 0 \tag{9}$$

2.2. *Steady-State Conduction Case*

The governing equation for the steady-state conduction case is determined by starting with Eqs. 1 and 2 and introducing Ohm's law, which is

$$\mathbf{I} = \sigma \mathbf{E}. \tag{10}$$

In this equation, σ is the electrical conductivity of a material and $\mathbf{I}$ represents the electrical current density. The differential law necessary is the conservation of free charge density i.

$$\nabla \cdot \mathbf{I}_u + \frac{\partial \rho_{vu}}{\partial t} = 0 \tag{11}$$

For steady ohmic conduction it is assumed that the accumulation of charge at any given location has a negligible influence on the distribution of the current density. Under this assumption the time derivative of the unpaired charge density in Eq. 11 vanishes, implying that the current density is solenoidal. Thus the electric field in such situations is governed by

$$\nabla \cdot \sigma \mathbf{E} = 0. \tag{12}$$

Since the electric field is irrotational, for regions of uniform conductivity Eq. 12 reduces to Laplace's Eq. 4. Thus steady-state conduction and dielectric insulating situations are governed by the same relationship.

The material interface continuity equations for steady-state conduction are virtually the same as those for the dielectric insulating case (7) and (8). The only change that is required is the usage of conductivity rather than dielectric constant in Eq. 7, which leads to:

$$\sigma_a \frac{\partial \phi_a}{\partial n} = \sigma_b \frac{\partial \phi_b}{\partial n} . \tag{13}$$

The boundary conditions to be applied around the border of the unit cell for the steady-state conduction case are identical to those for the dielectric insulating situation.

2.3. Secondary Variable Determination

The successful solution of the completely posed mathematical problem described above leads to the determination of the scalar electric potential distribution throughout the unit cell. Secondary variables of interest such as electric field distribution, current density, and interparticle electrostatic force can be calculated once the electric potential distribution is known. The electric field and current density can be determined through the direct application of Eqs. 3 and 10 respectively. The electrostatic force acting on either cylinder is computed by first calculating the Maxwell stress tensor τ_{ij} as follows:

$$\tau_{ij} = \varepsilon_o \varepsilon_f \left[E_i E_j - \frac{1}{2} |\mathbf{E}|^2 \delta_{ij} \right] . \tag{14}$$

In this relationship, ε_o is the universal dielectric constant, ε_f is the dielectric constant of the carrier fluid, E_i and E_j are components of electric field in the i and j directions respectively, and δ_{ij} is the Kronecker delta function. With the appropriate stresses known, the interparticle force **F** can then be determined by integration over the surface of interest using

$$\mathbf{F} = \int_A \tau_{ij}\, \mathbf{n} \cdot d\mathbf{A} , \tag{15}$$

in which **dA** represents an element of the surface and **n** is the unit vector normal to this surface.

3. Solution Technique

The irregularity of the unit cell geometry along with the coupled nature of the particle and carrier fluid electric potential distributions lead to the need for a numerical solution approach to address such problems. During the present study, a hybrid technique involving the usage of a traditional finite difference method in particle regions and a numerical grid generation based boundary-fitted coordinate system method in the carrier fluid region was applied. An example hybrid mesh for a unit cell that results when such an approach is adopted is shown in Figure 2. As is shown in the figure, traditional cylindrical polar mesh regions were utilized to represent suspended particles and an irregularly shaped boundary fitted coordinate mesh region was used in the carrier fluid region. Details of the solution methodologies applied in these two regions follow.

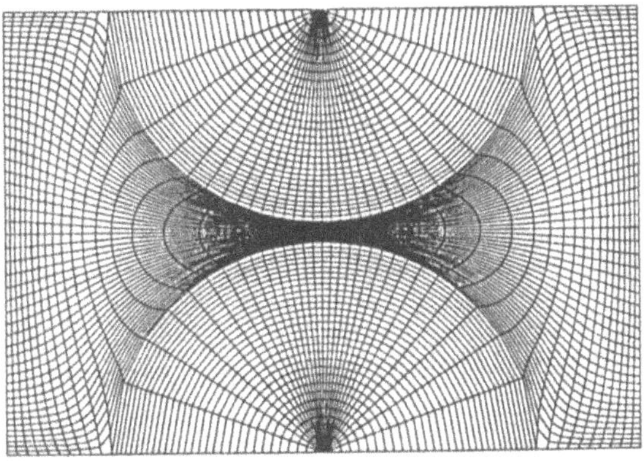

Figure 2: Example hybrid mesh utilized during the present study.

3.1. Particle Regions

Meshes were generated over particle regions using nodal points defined by the polar coordinates (r,θ). To allow for the concentration of nodal points in selected regions, the adjustable variation of both radial and angular nodal point spacing was included. This was accomplished using the following relations for radial and angular nodal point locations:

$$r_i = R\left[\frac{\tan^{-1}\left\{B_r\left(\frac{i-1}{M-1}\right)\right\}}{\tan^{-1}\{B_r\}}\right] \tag{16}$$

$$\theta_j = \theta_o\left\{0.5 + \left[\frac{\tan^{-1}\left\{B_\theta\left[\left(\frac{2j-2}{N-1}\right)-1\right]\right\}}{2\tan^{-1}\{B_\theta\}}\right]\right\} \tag{17}$$

In these equations, r_i and θ_j are the locations of the *i*th radial and *j*th angular nodal points, R and θ_o are the maximum coordinate values, B_r and B_θ are concentration constants, and M and N are the numbers of nodal points in the radial and angular directions respectively.

Throughout the regions representing suspended particles, the governing Laplace Eq. 4 was solved using its polar coordinate form. To assure solution stability, non-dimensional variables were introduced as follows

$$r^* = \frac{r}{R}, \qquad \theta^* = \frac{\theta}{\theta_o}, \qquad \phi^* = \frac{\phi - \phi_b}{\phi_t - \phi_b}, \tag{18}$$

where R was the total particle radius, $\theta_o = \pi$, ϕ_t is the potential applied along the top of the unit cell, and ϕ_b is the potential applied along the bottom of the unit cell. After inclusion of these variables, the polar form of the Laplace equation becomes

$$\frac{\partial^2\phi^*}{\partial r^{*2}} + \frac{1}{r^*}\frac{\partial\phi^*}{\partial r^*} + \frac{1}{\theta_0^2 r^{*2}}\frac{\partial^2\phi^*}{\partial\theta^{*2}} = 0. \tag{19}$$

To facilitate the determination of the electric potential distribution within suspended particles, Eq. 19 was discretized using a second-order accurate finite difference scheme.

3.2. Carrier Fluid Region

In order to handle the geometrical complexity of the irregularly-shaped carrier fluid region, a numerical grid generation technique encompassing a boundary-fitted coordinate system (BFCS) was selected. This technique, developed primarily over the last two decades, has been widely used to solve complex fluid flow and thermodynamic problems[12,13]. Coulter and coworkers have successfully applied BFCS approaches to the solution of laminar and turbulent convection processes[14-17] and composite material manufacturing processes[18-22]. The technique is advantageous in its ability to combine the geometric flexibility of finite element methods with the computational simplicity of finite difference techniques.

The underlying principle of the BFCS technique is the transformation of the irregular problem geometry in the physical x-y domain into a regular geometry in a computational ξ, η domain. A schematic representation of physical and computational domains is presented in Figure 3. As shown, the meshes were specified so as to have M_f nodal points in the ξ direction and N_f nodal points in the η direction.

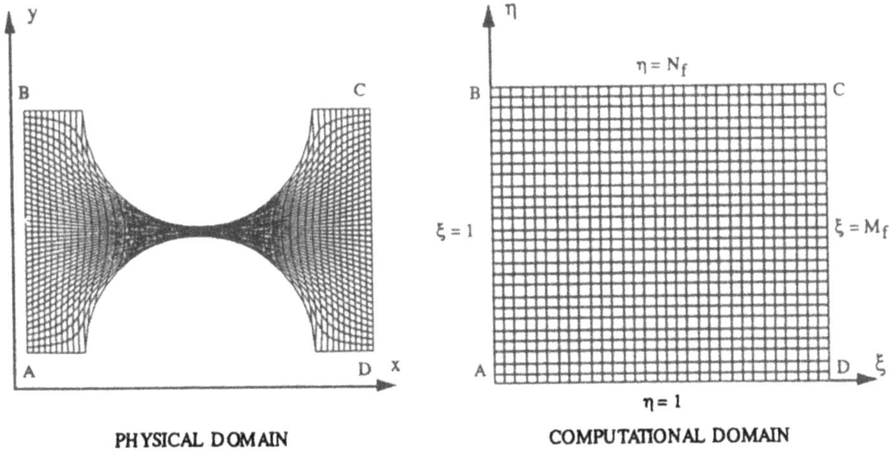

Figure 3: Physical and computational domains selected for the present study.

The BFCS transformation differs from more traditional transformations in that the relationships between the physical and computational coordinates are given in the form of differential equations, rather than as algebraic expressions. Generally, these generating equations can be of elliptic, hyperbolic, or parabolic nature. In the present analysis, elliptic type transformation equations of the Poisson form were chosen due to their inherent uniform mesh generation and characteristic smoothening of the boundary slope discontinuities in the physical domain. These mesh generating equations are given as

$$\frac{\partial^2 \xi}{\partial x^2} + \frac{\partial^2 \xi}{\partial y^2} = P(\xi,\eta) \tag{20}$$

and,

$$\frac{\partial^2 \eta}{\partial x^2} + \frac{\partial^2 \eta}{\partial y^2} = Q(\xi,\eta) \tag{21}$$

where the nonhomogeneous terms $P(\xi,\eta)$ and $Q(\xi,\eta)$ allow for grid control by regulating nodal point concentration in desired regions. In the present study, increased nodal resolution between the particles was accomplished using line and point attractor equations of the form

$$\begin{aligned} P(\xi,\eta) = &- \sum_{i=1}^{k_p} a_i \operatorname{sign}(\xi - \xi_i) \exp\left[-c_i \left|\xi - \xi_i\right|\right] \\ &- \sum_{j=1}^{L_p} b_j \operatorname{sign}(\xi - \xi_j) \exp\left[-d_j \sqrt{(\xi - \xi_j)^2 + (\eta - \eta_j)^2}\right] \end{aligned} \tag{22}$$

$$Q(\xi,\eta) = -\sum_{i=1}^{k_q} a_i \, \text{sign}(\eta - \eta_i) \exp\left[-c_i \, |\eta - \eta_i|\right]$$

$$- \sum_{j=1}^{L_q} b_j \, \text{sign}(\eta - \eta_j) \exp\left[-d_j\sqrt{(\xi - \xi_j)^2 + (\eta - \eta_j)^2}\right] \qquad (23)$$

In these expressions k_p and k_q are the number of points around which mesh concentration is desired, L_p and L_q are the number of nodal lines towards which concentration is desired, ξ_i, η_i, ξ_j, and η_j are concentration focal point coordinates, and a_i, b_i, c_j, and d_j are user-specified concentration coefficients. Introducing dimensionless coordinates of the form $x^*=x/R$ and $y^*=y/R$, Eqs. 20 and 21 may be rewritten with the physical (x,y) coordinates as the dependent variables as follows:

$$\alpha \frac{\partial^2 x^*}{\partial \xi^2} - 2\beta \frac{\partial^2 x^*}{\partial \xi \partial \eta} + \gamma \frac{\partial^2 x^*}{\partial \eta^2} + J^2\left(P\frac{\partial x^*}{\partial \xi} + Q\frac{\partial x^*}{\partial \eta} \right) = 0 \qquad (24)$$

$$\alpha \frac{\partial^2 y^*}{\partial \xi^2} - 2\beta \frac{\partial^2 y^*}{\partial \xi \partial \eta} + \gamma \frac{\partial^2 y^*}{\partial \eta^2} + J^2\left(P\frac{\partial y^*}{\partial \xi} + Q\frac{\partial y^*}{\partial \eta} \right) = 0 \qquad (25)$$

in which the coefficients are geometrical terms defined as

$$\alpha = \left(\frac{\partial x^*}{\partial \eta}\right)^2 + \left(\frac{\partial y^*}{\partial \eta}\right)^2, \qquad (26)$$

$$\beta = \frac{\partial x^*}{\partial \xi}\frac{\partial x^*}{\partial \eta} - \frac{\partial y^*}{\partial \xi}\frac{\partial y^*}{\partial \eta}, \qquad (27)$$

$$\gamma = \left(\frac{\partial x^*}{\partial \xi}\right)^2 + \left(\frac{\partial y^*}{\partial \xi}\right)^2, \qquad (28)$$

and,

$$J = \frac{\partial x^*}{\partial \xi}\frac{\partial y^*}{\partial \eta} - \frac{\partial x^*}{\partial \eta}\frac{\partial y^*}{\partial \xi}. \qquad (29)$$

Eqs. 24 and 25 can be solved in an explicit iterative manner to produce the desired transformation from any simply connected domain of general irregular shape into a rectangular computational domain containing square unit cells in which traditional finite difference techniques can be used to solve the problem. The Dirichlet boundary conditions required are precisely the locations of nodal points which describe the boundary of the physical domain presented in Figure 3a.

Corresponding to the transformation of geometry is a similar transformation of the governing equations from the physical to the computational domain. This is done by performing chain rule differentiations on the physical domain equations. In the computational domain the dimensionless form of Laplace's Eq. 4 becomes

$$\nabla^2\phi^* = \frac{\alpha}{J^2}\frac{\partial^2\phi^*}{\partial\xi^2} - \frac{2\beta}{J^2}\frac{\partial^2\phi^*}{\partial\xi\partial\eta} + \frac{\gamma}{J^2}\frac{\partial^2\phi^*}{\partial\eta^2} + P\frac{\partial\phi^*}{\partial\xi} + Q\frac{\partial\phi^*}{\partial\eta} = 0 \tag{30}$$

The vanishing normal derivative condition (9) which is applied along the unit cell side boundaries becomes

$$\frac{\partial\phi^*}{\partial n} = \frac{1}{J\sqrt{\alpha}}\left(\alpha\frac{\partial\phi^*}{\partial\xi} - \beta\frac{\partial\phi^*}{\partial\eta}\right) = 0 \tag{31}$$

for the $\xi = 1$ side, and

$$\frac{\partial\phi^*}{\partial n} = -\frac{1}{J\sqrt{\alpha}}\left(\alpha\frac{\partial\phi^*}{\partial\xi} - \beta\frac{\partial\phi^*}{\partial\eta}\right) = 0 \tag{32}$$

for the $\xi = M_f$ side.

3.3. Combining the Meshes

The governing Eqs. 19 and 30 were solved for ϕ^* in their respective interior regions using a central differencing discretization process with second order accuracy. The boundary conditions were calculated using single-sided differencing with second order accuracy. The two meshes were "married" together by allowing nodal point coordinates along the polar mesh semicircular boundaries to equal those coordinates of the boundary-fitted carrier fluid region mesh. The interfacial boundary condition (Eq. 7) can be expressed in terms of the combined mesh system as

$$\varepsilon_p\left(\frac{\partial\phi^*}{\partial r^*}\right)_p = \frac{\varepsilon_f}{J\sqrt{\alpha}}\left[\alpha\left(\frac{\partial\phi^*}{\partial\xi}\right)_f - \beta\left(\frac{\partial\phi^*}{\partial\eta}\right)_f\right] \tag{33}$$

where the superscript p denotes the particle region and f, the fluid region. The conductive boundary condition (Eq. 13) can be likewise expressed by substituting σ_p and σ_f for ε_p and ε_f, respectively.

4. Results

The model was initially tested for the trivial case in which the conductivities of the particles and fluid were equal. As expected, the equipotential contour lines throughout the physical cell for this case were parallel and straight. A second and more critical verification was made by comparing the effective conductivity of the chained cylinders to the analytical results of Chiew[23], who solved for the effective conductivity of a random dispersion of long cylinders in a continuous matrix phase. His computed effective conductivity values for a particle-to-fluid conductivity ratio of 50 and for varying particle area fraction ϕ are shown in Figure 4. The effective conductivity σ_{eff} is normalized here by the fluid conductivity σ_f. Plotted against his data are the results from calculating the same effective conductivity using the present model. The particle area fraction ϕ in the model was varied by changing the particle mesh size in the cell. As shown, the model is in close agreement with Chiew's calculations.

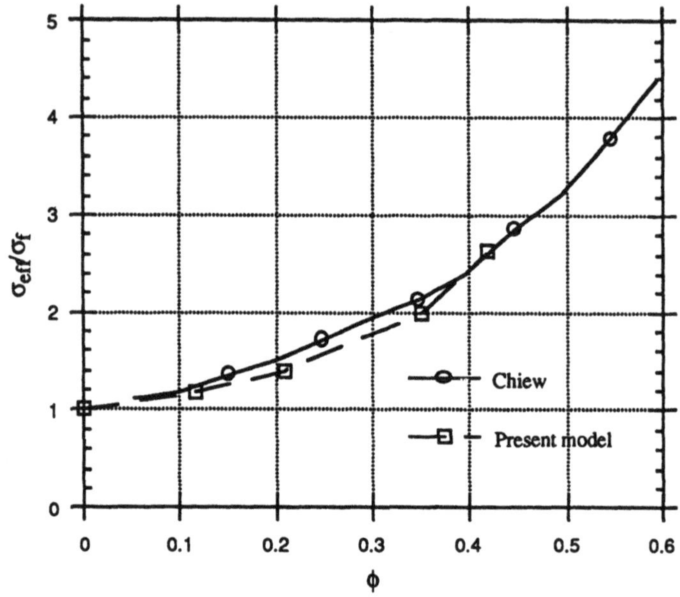

Figure 4: Comparison of effective conductivity of the present model with the analytical results of Chiew[23]. The effective conductivity is calculated for $\sigma_p/\sigma_f = 50$ and is normalized by σ_f.

For meshes such as that shown in Figure 2, the electric potential was calculated at every node. The resulting electric potential contours for a typical solution are shown in Figure 5. In this case, the particle-to-fluid conductivity ratio is 100 and the particles are spaced 1/10th of a radius apart. As expected for a higher particle conductivity, the potential changes most rapidly in the area directly between the particles, manifested by the closely packed contours.

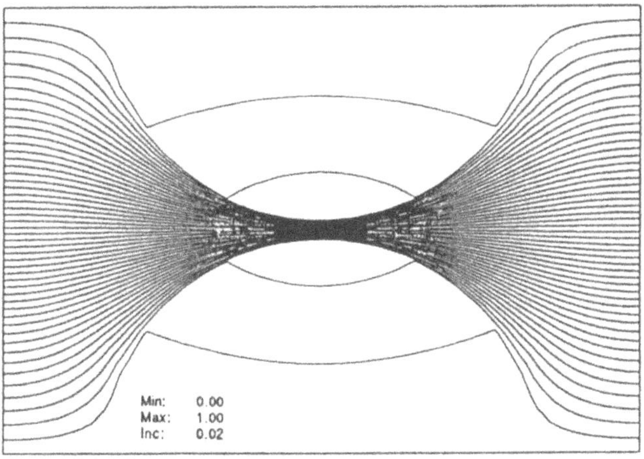

Figure 5: Equipotential contours for particles spaced 1/10 radius apart and $\sigma_p/\sigma_f = 100$.

Differentiating the potential over the cell yields the electric field. Figure 6 shows the electric field magnitude distribution corresponding to the electrical potential solution shown in Figure 5. High field levels are observed in the cap region where the particles nearly touch. Outside this area the electric field is greatly reduced.

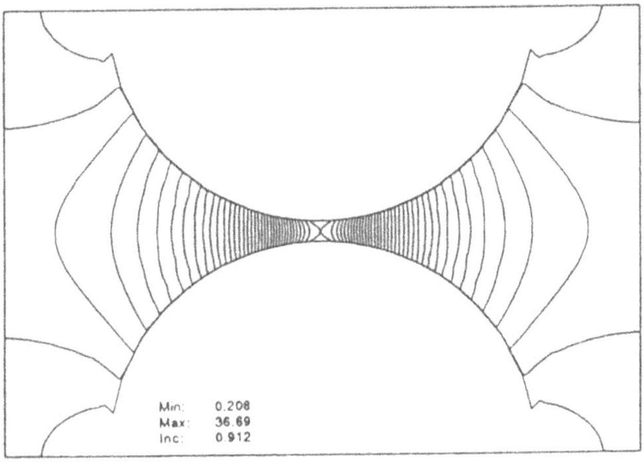

Figure 6: Electric field magnitude contours for the equipotential solution in Figure 5.

Another example of a resulting electric field concentration is given in the surface plot of Figure 7. The particles in this case are 1/4th of a particle radius apart and the conductivity ratio σ_p/σ_f is 10. As shown, the electric field between the particles is enhanced 5.6 times as compared to the equal conductivity case ($\sigma_p = \sigma_f$). For particles spaced 1/100th of a particle radius apart with the same conductivity ratio, the electric field is enhanced 17.5 times. There is also a saddle-like behavior of the electric field directly between the particles, implying the highest electric fields and stresses are located on the particle surfaces at the point of closest proximity.

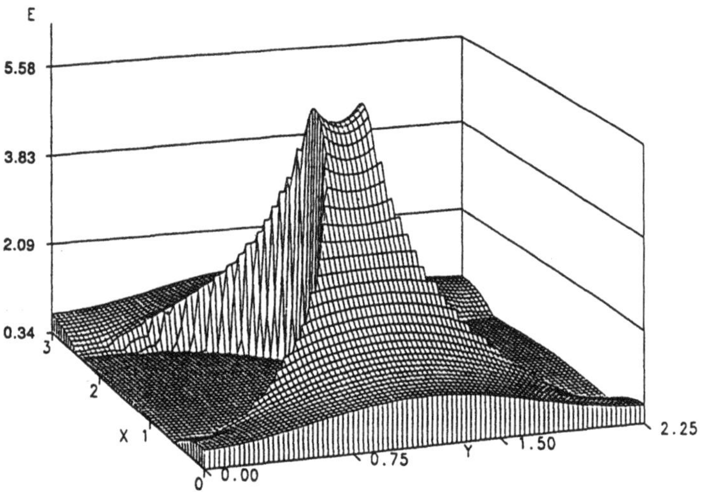

Figure 7: Electric field magnitude surface plot for particles separated 1/4 radius apart and $\sigma_p/\sigma_f = 10$. The electric field scale is normalized to the equal conductivity case where $\sigma_p = \sigma_f$.

Using Maxwell's stress tensor as described previously, the electrostatic force between the particles was calculated. The normalized electrostatic force as a function of particle separation in a chain is given in Figure 8. The points were calculated by solving meshes with varying particle separation distances. As expected the interparticle force increases dramatically as the particles are brought into close proximity. Moreover, from the previous electric field results it is seen that the high electric field in the narrow "cap region" between particles is responsible for the bulk of the electrostatic force between particles in a chain.

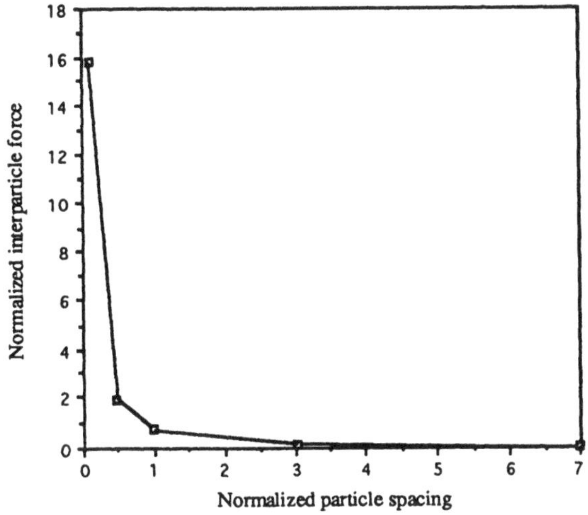

Figure 8: Interparticle force vs. particle spacing. Spacing is normalized to a particle radius.

5. Conclusions

A numerical model based on a hybrid finite difference/numerical grid generation approach has been developed to study the interparticle behavior in 2-dimensional electrostatic particle chains. Utilizing the boundary-fitted coordinate system transformation, a hybrid mesh was generated to simulate the chain cell geometry. Verification of the model was made through close agreement with calculations of effective conductivity by Chiew[23].

Although modest in its present scope of analysis, the model and associated technique for solution is expandable to more complex cases. In its present form, the model can solve for composite particles with arbitrary shapes. With further development, the model can be made to solve spherical particles in 3-dimensions. Transient governing equations as well as nonlinear interparticle constituent behavior may also be included. In addition, the modelling of shear geometries may also be accomplished to generate an effective yield stress for the chained system.

6. Acknowledgements

The authors wish to acknowledge the support of T. G. Duclos from Lord Corp. and the assistance of S. I. Güçeri from The University of Delaware. Partial funding for J. P. Coulter's involvement in the present study was provided by The National Science Foundation in the form of Grant No. MSS-9110909.

References

1. A. P. Gast and C. F. Zukoski, *Advances in Colloid and Interface Science* **30** (1989) 153.

2. T. C. Jordan and M. T. Shaw, *IEEE Transactions on Electrical Insulation* **24** (1989) 849.

3. H. Block and J. P. Kelly, *J. Phys. D: Appl. Phys.* **21** (1988) 1661.

4. P. A. Arp and S. G. Mason, *Colloid and Polymer Sci.* **255** (1977) 566.

5. M. H. Davis, *J. Mech. Appl. Math.* **17** (1964) 499.

6. D. J. Jeffery, *J. Mech. Appl. Math.* **22** (1978) 337.

7. J. B. Keller, *J. Appl. Phys.* **34** (1963) 991.

8. G. K. Batchelor and R. W. O'Brien, *Proc. R. Soc. Lond.* **355** (1977) 313.

9. D. J. Klingenberg, Masters Thesis, University of Illinois, Urbana (1988).

10. G. S. Smith and R. Barakat, *Appl. Sci. Res.* **30** (1975) 418.

11. S. R. H. Hoole, *IEEE Transactions on Magnetics* **25** (1989) 2831.

12. J. Hauser and C. Taylor (eds.), *Numerical Grid Generation in Computational Fluid Mechanics, Proceedings of the First International Conference*, (Pineridge Press Ltd., Swansea, U.K., 1986).

13. S. Sengupta, J. F. Thompson, P. R. Eiseman, and J. Hauser (eds.), *Numerical Grid Generation in Computational Fluid Mechanics, Proceedings of the Second International Conference*, (Pineridge Press Ltd., Swansea, U.K., 1988).

14. J. P. Coulter, S. D. Gilmore, and S. I. Güçeri, in *Numerical Grid Generation in Computational Fluid Dynamics*, eds. J. Hauser and C. Taylor, (Pineridge Press, Swansea, U. K., 1986).

15. J. P. Coulter and S. I. Güçeri, in *Solar Energy Utilization: Fundamentals and Applications*, ed. H. Yüncü, (Martinus Nijhoff Publishers, Dordrecht, The Netherlands, 1987).

16. J. P. Coulter and S. I. Güçeri, *Numerical Heat Transfer* **12** (1987) 211.

17. M. M. Elshamy, M. N. Ozisik, and J. P. Coulter, *Numerical Heat Transfer; Part A*, **18** (1990) 95.

18. J. P. Coulter, B. F. Smith, and S. I. Güçeri, in *Proceedings of the Second Technical Conference of the American Society for Composites*, (Technomic Publishing Co., Inc., Lancaster, Pennsylvania, 1987).

19. J. P. Coulter and S. I. Güçeri, *Journal of Reinforced Plastics and Composites* **7-3** (1988) 200.

20. J. P. Coulter and S. I. Güçeri, in *The Manufacturing Science of Composites*, ed. T. G. Gutowski, (The American Society of Mechanical Engineers, New York, 1988).

21. J. P. Coulter and S. I. Güçeri, *Composites Science and Technology* **35-4** (1989) 317.

22. J. P. Coulter and S. I. Güçeri, in *Advances in Thermoplastic Matrix Composite Materials*, ed. G. M. Newaz, (American Society for Testing and Materials, Chelsea, Michigan, 1989).

23. Y. C. Chiew, *J. Appl. Phys.* **67** (1990) 1684.

III. PROPERTIES OF ER FLUIDS

THE STRENGTH OF ELECTRORHEOLOGICAL (ER) FLUIDS

H. Conrad
Materials Science and Engineering Department
North Carolina State University
Raleigh, NC 27695–7907

and

Y. Chen and A. F. Sprecher
Materials Science and Engineering Department
North Carolina State University
Raleigh, NC 27695–7907

ABSTRACT

The definition of the strength of an ER fluid is discussed. Studies on the electrorheology of ER fluids containing zeolite particles in various oils indicate that the order of magntiude difference between the measured values of the yield stress and those calculated based on the axial force of interaction between particles in a single-row chain can be explained by an enhancement of the force due to the observed clustering of particles into multi-row chains. The force enhancement factor varied with the shear rate and the concentration of particles, but was relatively independent of the electric field, temperature and host fluid. Reasonable agreement existed between the predicted and the measured shear stress-shear strain curves and the concentration dependence of the yield stress when the appropriate force enhancement factor was taken into account. The present theoretical-experimental considerations suggest that ER fluids may attain a yield strength of ~ 50 kPa.

1. Introduction

An important objective of the research on ER fluids is to be able to predict their strength in terms of fundamental physical properties and atomic or molecular mechanisms. In this paper we will present recent work by the authors on this subject. The paper focuses on comparing the measured yield (flow) stress and shear modulus of ER fluids containing zeolite particles with those predicted by a model of the force of interaction between particles in a chain consisting of a single row of spherical particles. It will be shown that reasonable agreement exists between the predicted and measured values if one takes into account that the interaction force may be enhanced by an order of magnitude in real ER fluids as a result of the fact that the chains consist of multi-rows or clusters of particles.

2. Definition and Measurement of Strength

To understand the strength of ER fluids in terms of fundamental physical properties and mechanisms we must first define what we mean by strength and how we measure it. Three types of equipment are generally employed in the laboratory to measure the rheological (mechanical) properties of an ER fluid: (a) lateral shearing device[1–4], (b) couette cell[5,6] and parallel plate shearing device[7]. The lateral shearing device is most generally employed at relatively small shear strains (< 10) and shear rates (< $1s^{-1}$) and the other two at higher strains and rates. However, either type of equipment can be modified to cover a wide range of strains and rates; for example, the shear modulus of an ER fluid is frequently determined from small strain oscillations in couette flow[8].

Idealized plots of data obtained with the lateral shear device and the couette cell and the terms proposed (taken from texts such as that of Ref. 9) are illustrated in Figs. 1 and 2. The term flow stress τ_{fl} is here used to define the instantaneous stress at a given shear strain γ, shear rate $\dot{\gamma}$,

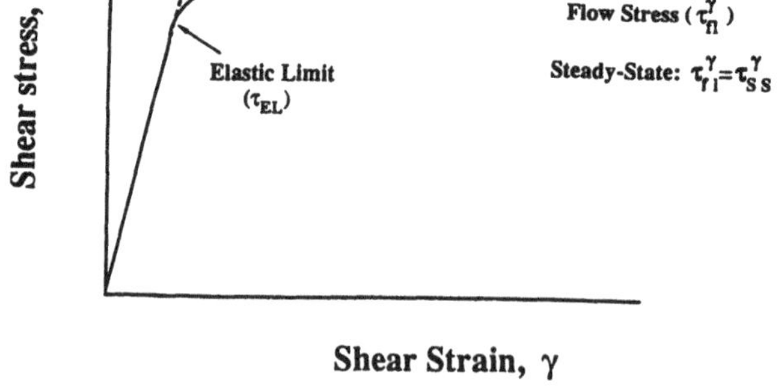

Fig. 1 Schematic of an idealized shear stress-shear strain curve obtained with a lateral shearing device.

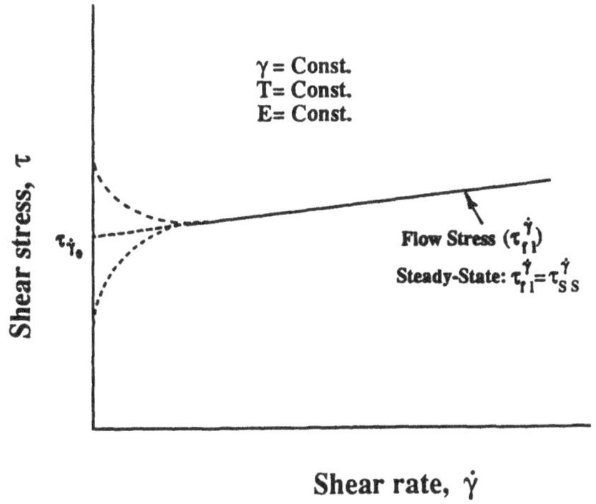

Fig. 2 Schematic of an idealized shear stress-shear rate curve obtained with a rotating rheometer (couette cell).

temperature T and electric field E. The magnitude of τ_{fl} depends on these test conditions along with the prior strain history, which determines the existing or instantaneous "structure" of the material, i.e.,

$$\tau_{fl} = f\,(\text{Str.}\ \dot{\gamma},\ T,\ E)$$

and

$$\text{Str.} = f\,(\gamma,\ \dot{\gamma},\ T,\ E,\ t)$$

where t is time. In the discussion to follow, we will restrict ourselves to a d.c. electric field. For an a.c. electric field, the frequency of the field needs to be taken into account as well as its magnitude.

The following should be considered when attempting to interpret or compare the results in Figs. 1 and 2. In the case of ER fluids, one expects that for a given $\dot{\gamma}$ and E the structure remains relatively constant in the plateau stress region of Fig. 1, giving the steady-state flow stress τ_{ss}^{γ}, which in turn is equal to the extrapolation τ_{γ_0}.If the data points in Fig. 2 were obtained at shear strains, strain rates and strain histories identical to those in the plateau region of Fig. 1, then the structure for each shear rate in Fig. 2 should be the same as that in Fig. 1 for the same test conditions. Hence, for this case, each steady-state flow stress $\tau_{ss}^{\dot{\gamma}}$ along the τ vs $\dot{\gamma}$ curve of Fig. 2

should be equal to $\tau_{ss}{}^{\gamma}$. The extrapolation of the $\tau_{fl}{}^{\gamma}$ vs $\dot{\gamma}$ curve to $\dot{\gamma} = 0$ only has meaning when the following two conditions are met: (a) each $\tau_{fl}{}^{\dot{\gamma}}$ is a steady-state flow stress and (b) $\tau_{ss}{}^{\dot{\gamma}}$ is linear with $\dot{\gamma}$. If these two conditions are met, then $\tau_{\dot{\gamma}_0} = \tau_{\gamma_0}$ ($\dot{\gamma} \rightarrow o$). Otherwise, there may be no direct correspondence between $\tau_{\dot{\gamma}_0}$ and τ_{γ_0}.

For structural materials the yield stress is generally defined as the flow stress associated with a fixed plastic deformation of 0.2%. Drawing upon this, the yield stress of ER fluids might also be defined as the flow stress at some fixed value of plastic deformation. In view of the form of the τ vs γ curve for ER fluids suggested in Fig. 1, one likely choice for the yield stress might be the flow stress at a fixed plastic strain in the plateau region. For lack of a better definition, this was the definition of yielding employed by the present authors. Finally, in keeping with common terminology[9], the flow stress at $\dot{\gamma} \leq 10^{-1}$ s^{-1} is considered **static or quasi-static** and that for higher shear rates as **dynamic**.

To examine some of the above-mentioned points we investigated the rheology at 25°C of an ER fluid consisting of 20 vol.% Sigma Chemical Co. Z3125 zeolite (8 wt.% H_2O) particles (2 µm dia.) in a mineral oil (Exxon N–61, η = 8.87 cst) employing: (a) the lateral shearing device described in Ref. 4 and (b) a Physica universal rheometer. Three types of test were conducted: (a) shear stress vs shear strain at a constant shear rate in the range of 0.0085–500 s^{-1}, (b) shear stress vs shear rate (2–500 s^{-1}) with a continuously increasing shear rate $d\dot{\gamma}/dt$ = 1.25 s^{-2} and (c) shear stress vs shear rate with a continuously increasing stress rate of 1 Pa-s^{-1}.

The τ vs γ curves at a constant shear rate obtained with our lateral shear device and with the Physica rheometer are given in Figs. 3 and 4, respectively. The shear rate range in lateral shear is 0.0085 to 2.12 s^{-1}; that in couette shear is 2 to 500 s^{-1}, giving an overlap at $\dot{\gamma}$ = 2 s^{-1}. To more fully illustrate the behavior, the τ vs γ data points are plotted in both linear and semilog form. Worthy of note regarding Fig. 3 is that the shear strain for the beginning of the stress plateau increases with increase in shear rate. Also, the curve for the lowest shear rate exhibits an upper yield (flow) stress. To be noted in Fig. 4 is that except for the shear rate of 200 s^{-1} shear weakening occurs, i.e., τ decreases with γ at $\gamma > \sim 10$. The results of Figs. 3 and 4 indicate that for a fixed shear rate throughout the test the flow stress (and presumably the structure) remains approximately constant in the strain range $\gamma \approx$ 1–10. This relatively constant stress was chosen to be termed the yield stress τ_y. For the present tests τ_y was taken as the flow stress at $\gamma = 5$, which is in the plateau stress regime.

Linear and semilog plots of τ vs $\dot{\gamma}$ for couette flow are given in Fig. 5 for the shear rate-control and stress rate-control tests, respectively. To be noted is that at the higher shear rates, the two curves are essentially parallel, that for the stress rate control being lower by an amount equivalent to a shift to a lower shear rate by a factor of about two. The lower

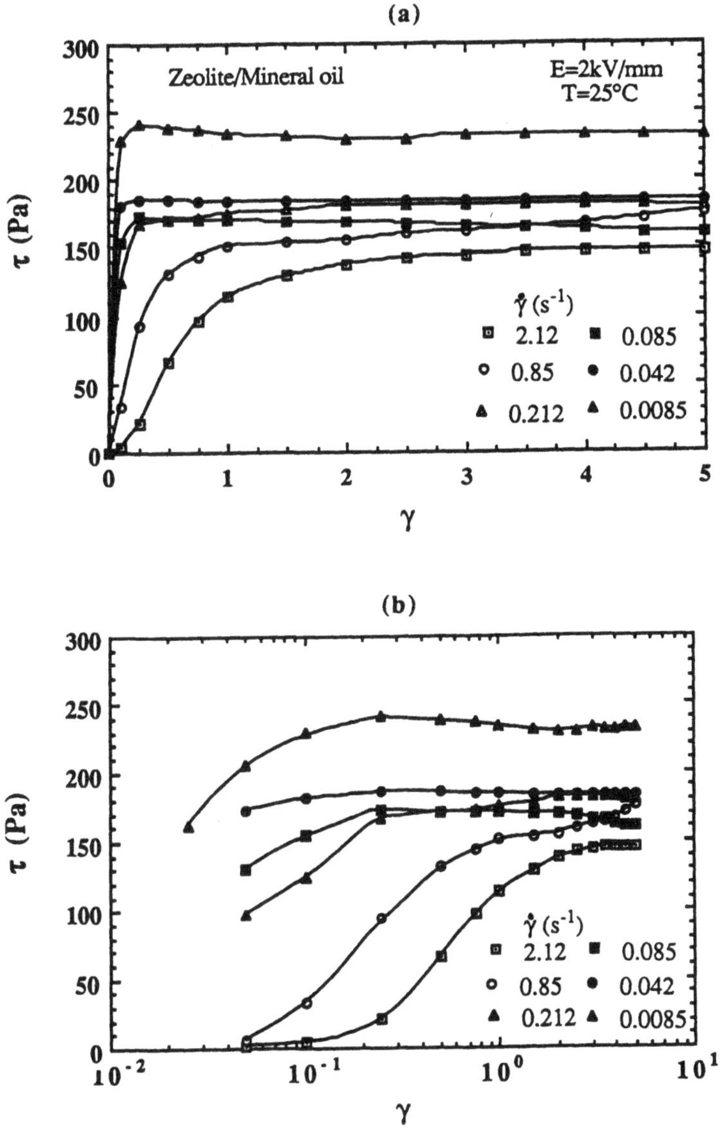

Fig. 3 **Shear stress τ vs shear strain γ curves as a function of shear rate $\dot{\gamma}$ for an ER fluid consisting of 20 vol % Sigma Z3125 zeolite particles (8 wt.% H_2O, 2 μm dia.) in mineral oil and tested in lateral shear with E = 2 kV/mm (d.c.) and T = 25°C: (a) linear plot and (b) semilog plot.**

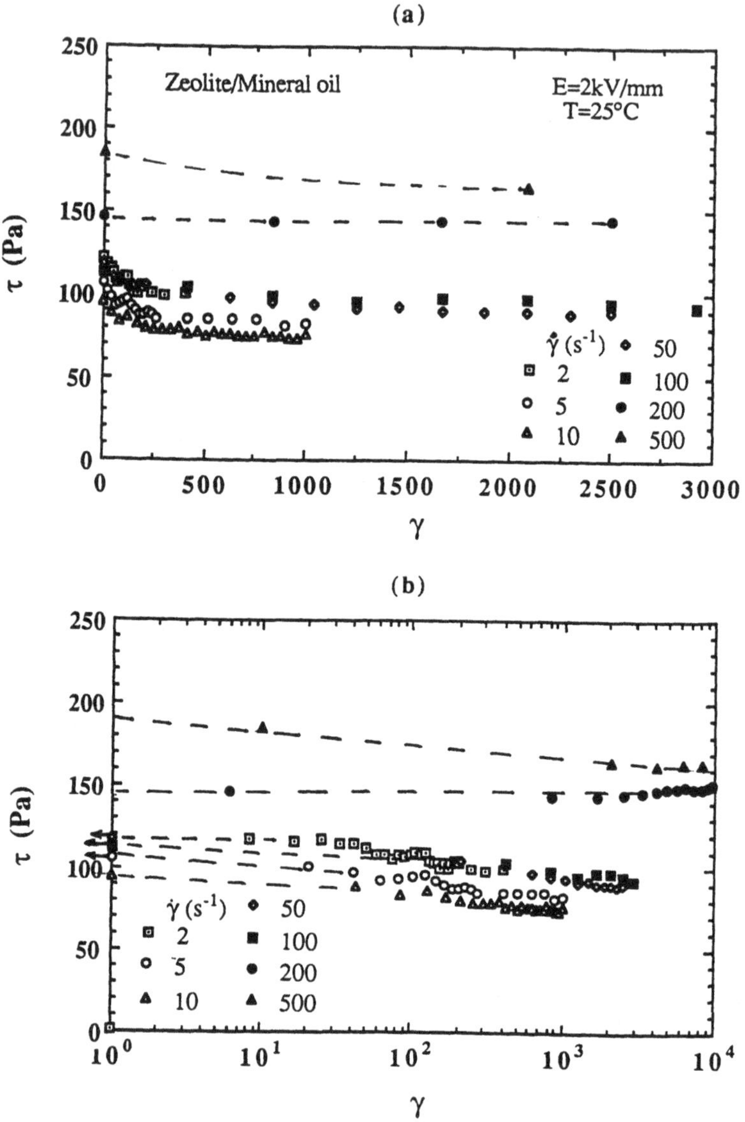

Fig. 4 Shear stress τ vs shear strain γ curves as a function of shear rate $\dot{\gamma}$ for the same ER fluid as in Fig. 3 but tested in a rotating rheometer with E = 2 kV/mm (d.c.) and T = 25°C: (a) linear plot and (b) semilog plot.

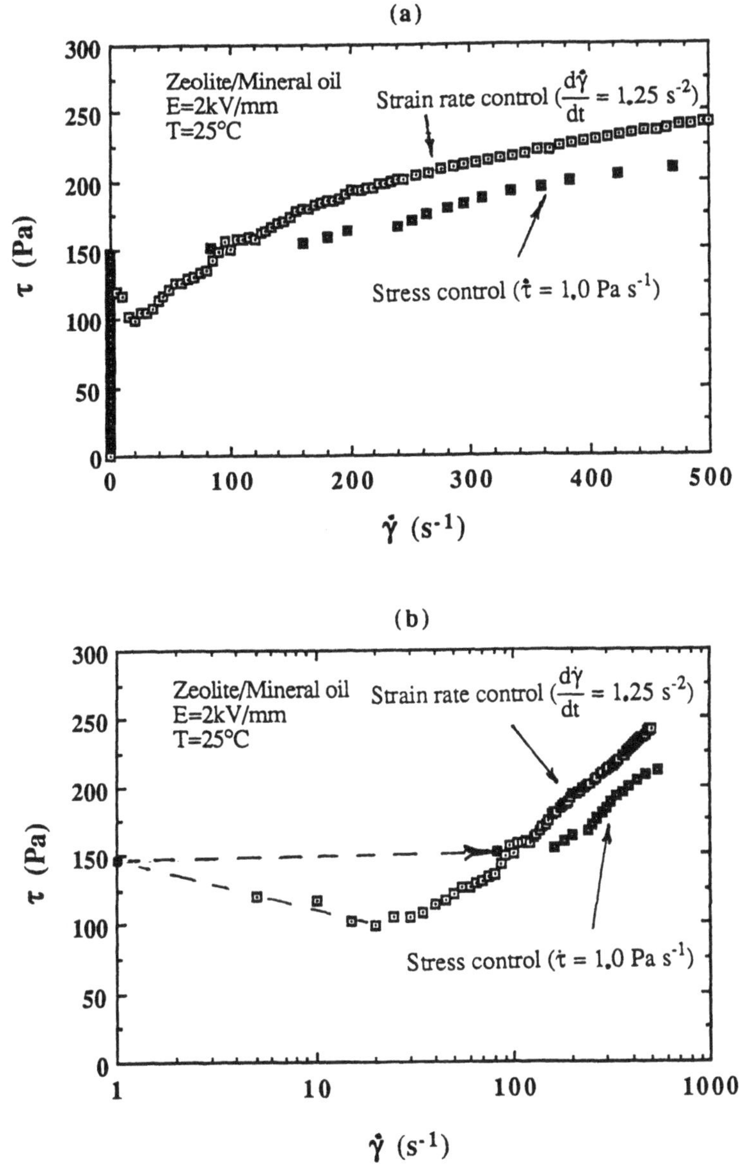

Fig. 5 Shear stress τ vs shear rate $\dot{\gamma}$ for the same ER fluid, electric field and test temperature as in Fig. 3 but under shear rate-control and stress rate-control modes: (a) linear plot and (b) semilog plot.

stress for the stress-control test at the same shear rate is considered to reflect a difference in structure produced by the prior strain history.

The effect of shear rate on the yield stress τ_y of the zeolite/mineral oil ER fluid obtained with the various test methods is shown in Fig. 6. For the constant shear rate (τ vs γ) tests τ_y was taken as the flow stress at $\gamma = 5$. This value of the stress was approximately equal to τ_{γ_0} or the maximum stress recorded for each rate. For the shear rate-control and the stress rate-control tests τ_y was taken as the flow stress at each shear rate. To be noted is that there is reasonable agreement between the values of τ_y obtained with the lateral shear device and the couette cell at the constant shear rate of 2 s^{-1}. The values of τ_y for the stress rate-control are nearly in accord with those for the constant shear rate tests in the couette flow, indicating that the structure under these test conditions was similar. τ_y for the shear rate-control tests are slightly higher (especially for E = 2 kV/mm), suggesting that the corresponding strain history produced a stronger structure. However, this may only apply to this particular ER fluid system and not represent general behavior. Worthy of mention is that the initial decrease in τ with $\dot{\gamma}$ followed by an increase at higher shear rates which occurred in the $\dot{\gamma}$-control test in Fig. 5 is in keeping with the effect of

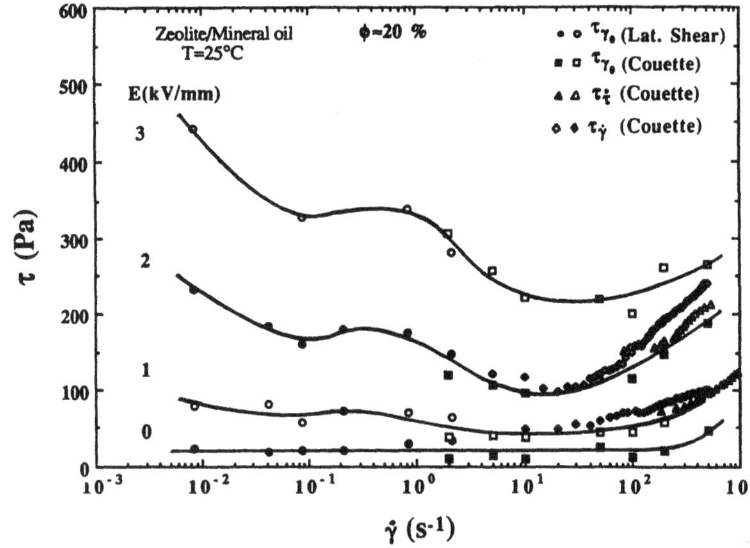

Fig. 6 Comparison of the effect of shear rate $\dot{\gamma}$ on the yield stress determined by various test methods. The ER fluid is that in Figs. 3–5.

shear rate on τ_y obtained in the constant shear rate tests of Figs. 3 and 4 and plotted in Fig. 6. An explanation for this reversal in the effect of $\dot{\gamma}$ on τ_y is given below. Finally, worthy of note is that a plateau or hump in the τ vs $\dot{\gamma}$ plot of Fig. 6 occurs in the strain rate range of 0.1–1 s^{-1}.

3. Structure

The above suggests that an important aspect regarding the strength of ER fluids is the nature of the structure. It is now generally known that upon application of an electric field the particles in an ER fluid align in the direction of the field forming a chain-like or fibrous structure; see for example Figs. 7 and 8. Some of the more important stereological characteristics of this structure which were determined for a model ER fluid of glass beads in silicone oil included the following[10–12]:

a. Upon increasing the electric field the degree of alignment of the particles initially increased rapidly with the field to about 0.5 kV/mm and then leveled off with further increase in field; see Fig. 9. Also to be noted is that the degree of alignment in the plateau region of the plot decreased slightly with increase in particle concentration.

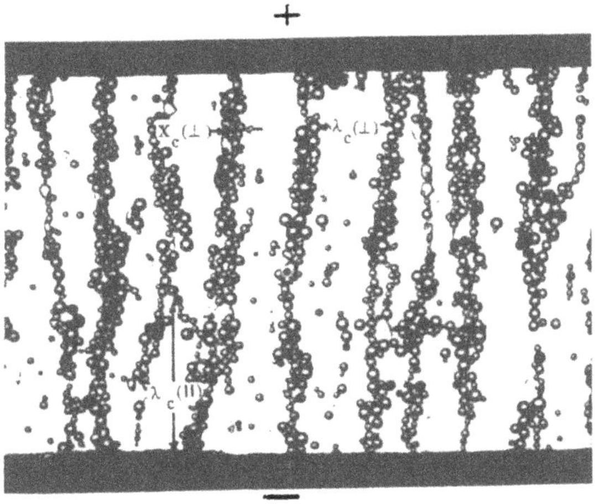

Fig. 7 Structure of a model ER fluid with 0.4 area fraction (vol. fraction) of 27 μm dia. glass beads in silicone oil following application of an electric field of 2 KV/mm (d.c.). From Conrad et al[10].

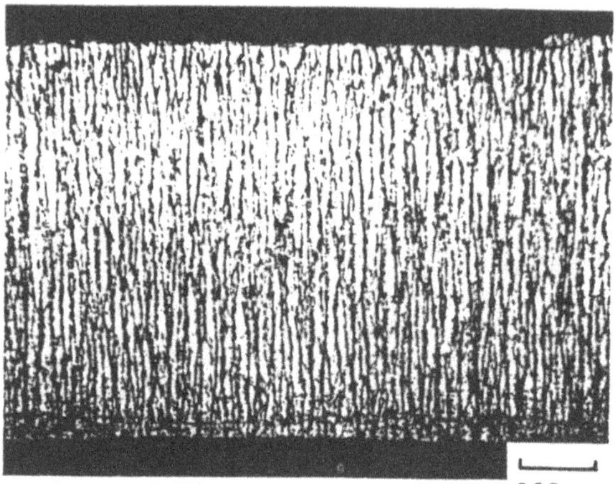

Fig. 8 Structure of a 20 vol.% zeolite (8 wt.% H_2O, 2 μm dia.) /silicone oil ER fluid with E = 2 kV/mm (d.c.).

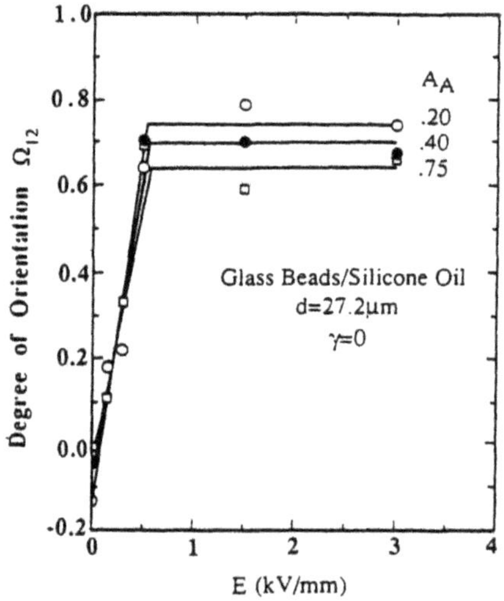

Fig. 9 Effect of electric field on the degree of alignment of the particles in an ER fluid consisting of 27 μm glass beads in silicone oil. The orientation parameter $\Omega_{12} = [P_L(\perp) - P_L(\parallel)]/[P_L(\perp) + 0.571\ P_L(\parallel)]$, where $P_L(\perp)$ is the number of intersections per unit length of chain structure taken in a direction perpendicular to the field and $P_L(\parallel)$ is that taken parallel to the field. From Conrad et al[10].

b. Fully-developed chains extending from one electrode to the other did not occur until a critical field of ~ 0.5 kV/mm was exceeded.

c. For fully-formed chains, their average spacing decreased and their thickness increased with concentration of particles, Fig. 10. At the lowest values of A_A (equivalent to volume fraction), the thickness of the chains was of the order of one particle diameter, i.e., the chains consisted mostly of single rows of particles. At higher concentrations, the chains invariably consisted of clusters of particles or multi-rows (Fig. 7). Also shown is that the average spacing between chain cross-links decreased; i.e. the number of cross-links increased with concentration.

d. The chain-like structure was imperfect, i.e. there occurred chain cross-links and debris of chain segments in addition to the chains extending from one electrode to the other. The degree of perfection was of the order of $\pi/4$ to $\pi/6$ for the particle diameters (7–27 μm) and size distributions (normal and log-normal) studied.

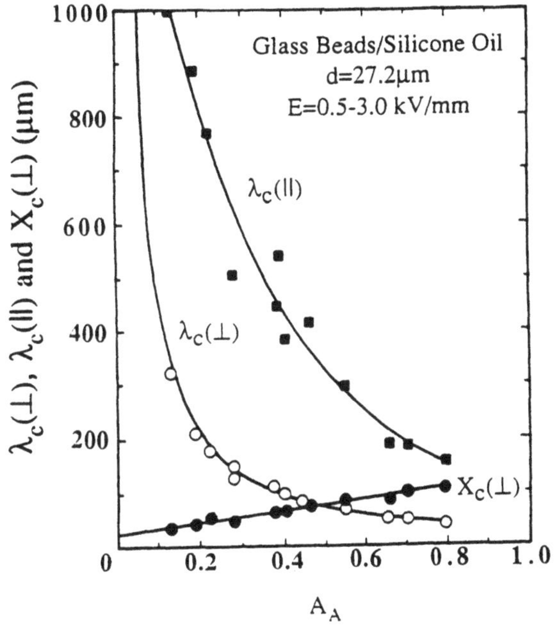

Fig. 10 Mean free spacing between chains λ_c (⊥), mean spacing between chain cross-links λ_c (||) and thickness of chains X_c (⊥) in an ER fluid consisting of 27 μm glass beads in silicone oil. From Fisher et al[12].

e. The dominant features of the chain structure formed faster than the eye could follow. However, continuous readjustments of the particles occurred for an extended period of time following the initial chain structure formation.

4. Particle Interaction Force and Strength

Accepting that a chain-like structure is generic to ER fluids, one expects that the interaction force between the particles in the chains is important to their strength. With this in mind, the present authors calculated the force between neighboring particles in a chain consisting of a single row of spherical particles[13]. The axial force F between the particles varied with the dielectric mismatch parameter β and the ratio of the spacing between particles R to their radius a as shown in Fig. 11, which is approximated by the following expression for $R/a \geq 2.05$

$$F/f_d = 1.56 \exp\left\{\left[14.84 - 6.16\left(\frac{R}{a}\right)\right]\beta^2\right\} \tag{1}$$

where

$$f_d = 24\,\pi\, a^6\, \varepsilon_o\, K_f\,(\beta E)^2/R^4 \tag{2}$$

f_d is the force between two spherical particles in the classical point dipole approach[14]. a is the radius of the particles, R their center-to-center spacing, ε_0 the dielectric constant of free space, E the electric field and β the dielectric mismatch parameter given by

$$\beta = \left(\frac{K_p - K_f}{K_p + 2K_f}\right) \tag{3}$$

$K = \varepsilon/\varepsilon_0$ is the total permittivity, the subscripts p and f referring to the particle and host fluid, respectively. Substituting f_d given by Eqn. 2 into Eqn. 1, we obtain

$$F = \frac{37.44\,\pi\, a^2\, \varepsilon_o\, K_f\,(E\beta)^2}{(R/a)^4} \exp\left\{\left[14.84 - 6.16\left(\frac{R}{a}\right)\right]\beta^2\right\} \tag{4}$$

Since the rheology of ER fluids is generally determined in shear, we need to consider the axial force between neighboring particles in a chain undergoing shear; see Fig. 12. Geometric considerations give for the shearing force

$$F_s = F_a \sin\theta \tag{5}$$

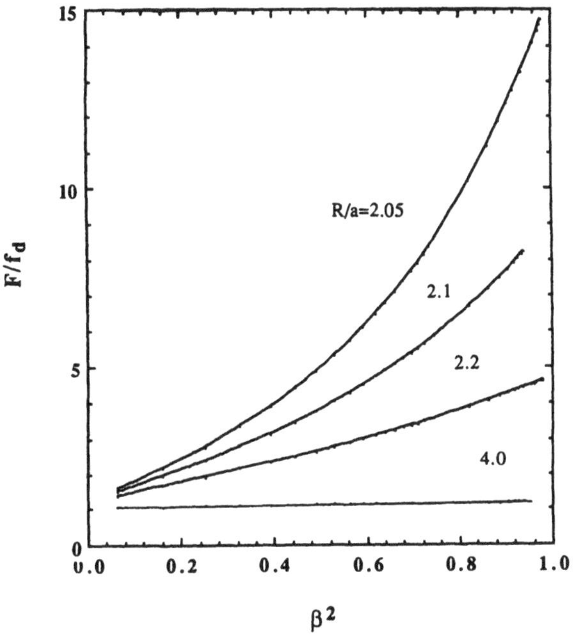

Fig. 11 Ratio of the calculated interaction force F between particles in a single-row chain to that of the classical dipole approach f_d vs the square of the dielectric mismatch parameter β. From Chen et al[13].

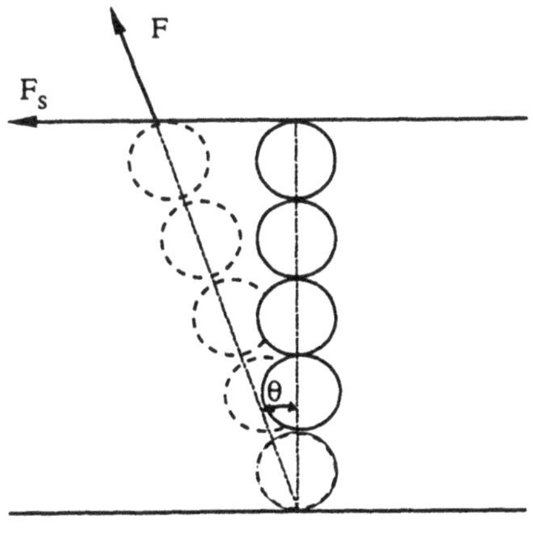

Fig. 12 Schematic of the forces involved in the shear of a chain.

where F_a is the axial force along the chain and $\sin\theta = \sqrt{1-4/(R/a)^2}$. For small values of θ we expect that $F_a = F$ (Eqn. 4), so that

$$F_s = 37.44\pi a^2 \varepsilon_o K_f (E\beta)^2 \exp\left\{\left[14.84 - 6.16\left(\frac{R}{a}\right)\right]\beta^2\right\} \sqrt{\frac{1}{(R/a)^8} - \frac{4}{(R/a)^{10}}} \qquad (6)$$

Plots of F_s/E^2K_f for three values of the ratio K_p/K_f [$=(1+2\beta)/(1-\beta)$] are given in Fig. 13. To be noted is that F_s goes through a maximum in each case. The maximum force $F_{s,\ max}$ represents an unstable condition in that the chain will break at a higher force. That Eqn. 5 gives a reasonable estimate for the force between particles in a single-row chain undergoing shear is shown in Fig. 14.

Knowing the interaction force between particles, we can derive an expression for the strength of an ER fluid. As a first approximation we expect that the yield stress τ'_y (i.e. the maximum or steady state flow stress τ_{ss}) is proportional to the force $F_{s,max}$ to rupture a chain times the number of chains per unit area N_A, i.e.

$$\tau_y = A_s N_A F_{s,max} \qquad (7)$$

where A_s is a constant which represents the structure, i.e., the departure of the geometric arrangement of the particles from single row chains. For

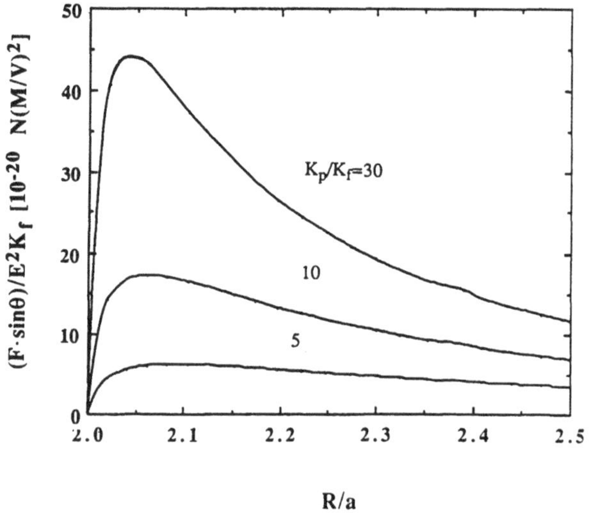

Fig. 13 The normalized shear force $F_s = F\sin\theta$ vs R/a for three values of the ratio K_p/K_f.

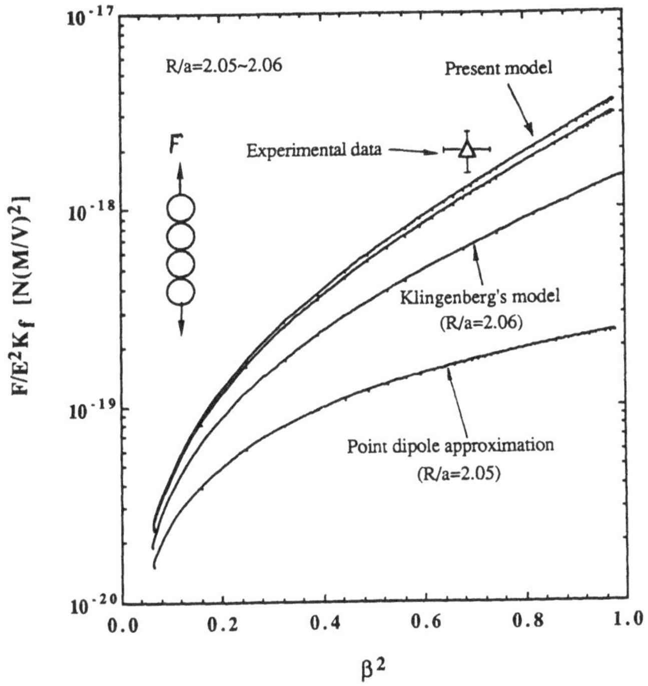

Fig. 14 Comparison of the normalized interaction force vs β^2 between that predicted and that measured directly for a single-row chain. The two curves for the present model are for R/a = 2.05 and 2.06. From Chen et al[13]. Experimental data point is from Sprecher et al[15].

complete chains of equisized spherical particles it can be shown that

$$N_A = 3\,\chi\,\phi/2\,\pi\,a^2 \tag{8}$$

where χ is the fraction of particles that lie in the chains parallel to the electric field and ϕ is the volume fraction of particles. As mentioned above, for the model ER fluid consisting of glass beads in silicone oil it was found that $\chi = \pi/4$ to $\pi/6$ depending on the size and size distribution of the particles. Taking $\chi = \pi/4$ and substituting Eqns. 6 and 8 into Eqn. 7 gives

$$\tau_y' = 44.1\,A_s\phi\epsilon_o K_f(\beta E)^2 \left| \left\{ \exp\left[\left(14.84 - 6.16\,\frac{R}{a} \right) \beta^2 \right] \right\} \sqrt{\frac{1}{(R/a)^8} - \frac{4}{(R/a)^{10}}} \right|_{max} \tag{9}$$

In Fig. 15 the measured yield stress (taken here to be $\tau'_y = \tau_y$ (E) − τ_y (E = 0) at $\gamma = 5$) for ER fluids consisting of 20 vol % Z3125 zeolite particles (8 wt.% H_20, 2 μm dia.) in various host oils is compared with that predicted by Eqn. 9. Included are data points for a number of conditions: (a) two shear rates and test methods (lateral shear with $\dot{\gamma} = 0.085\ s^{-1}$ and couette flow with $\dot{\gamma} = 2\ s^{-1}$), (b) a range in electric fields (0.5–4 kV/mm) and test temperatures (25°–150°C) and (c) various oils (silicone, mineral, castor, DOP, corn). To be noted in Fig. 15 is that the data points for the various test conditions can be considered to lie on two straight lines through the origin, one for the shear rate of 0.085 s^{-1} and the other for the shear rate of 2 s^{-1}, their slopes being 15 and 11 respectively. Thus, the measured yield stress agrees with the predicted stress by taking with $A_s = 15$ for $\dot{\gamma} = 0.085\ s^{-1}$ and $A_s = 11$ for $\dot{\gamma} = 2\ s^{-1}$, independent of the other test conditions. This decrease in the structure parameter A_s with increase in shear rate is in keeping with the effect of $\dot{\gamma}$ on the flow stress in Fig. 6.

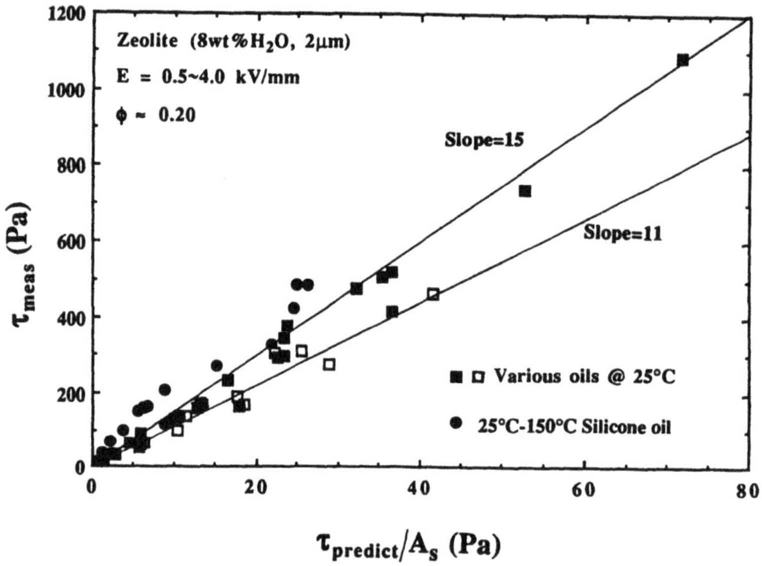

Fig. 15 Comparison of the measured yield stress ($\tau'_y = \tau_y$ (E) − τ_y (E = 0)) with that predicted by Eqn. 9 of the text for ER fluids containing ~ 20 vol.% zeolite (8 wt.% H_20, 2 μm dia.) particles in various oils (silicone, mineral, caster, corn, DOP). Open symbols are for measurements with couette cell and $\dot{\gamma} = 2\ s^{-1}$; closed symbols are with lateral shear and $\dot{\gamma} = 0.085\ s^{-1}$.

It was found that τ'_y obtained from Fig. 6 was essentially proportional to E^2 at all shear rates. Hence, the three curves in Fig. 6 scaled to a single curve by dividing each measured yield stress at a given electric field by a standard reference stress $\tau'_y{}^\circ$ for that field; see Fig. 16. The reference stress chosen for each field was the stress at a shear rate of 1 s^{-1}. If we now assume that Eqn. 9 applies over the entire strain rate range considered in Fig. 16, then A_s will vary with shear rate in the same manner as the ratio $\tau'_y/\tau'_y{}^\circ$. The variation of A_s with shear rate deduced in this manner (taking $A_s = 14$ at $\dot{\gamma} = 1\ s^{-1}$) is depicted by the ordinate on the right side of Fig. 16. In keeping with the variation of τ_y with shear rate, A_s decreases with increase in $\dot{\gamma}$ to about 30 s^{-1} and then increases at higher rates. Again, a plateau or hump is indicated at $\dot{\gamma} = 0.1$–$1\ s^{-1}$.

The decrease in A_s with shear rate at the lower shear rates is considered to result from insufficient time for the particles to reform the geometric arrangement which gives the strongest interaction force. The increase in τ'_y and in turn the deduced A_s at the highest shear rates probably reflect the observed degradation at high shear rates of the uniform chain structure extending from one electrode to the other. The degraded structure consists of solid-like regions near the electrodes separated by a

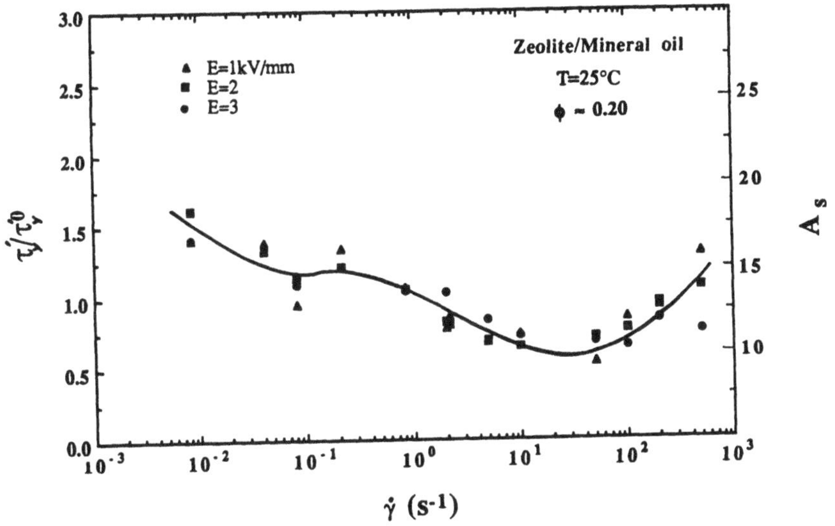

Fig. 16 Semilog plot of the scaled yield stress $\tau'_y/\tau'_y{}^\circ$ and the structure parameter A_s vs shear rate for 20 vol.% zeolite (8 wt.% H_2O, 2 μm dia.) /mineral oil fluid at T = 25°C and E = 1–3 kV/mm (d.c.). Left ordinate is for the stress ratio; right is for A_s.

fluid-like central region in which most of the shearing takes place[1,16]. With increasing shear rate the thickness of the fluid-like region expands and the rheological behavior of the suspension becomes more-and-more Newtonian. Eqn. 9 does not apply to this condition and therefore for $\dot{\gamma} > \sim 20\ s^{-1}$ the deduced values of A_s in Fig. 16 do not have the same meaning as for the lower shear rates. The fact that in the high shear rate region the flow stress depends on the test method suggests that the two-phase structure which forms in this region depends on the strain history, as is also expected for the chain structure in the lower shear rate region.

Besides depending on the shear rate, it is expected that the structure parameter A_s may depend on the concentration of the particles in the ER fluid. This was investigated by considering the effect of the volume fraction of particles ϕ in the zeolite/mineral oil fluid on the yield strength; see Fig. 17. The form of the τ_y vs ϕ curve shown here is similar to those given by Klingenberg and Zukoski[16] for an ER fluid consisting of lithium poly(methacrylate) in a chlorinated hydrocarbon oil, whereby the strength at the lower concentrations is less than expected from an extrapolation through the origin of that at the higher concentrations. The values of A_s derived from the results in Fig. 17 using Eqn. 9 are plotted vs ϕ in Fig. 18.

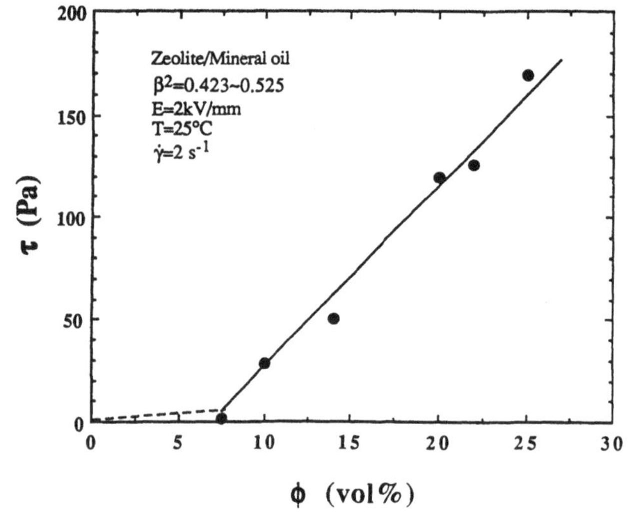

Fig. 17 Yield stress of a zeolite (8 wt.% H_2O, 2 μm dia.)/mineral oil ER fluid vs volume fraction of zeolite particles. Couette flow at 25°C, E = 2kV/mm (d.c.) and $\dot{\gamma} = 2\ s^{-1}$.

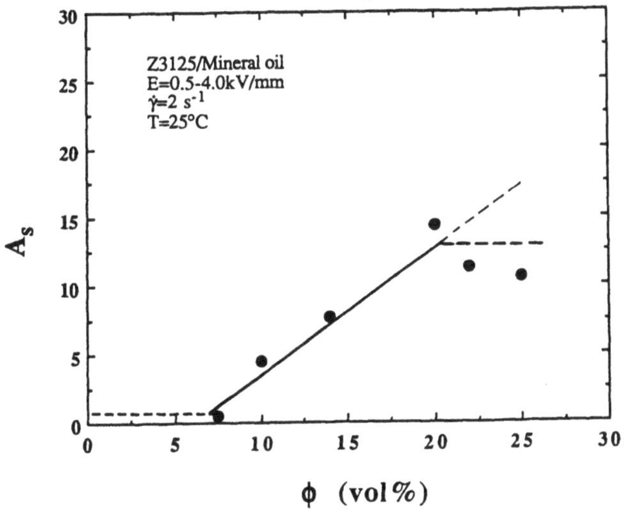

Fig. 18 Structure parameter A_s vs volume fraction of zeolite particles in mineral oil derived from the data in Fig. 17 and Eqn. 9 in the text.

The straight line through the data points gives for the concentration dependence of A_s

$$A_s = 92\,(\phi - 0.06) \tag{8}$$

Worthy of note is that $A_s = \sim 1$ at the lowest concentration investigated, where the structure is expected to consist of single-row chains[10,11]. Thus, the form of the τ_y vs ϕ curve of Fig. 17 can be explained in terms of the combined effect of ϕ on the structure factor A_s and on the number of chains per unit area N_A (Eqn. 8). It is expected that A_s may saturate as the concentration of particles is increased. The results in Fig. 18 do not permit a positive statement whether or not saturation of A_s occurs at $\phi = 0.2$ for the present ER fluids, since the decision depends mainly on the single data point at $\phi = 0.25$.

The variation of A_s with ϕ along with the observations on the chain-like arrangement of the particles in Figs. 7 and 8 provide some insight into why A_s is greater than unity as ϕ increases. The theoretical calculation of the force F in Eqn. 4 is for a chain consisting of a single row of particles. On the other hand, for $\phi > \sim 0.1$ the chains in real fluids consist of clusters of particles or multi-rows (Figs. 7 and 8), whose thickness increases with ϕ. The force between individual particles in such clusters is expected to be larger than that between particles in a single-row chain. This is supported by the finding that A_s approaches unity at low ϕ where the structure consists mostly of single-row chains and its magnitude increases at higher concentrations where multi-row chains occur. Further, recent calculations by Shih et al[17] give an order of magnitude enhancement of the force between particles in a cluster consisting of two rows of particles in close-packed arrangement compared to a single row; see Fig. 19.

5. Comparisons and Predictions

Having established the magnitude of A_s, it is of interest to compare the measured τ vs γ curve with that predicted by taking $\tau_y = A_s N_A F_s$ and $\gamma = \tan\theta$ $(=\sqrt{(R/a)^2/4-1})$. Such a comparison is made in Fig. 20 for a zeolite/mineral oil ER fluid tested at two shear rates at 25°C and E = 2 kV/mm and taking A_s from Fig. 16. Reasonable agreement occurs for the maximum stress. However the stress drops off with further strain in the predicted curve, whereas for the measured curve the stress remains relatively constant giving the plateau or steady-state flow stress. It is proposed that the reason the measured stress does not drop off in the real fluid is that the chains which have been broken are continually reformed as shearing continues, thereby avoiding the decrease in force which occurs in the theoretical calculations pertaining to the stretching of a single chain.

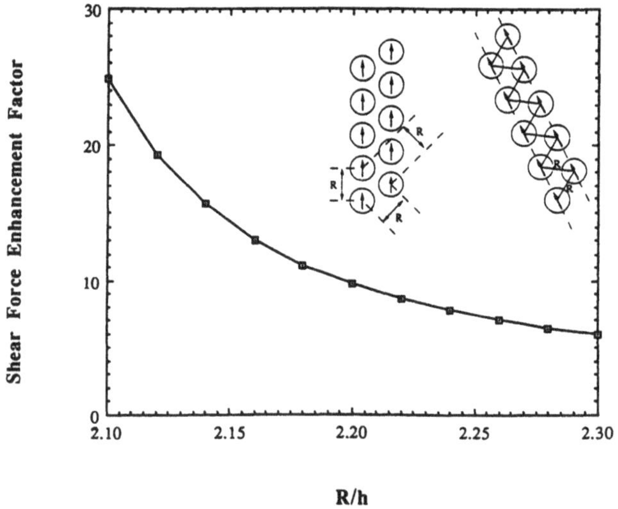

Fig. 19 Shear force enhancement factor for a two-row chain of particles in hexagonal array compared to a single row chain vs the ratio of center-to-center spacing R of the particles to the dipole separation distance h in each particle (≈ a). From Shih et al[17].

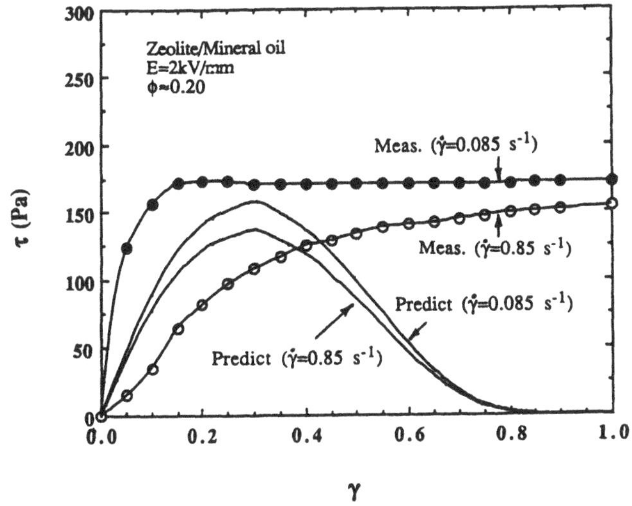

Fig. 20 **Comparison of the measured shear stress-shear strain curves at two shear rates (lateral shear) with those predicted using the appropriate values of A_s.**

The agreement between the measured stress-strain behavior in the pre-yield region and that predicted is reasonable but not as good as that for the yield stress. This is not surprising, since the values of the structure parameter employed in the calculated τ vs γ curves were derived from post-yield measurements, where the geometric arrangement of the particles is expected to differ from that in the pre-yield region. For a better comparison of the stress-strain curves in the pre-yield region (and in turn the shear modulus) one needs to establish the values of A_s for this region.

In view of the agreement between the measured values of the flow stress and those predicted by Eqn. 9 with the appropriate value for A_s, it is tempting to estimate the maximum strength attainable with ER fluids at room temperature and a shear rate of 1 s^{-1}. Theoretical considerations give that the maximum force of interaction between two particles in a single-row chain occurs for $K_p/K_f \approx 15$[13]. Taking the reasonable values of K_f = 12.5, K_p = 187.5, ϕ = 0.5, E = 5 kV/mm, R/a = 2.05 and A_s = 15 one obtains

$$\tau_y \approx 49 \text{ kPa}$$

which is more than 3 times the maximum strength of 15 kPa reported to-date[18,19]. To realize this predicted strength the local dielectric breakdown strength of the host fluid would need to be of the order of 65 kV/mm, because of the electric field concentration between the particles[13].

6. Summary and Conclusions

A. *General*

1. It is recommended that the strength of an ER fluid be defined as the flow stress at a constant structure, shear rate, temperature and electric field, recognizing that the structure is dependent on the prior strain history.
2. Similar to the case of structural materials, the yield stress τ_y of ER fluids in a constant shear rate test is best taken at some fixed value of the permanent shear strain beyond the elastic limit. A reasonable choice for ER fluids appears to be a strain in the regime of the stress plateau, i.e., in the steady-state flow stress regime.
3. The flow stress in shear rate-control and stress-rate control tests depends on the control rate (i.e. on the strain history) and therefore such tests may not give unique or unambiguous values for the yield strength.

B. *Zeolite-Based ER Fluids*

1. A shear stress plateau occurred in the shear strain range γ = 1–10. Therefore the yield stress τ_y was taken as the flow stress at $\gamma = 5$.
2. τ_y at 25°C first decreased with rise in shear rate $\dot{\gamma}$ to about 20 s^{-1} and then increased with further increase in rate. A plateau or hump in the τ_y vs $\dot{\gamma}$ curve occurred at shear rates of 10^{-1} to 1 s^{-1}.
3. The measured values of τ_y were about two orders of magnitude larger than predicted by the point dipole approximation and about an order of magnitude larger than those calculated based on the force of interaction between particles in a chain consisting of a single row of particles. This latter difference was explained by an enhancement of the interaction force due to the existence of clusters of particles or multi-row chains rather than single-row chains.
4. The force enhancement factor A_s varied with shear rate and concentration of particles, but was relatively independent of electric field, temperature and host fluid.
5. There was reasonable agreement between the predicted and the measured shear stress-shear strain curves and the concentration dependence of the yield strength when the force enhancement factor was taken into account.

C. Theoretical Strength

1. Present theoretical-experimental considerations suggest that ER fluids may attain a yield strength approaching 50 kPa.

7. Acknowledgements

This research was jointly sponsored by NSF Award #CBT–8714515, the Ford Motor Company and the NCSU University/Industry Consortium on ER fluids.

8. References

1. A. F. Sprecher, J. D. Carlson and H. Conrad, *Mat. Sci. Engr.* **95** (1987) 187.
2. N. G. Stevens, J. L. Sproston and R. Stanway, *Trans. ASME* **54** (1987) 456.
3. D. A. Brooks, in *Proc. 1st Int. Symp. on Electrorheological Fluids*, H. Conrad, A. F. Sprecher and J. D. Carlson, eds., (NCSU Engineering Publications, Raleigh, NC, 1989) p. 63.
4. H. Conrad, Y. Chen and A. F. Sprecher, in *Proc. 2nd Int. Conf. Electrorheological Fluids*, J. D. Carlson, A. F. Sprecher and H. Conrad, eds. (Technomic, Lancaster, PA, 1990) p. 252.
5. D. L. Klass and T. W. Martinek, *J. Appl. Phys.* **38** (1967) 67.
6. H. Uijima, *Jap. J. Appl. Phys.* **11** (1972) 319.
7. T. C. Jordon and M. T. Shaw, in *Proc. 2nd Int. Conf. Electrorheological Fluids*, J. D. Carlson, A. F. Sprecher and H. Conrad, eds. (Technomic, Lancaster, PA 1990) p. 231.
8. Y. Choi, A. F. Sprecher and H. Conrad, "Response of a ER Fluid-Filled Laminate Composite to Forced Vibration", submitted to *J. Int. Mater. Sys. and Str.* (1991).
9. G. Dieter, *Mechanical Metallurgy*, 3rd Ed., (McGraw Hill, N. Y., 1986).
10. H. Conrad, M. Fisher and A. F. Sprecher, in *Proc. 2nd Int. Conf. on ER Fluids*, J. D. Carlson, A. F. Sprecher and H. Conrad, eds., (Technomic, Lancaster, PA, 1990) p. 67.
11. M. D. Fisher, A. F. Sprecher and H. Conrad, in *Metallography 75 Years Later*, ASTM, to be publ.
12. M. D. Fisher, A. F. Sprecher and H. Conrad, unpublished results, NCSU (1989).
13. Y. Chen, A. F. Sprecher and H. Conrad, *Jnl. Appl. Phys.*, in print.
14. C. A. Coulson and T. S. M. Boyd, *Electricity* (Longman, N. Y., 1979) p. 43.

15. A. F. Sprecher, Y. Chen and H. Conrad, in *Proc. 2nd Int. Conf. ER Fluids*, J. D. Carlson, A. F. Sprecher and H. Conrad, eds. (Technomic, Lancaster, PA, 1990) p. 82.
16. D. J. Klingenberg and C. F. Zukoski, *Langmuir* **6** (1990) 15.
17. Y-H. Shih, A. F. Sprecher and H. Conrad, to be submitted *J. Rheology*.
18. American Cynamid Co. Bulletin, "Electrorheological Fluids", NBO–001–86–2507–4.5M 1086 (1986).
19. H. Block and J. P. Kelley, in *Proc. 1st Int. Symp. ER Fluids*, H. Conrad, A. F. Sprecher and J. D. Carlson, eds. (NCSU Engr. Publ., Raleigh, NC, 1989) p. 1.

EXPERIMENTAL MEASUREMENTS OF THE DYNAMIC TORQUE RESPONSE OF AN ELECTRO-RHEOLOGICAL FLUID IN THE SHEAR MODE

A. HOSSEINI-SIANAKI
Department of Mechanical and Process Engineering,
University of Sheffield, P. O. Box 600,
Sheffield, S1 4DU, U.K.
and
W. A. BULLOUGH, R. FIROOZIAN, J. MAKIN, R. C. TOZER
University of Sheffield.

ABSTRACT

Experimental measurements of the time response of an electro-rheological fluid in a high natural frequency Couette type of rheometer system are reported for both step and variable frequency, DC biased small amplitude sine wave voltage excitation.

The effects of both the speed of the rotating element of the rheometer and the magnitude of its voltage relative to the stationary element as well as the fluid temperature level are presented for time, frequency and steady domains. For a given DC voltage level and over the range of tests carried out, the fluid response time improves considerably with both increase in the temperature and speed, whereas the effect on the magnitude of the steady state ER torque value is marginal. Very encouraging results are obtained when the performance of the fluid with regard to the step tests are compared with results of transformed data taken from a larger and more practical clutch system and those acquired from the same fluid but in the flow mode valve arrangement.

1. INTRODUCTION

The factor which makes ER force transmission systems distinct is their capability to perform a function quickly. This has meant that experimental measurements of the speed of torque response of these fluids has proved to be rather a difficult task. This is because most available torque measuring systems (torque transducers) when coupled to an ER device produce a system with a torsional natural frequency below that of the response time of the E.R. device under test. The consequence is that the actual fluid response would be concealed by the dynamic oscillations on the torque transducer. In some instances[1] attempts have been made to estimate the actual fluid response by implementing an identification and digital simulation software. Useful results were achieved albeit without the confidence in the results that would accompany a direct measurement.

The work reported here was carried out on a test rig with a torsional natural frequency much greater than the expected speed of response of the fluid under test, thus enabling direct measurement of response time. However, this test rig differed from a practical, industrial clutch arrangement in three ways: (a) one of the electrodes was held stationary; (b) the small diameters of electrodes resulted in shear rate and field strength varying across the gap and; (c) a much lower shear rate than the clutch test rig.

In order to obtain meaningful results (on the speed of response of the fluid) from step and small amplitude sine wave tests, the power supply unit should be able to produce a good quality square voltage pulse for the step voltage tests and undistorted sine waves at high frequencies - for the frequency response trials. Good quality signals in conjunction with a high frequency viscometer system and a high quality computer controlled data logging system facilitate the acquisition of accurate data in the steady state, time and frequency domains.

Amongst the variables affecting the fluid performance are: bias DC voltage amplitude, temperature, shear rate and rate of imposition (as well as the magnitude) of the step voltage across the electrodes. The aim of the current work is to investigate these effects and compare the results, where possible, with those from a flow mode test rig[2,3].

2. TEST RIG FACILITIES

A schematic view of the experimental apparatus is shown in Figure 1. The rig was designed and developed by Stangroom et al[4] and details are reported in that reference. It is nevertheless necessary briefly to explain some of the main features of this rig.

Given that the objective is to measure the response of the ER fluid without any interference from the measuring system and since the fluid is believed to have a high speed of response, the measuring viscometer system must have a much greater natural frequency, i.e. more than five times that of the highest measurement frequency anticipated. Also, in order to eliminate any resonance of the main rheometer frame within the measuring bounds and because it is impracticable to achieve very high natural frequency (i.e. order of kilohertz) the frame torsional natural frequency had to be decreased (by an increase in moment of inertia and reduction in torsional stiffness) to as low a value as possible. With the arrangement shown in Figure 1, this is well below 0.5 Hz. The air bearings further reduce the torsional stiffness in the mounting system and prevent the introduction of noise which would be present in the case of conventional bearings.

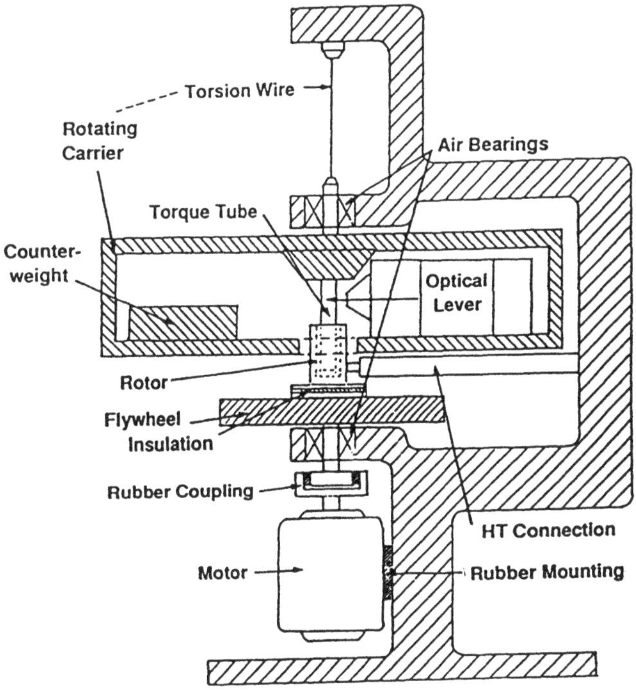

Fig. 1. Schematic layout of the experimental apparatus.

The viscometer consists of two parts : the rotor and the torque tube, both made from stainless steel. The torque tube constitutes the stationary element of the system and is kept at earth potential. Sizes are chosen so that a natural frequency of about 10 kHz is achieved. The inner and outer diameters of the tube are 5 mm and 6 mm respectively and the active length is about 80 mm. The tube is immersed for about 40 mm in the fluid and the torsion in it is measured at a point about 10 mm above the fluid surface.

The main part of the rotor is the cup section which has a solid rod at the centre so that both the torque tube outer and inner surfaces are active. The rotor is also very stiff and acts as the high potential electrode. Hence it is electrically insulated from the rest of the viscometer. Power is supplied via a carbon brush which is in contact with the lower part of the rotor cup section.

The torque developed in the form of torsion in the torque tube after the energisation of the electrodes is picked up by a very sensitive optical lever system. The basic design of this equipment and its associated electrical instrumentations are based on the work by Jones and Richards[5]. Basically a differential photocell arrangement is used with a standard optical grid system to observe the displacement of light beam (from a small 12v bulb) reflected by a 2 mm square mirror which is an integral part of the torque tube.

The composition of the electro-rheological fluid used throughout the tests reported in this paper was dielectric base oil with some 30% by volume of lithium polymethacrylate solid. The porous solid contained approximately 10% water and were generally in the size range 10-20 μm. This fluid was from the same batch as that employed in earlier valve and clutch test rigs[1,2,3] and that is the principal reason for using it here - so the results could be compared with one another. Further details of the fluid are given in Ref. 6.

The motor employed to drive the rotating member of the rheometer was a printed armature DC servo type with rated speed and output power of 2700 rpm and 41 W respectively. The motor was mounted under the support plate using rubber mounting and was connected to the flywheel via a flexible coupling to eliminate undesired dynamic effects. The rotational speed was measured with a small tacho-generator.

The power supply arrangement comprised two units; (a) the DC supply unit and (b) the facility for superimposing the small amplitude sine wave voltage excitation.

The high DC voltage across the electrodes was supplied by a Wallis high tension unit which was able to develop 2kV at a maximum current of 1A with an overload protection of 50mA. With this arrangement it was possible to generate a step voltage with insignificant regulation and time delay over the maximum predicted current range of 0-50mA.

Further particulars of the current and voltage monitor circuits and details of imposition of the step voltage across the electrodes are reported elsewhere[3]. For these experiments the supply system was modified to include a high performance supply voltage modulator specially designed and built for the superposition of a small amplitude signal AC component onto the main DC bias supply voltage. A block diagram of this system is shown in figure 2.

The "small signal" amplifier, A, is a standard Class B power amplifier circuit with its ground line connected to the high voltage DC output from the main supply. The term "small signal" in this context describes the relative amplitude of superimposed AC to the DC value of the high supply voltage. The input to the amplifier is derived from an 8 bit digital to analogue (D/A) converter, B, which is driven by an analogue to digital (A/D) converter connected to the input signal, D, via a 9 channel opto-isolator, C. The sampling rate of the A/D converter, D, is 500kHz and both the opto-isolator, C, and D/A converter, B, have been designed to operate at this rate. Since the required system bandwidth of 5kHz is much smaller than the limit of 250kHz set by the sampling rate, it is not necessary to filter the conversion noise which occurs at harmonics of 500kHz.

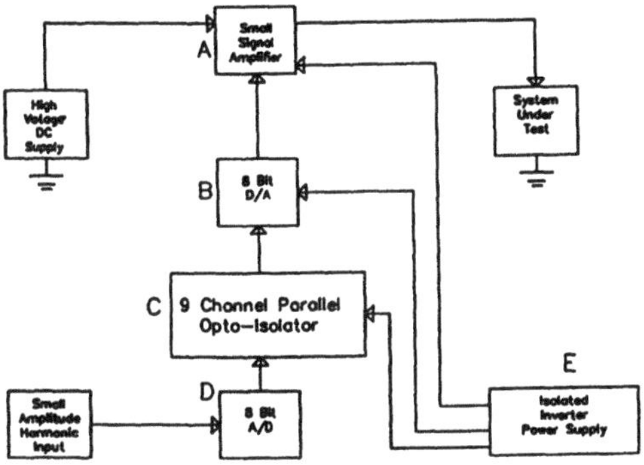

Fig. 2. Supply voltage modulator for superimposing a small amplitude sine wave signal onto the bias DC voltage supply.

The amplifier, A, 8-bit D/A, B and the high voltage side of the opto-isolator are supplied by a conventional push-pull inverter. Both the bobbin of the inverter transformer and the opto-isolator are designed to withstand a low voltage to high voltage difference of ±5kV.

For logging the experimental data a high quality computer controlled data acquisition system (Microlink 4000) was commissioned. This system had six input channels with each channel having its own programmable gain amplifier, analogue to digital converter and on-board memory with 8 K storage capacity. The analogue to digital converter had a 12 bit resolution and a sampling rate of up to 1 MHz was possible on all channels. The sampling rate was not restricted by the controller, i.e. the computer. However, a single clock source ensured synchronous sampling on all channels. A distinctly useful facility of the data recorder system was its capability to capture and store data prior to receiving a trigger signal. Data capture could be commenced by some event taking place on one of the input signals, e.g. a voltage crossing some predefined threshold.

The data logging process was managed by the software package installed in the host computer. After an initial inspection, the recorded data was transferred to the mainframe computer (IBM 3083) for further detailed analyses.

3. EXPERIMENTAL TEST PROCEDURE

To attain the objective of the present exercise that is, to determine the time response characteristics of an ER fluid under various operating conditions, two types of tests, namely step and harmonic excitation were carried out. This section describes the scheme of these tests, whereas, in the next section results are presented, discussed and compared with previous works where possible.

3.1 Step Tests

For these tests, with the bottom air bearing activated, the motor was set to the required speed and together with the flywheel was left running to achieve the desired steady level. The trigger facility on the data logging system was employed to capture some pre-trigger data (i.e. at zero applied voltage) which was used as a reference level for the measurement of the E.R. torque. The electrodes were then energised by a steep imposition of the applied voltage across them. This also conveyed a trigger signal to the data acquisition system to initiate simultaneous recordings of four channels of data, i.e. excitation voltage and current, motor speed and the signal from the torque measuring equipment. A sampling rate of 100 kHz was adjudged to be the most suitable for these series of trials.

At the beginning and end of each individual test the fluid temperature was measured by lowering a thermocouple through the torque tube into the fluid. Since the period during which each test was performed was very short no significant rise in temperature was observed. The above test procedure was repeated for motor speeds of 250 and 1000 rpm (these are equivalent to shear rates of $160s^{-1}$ and $640s^{-1}$) with excitation voltages of 1200, 1500 and 1800 volts.

In an attempt to discern the influence of temperature on the fluid performance (i.e. steady state torque output, response time etc.) a similar technique to that described above was carried out. However, after completing the tests at room temperature, the fluid was warmed up by blowing warm air at the rheometer until the desired higher steady temperatures were achieved. For a chosen excitation voltage of 1500 volts tests were repeated for speeds of 250 and 1000 rpm at temperatures of 23, 28, 32, 38 and 42°C.

It has been argued in the literature[7] that change in electric field strength rather than its polarity reversals are responsible for the heightening of the ER Effect and[8] that the rate of polarisation is important. As a contribution to this area of interest in the last set of the step trials, a step voltage with different slew rates, was applied across the electrodes. The test method was similar to those explained earlier, however, on this occasion time delays were introduced in order to produce step voltages with different rise times. Here, again tests were repeated for different motor speeds and steady state voltage values.

For all the various testing conditions mentioned so far, each test was repeated at least twice to check for repeatability of the results.

3.2 Harmonic Excitation Tests

Another approach for determining the response characteristics of ER fluids is by using frequency or harmonic response techniques. For these tests, the operational setting of air bearings, motor and other mechanical arrangements were similar to the step tests. The electrical excitation across the electrodes, however, comprised a base DC voltage over which a sinusoidal voltage of small amplitude compared to the base voltage was superimposed. The sinusoidal voltage amplitude was kept low to avoid voltage dependent non-linearity problems. After all initial transients had died out the torque signal in addition to the excitation voltage and current was recorded on the data capture system. The sampling rate was chosen such that there were at least 200 sample points per cycle.

The high gain required in the optical lever to measure the very small change in angle of the mirror due to the response of the fluid to the harmonic driving signal resulted in a torque response contaminated with large amplitude high frequency noise. Some low frequency unwanted signals were also present, probably due to slight misalignment of the electrodes. Thus, either a digital filter had to be used on the recorded data (after the event) or an analogue filter prior to logging the data. It was decided that the analogue filter was more appropriate. However, in order to avoid problems with regard to signal attenuation and phase change when filtering the data, both the voltage and torque signals were passed through filters with the same gain and bandwidth settings.

Experimental results were obtained for a range of frequencies between 1Hz and 3kHz. To monitor the effect of fluid temperature, motor speed and bias excitation voltage, two of the parameters were held constant whilst the third was varied over a suitable range.

4. RESULTS AND DISCUSSION

4.1 Step Test Response

Table 1 shows the effect of temperature and speed (shear rate) on the steady state ER torque value, the time taken to reach the onset of this value and the current drawn, whereas Table 2 presents the influence of DC voltage level on the same parameters. The time response apparently comprised three sections, a zero effect initial delay followed by two subsequent rises which lead to the steady state level. The time given in the tables are the sum of these three periods.

Table 1. Effect of temperature and motor speed on the steady state ER torque, current and time response. Steady state voltage = 1500 volts

Temp.	23°C		32°C		42°C	
Speed	250 rpm	1000 rpm	250 rpm	1000 rpm	250 rpm	1000 rpm
Response Time (ms)	3.47	2.7	2.6	1.54	1.71	0.98
Torque, τ_e (Nm)x10^{-3}	10.23	8.85	10.35	9.38	10.43	10.33
Current (m Amps)	0.32	0.28	1.58	1.66	8.41	10.35

Table 2. Effect of various DC voltage levels on the steady state ER torque, current and time response. Motor speed = 1000 rpm, Fluid Temp. = 23°C

Voltage (V)	Torque, τ_e (Nm)x10^{-3}	Current (m Amps)	Response Time (ms)
900	3.9	0.11	6.4
1200	6.05	0.21	3.75
1500	8.85	0.28	2.7
1800	12.44	0.50	2.3

As reflected in Table 2 the steady state ER torque value, for a given motor speed and fluid temperature, increases substantially with increase in the step voltage level. However, for a constant temperature and applied voltage, as the speed of the rotating member of the rheometer increases, thus producing more shear, the output torque decreases somewhat (see Table 1). This characteristic has also been verified in the flow mode arrangement[2] where electro or yield pressure decreases with increase in the fluid flow rate through the valve. Similar results were obtained from a catch device[9]. The temperature seems to have an opposite impact on the steady state torque output. Table 1 demonstrates that its value rises as the temperature increases especially at the higher shear rate. It is however important to point out that the sensitivity of the torque measuring system and the noise on the signal may introduce some discrepancies in the results. Thus if the change in the torque value is not substantial for a considerable change in the variables; speed, voltage level and temperature, small changes may be attributed to measurement errors. Nevertheless the span of error can be seen to be restricted - figure 3 shows three measurements superposed, i.e. as an indication of repeatability.

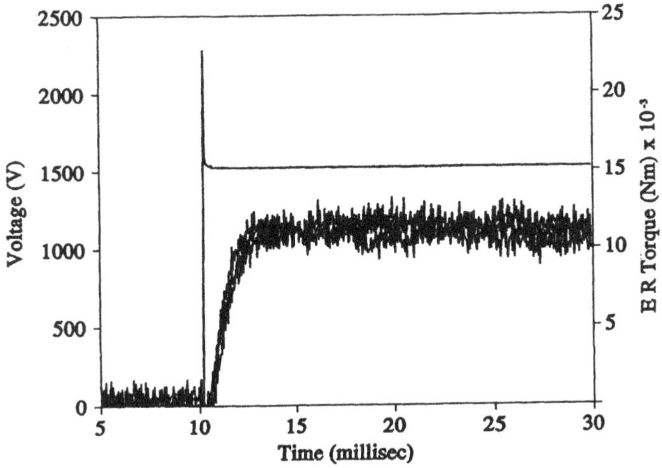

Fig. 3. Repeatability tests, results of three tests superimposed. Motor speed = 1000 rpm, Fluid temperature = 18°C

The response time of the fluid, as it is defined here (i.e. total time taken from when the step voltage is applied to the onset of steady state level), takes an entirely different path. Table 2 demonstrates that this period reduces as the voltage excitation level is enhanced. Similarly, as may be perceived from Table 1, increases in the shear rate and fluid temperature lead to a substantial reduction in the response period. For example, an increase of 19°C in the fluid temperature has halved the response time for a speed of 250 rpm and has reduced this to about a third for a speed of 1000 rpm.

These are all important results since a main stream of research on the application of ER fluids is directed towards high speed computer controlled machines. If required, maintaining a high temperature would not be a difficult task. The only drawback, with regard to high temperature operation, is that more current is drawn by the fluid thus increasing the power consumption. However, the ER torque has also increased although not at the same rate.

Some typical graphical representation of the results discussed above are shown in Figures 4 and 5. Figure 4 displays the effect of temperature on the response time and steady state torque value for a motor speed of 1000 rpm and step voltage of 1500 volts. Figure 5, on the other hand reveals the influence of speed at a constant temperature of 42°C and step voltage of 1500 volts. It is worth pointing out that the noise on the voltage signal was only observed at high temperatures and is due to poor smoothing (by the power supply) when delivering power at high current.

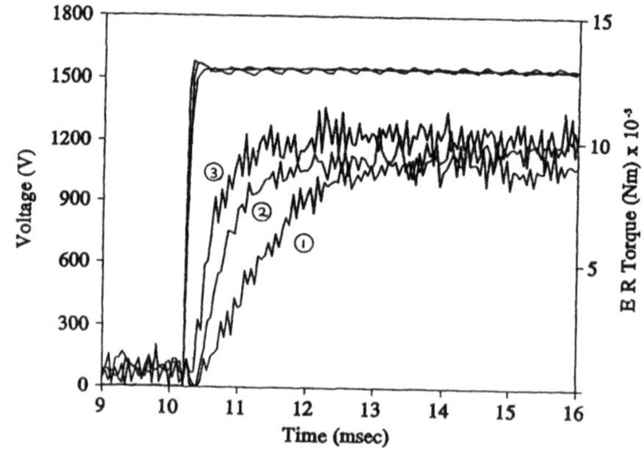

Fig. 4. Effect of fluid temperature on the response time and steady state torque value.
Motor speed = 1000 rpm. 1-23°C, 2-32°C, 3-42°C.

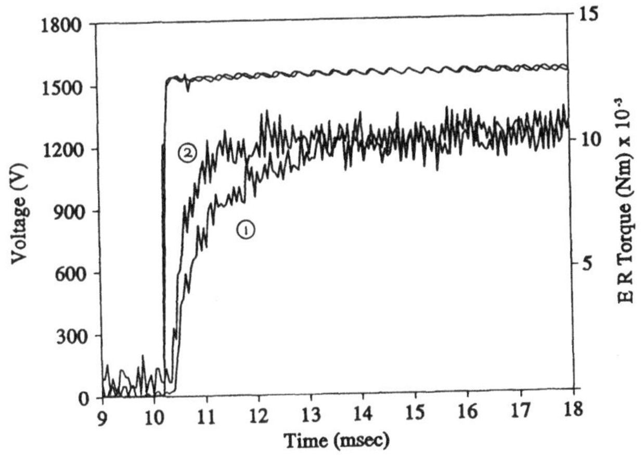

Fig. 5. Effect of speed on the response time and steady state torque value.
Fluid temperature = 42°C. 1-250 rpm, 2-1000 rpm.

Another objective of the present work has been to compare the step test with those taken on the same fluid but in a valve arrangement[2] and a clutch device[1]. Figure 6 demonstrates a typical result of such a comparison. In the clutch device, a low stiffness torque transducer was used for torque measurements thus giving rise to a system with a low natural frequency of about 110 Hz. This obscured the true response of the ER fluid, therefore the recorded data had to undergo significant processing due to the presence of dynamic oscillation on the torque meter shaft.

Although the results shown in Figure 6 have been obtained under different operating conditions, they possess nearly all of the same features. That is the response apparently constitutes of three distinct parts: a zero effect initial delay, a rapid first rise and a slower second rise. With regard to the effect of fluid temperature, no information is available from the clutch and valve arrangements. However, the rheometer results indicated that the duration of each period decreases with increase in temperature for the same bias DC voltage and fluid shear rate. For instance, for 1500 volts and a speed of 1000 rpm at 23°C durations of 0.38, 0.77 and 1.24 milliseconds were noted for the delay, first rise and second rise respectively. Whereas for the same conditions but at 42°C these durations were 0.13, 0.3 and 0.55 milliseconds. The only distinct difference between the various graphs in Figure 6 is that the flow mode results (i.e. Figure 6c) shows some kind of overshoot which is not present in the other two cases.

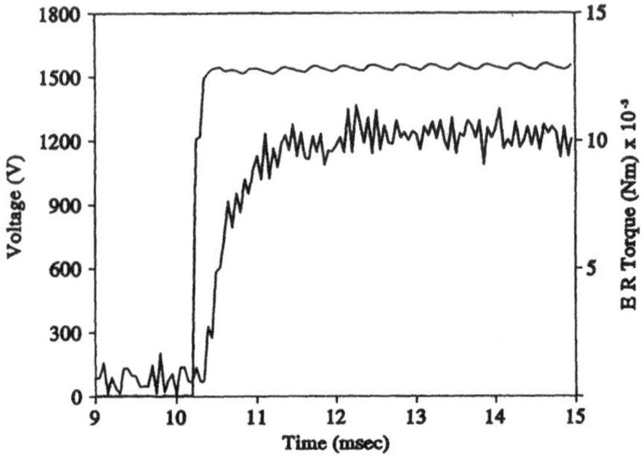

Comparison of the fluid response time from the rheometer, a clutch device and a valve arrangement.

Fig. 6 (a) Raw data taken from the present rheometer (1000 rpm, 42°C).

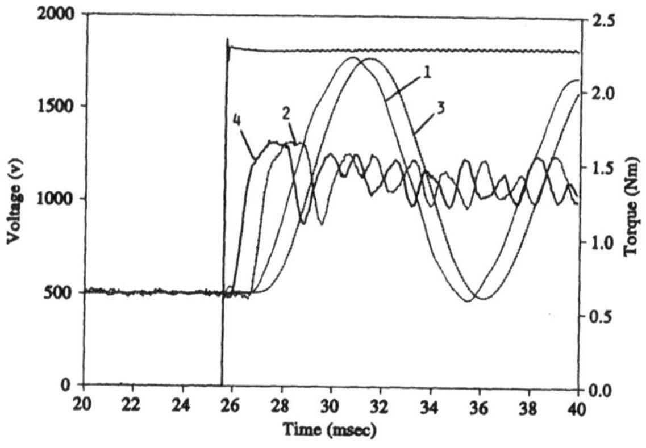

Comparison of the fluid response time from the rheometer, a clutch device and a valve arrangement.

Fig. 6 (b) Results of transformed data taken from a high speed clutch device,

1. Measured signal.
2. First estimate.
3. Predicted response for the first estimate.
4. Final estimate of torque response.

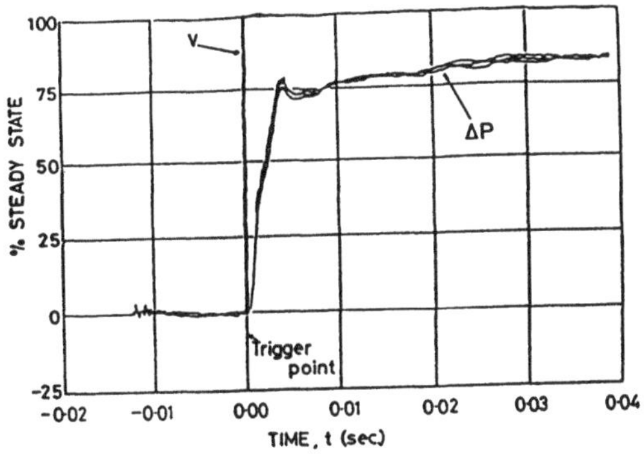

Comparison of the fluid response time from the rheometer, a clutch device and a valve arrangement.

Fig. 6(c) Flow mode (valve) results.

At this stage no quantitative explanation can be put forward for this multi part response behaviour. Further research is required on the effect of fluid composition and the electrodynamic basis of the electro-rheological effect before optimal response can be obtained.

Figure 7 exhibits the outcome of the last set of step trials in which a near step in voltage was applied across the rheometer electrodes. In these tests, the rate of imposition of the voltage was altered to monitor its influence on the final torque level. As can be discerned, the initial delay and the subsequent rate at which the torque was developed varies with the rate of voltage application but the eventual torque level is the same in each case studied. The example shown is for the case of 250 rpm, 1500 volts. Similar results were achieved for various other speeds and voltage levels. The case for rapid voltage applications is obvious.

4.2 Frequency Response

One way of efficiently presenting the frequency response readings is the Bode plot. This is composed of two plots usually drawn on a semi-logarithmic graph paper: an amplitude ratio or gain plot and a phase plot, both on a linear scale, versus the excitation frequency plotted on a logarithmic scale. The amplitude ratio is normally plotted in decibels, that is 20 log(M). Where M is the ratio of the output amplitude (in this case torque) to the input amplitude (in this case voltage) and the log is to base 10. For the results presented here, it should be pointed out that M at every frequency was first normalised with respect to the M from the lowest frequency tested.

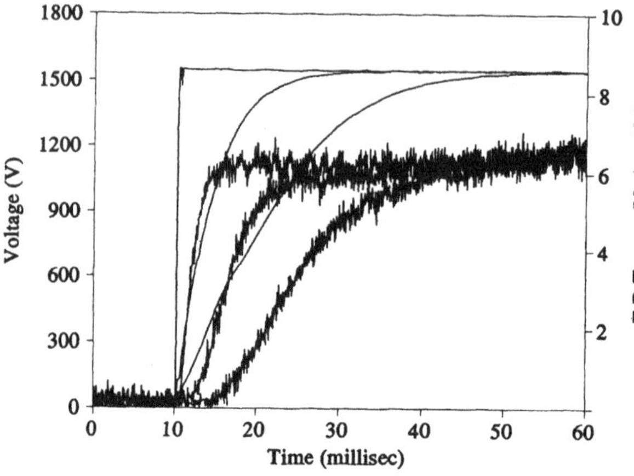

Fig. 7. Effect of slew rate of voltage on the steady state ER torque level.
Motor speed = 250 rpm
Fluid temperature = 20° C.

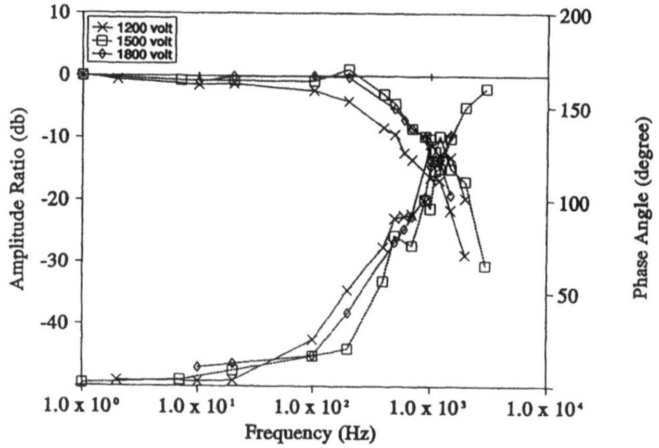

Fig. 8. Influence of the DC bias excitation voltage level on the frequency response.
Rotational speed = 250 rpm
Fluid temperature = 23° C.

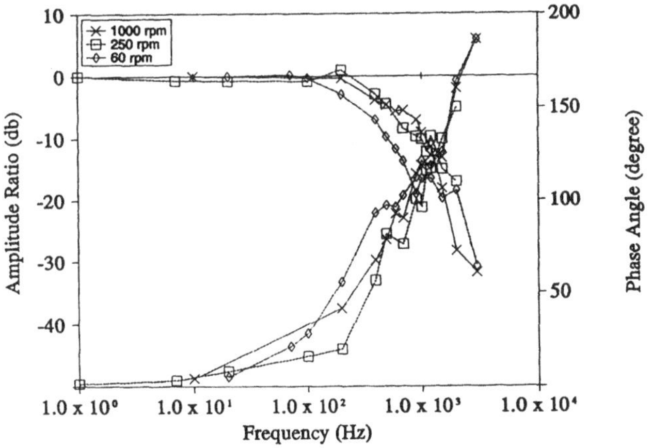

Fig. 9. Influence of the shear rate on the frequency response.
Bias DC amplitude = 1500 volts.
Fluid temperature = 23°C.

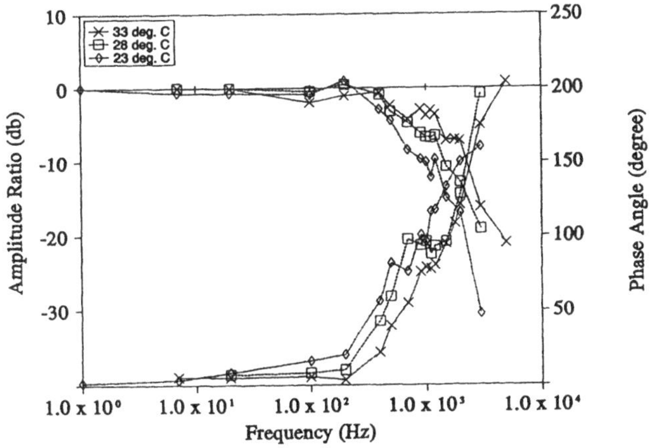

Fig. 10 Influence of fluid temperature on frequency response.
Bias DC level = 1500 volts.
Rotational speed = 250 rpm.

As with the step response tests, effects of the speed of the rotating member of the rheometer and the DC amplitude of its voltage relative to the fixed member and also fluid temperature influence were investigated. Results are displayed in Figures 8, 9 and 10.

Figure 8 shows how the bias DC voltage amplitude influences the fluid performance. The motor speed in these tests was 250 rpm and three voltages of 1200, 1500 and 1800 volts were tried at 23°C. Experimental results suggest that a complex model is required to describe the fluid characteristics. However, if a first order linearised approximation is assumed, the parameter of interest is the break point or corner frequency, that is, the frequency at which the amplitude ratio drops by 3 dB and from which the system time constant is calculated. This point also coincides with a phase angle of 45°.

With reference to the definition of corner frequency, Figure 8 shows that the fluid performance improves as the bias DC level is increased. In fact for a mean level of 1200 volts this frequency is about 160 Hz whereas for higher voltages this figure is between 300 and 400 Hz. These conclusions are also consistent with step test results presented in Table 2 in which it was shown that the response time decreased by raising the DC excitation level.

Similar graphs demonstrating the influence of rotational speed and fluid temperature are shown in figures 9 and 10 respectively. In both cases the bias DC voltage was 1500 volts. Figure 9 shows that for higher rotational speed (i.e. higher shear rate) the fluid performs better and that the corner frequency increases from about 150 Hz at 60 rpm to around 300 Hz for 250 rpm and over. Figure 9, on the other hand shows most importantly that increasing the temperature of the fluid (for constant DC voltage level and rotational speed) by about 9°C has doubled the corner frequency from 300 Hz at 23°C to about 600 Hz at 32°C. Results in both figures (i.e. 9 and 10) are also compatible with the step response results (see Table 1) in that both augmentations in shear rate and temperature boost the fluid functioning.

5. CONCLUSIONS

Results have been presented from step and biased harmonic voltage excitation tests carried out on an ER fluid in a high frequency Couette type of rheometer arrangement. Effects of temperature, shear rate and the bias DC voltage amplitude on the fluid performance have been studied. In general, an increase in each of these variables can significantly enhance the time response characteristics of the fluid and time constant of less than one millisecond is readily achieved. This has important implications on say the operation of an industrial clutch.

The acquired multi part time response compares well with transformed results of readings taken from a low stiffness torque transducer in a clutch system, thus adding confidence to that indirect process. Both the frequency response and step test results suggest that the fluid may not be modelled as a simple (linearized) first or second order system but possibly as a combination of several modes of these two and that further research is necessary to identify a linear/non-linear mathematical model which describes the response characteristics of the fluid over the operating range of interest.

6. ACKNOWLEDGEMENTS

The authors wish to express their thanks to the University of Sheffield for funding this work under the Research Fund Scheme.

7. REFERENCES

1. R. Firoozian, et al. *Dynamic Torque Response of an Electro-Rheological Clutch*, A.S.M.E. Publication, DE-Vol.**35**, Miami, pp. 295-301.

2. W. A. Bullough and D. J. Peel, *The Field-Controlled Liquid State, : A Conceptual Means for the Implementation of Advanced Mechatronics Systems*, Proc. I.Mech.E. Conference on Mechatronics - Designing Intelligent Machines, Cambridge, September 1990, pp. 171-177.

3. D. J. Peel, R. Firoozian and W. A. Bullough, *Derivation of the Electrical Characteristics of an Electro-Rheological Fluid from Biased Sine Wave Excitation Tests*, Proc. 8th Intl. Fluid Power Symposium, NEC Birmingham, 1988, pp. 527-545.

4. J. E. Stangroom, *The Measurement of the Speed of Response of ER Fluids*, Proceedings First International Symposium on Electro-Rheological Fluids, Boston, Mass., 1987, pp. 81-98.

5. R. V. Jones and J. C. S. Richards, *Recording Optical Lever*, Journal of Scientific Instruments, **36**, 1959, pp. 90-94.

6. J. E. Stangroom, *Improvements in or Relating to Electric Field Responsive Fluids*, UK Patent 1,501,635, published 22 Feb. 1978.

7. A. Inoue, *Study of New Electro-Rheological Fluid*, Proceedings Second International Conference on Electro-Rheological Fluids. Raleigh NC, 1989, pp. 176-183.

8. H. Block, *Electro-Rheology*, **21**, Chemtech, American Chemical Society - in press.

9. A. Hosseini-Sianaki, et. al., *Operational Considerations in the Use of an Electro-Rheological Clutch Device*, Proc. China Fluid Power Soc., 1st Int. Symp. Fluid Power Trans. and Control, Beijing, 1991, pp. 591-595.

PERFORMANCE CHARACTERIZATION OF ER FLUIDS: DURABILITY

F.P. BOYLE
RheoActive, Inc., 35585 Curtis Blvd., Eastlake, OH 44095 USA

ABSTRACT

ER fluids are specialty chemicals with performance characteristics that are being developed for commercial applications. Until recently, the only ER fluid performance characteristics generally discussed in the technical and popular literature was maximum yield stress. Now researchers understand that zero field viscosity, power density required to achieve the stated yield stress, and response time of the fluid are also important performance characteristics. In particular, documentation of fluid performance as a function of temperature is now recognized as being essential for proper design of an ER device. For long term operation, however, the fluid's durability must also be considered. This paper presents data that show the changes in performance characteristics of two ER fluids as a function of time in an ER shock absorber durability test. The prototype shock absorber used in the test is designed for an automobile with an electronically-controlled suspension system, and the test conditions are typical for commercial shock absorbers. The data demonstrate that realistic damping forces can be consistently achieved over the life of the test using appropriate ER fluids.

Introduction

Electrorheological (ER) fluids are electric field sensitive materials[1] that have significant commercial application.[2] To achieve that potential, the performance characteristics of individual fluids must be tailored for specific applications, and those performance characteristics must not change significantly over the designed life of the device.

As is well known, the shear stress versus shear rate performance of an ER fluid varies as a function of applied electric field. To a first approximation, ER fluids can be considered Newtonian fluids when zero field is applied, and Bingham plastics when non-zero electric field is applied. The shear stress for this approximation is given by:

$$\tau = \dot{\theta}\eta + \tau_{ER} ,$$

where $\dot{\theta}$ is the shear rate applied to the fluid, η is the temperature dependent viscosity and τ_{ER} is the electric field dependent ER yield stress. For a fluid described by this model, the total shear stress is the sum of the viscous shear stress and the field dependent ER yield stress.

Actual ER fluids do not perfectly fit this model. In practice, ER fluids have τ_{ER} that is dependent on shear rate, temperature, and flow-geometry and type in addition to being dependent on the applied electric field. Figure 1a shows the shear stress versus shear rate performance of an ER fluid measured in steady-state Couette-flow geometry, and Figure 1b shows the shear stress versus shear rate performance of the same fluid measured in oscillating Poiseuille-flow geometry. Couette flow is characterized by the fluid being sheared between two essentially parallel plates moving relative to each other, and Poiseuille flow (sometimes referred to as duct flow) is characterized by shear due to fluid moving relative to parallel plates that are stationary relative to each other. The zero-field viscosity measured by both methods is aproximately the same. The τ_{ER} values measured by the two methods as a function of electric field, however, vary significantly. For the purposes of this paper, only the Poiseuille-flow performance of the ER fluids is considered and the Bingham model is used to approximate the performance of the fluid.

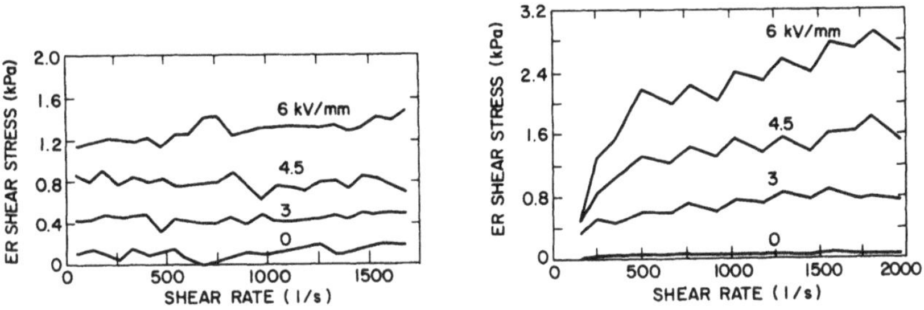

Figure 1: a) Room temperature ER shear stress versus shear rate performance of a typical ER fluid measured in a Couette-flow test stand.

b) Room temperature ER shear stress versus shear rate performance measured in an oscillating Poiseuille-flow test stand for the same fluid in Fig. 1a.

Force transfer in a fluid-coupled device in laminar flow is a function of the shear stress of the fluid. When the device is in the "off" state, zero applied electric field, the force transfer is proportional to the fluid viscosity ($F \propto \eta$). When the device is in the "on" state, non-zero applied electric field, the ER stress is normally much greater than the viscous stress; hence, the force transfer is approximately proportional to the ER yield stress ($F \propto \tau_{ER}$). This implies that the "on/off" ratio of the device for fixed geometry and shear rate, is proportional to the τ_{ER}/η ratio.

If the device design is optimized for a particular "on/off" ratio the electrode separation and surface area are selected in such a manner that the volume of the fluid between the electrodes is minimized. The volume of fluid between the electrodes in the optimized design, within manufacturing constraints, can be shown to be approximately proportional to η/τ_{ER}^2.[3]

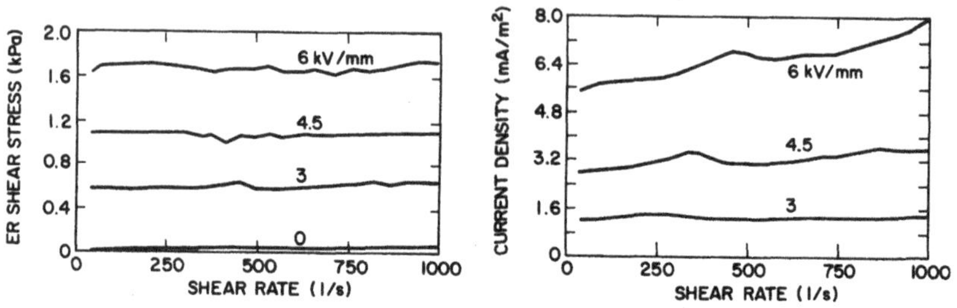

Figure 2: a) Room temperature ER shear stress versus shear rate performance of a typical ER fluid.

b) Current density versus shear rate measured during the test in Fig. 2a.

ER fluids conduct current when an electric field is applied. Figure 2a shows the shear stress versus shear rate characteristics of a typical ER fluid and Figure 2b shows the attendant DC room temperature current density versus shear rate characteristics measured during test. An ER device is essentially one or more pairs of parallel plates that are separated by some distance "h" and have opposed surface area "A". In DC operation, to turn the device "on", an electrical field "E" with a resulting current density "j" must be applied between the plates. This requires that the power supply be capable of providing voltage $V = E \bullet h$ and current $i = j \bullet A$. Since high voltage power supplies are essentially sold by the "Watt," the power required by an ER device is of particular concern. The power requirement is voltage times current which is:

$$\text{POWER} = (Ej) \bullet (hA) .$$

Defining ϕ as the fluid's power density, that is, the power required per unit volume to achieve the desired ER yield stress, then the power required to operate the device is the fluid's power density times the volume,

$$\text{POWER} = \phi\eta/\tau_{ER}^2 = W_n^{-1} .$$

The Winslow number, W_n, is a dimensionless number that is electric field, temperature and flow geometry dependent. The number is useful as a cursory means for comparing ER fluids when evaluated in a standardized test.

In addition to requiring a high Winslow number, many applications have limits on the time required for the fluid to respond to turning the electric field "on" or "off". Shown in Figure 3 is the room temperature response of two ER fluids to a pulsed DC electric field in an oscillating Poiseuille-flow test stand. Both fluids shown in the figure have the same ultimate ER shear stress value and have approximately the same Winslow number; however, one fluid responds significantly faster than the other.

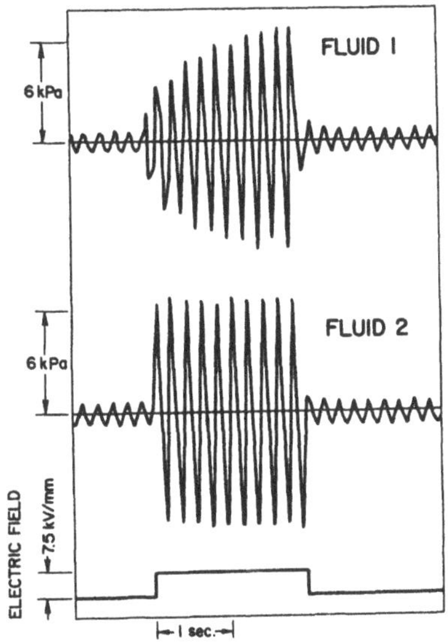

Figure 3: Room temperature response to a pulsed DC electric field for two ER fluids measured in an oscillating Poiseuille-flow stand.

Each of the performance characteristics of the ER fluids has temperature dependence. In general, zero-field viscosity and response times decrease while power density increases as the fluid temperature increases. ER shear stress can either increase or decrease as a function of temperature. In most applications, ER fluids with the least temperature dependence are preferred.

Stable performance of ER fluids under extreme operating conditions is also a desired characteristic. Any time-dependent change of the performance characteristics limits the fluid's durability. An understanding of this durability is essential to predicting the service life of an ER system. Since there is no documentation in the technical literature of durability for any of the known ER fluids, a program was initiated to investigate the time-dependent change of ER fluid performance in a simulated-service-life test.

Test Program

An ER shock absorber prototype[4] was selected as the test device for the durability investigation. The prototype is designed as a retrofit for the shock absorbers on an automobile that comes equipped with an electronically-controlled damping system. Picture 1 shows the prototype ER shock absorber and the production electro-mechanical shock absorber that the ER unit is designed to match. To achieve the desired damping performance, the ER prototype requires an fluid with a room temperature Winslow number of greater than 5000 when tested in RheoActive's standard oscillating Poiseuille-flow test stand. In particular a room temperature ER shear stress greater than 4 kPa, and a room temperature viscosity less than 75 mPa-s is required. Also a reltively stable performance over a wide temperature range is preferred. In the ultimate application of an active damper, a fast response time is required. In the initial application, however, response time is not as critical since the commercial shock absorber and controller are relatively slow.

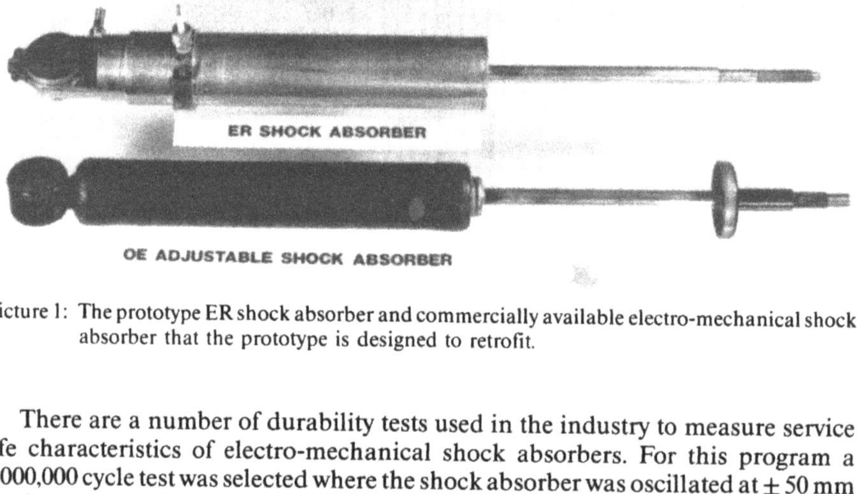

Picture 1: The prototype ER shock absorber and commercially available electro-mechanical shock absorber that the prototype is designed to retrofit.

There are a number of durability tests used in the industry to measure service life characteristics of electro-mechanical shock absorbers. For this program a 2,000,000 cycle test was selected where the shock absorber was oscillated at ± 50 mm stroke and 1.5 Hz frequency, and was maintained at 80°C. The shock absorber alternated between five cycles at high damping and five cycles at low damping throughout the test. For the ER prototype, low damping occurred when zero electric field was applied to the ER fluid, and high damping occurred when an electric field of the appropriate magnitude was applied to the fluid.

Test Fluids

Two ER fluids, which will be referred to as fluid "A" and fluid "B", were selected for the test based on the prototype shock absorber requirements. Both fluids were tested at 20°C increments in a standard RheoActive test at 6 kV/mm before the start of the durability test. Fluid "A" was tested from 0 to 100°C, and fluid "B" was tested from 0 to 120°C. Power supply limitations of the test stand prevented fluid "A" from being tested at 120°C.

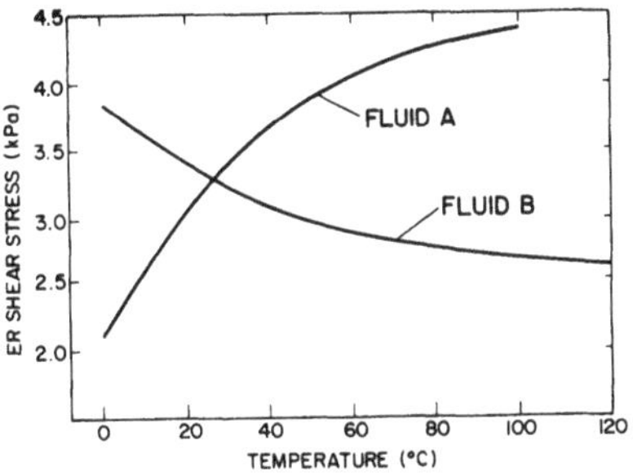

Figure 4: ER shear stress at 6 kV/mm versus temperature for fluids "A" and "B" measured in a standard oscillating Poiseuille-flow test stand.

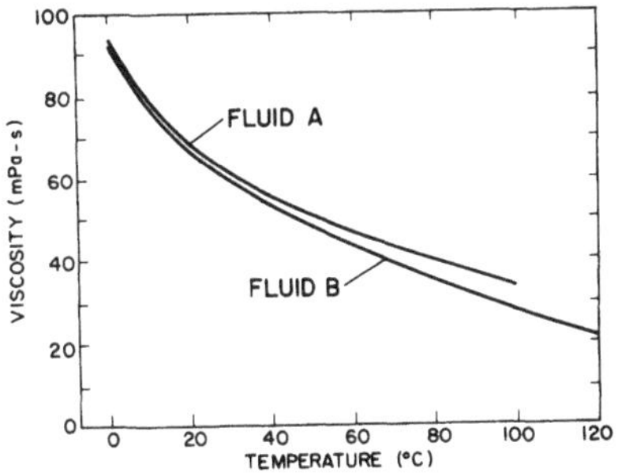

Figure 5: Zero-field viscosity versus temperature for fluids "A" and "B" measured in a standard oscillating Poiseuille-flow test stand.

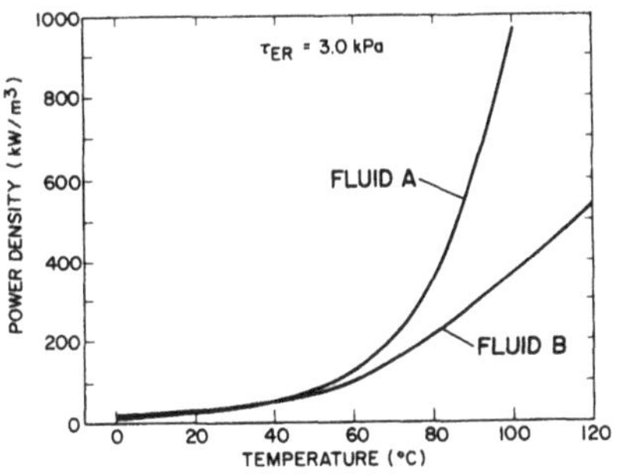

Figure 6: Power density required to achieve an ER shear stress of 3.0 kPa versus temperature for fluids "A" and "B" measured in a standard oscillating Poiseuille-flow test stand.

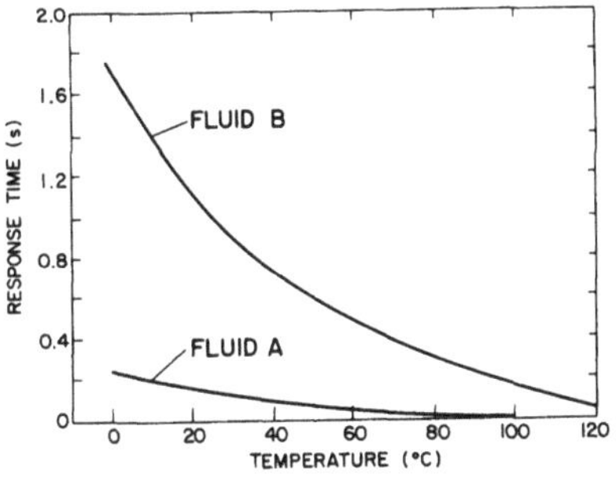

Figure 7: "Turn-on" response time to achieve 90% of the ER shear stress shown in Fig. 4 versus temperature for fluids "A" and "B" measured in a standard oscillating Poiseuille-flow test stand.

Figures 4, 5, 6 and 7 show the ER shear stress, zero field viscosity, power density, and "turn on" response time respectively for fluids "A" and "B". The ER shear stress increases with temperature for fluid "A", and decreases with temperature for fluid "B". Fluids "A" and "B" have similar zero-field viscosity versus temperature characteristics. Fluid "A" has power density that increases very rapidly as a function of temperature when compared to fluid "B". At low temperature, fluid "B" responds very slowly. Even at 120°C, fluid "B" would not be considered a fast fluid. Within the time resolution of the standard test, response time of fluid "A" at 100°C is less than 10 ms. For comparison, Figure 8 shows the 6 kV/mm Winslow number for both fluids in the standard test as a function of temperature. Fluid "B" is the better fluid at the temperature extremes, and fluid "A" is better in the mid-temperature range.

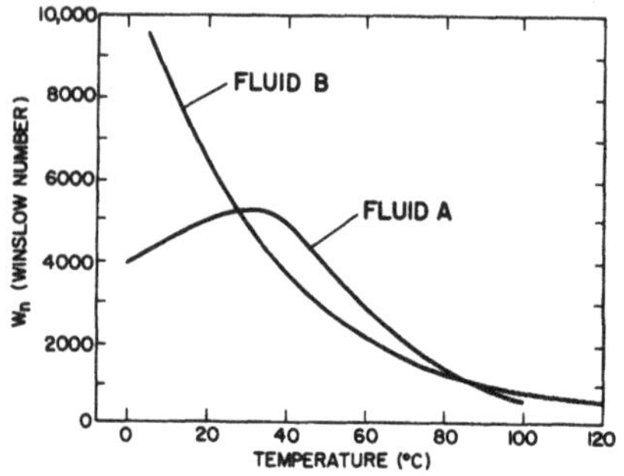

Figure 8: Winslow number at 6 kV/mm versus temperature for fluids "A" and "B" measured in a standard oscillating Poiseuille-flow test stand.

Test Results

The durability test for each fluid required approximately 370 hours to complete. The shock absorber performance was automatically recorded at fixed intervals during the test. The high damping for the tests was achieved with an electric field of 8 kV/mm. Figures 9, 10, 11 and 12 show the performance of the fluids in the shock absorber as a functon of time at the 80°C operating temperature. The initial ER fluid performance characteristics measured in the shock absorber do not reproduce exactly the values measured in the standard test at 80°C due to differences in the magnitude of the electric field used in the two tests, and due to flow-path geometry differences between the test cell and the prototype shock absorber.

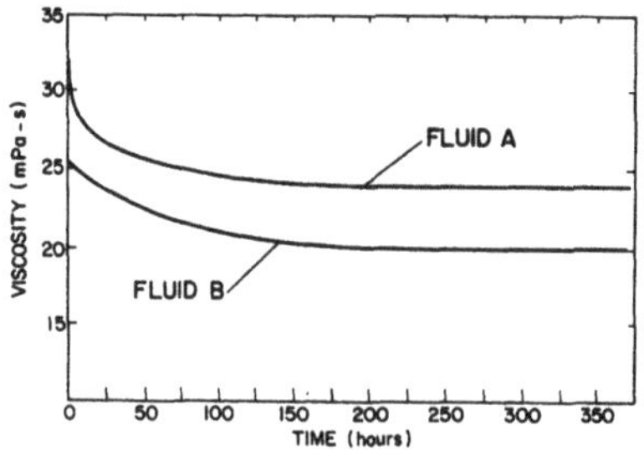

Figure 9: ER shear stress at 8 kV/mm and 80°C versus durability-test-time for fluids "A" and "B" tested in the prototype ER shock absorber.

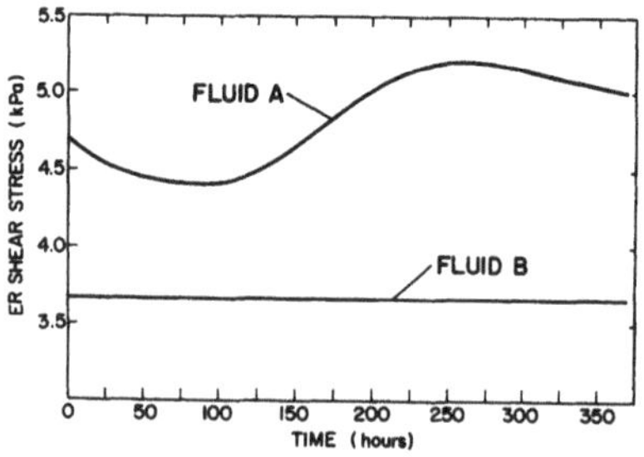

Figure 10: Zero-field viscosity at 80°C versus durability-test-time for fluids "A" and "B" tested in the prototype ER shock absorber.

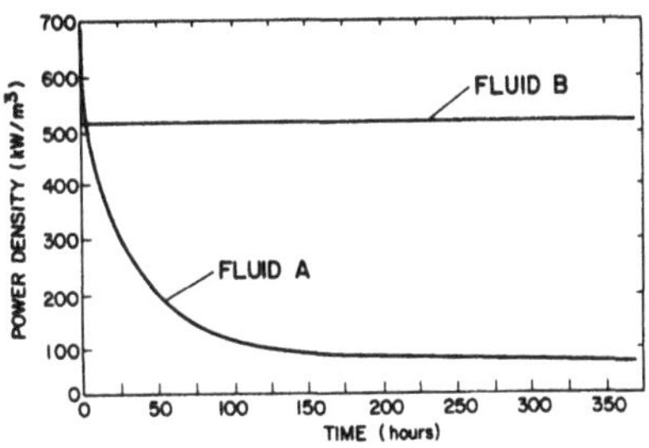

Figure 11: Power density versus durability-test-time for fluids "A" and "B" to achieve the ER shear stress shown in Fig. 10.

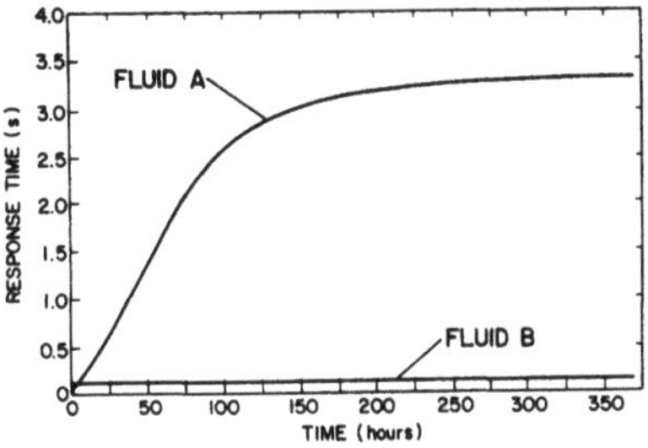

Figure 12: "Turn-on" response time to achieve 90% of the ER shear stress shown in Fig. 9 versus durability-test-time for fluids "A" and "B" tested in the prototype ER shock absorber.

After the start of the test, fluid "A" showed a rapid decrease in viscosity and current density. Fluid "A"'s ER shear stress initially decreased as a function of time and then increased. The ER shear stress decrease after 275 hours is within the error limits of the test and may not be significant.

The only change that fluid B showed during the test was a decrease in the zero field viscosity. Even that change was relatively gradual. Figure 13 shows the 80°C shock absorber damping as a function of time for fluid "B" at the start of the test and after 2,000,000 cycles. The curves are essentially indistinguishable.

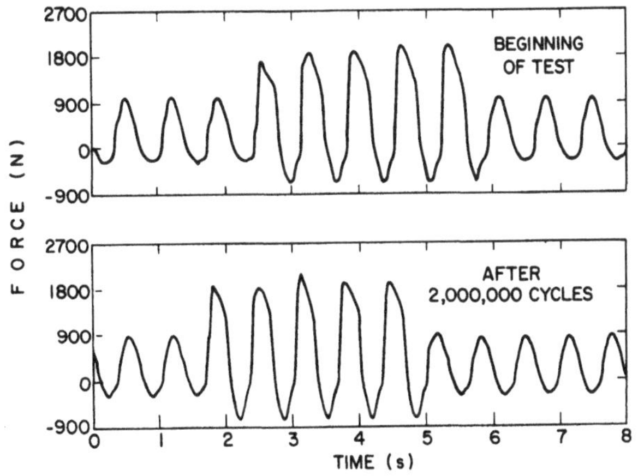

Figure 13: Damping force versus time for the ER shock absorber at 80°C with fluid "B" at the beginning and end of the 2,000,000 cycle durability test.

At the end of the durability test, each ER fluid was taken from the shock absorber and evaluated in the standard test at room temperature. The initial and final room temperature performance characteristics are listed in Table 1.

	FLUID A		FLUID B	
	BEFORE	AFTER	BEFORE	AFTER
ER Shear Stress (kPa)	3.2	1.2	3.3	3.3
Zero-Field Viscosity (mPa-s)	65	55	64	50
Power Density (kW/m³)	30	20	30	30
"Turn-on" Response Time (s)	0.2	< 0.05	1.2	1.0
Winslow Number	5,200	1,300	5,700	7,300

Table 1: Room temperature performance of fluids "A" and "B" measured in a standard oscillating Poiseuille-flow test stand before and after the 2,000,000 cycle durability test.

Discussion

Fluid "A" showed significant performance change during the durability test. While the ER shear stress and power density remained within acceptable limits during the test, the dramatic increase in response time at 80°C was unacceptable. The room temperature data measured in the standard test (Table 1) show that the final ER fluid was unacceptable as evidence by the low Winslow number. An interesting comparison is the final performance of fluid "A" as measured in the 80°C shock absorber and the room temperature standard test. In the shock absorber, the fluid still had a high ER shear stress, but was slower at the end of the test. In the standard test the fluid had lower ER shear stress, but appeared faster. Actually, the faster response time is somewhat misleading since in both the "before" and "after" tests the fluid reached 1.2 kPa shear stress within 50 ms. In the "before" test, however, the ER shear stress continued to rise until the 3.2 kPa value was reached. What these results may imply is that fluid "A" has two or more mechanisms that contribute to its ER effect. Each of the mechanisms may be temperature dependent, and may improve or degrade independently as a function of time. Hence, the mechanism that resulted in the fast 1.2 kPa ER shear stress may not have changed during the test, while the mechanisms that resulted in the slower, higher ER shear stress changed significantly. Nonetheless, fluid "A" would not be an ideal fluid for a shock absorber application due to the changes that did occur during the test.

The changes in fluid "B" during the test were insignificant to the performance of the prototype shock absorber. That is not to imply that fluid "B" is an ideal fluid for the application since the response time characteristics are less than ideal. This fluid does show, however, that ER fluids can provide adequate performance during a durability test simulating the service life of an actual application.

Conclusion

In a durability test using a prototype ER shock absorber that is designed as a replacement for a production electro-mechanical shock absorber, results demonstrate that an appropriate ER fluid can be used to consistently achieve realistic damping forces over the life of the test. This implies that the durability of ER fluids need not be the limiting factor in the service life of an ER shock absorber.

Acknowledgement

The ER fluids used in this study were provided by The Lubrizol Corporation.

References

1. W.M. Winslow, *J. Appl. Phys.* 20: 1137 (1949).

2. D.A. Brooks, *Electrorheological Fluids: Proceedings of the Second International Conference on ER Fluids*, (Technomic Publishing Co., Inc., 1990) p. 371.

3. T.G. Duclos, *Design of Devices Using Electrorheological Fluids*, Society of Automotive Engineers, Paper No. 881134 (1988).

4. N.K. Petek, *An Electronically Controlled Shock Absorber Using Electrorheological Fluid*, Society of Automotive Engineers, (To Be Published).

Linear/Non-linear Mechanical Properties of Electrorheological Materials

Daniel R. Gamota
Department of Materials Science and Engineering
The University of Michigan, Ann Arbor, Michigan 48109

and

Frank E. Filisko
Department of Materials Science and Engineering
The University of Michigan, Ann Arbor, Michigan 48109

ABSTRACT

The stress response of an electrorheological (ER) material is modified by the application of an electric field. Various studies have shown that the ER material can behave as a linear viscous, linear viscoelastic, nonlinear viscoelastic, plastic, or viscoelastic-plastic body. Furthermore, several different experimental techniques are conducted to observe the ER material's behavior as a function of strain, strain frequency, field strength, and ER material concentration. Small amplitude dynamic studies are used to observe the ER material's linear viscoelastic properties, while moderate and large amplitude studies are used to observe the material's fundamental nonlinear dynamic properties. Finally, constant shear rate experiments are performed to observe the apparent viscosity of the ER material during flow conditions.

I. Introduction

The rheological properties of electrorheological (ER) materials can be modified by the application of an electric field. Furthermore, the response of the materials can be discussed in terms of three rheological regimes: a pre-yield, a yield, and a post-yield regime.

The behavior of ER materials under the influence of an electric field is commonly discussed in terms of the ideal Bingham fluid model. An illustration of the ideal shear stress - shear rate behavior for an ER material is illustrated in Figure 1. It has been suggested that these materials behave as fluids (as opposed to greases or gels) under zero field conditions when subjected to unidirectional shearing of constant rate, but under similar flow conditions, the shear stress values increase when the material is subjected to a nonzero electric field. Also, under a nonzero field, the materials behave as solids up

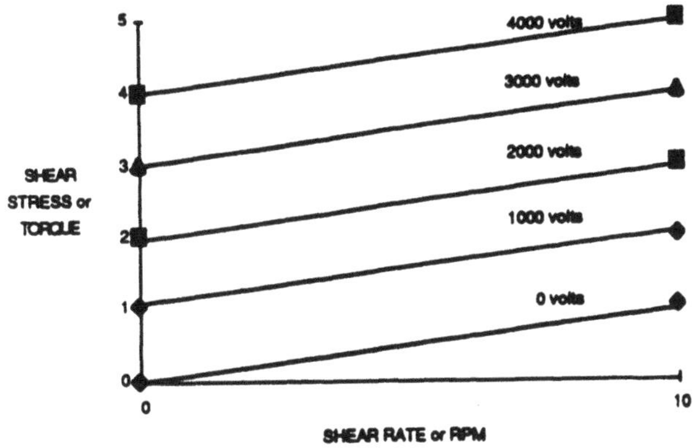

Figure 1: Illustration of shear stress versus shear rate data for an ER material at various electric field strengths (Ideal Bingham behavior).

to a certain critical shear stress (S_c) and liquids at stresses above S_c. The critical shear stress dependence is commonly observed as being proportional to the field squared (Uejima, 1972) but many other types of behavior are observed (Block and Kelly, 1988). While the Bingham model has been frequently applied to ER materials, it completely overlooks the properties of the material at stresses less than S_C.

A more complete description of the behavior of these materials may be illustrated by a plot of stress versus strain as illustrated in

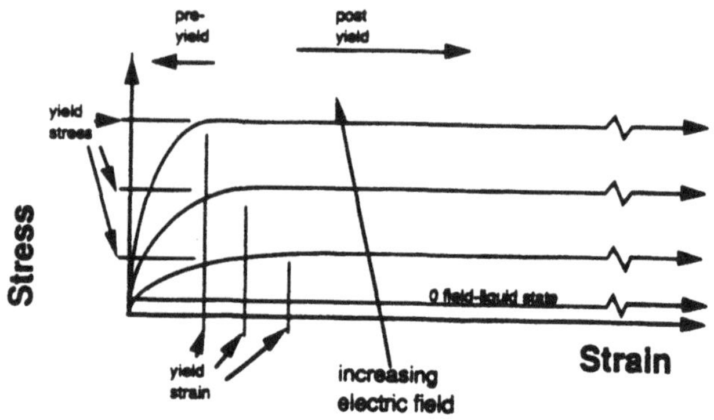

Figure 2: Illustration of shear stress versus shear strain properties of a typical ER material under zero electric field (i.e. liquid state) and under increasing electric field strengths.

Figure 2. The materials under a nonzero field can be described as viscoelastic solids below a certain yield stress and yield strain and as viscous liquids for stresses at or above the yield stress or strains greater than the yield strain. When characterized in this fashion, the materials may be described as viscoelastic-plastic materials whose yield stress and yield strain are strong functions of electric field. Furthermore, the yield stress is highly variable and can be strongly dependent on numerous factors, including the ER fluid composition.

The pre-yield region is observed experimentally by conducting studies using (a) constant stress rheometers and applying stresses below the "critical stress", (b) constant strain rate rheometers and applying low strain rates while maintaining a strain lower than the "critical strain", and (c) dynamic mechanical spectrometers and applying a sinusoidal strain of low amplitude. Studies conducted within this region have suggested that ER materials behave as linear viscoelastic solids whose loss modulus, storage modulus, and phase angle are functions of the applied field strength.

The yield region is observed experimentally during sinusoidal straining experiments when the stress response deviates from sinusoidal behavior or in the transient portions of steady state shear tests as the material begins to flow. The ER material's stress response may be analyzed using Fourier analysis when it deviates from sinusoidal behavior and the fundamental loss modulus, fundamental storage modulus, and fundamental phase angle can be calculated. In addition, shear stress - shear strain loops (hysteresis loops) may be generated for the material and the energy dissipated by the material per volume, per cycle may be calculated.

The post-yield region is observed experimentally when the ER material is sheared at a constant rate to induce flow. Studies concerning this region are abundant in the literature and generally the experimental results suggest that the apparent viscosity of the ER material increases with applied field strength raised to the second power. Moreover, post-yield region studies are often performed using constant shear rate viscometers and subjecting an ER material to flow conditions.

II. Pre-Yield Region

Dynamic studies conducted with sufficiently small sinusoidal shear strains suggest that ER materials deform as viscoelastic bodies (Brooks et al. 1986; Gamota and Filisko, 1991; Shulman et al., 1989; Vinogradov et al., 1986). Moreover, it is generally accepted that under a dynamic stimulus of low amplitude the structure formed in response to the electric field is virtually undisturbed. These studies also report that the viscoelastic parameters, i.e. storage modulus, loss modulus, and phase angle, are functions of the field strength, ER material composition, strain frequency, and strain amplitude. The ER material behavior as related to the applied field is unique in that as the field is increased, the loss modulus increases in magnitude while the tangent of the phase angle decreases. The tangent of the phase angle is equal to the ratio of the loss modulus to the storage modulus and thus a decreasing phase angle would suggest that the material is becoming more elastic in nature. However, because the stress amplitude is increasing with field simultaneously as the phase angle is decreasing, an increasing loss modulus is recorded.

A theoretical study conducted to correlate the viscoelastic material parameters, i.e. elastic shear modulus and dynamic viscosity, to the particle volume fraction and dipole strength was reported by Adriani and Gast (1988). They found that both parameters vary with ER material particle volume fraction and that the elastic modulus shows a strong dependence on the dipole strength while the dynamic viscosity appears to be insensitive to the dipole strength.

Evaluation of some experimental dynamic studies suggests that the dependence of the loss modulus and storage modulus on the field strength is ER material specific. Vinogradov et al. (1986) and Shulman et al. (1989) report that both the loss modulus and the storage modulus increase with field for a diatomite and transformer oil ER material subjected to sinusoidal strains in the frequency range 0.1 Hz to 1.0 Hz. They also report that the increase in moduli magnitude is dependent on the particle volume fraction, field strength, strain amplitude, and strain frequency.

A dynamic study conducted at a frequency of 190/s using an ER material composed of lithium poly(methacrylate) dispersed in a chlorinated hydrocarbon oil suggests that the dynamic moduli initially increase with increasing field, reach a maximum and subsequently decrease with increasing field strength (Brooks et al., 1986). Moreover, at field strengths lower than 0.2 kV/mm, the storage modulus is greater in magnitude than the loss modulus, and at field strengths greater than 0.2 kV/mm the loss modulus appears to be greater in magnitude than the storage modulus. They further state that the occurrence of a crossover point where the loss modulus is equal to the storage modulus could not be unambiguously identified.

An ER material composed of alumino-silicate powder dispersed in paraffin oil which was subjected to 300 Hz to 400 Hz sinusoidal strains appears to show a similar dependence of the viscoelastic parameters on the field strength as reported by Shulman et al. (1989). Both the storage and the loss moduli increase in magnitude with increasing field: the storage modulus is a function of the field raised to the 2.1 power while the loss modulus is a function of the field raised to the 1.1 power (Gamota and Filisko, 1991). Figure 3 presents the loss and storage modulus as a function of field for strain of frequency 300 Hz and amplitude 0.001 rad.

ER material pre-yield behavior was also observed during shear stress - shear strain experiments performed using a tensile testing configuration (Conrad et al., 1991). Conducting studies with an ER material composed of Zeolite 3125 particles and silicone oil, they observed an initial increase in shear stress with increasing strain, which can be referred to as the pre-yield or linear viscoelastic region. The slopes of the shear stress - shear strain data are dependent on the field strength, with the slopes increasing with increasing field within the pre-yield region. However, there was a stress value above which the stress became constant, i.e. independent of strain, which they refer to as a yield stress or flow

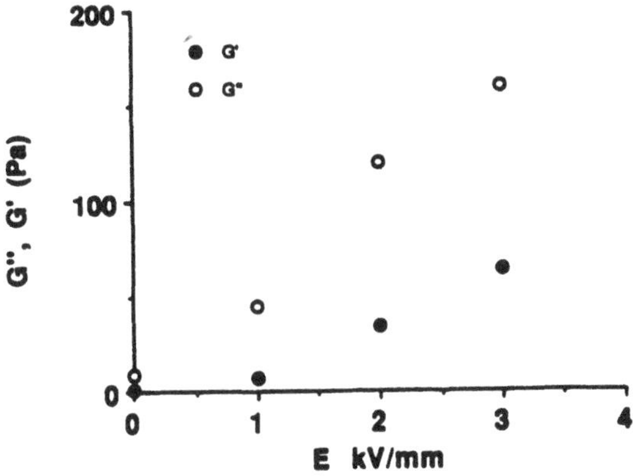

Figure 3: Storage modulus and loss modulus as a function of applied field strength (E) during sinusoidal straining of frequency 300 Hz and strain amplitude of 0.001 rad.

stress. Furthermore, it is suggested that the appearance of a yield stress is due to instabilities associated with the fibrated structure i.e. the onset of yield.

A. Modelling of ER material behavior

The rheological interpretation of the pre-yield state is important for the understanding of the properties of an ER material. Modelling of the ER material behavior in the pre-yield region was performed, and a few rheological models were studied for their experimental accuracy. Brooks et al. (1986) used the constitutive equations for a Maxwell Fluid to model an ER material. They found that the the predicted behavior of the storage modulus as a function of field strength correlated well with experimental results; however, the behavior of the loss modulus was different than the experimental results. They stated that the lack of agreement between the theoretical loss modulus data and the experimental data was possibly due to the need for the inclusion of an inter-particle force term within the loss modulus equation.

The standard linear solid model, which is a spring and dashpot in parallel placed in series with a spring, has also been suggested as a model for the ER material when its behavior is restricted to the pre-yield region (Gamota and Filisko, 1991). This model was suggested due to the observation that the yield stresses increased while the yield strains decreased when the field was increased.

A preliminary analysis of the model was performed, and it appears that there is agreement between theoretical and experimental results. Observing the boundary conditions of the loss modulus, i.e. the loss modulus becomes zero for frequencies approaching zero and infinity, a nonlinear correlator program was used to simulate the behavior of the ER material as a function of field. The computer-simulated loss modulus behavior as a function of frequency showed that the modulus initially increased at an exponential rate with increasing strain frequency, reached a maximum value at a strain frequency of about 500 Hz and then slowly decayed in magnitude as the strain frequency was increased further. In addition, it is important to state that all the mechanical elements must exhibit some field dependence since both the storage and loss moduli show field dependency.

B. ER material damping properties

The damping properties of an ER material can be studied by calculating the loss modulus when the ER material is deforming as a viscoelastic body. In addition, the energy dissipated by the material during a strain cycle can be calculated from the area enclosed within the hysteresis loop. When the ER material is deforming as a viscoelastic material, the loop will be elliptical in shape, and the energy dissipated by the ER material is related to the strain amplitude raised to the second power. These damping conditions are referred to as linear damping, and modelling of the energy dissipation mechanism is performed by the use of a viscous damper or dashpot. It is apparent by the increase in area of the elliptical hysteresis loops that with increasing field the energy dissipated by an ER material increases and that the ER material energy dissipation mechanism can be represented as a linear damper. In addition, the angle between the major axis of the ellipse and the abscissa, which is related to the elastic properties of the material, increases with increasing field.

III. Yield Region

A representative series of stress responses to a sinusoidal strain of frequency 20 Hz and strain amplitude of 0.1 for an alumino-silicate in paraffin oil ER material is presented in Figure 4. Curve B is the stress response of the ER material when it is subjected to a zero electric field. The stress response appears to be sinusoidal and 90° out of phase with the applied strain (Curve A), suggesting that the

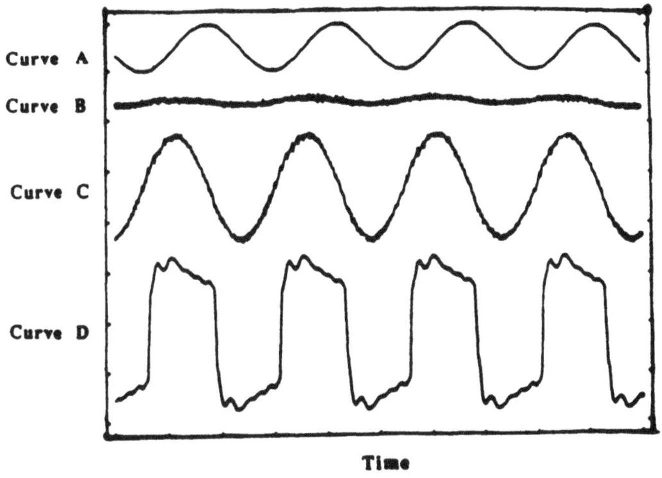

Figure 4: Stress response to a constant strain amplitude during the application of an electric field. Curve A is the applied strain. Curve B is the stress when E=0.0 kV/mm. Curve C is the stress when E=1.0 kV/mm. Curve D is the stress when E=3.0 kV/mm.

material is behaving as a viscous material. If the material is subjected to a field strength of 1.0 kV/mm, the stress appears sinusoidal and the amplitude has increased while the phase angle has decreased, suggesting that the ER material is deforming as a viscoelastic material (Curve C). If the applied field strength is increased to 3.0 kV/mm the material stress response deviates from sinusoidal behavior i.e. it exhibits non-linear viscoelastic behavior (Curve D). In addition, the amplitude of the fundamental harmonic has increased, and the fundamental phase angle has decreased. Thus, the magnitude of the fundamental harmonic is an increasing function with increasing applied field strength while the fundamental phase angle appears to be a decreasing function with increasing field strength. From these results it is clear that the ER material behavior is modified by the application of an electric field; furthermore, it is suggested that increasing the field progressively transforms the ER material from a viscous to a viscoelastic to a nonlinear viscoelastic body. (Note that the scales of the stress response signals in Fig. 4 are of different magnitudes. Curve B: a unit on the ordinate is equal to 25.0 Pa. Curve C: a unit on the ordinate is equal to 7.5 Pa. Curve D: a unit on the ordinate is equal to 50.0 Pa.)

The deviation from sinusoidal behavior suggests a transition in the mode of deformation away from viscoelastic behavior. This deformation transition may be related to the yielding of the material which is suggested by the authors of this paper to be a gross planar movement within the electrically induced structure. The yielding phenomenon is believed to be a long-range coordinated shearing of planes which is a consequence of a critical activation stress having been reached. Electrorheological materials appear to have a field-induced deformation transition which is observed when the stress response deviates from sinusoidal behavior.

Analysis of the viscous and elastic properties of the ER material during this transition may be performed by the use of Fourier analysis. Fourier analysis correlates the fundamental (first) harmonic of the imposed strain to the fundamental harmonic of the stress response, and thus allows the fundamental nonlinear dynamic properties (fundamental storage modulus, fundamental loss modulus, and fundamental phase angle) to be calculated. If the ER material is subjected to a sinusoidal strain history

$$\gamma(t) = \gamma_0 \sin(\omega t) \tag{1}$$

then the resulting stress response can be represented as a Fourier series:

$$\tau(t) = \tau_0 + \sum_{n=1}^{\infty} \tau_n \sin(n\omega t + \delta_n) \qquad (2)$$

where τ_0 is the mean stress value, and $\tau_1, \tau_2, \tau_3, \ldots, \tau_n$ are the amplitudes of the harmonics of the stress response, and $\delta_1, \delta_2, \delta_3, \ldots, \delta_n$ are the phase angles of the harmonics. If fluid inertia is negligible, the

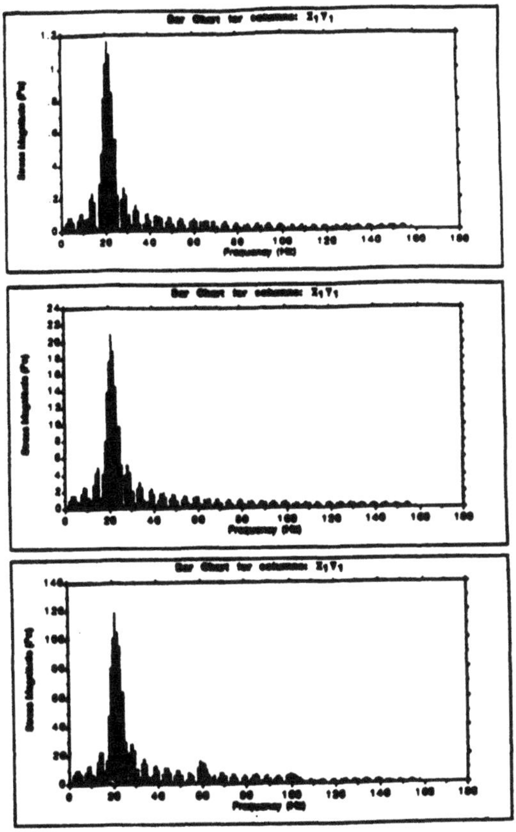

A

B

C

Figure 5: Fourier spectra of the stress response for an ER material. Spectrum A: E=0.0 kV/mm. Spectrum B: E=1.5 kV/mm. Spectrum C: E=3.0 kV/mm.

fundamental nonlinear dynamic properties may be evaluated directly from the fundamental amplitude ratio, τ_1/γ_0 and the phase angle δ_1 (MacSporan and Spiers, 1982).

A representative series of Fourier spectra of the stress response to a sinusoidal strain of frequency 20 Hz and amplitude 0.1 for an alumino-silicate in paraffin oil ER material is presented in Figure 5. The Fourier spectrum of the stress response when a zero field is applied to the material contains the fundamental harmonic, i.e. a harmonic at 20 Hz (5A). If a field of strength 1.5 kV/mm is applied to the material, the Fourier spectrum appears to contain only the fundamental harmonic; however, the magnitude of the harmonic has increased (5B). Application of a 3.0 kV/mm field causes the Fourier spectrum of the stress response to contain higher harmonics, i.e. third harmonic at 60 Hz and fifth harmonic at 100 Hz, and the magnitude of the fundamental harmonic to increase further (5C). Thus, increasing the strength of the field increases the harmonic content, as well as the magnitudes of the harmonics.

A study was conducted in which the fundamental storage and loss moduli for an alumino-silicate and paraffin oil ER material, subjected to a sinusoidal strain of 0.1 and strain frequency of 10 Hz, were calculated (Gamota and Filisko, submitted). Both fundamental moduli increase as the field strength is increased, with the fundamental storage modulus showing a stronger dependence on the electric field. Approximate relationships between the fundamental storage modulus and the fundamental loss modulus and the electric field strength are given by

$$G_1' = \text{constant} * E^{1.8} \qquad (3)$$

$$G_1'' = \text{constant} * E^{1.2} \qquad (4)$$

Conrad et al. (1991) also observed the yield region during tension experiments performed on a Zeolite 3125 and paraffin oil ER material. They observed the yield stress or flow stress, which is defined as the stress above which the shear stress became independent of the applied shear strain, during constant-rate straining. Furthermore, they report that the yield stress increases with increasing applied field strength and that a quadratic relationship between the two experimental parameters is observed.

A. Modelling of ER material behavior

A rheological mechanical model which has been suggested to explain qualitatively the observed behavior of an ER material deforming within the yield region consists of the standard linear solid model placed in series with a Coulomb element and a dashpot (Figure 6). During steady state conditions, the standard linear solid is noncontributing, and the rheology of the ER material may be modelled by the Coulomb element in series with the viscous element i.e. post-yield region. The Coulomb element is used to simulate the critical shear stress for the onset of flow, while the viscous element

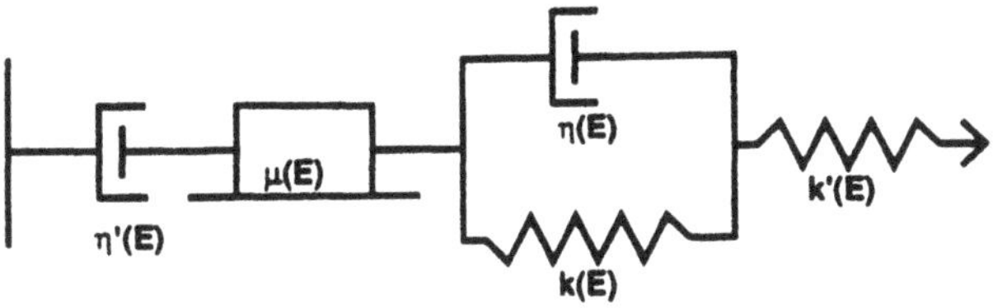

Figure 6: ER material rheological mechanical model.

accounts for the viscosity of the material. As mentioned earlier, the standard linear solid model is used to represent the linear viscoelastic behavior of the material, and all elements exhibit some dependence on the strength of the applied field.

During transient conditions or non-constant flow situations, the Coulomb element, viscous element, and the standard linear solid model must be considered. Furthermore, the contributions of the elements are field-dependent and rate-dependent as dictated by the characteristic times of the elements. Thus, a nonsinusoidal stress response, as shown in Figure 4 (Curve D), suggests nonlinear viscoelastic response, implying that the Coulomb element is moving during the cycle of straining, which results in the involvement of the viscous element.

B. ER material damping properties

As mentioned earlier, the energy dissipated by an ER material is related to the area enclosed within the loop. A series of hysteresis loops for the material during sinusoidal straining of 20 Hz and amplitude 0.5 are presented in Figure 7. The hysteresis loop for the

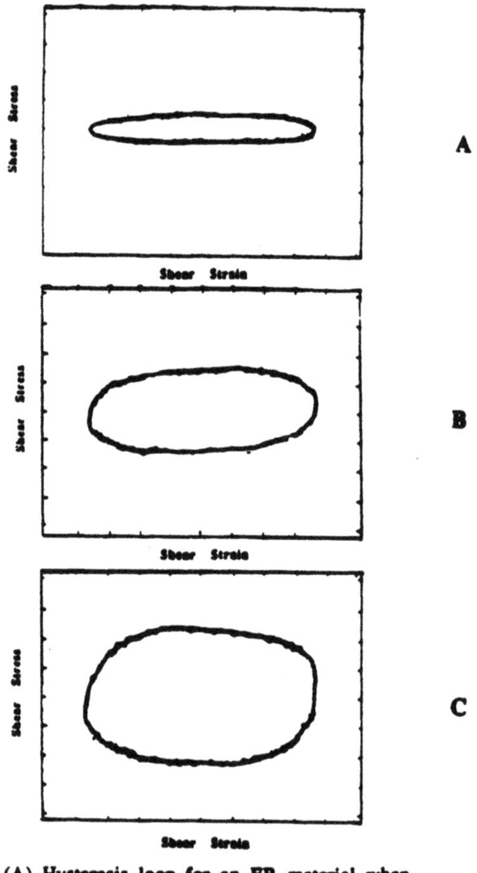

Figure 7. (A) Hysteresis loop for an ER material when E=0.0.kV/mm, (B) Hysteresis loop for an ER material when E=1.0 kV/mm, (C) Hysteresis loop for an ER material when E=2.0 kV/mm.

material when it is subjected to a zero field appears to be elliptical, with the major axis of the ellipse parallel to the abscissa, suggesting the material is behaving as a viscous body (7A). Applying a field strength of 1.0 kV/mm causes the ellipse to increase in area and the angle between the major axis of the ellipse and the abscissa to increase, i.e.viscoelastic behavior (7B). The amount of energy dissipated by the material has increased simultaneously with the elastic properties of the material. This behavior suggests that the energy dissipation mechanism of the ER material may be modelled by a linear damper, i.e. a viscous damper. If a 2.0 kV/mm field is applied to the material, the hysteresis loop deviates from elliptical shape, suggesting that there has been a transition in the energy dissipation mechanism from a linear damping mechanism to a nonlinear damping mechanism (7C). Moreover, the area within the loop has increased, thus the amount of energy dissipated by the ER material has increased. (Note that the scales of the hysteresis loops in Fig. 7 are of different magnitudes. Loop A: a unit on the ordinate is equal to 10.0 Pa. Loop B: a unit on the ordinate is equal to 15.0 Pa. Loop C: a unit on the ordinate is equal to 30.0 Pa.)

A moderate frequency dynamic study conducted with an alumino-silicate and paraffin oil ER material suggests that the energy dissipated by the material is a function of the strain amplitude, strain frequency, and applied field strength (Gamota and Filisko, 1991). For a constant strain amplitude, the amount of energy dissipated increased when the strain frequency was increased over the strain frequency range of 10 Hz to 50 Hz.

Damping studies observe the relationship between the dissipated energy and the applied strain amplitude to model the damping mechanism as either a linear (viscous) or a nonlinear (Coulombic) damper. Gamota and Filisko (1991) present data that suggests that the ER material's dissipation mechanism can be modelled as a viscous damper for low applied strain amplitudes and low applied fields and as a Coulombic damper for high applied fields and high strain amplitudes. In addition, this transition in damping mechanisms is related to the yielding of the ER material.

Also they relate the energy dissipated by the ER material to the field strength raised to the second power (E^2) and find that the relationship appears to be sigmoidal in nature. However, they do comment that an apparent E^2 relationship is observed at intermediate field strengths with stronger field relationships occurring for low fields and high field (before dielectric breakdown).

IV. Post-Yield Region

The post-yield behavior of an ER material may be characterized by an apparent viscosity (η_a) Under steady state flow conditions these parameters are related by

$$\eta_a = [S_c(E) + S_o(\gamma)] / \gamma_a \qquad (5)$$

where S_c (critical shear stress) is a function of the electric field (E) and S_o is a function only of the shear rate and temperature.

Two common criteria for evaluating an ER material in flow are the magnitude of the increase in apparent viscosity when the material is subjected to an electric field and the relative viscosity ratio, i.e. ratio of the apparent viscosity of the material when it is subjected to a nonzero field ($\eta_a(E)$) to the apparent viscosity of the material when it is subjected to a zero electric field (η_o). The relative viscosity ratio is the more important criterion since it relates how effective an electric field is to the rheology of the material. Moreover, if we define an ER material effectiveness factor, K, then the following relationship can be derived

$$K = \eta_a(E)/\eta_o = [(S_o+S_c)/\gamma] / (S_o/\gamma) = 1 + S_c/S_o \qquad (6)$$

Experimental observations suggest that S_c is essentially constant for a certain field strength but that S_o increases with increasing shear rate. Therefore, Equation 6 predicts that K will decrease towards unity as the shear rate increases. This is an important first order relationship for understanding the characteristics of the materials under flow since it implies that making a more effective ER material requires making S_c as large as possible while keeping S_o as small as possible.

The second parameter, S_o, is a function of the ER material composition and flow conditions. Thus, the maximum shear stresses can be increased by making the zero field ER materials more viscous; however in doing so the K factor can become extremely small so that the field induced change in rheology, the critical shear stress will become insignificant.

Torque - RPM data for an alumino-silicate and paraffin oil ER material subjected to fields of various strengths is presented in Figure 8. The torque - RPM data is linear and the slopes of the lines

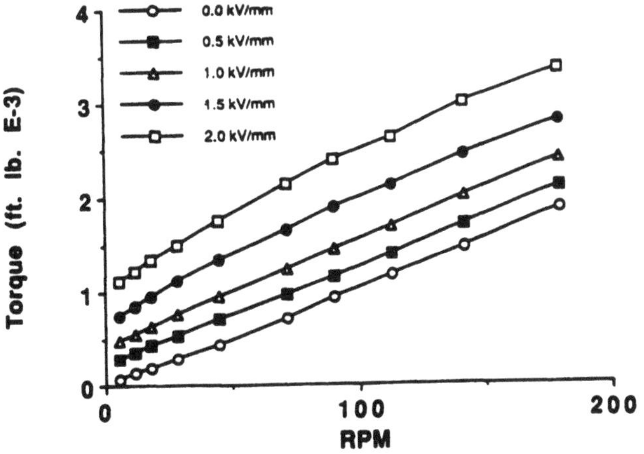

Figure 8: Torque versus RPM for an ER material under various field strengths.

appear to be similar as portrayed by an ideal Bingham fluid (Figure 1). Moreover, the critical torque may be calculated by simply extrapolating the data back to zero RPM which will allow the field dependency of the Coulomb element to be calculated.

A. Modelling of ER material behavior

Modelling of the ER material when deformation is restricted to the post-yield region is generally accomplished by using the Bingham model. This model suggests that the critical shear stress for the onset of flow can be simulated by the Coulomb element and the viscous element accounts for the viscosity of the material. In some studies the slopes of the shear stress versus shear rate data is the same for various applied field strengths, which suggests that the viscous element shows minimum field dependence. Moreover, if variations in the slopes exist for the various field strengths, the behavior may be accounted for by having the viscous element act as a non-Newtonian or shear thinning element.

IV. Conclusion

The rheological behavior of ER materials can be discussed in terms of three rheological regimes; pre-yield, yield, and post yield. Deformation of the ER material restricted to the pre-yield regime may be interpreted using linear visoelastic theory, and calculating the ER material's storage and loss moduli. Moreover, studies suggest that in this region of deformation the ER material's energy storing and energy dissipating properties are increased by the application of an electric field.

Yield studies have shown that the shear stress or shear strain for the onset of the yield region is a function of the field strength. Analysis of the rheological properties of an ER material within this region may be performed by either the use of Fourier analysis to calculate the fundamental nonlinear dynamic properties or by generating hysteresis loops and determining the energy dissipated by the ER material per volume, per cycle.

Finally, the post-yield region is observed under continuous shearing at a constant rate so as to initiate the "flow" of the ER material. ER materials are often studied under these conditions, and it is common to observe an increase in the apparent viscosity of the material when it is subjected to an electric field. In addition, some studies have reported a quadratic relationship between shear stress and shear rate.

V. References

Adriani, P., M, and A. P. Gast. "A Microscopic Model of Electrorheology." Phys. Fluids 31 (10), 2757- 2768 (1988).

Block, H., and J.P. Kelly. "Electro-rheology." Journal of Physics D: Applied Physics 21, 1661- 1677 (1988).

Conrad, H.,. Sprecher, A., F.,. Choi, Y., and Y. Chen. "The Temperature Dependence of the Electrical Properties and Strength of Electrorheological Fluids." J. Rheol. 35(7), 1393-1410 (1991).

Brooks, D., J., Goodwin, C. Hjelm, L. Marshall, and C. Zukoski. "Viscoelastic Studies on an Electro-Rheological Fluid." Colloids and Surfaces **18**, 293-312 (1986).

Gamota, D. R. and F. E. Filisko. " Dynamic Mechanical Studies of Electrorheological Materials: Moderate Frequencies." J. Rheol. **35**(3), 399-425 (1991).

Gamota, D. R. and F. E. Filisko. " High Frequency Dynamic Mechanical Study of an Aluminosilicate Electrorheological Material." J. Rheol. **35**(7), 1411-1426 (1991).

Gamota, D. R. and F. E. Filisko. " Fourier Analysis: Fundamental Response of an Electrorheological Material." J. Rheol. (submitted).

Shulman, Z.P., E.V. Korobko, and Yu. G. Yanovskii. "The Mechanism of the Viscoelastic Behaviour of Electrorheological Suspensions." Journal of Non-Newtonian Fluid Mechanics **33**, 181-196 (1989).

Uejima, H." Dielectric Mechanism and Rheological Properties of Electro-fluids." Japanese Journal of Applied Physics **11** (3), 319-326 (1972).

Vinogradov, G. V., Z. P. Shulman, Yu. G. Yanovskii, V. V. Barancheeva, E. V. Korobko, and I. V. Bukovich." Viscoelastic Behavior of Electro-Rheological Suspensions." Inzhenerno-Fizicheskii Zhurnal **50** (4), 605-609 (1986).

MACROSCOPIC BEHAVIOR OF ELECTRORHEOLOGICAL FLUIDS: TECHNIQUES FOR MEASURING RESPONSE TIME

KEITH D. WEISS and J. DAVID CARLSON

Thomas Lord Research Center, Lord Corporation, 405 Gregson Drive, Cary, North Carolina 27511-7900, USA

ABSTRACT

Two experimental methods, a dielectric spectroscopic technique and a channel flow technique, have been found to provide reasonable estimates of an ER fluid's response time. The dielectric data obtained for an ER fluid was adequately modeled using a well established Cole-Cole relationship for a symmetric distribution of relaxation times. A calculation of the particle's conductivity in the ER fluid also was found to be possible using the measured dielectric data. The response time obtained with the channel flow technique was in good agreement with the measured response time of a working ER fluid device.

[This paper also has been submitted for publication in the *International Journal of Modern Physics B*.]

1. Introduction

Industrial interest in the electrorheological (ER) phenomenon has centered around the potential utilization of these fluids in systems that require continuous and variable controllability. Both the evaluation and optimization of post-yield material properties, such as yield stress, viscosity and current density, for a variety of ER fluid formulations have been described in the literature.[1,2,3] However, one material parameter that has been only sporadically evaluated is the time required for the components of the fluid to respond to an electric field. In order for an ER fluid to be utilized in continuously variable, controllable devices a response time on the order of several milliseconds is required. It has been assumed that the response times associated with all known ER formulations reside within this time limit. The origin of this assumption can be traced back to the results disclosed in the initial work by Winslow.[4] Further validation for this assumption arose from the early investigations performed by Klass and Martinek.[5] It usually is not recognized that these results [4,5] are only applicable to the specific material compositions that were evaluated. A variety of ER fluids prepared in this laboratory have demonstrated response times that range from less than a millisecond to greater than one second. Thus it is important that the response times associated with available ER formulations be measured prior to selection for use in a controllable device.

One problem encountered upon attempting to compare the mechanical and electrical properties reported for ER fluids is the lack of uniformity in both test methodology and property definitions used by different investigators. This same problem is beginning to occur in the measurement of response times associated with ER fluids. For instance, should response time be defined as the period of time required for an ER fluid to react to an incremental voltage change, form the first fibril bridging the electrodes or reach final rheological equilibrium. Although optical imaging techniques allow for the measurement of time necessary for fibril formation in dilute ER fluids,[6] device design requirements dictate the practicality of defining a fluid's response time as a function of both a change in interparticle force and the final rheological equilibrium. In devices that are designed for on/off applications it is essential to know the time required to reach the maximum control (shear stress) ratio. In continuously, variable devices the establishment of this final rheological equilibrium is important only upon initial system start-up. It is more important in the design of these devices to know how fast a fluid can respond to an incremental change in applied voltage.

A description of only one experimental technique applicable to the measurement of an ER fluid's response to small variations in applied voltage has been reported.[7] This technique is based upon the optical measurement of small angular displacements induced in a tube by varying the voltage applied to a surrounding electrorheological medium. Two experimental methods that can be used to obtain an estimate of the time required for an ER fluid to reach a rheological equilibrium have been reported.[8,9] One of these techniques uses light scattering to measure the turbidity of an ER fluid upon the application of voltage.[8] The second technique is based upon the measurement of pressure build-up as a function of time for an ER fluid flowing through a valve.[9] It is the objective of this paper to demonstrate two additional methods that can be used to estimate the response times associated with ER fluids. The first technique, which is based upon dielectric or impedance spectroscopy, allows for the measurement of the time necessary for individual particles to respond to a small change in electric field. This measurement provides a lower limit for the time interval in which an ER fluid response can be expected. The second technique, which is based on a custom designed, fixed-electrode, channel-flow rheometer, is an adaptation to the previously described valve technique.[9] The measurement obtained by this technique corresponds to an upper limit for the time interval necessary for an ER fluid to reach its maximum control ratio.

2. Impedance Spectroscopic Technique

Dielectric or impedance spectroscopy (IS) is a technique capable of obtaining data related to the interactions that exist between different components in heterogeneous materials. Data describing the electrical response of a material is obtained through the application of a small A.C. signal to the sample over a broad frequency range. Analysis of this data using circuit theory provides a description of the material's dielectric properties. This technique is applicable to evaluation of a wide variety of interfacial reactions, such as conductivity effects in ionic solids, defects in semiconductors, electrochemical reactions (i.e., metal corrosion, etc.), and solid particles in liquid electrolytes (i.e., chemical sensors, etc.). The measurement of the dielectric spectra associated with an ER fluid is not a new concept.[5] Although it has been suggested that Debye theory is applicable to ER fluids [5,10], no analysis of relaxation times determined from the dielectric properties of ER fluids has been reported. Several general overviews describing recent developments in impedance spectroscopy can be found in the literature.[11,12]

2.1. Dielectric Theory

By definition the permittivity, ε^*, of a homogeneous material describes the ability of this material to polarize in the presence of an electric field. Reorientation of the electrical charge distributions within polar molecules is the basic phenomenon underlying the particle rotation necessary for the observation of an ER effect. In the presence of an alternating current a homogeneous material's permittivity is known to vary with frequency. When the frequency of the A.C. field becomes faster than the ability of the material's dipoles to reorient themselves to the field direction, the appearance of dielectric loss is observed. This phenomenon is visualized as a complex representation of the material's permittivity as shown in Equation 1, where ε' and ε'' represent the relative permittivity (dielectric constant) and the imaginary dielectric loss factor, respectively. The ratio of $\varepsilon''/\varepsilon'$ defines the loss tangent, tan δ, of the material where δ is the loss angle. It should be acknowledged that conductivity processes also effect the dielectric loss of a material.

$$\varepsilon^* = \varepsilon' - i\,\varepsilon'' \qquad (1)$$

If the polarization of the homogeneous material is considered as a function of time, the complex permittivity can be redefined in terms of a dielectric (Debye-like) relaxation process (Equations 2, 3 and 4).[13,14,15,16,17] In these equations ε_∞, ε_s, ω and τ represent the relative permittivity of the material at infinite frequency, the static field permittivity of the material, the angular frequency of the alternating field and the relaxation time associated with the material, respectively. The dielectric loss factor, ε'', is observed to reach a maximum when $\omega\tau = 1$, while the loss tangent, tan δ, reaches a maximum when $\omega\tau = (\varepsilon_s/\varepsilon_\infty)^{1/2}$.

$$\varepsilon' = \varepsilon_\infty + \frac{\varepsilon_s - \varepsilon_\infty}{1 + \omega^2\tau^2} \qquad (2)$$

$$\varepsilon'' = \frac{(\varepsilon_s - \varepsilon_\infty)\,\omega\tau}{1 + \omega^2\tau^2} \qquad (3)$$

$$\tan\delta = \frac{(\varepsilon_s - \varepsilon_\infty)\,\omega\tau}{\varepsilon_s + \varepsilon_\infty\,\omega^2\tau^2} \qquad (4)$$

The time dependent dielectric behavior associated with ER fluids is expected to be a more complex phenomenon than indicated in the previous discussion for molecular relaxation processes. Since ER fluids are heterogeneous materials consisting of conductive particles dispersed in a dielectric medium, their polarization process relies upon charge build-up at the particle/oil interface. In the simplest case, a heterogeneous material can consist of particulates with a relative permittivity, ε'_2, and a conductivity, $\sigma_2 = \omega\varepsilon_o\varepsilon''_2$, dispersed in a medium with a relative permittivity, ε'_1, and conductivity, $\sigma_1 = \omega\varepsilon_o\varepsilon''_1$. Wagner has provided an in depth analysis of this situation for a material containing a dilute particle concentration.[16, 18] The derived relaxation expressions for a heterogeneous system displaying a Maxwell-Wagner polarization process are similar to those previously described for a simple Debye relaxation process (see Equations 2-4). In this case, the parameters, ε_∞, ε_s and τ, can be described in terms of the relative permittivites and conductivities associated with the particulates and dielectric medium as shown in Equations 5-7.[14] Although the validity of this treatise by Wagner has been confirmed in systems containing small particle volume fractions, ϕ_2, an upper limit for the particle concentration has not been established.[19,20,21] The major difficulty in determining a Maxwell-Wagner relaxation time for a heterogeneous material is the difficulty encountered in obtaining an independent measurement of the conductivity associated with the dispersed particles.

$$\varepsilon_\infty = \frac{\varepsilon'_1\left[2\varepsilon'_1 + \varepsilon'_2 + 2\phi_2(\varepsilon'_2 - \varepsilon'_1)\right]}{\left[2\varepsilon'_1 + \varepsilon'_2 - \phi_2(\varepsilon'_2 - \varepsilon'_1)\right]} \tag{5}$$

$$\varepsilon_s = \frac{3\phi_2\sigma_1\left[(2\sigma_1 + \sigma_2)(\varepsilon'_2 - \varepsilon'_1) - (2\varepsilon'_1 + \varepsilon'_2)(\sigma_2 - \sigma_1)\right]}{\left[2\sigma_1 + \sigma_2 - \phi_2(\sigma_2 - \sigma_1)\right]^2} + \frac{\varepsilon'_1\left[2\sigma_1 + \sigma_2 + 2\phi_2(\sigma_2 - \sigma_1)\right]}{\left[2\sigma_1 + \sigma_2 - \phi_2(\sigma_2 - \sigma_1)\right]} \tag{6}$$

$$\tau = \frac{\varepsilon_0\left[2\varepsilon'_1 + \varepsilon'_2 - \phi_2(\varepsilon'_2 - \varepsilon'_1)\right]}{\left[2\sigma_1 + \sigma_2 - \phi_2(\sigma_2 - \sigma_1)\right]} \tag{7}$$

2.2. *Experimental Design*

All measurements were made using a Hewlett-Packard model 4192A LF impedance analyzer. All measurements performed by the analyzer and subsequent data analysis were controlled through an interface with an IBM 5150 laboratory computer system. The impedance parameters measured were capacitance and conductance from which ε', ε'', $\tan\delta$ and σ were calculated. Frequency was stepped from 5 Hz to 1 MHz with 20 increments per decade. A complete spectrum could be completed in approximately 2 minutes. The excitation signal delivered to the sample in these measurements had an amplitude of 1.0 volt. All measurements were taken at 25 °C.

The sample cell designed for the measurement of the dielectric properties of ER fluids is shown in Figure 1. This cell consists of a hollow aluminum bob (80 mm in diameter) inserted into a cup sized to provide a 0.25 mm gap at the bottom and sides of the bob. The ER fluid sample is placed within this gap. The conical shape at the bottom of the bob insured that air bubbles were not trapped. The upper portion of the bob had a reduced diameter so that the measurements were not sensitive to small variations in the actual volume of fluid placed in the cell. The bob was purposely made hollow (wall thickness ~ 1 mm) in order to reduce the thermal mass of the cell and allow rapid heating and cooling. All aluminum surfaces in contact with the ER fluid were highly polished and all inside corners were provided with a generous radius to facilitate cleaning between samples. The insulating washer and spacer were constructed of Ryton, while the heating/cooling jacket was made of Delrin. The bob and insulators were connected to the lid by a single, large knurled nut. The lid and bob assembly was secured to the cup by several thumb screws. The impedance analyzer was connected to the sample cell by a pair of wires that plugged into small holes in the bob stem and lid.

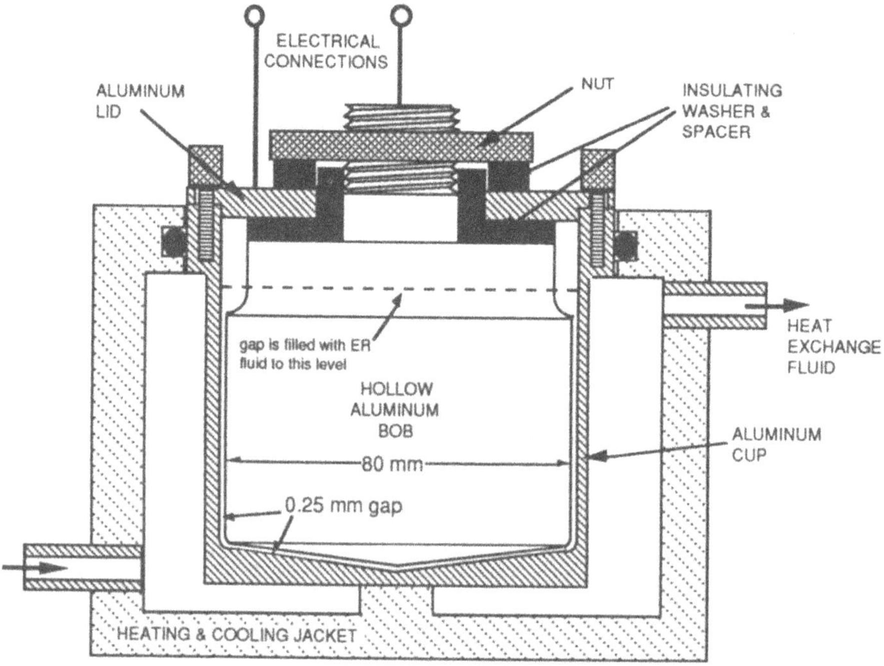

2.3. *Analysis of Dielectric Measurements*

The dielectric spectra obtained for a specific Lord ER fluid (#6533-16A) is shown in Figure 2. The maxima observed in both the dielectric loss and loss tangent data when plotted as a function of frequency demonstrates that the ER fluid is undergoing a Debye-like relaxation process. The dielectric loss, ε", is observed to go through a maximum at a frequency f_{max} = 309 Hz. This frequency also is observed to correspond to the midpoint of the curve observed for the relative permittivity, ε', of the fluid. According to simple Debye theory (Equations 2-4) this data can be used to calculate the relaxation time associated with the polarization decay process in the ER formulation. The frequency of the maxima observed in the dielectric loss factor corresponds to a relaxation time for this ER fluid equivalent to 0.51 ms ($\tau = 1/[2\pi f_{max}]$). The maximum, δ_{max}, observed for the loss tangent δ also is a measure of the relaxation time associated with the material. However, this maxima is displaced to a higher frequency such that $f_{\delta max} = f_{max} (\varepsilon_s / \varepsilon_\infty)^{1/2}$.

A measure of the range associated with the distribution in relaxation times in a material can be obtained by plotting the dielectric loss data as a function of the relative permittivity. The resulting semi-circular arc, which is termed a Cole-Cole diagram [22], corresponds to a Nyquist plot of a RC circuit. A very narrow distribution of relaxation times is indicated when this graphical representation is a true semicircle whose center of origin lies on the abscissa.[14] A broader distribution in relaxation times is indicated when the semicircular arc is either depressed (center lies below the ε'-axis)[22] or skewed towards lower frequencies.[23] The corresponding Cole-Cole diagram (Figure 3) for the ER fluid #6533-16A was found to yield a depressed semicircular arc. The increase in the dielectric loss factor at high values of relative permittivity observed in the diagram is an indication of dielectric loss at low frequencies due to conduction. The actual range associated with this symmetrical distribution in relaxation times is defined as the exponent, α, in the derivation of complex permittivity shown in Equation 8.[14,22] This exponent was determined from the semicircular arc diagram (Figure 3) for ER fluid #6533-16A to be equal to 0.43. The values of ε_∞ and ε_s also were determined from the diagram to be 13.82 and 2.81, respectively. Through substitution of this data into the Cole-Cole expression (Equation 8), it has been verified that this relationship is an adequate model for the dielectric properties measured for this ER fluid (see Figure 2).

$$\varepsilon^* = \frac{\varepsilon_s - \varepsilon_\infty}{1 + (i\omega\tau)^{1-\alpha}} \qquad 0 \le \alpha \le 1 \tag{8}$$

An additional check to insure the validity of using the Cole-Cole relationship to describe the distribution in relaxation times for this ER fluid is to measure the length of the chords U and V from the semicircular arc diagram. By definition, chord U stretches from the $\varepsilon' = \varepsilon_\infty$ intercept to a point on the ε" versus ε' arc, while chord V stretches from the $\varepsilon' = \varepsilon_s$ intercept to the same point on the arc. A plot of log(V/U) against log(ω) should result in a straight line with a slope equal to (1-α) and an intercept equal to (1-α)log[τ].[16] This linearity test when applied to the data shown in Figure 3 was found to yield a straight line (y = 0.59x - 1.92). The values for α = 0.41 and τ = 0.56 ms calculated from these test results provide an excellent correlation with the previously described data.

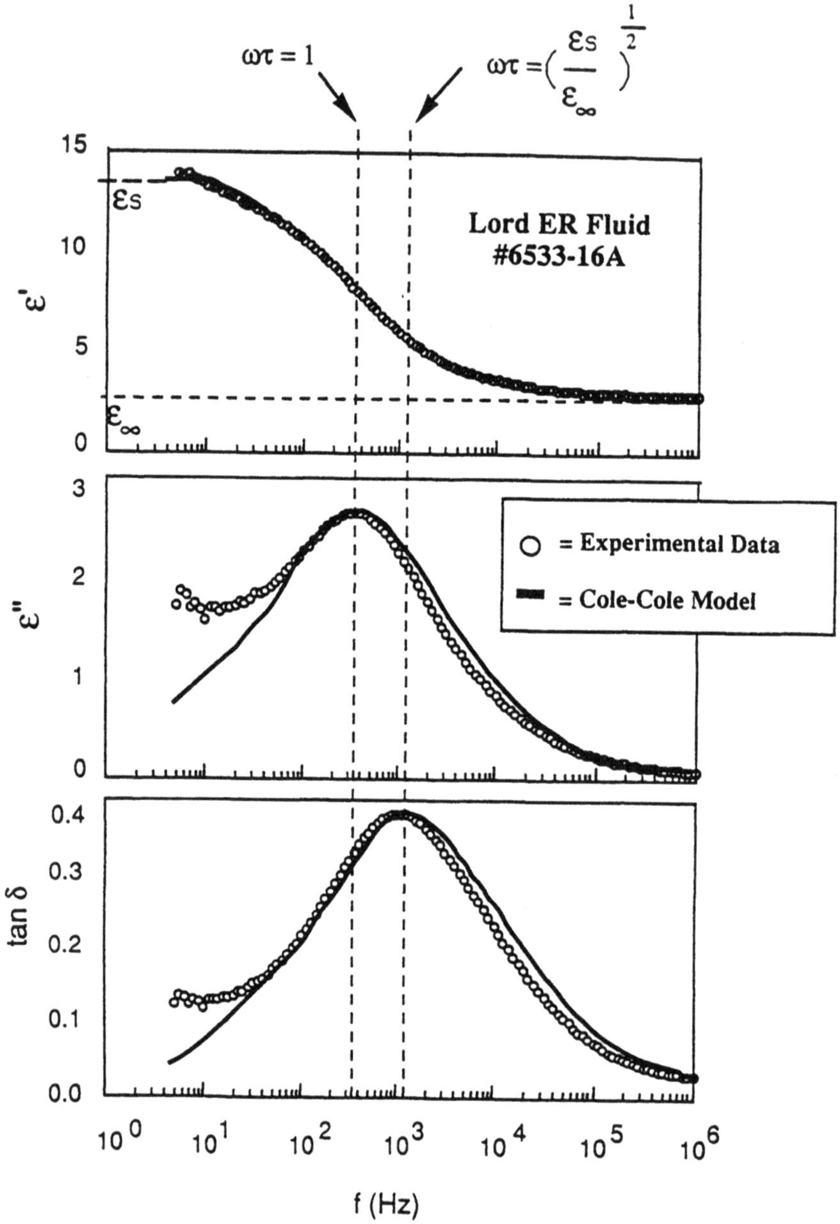

Figure 2. Dielectric Spectra Obtained for Lord ER Fluid #6533-16A.

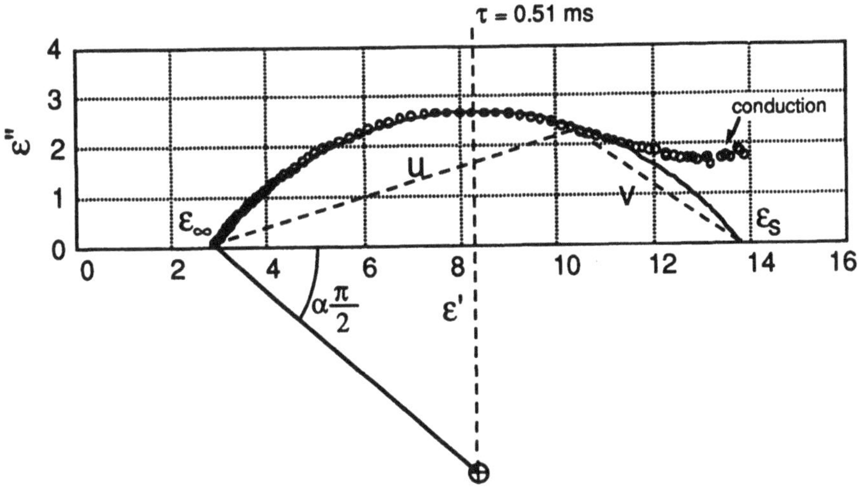

Figure 3. Cole-Cole Diagram Corresponding to Dielectric Spectra for Fluid Shown in Figure 2 (#6533-16A).

It is possible to obtain a measure of particle conductivity by substituting the relaxation time obtained from the dielectric data into Equation 7 for the Maxwell-Wagner effect provided all other parameters are known. For this ER fluid (#6533-16A) the particle conductivity was calculated to be 2.41 x 10^{-8} S/m. In order to determine the accuracy of this calculation the ER fluid was successively diluted with carrier oil and the impedance spectra remeasured. The particle volume fraction after the final oil dilution corresponded to 20% of its original value. It was assumed that the particle conductivity would remain constant through this dilution process. The Cole-Cole diagram describing the dielectric properties of the ER fluid after each oil dilution is shown in Figure 4. A decrease in dielectric loss at low frequencies due to conduction was observed upon decreasing the particle volume fraction. In addition, a slight increase in the width of the relaxation time distribution (α) was also observed upon decreasing the particle volume fraction. The dielectric data for each fluid (A-E) was found to pass the previously described linearity test. The average particle conductivity for this ER fluid #6533-16A was determined to be (2.57 ± 0.41) x 10^{-8} S/m (95% confidence limit, 5 degrees of freedom).

3. Fixed Plate Flow Fixture Technique

Two electrode geometries, the fixed plate and the sliding plate, are commonly used in most electrorheological fluid devices. A fixed plate device is in essence a valve in which the fluid flows between fixed electrodes. Sliding plate devices, on the other hand, involve a layer of fluid sheared between a pair of electrodes which move parallel to one another, e.g. a clutch. A fixed electrode flow fixture was designed by Lord Corporation to provide

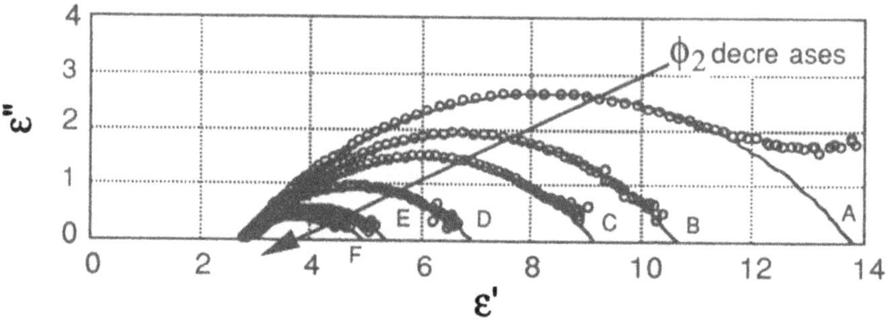

Fluid	φ2	α	ε_s	ε_∞	τ (s)	σ2 (S/m)*
6533-16A	1.00 φ	0.43	13.82	2.81	0.00051	2.41x10-8
-B	0.84 φ	0.42	10.65	2.91	0.00036	3.04x10-8
-C	0.72 φ	0.43	8.72	3.40	0.00034	2.96x10-8
-D	0.56 φ	0.46	6.93	2.77	0.00034	2.65x10-8
-E	0.40 φ	0.48	5.36	2.81	0.00035	2.33x10-8
-F	0.20 φ	0.50	4.82	2.76	0.00037	2.61x10-8

(* calculated from Equation 7)

Figure 4. Cole-Cole Diagrams and Analysis for Lord ER Fluid #6533-16A Diluted to Different Volume Fractions.

both mechanical and electrical measurements of ER fluids in an environment similar to that expected for a variety of valve type devices. The actual construction and test capabilities of this flow apparatus have only been briefly described in the literature.[24] The test most commonly performed with this instrument is the measurement of yield stress and current density as a function of applied voltage. The appearance of any particle/oil hysteresis or particle filtering during channel flow can easily be detected during this routine test.[24] One test that has not been previously mentioned in conjunction with this instrument is the measurement of an ER fluid's response time.

3.1. *Theoretical Background*

The theory behind the development of the fixed plate flow fixture is based upon the establishment of a steady state fluid flow along a channel having a rectangular cross section and the subsequent measurement of the pressure drop along the channel resulting from this flow state. For a Bingham plastic material, the pressure drop P, yield strength T_y, viscosity η and flow rate Q are related by a cubic equation derived by Phillips.[25] The solution to Phillips cubic equation can be written as shown in Equation 9, where b is the

$$T_y = \frac{1}{c}\left[\frac{h}{L}\right]P - \frac{4Q\eta}{bh^3} \qquad (9)$$

channel width, h is the channel height, L is the channel length and c is a parameter which depends on the specific flow conditions and has a value between 2 and 3. If $T_y/\eta < (20\ Q/bh^2)$, e.g. very high flow rates, then to a good approximation $c \approx 3$. On the other hand for low flow rates where $T_y/\eta > (200\ Q/bh^2)$ then $c \approx 2$. The flow rates and geometry used in the Lord flow fixture are such that the $c \approx 2$ works well. Errors associated with this approximation are generally less than 5% for most fluids tested on the flow fixture. Yield strength values obtained from the Lord flow fixture using this approximation have been observed to be in agreement with data obtained using a concentric cylinder cell geometry on a couette rheometer.

The determination of a yield stress for an ER formulation only establishes the presence of an unsheared region in the fluid. For all practical purposes it is the viscosity of the material that determines the velocity of the fluid in sheared regions. This relationship between viscosity and velocity becomes more apparent when one considers a change in the behavior of an incompressible fluid in going from Newtonian flow to a plugged flow condition. If the fluid's flow rate is held constant, the derivation of the valve response time, τ_v, leads to Equation 10, where ρh corresponds to the fluid's inertia and η/h represents viscous shear. In addition to the viscosity of the ER fluid, it also is necessary for the estimation of an ER fluid's response time in channel flow to take into consideration the compliance of the materials used to construct the test fixture, the compressibility of the fluid and the necessary particle structure building to reach an equilibrium yield stress.

$$\tau_v = \frac{\rho h^2}{12\eta} \qquad (10)$$

3.2. *Experimental Design*

The construction of the fixed plate flow fixture as shown in Figure 5 includes two facing conductive walls along the length of the flow channel. The flow channel was 100 mm long, 10 mm wide and the electrodes could have a separation of either 1 or 2 mm. High voltage was applied to the fixture using a Trek model 610C high voltage operational amplifier controlled by a Wavetek signal generator which provided the appropriate control voltage signal. This power supply is adjustable over the range of ± 10 kV and ± 2 mA with a bandwidth of 12 kHz and a slew rate of 20 V/μs. The flow fixture was constructed out of a heavy aluminum block mounted on a rigid aluminum channel in order to provide the stiffness necessary to maintain alignment of the fixed electrodes under pressure.

During testing the plunger of the syringe is operated with a Velmex Uni-Slide linear drive powered by an Electro-Craft motor and controller. The pressure is monitored with a 100 psi pressure transducer. A total of 15 mL of ER fluid is normally required for effective operation of the instrument. All test operations, data collection and analysis are controlled through the use of a MicroVax computer system.

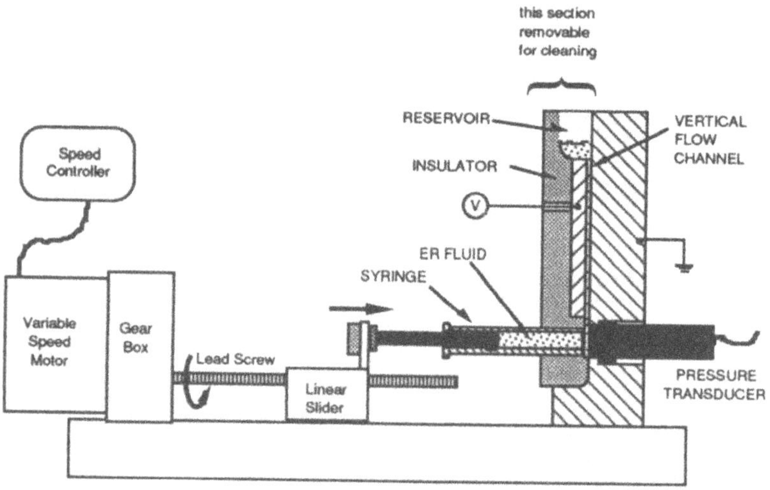

Figure 5. Simplified Diagram of Lord Fixed Electrode, Channel Flow Fixture.

For testing of an ER fluid's response time the motor controller was set to move the plunger at a linear speed of 44 mm/s or a flow rate $Q = 1.04\mathrm{x}10^{-6}$ m^3/s. This corresponds to approximate shear rates of 537 s^{-1} and 134 s^{-1} for 1 mm and 2 mm gap heights, respectively. A 20 Hz, unipolar square wave with a peak voltage of 4 kV was then applied to the flow fixture. Pressure readings were then taken at a rate of 1000 Hz. This test is appropriate for ER fluids that exhibit response times between about 1 ms and 25 ms. For slower ER fluids it is necessary to decrease the frequency of the high voltage signal. In addition, the speed of the plunger must be decreased so that at least several cycles of the high voltage signal can be applied within the allowable travel range of the plunger. For example, a plunger speed of 4 mm/s and a 0.125 Hz unipolar square wave can be used to test fluids having a response time on the order of several seconds.

3.3. *Analysis of Flow Fixture Measurements*

Experimental data obtained for Lord ER fluid #6533-16A using this flow fixture technique is shown in Figure 6. The time necessary for this ER fluid to reach an equilibrium yield stress was observed to be about 15 ms. Note, this is the same fluid as that whose impedance results are shown in Figures 2-4. The actual time constant τ

associated with this flow measurement is taken to be the amount of time necessary for the pressure to increase from the voltage-off base line pressure to 63% of the final, voltage-on equilibrium pressure or about 8 ms for the fluid shown in Figure 6. This assumes that the system behaves as a first order system and that the pressure rise when the fluid is subjected to a step input voltage can be written as shown in Equation 11. While this time constant is

$$p(t) = p_o\left(1-e^{-\frac{t}{\tau}}\right). \tag{11}$$

related to the electrical time constant for the fluid determined from the impedance measurements it includes additional time necessary for macroscopic structure formation in the fluid as well as any time needed to accommodate mechanical compliance in the test fixture. The overall "turn-on" time of an ER fluid is generally taken to be 2-3 time constants.

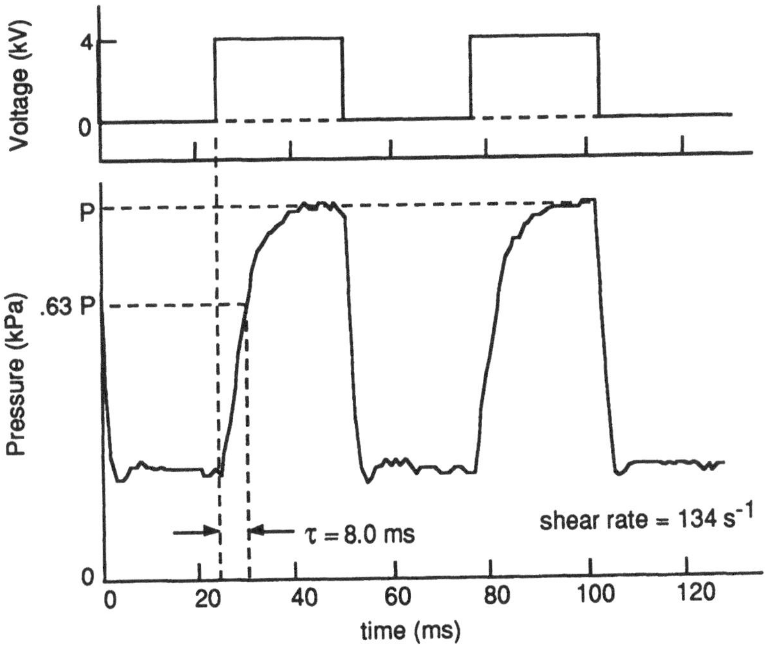

Figure 6. Response Time Data Obtained on Fixed Electrode Flow Fixture for Lord Fluid #6533-16A.

As was observed to be the case with the impedance measurements, the pressure rise times determined in the flow fixture were found to decrease as particle volume fraction was decreased. Decreasing the particle volume fraction by a factor of two was found to decrease the flow fixture time constant of the fluid shown in Figure 6 from 8 ms to 4 ms. Pressure rise times were always observed to be longer than pressure fall times. The time required to reestablish an off-state baseline for the fluid shown in Figure 6 was approximately 3 msec. Fall times, however, were observed to be nearly independent of particle volume fraction.

The main difficulty with channel flow devices is the potential for ER fluids to undergo particle/oil separation and/or filter bed formation and jamming. These can be particularly serious problems if the particles are large and the oil viscosity is low. Such effects can be readily detected, however, during a voltage sweep with the flow fixture if the voltage is allowed to sweep up to a maximum value and then back to zero while monitoring the pressure. Problematic effects such as separation and jamming show their presence by the formation of a hysteresis loop in the pressure/voltage curve. It is essential that channel flow test equipment be used to identify the presence of this hysteretic behavior in specific ER formulations prior to their utilization in fixed electrode devices. The Lord flow fixture has been found to be well suited to identify this sort of behavior in ER fluids.[24] These same problematic effects can also have an impact on response time measurements made with the flow fixture. Fluids that exhibit hysteretic behavior in their pressure/voltage sweep will also show a gradual increase in pressure over the course of the response time measurement as demonstrated in Figure 7.

In order for response time measurements on the flow fixture to be meaningful the time interval for which voltage is applied must be longer than several time constants of the ER fluid. If the fluid never approaches an equilibrium peak pressure then a time constant cannot be determined. In fact, if the fluid response time is much longer than the applied voltage period, the fluid will act as a low pass filter and respond only to the low frequency Fourier components of the high voltage signal. In the extreme case, a fluid responding to only the DC component of the high voltage signal will exhibit a small ripple at the fundamental signal frequency. This phenomena is illustrated in Figure 8. The time constant for this ER fluid was so long that the measured dynamic pressure exhibited only a small 0.125 Hz ripple superimposed on an average pressure due to the DC component of the voltage signal.

4. Comparison of Response Time Measurement Techniques

The ER fluid response time measured by impedance spectroscopy represents a lower limit in time required for the ER fluid to respond to the application of voltage. The response time obtained with the flow fixture technique is generally both longer and more likely to approximate the value one would expect to see in an actual device.

It is appropriate to determine how closely the measurement of a response time using the flow fixture technique will approximate the value obtained in a working device. To this end the response time of an ER fluid #6184-38A was measured using both the impedance and flow fixture techniques, as well as in an experimental ER fluid damper. The response time of the damper was measured as the time necessary to reach an equilibrium damping force at maximum voltage. The single time constant relaxation time for this ER fluid

(#6184-38A) was found by impedance spectroscopy to be 6.7 ms. The flow fixture technique gave a time constant of 14 ms and a time to reach peak pressure of about 25 ms. This response time is very similar to the average 28 ms response time measured for the ER damper to switch from its full off state to full on state.[26]

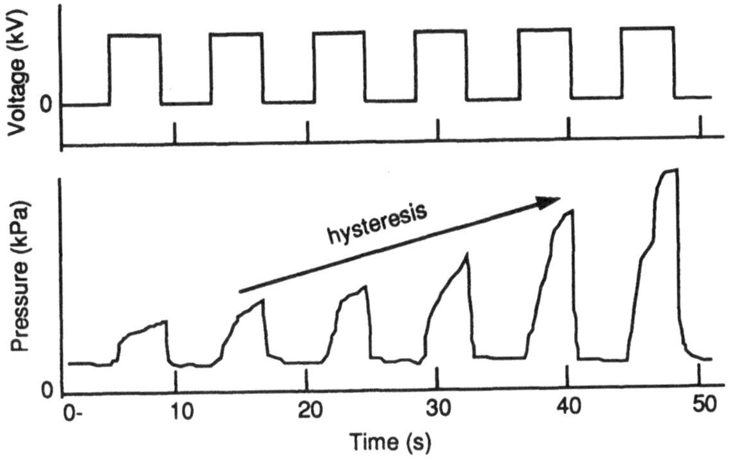

Figure 7. Hysteretic Behavior of ER Fluid #6184-77M During Measurement of Response Time on Flow Fixture.

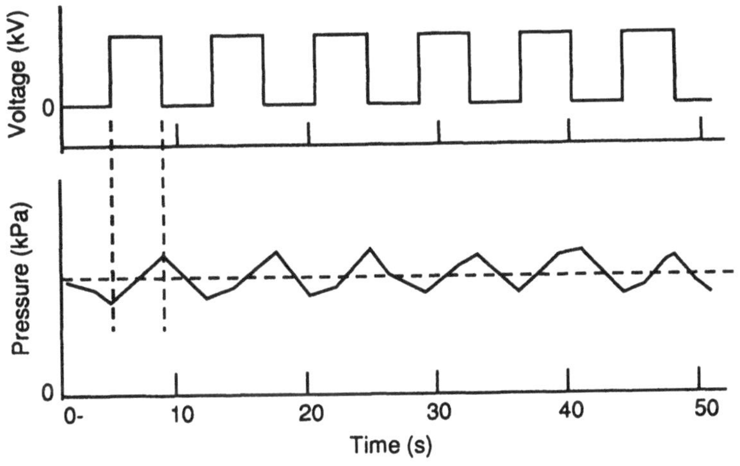

Figure 8. High Frequency Filtering Effect Observed for ER Fluid #6184-77K During Measurment of Response Time on Flow Fixture.

5. Conclusions

Two experimental procedures, the impedance spectroscopic technique and the fixed plate flow fixture technique, have been shown to provide reasonable estimates of an ER fluids response to an applied voltage. The dielectric data obtained for ER fluid #6533-16A through the impedance spectroscopic technique was observed to resemble a Debye-like relaxation process. This dielectric data was adequately modeled using a well established Cole-Cole relationship for a symmetric distribution in relaxation times. The relaxation time obtained from the dielectric data represents a lower limit for the time interval in which an ER fluid response can be expected. This response time measurement also allowed for the calculation of the particle's conductivity in the ER fluid based on the Maxwell-Wagner relationship. The response time measured in the fixed plate flow fixture technique correlates with the time required for a fluid to reach a final rheological equilibrium. This technique also can be used to identify the susceptibility of an ER fluid to hysteretic particle/oil separation and jamming. The response time measured through this technique has been found to provide an excellent correlation with the measured response associated with working ER fluid devices. The extension of both these techniques to other ER fluid systems, as well as the effect of temperature on the measured response times, is currently in progress.

6. Acknowledgements

The authors would like to acknowledge the support of Theodore G. Duclos through out the development of these response time measurement techniques. We also are indebted to both Mike Chrzan and Scott Grady for the design and construction of an ER damper, as well as for collecting and analyzing the data associated with this device. Finally, the efforts of Don Nixon and Jeannine Jones in obtaining the experimental data associated with both the impedance spectroscopic and flow fixture techniques need to be recognized.

7. References

1 H. Block,and J. P. Kelly, *J. Phys. D: Appl. Phys.* 21 (1988) 1661.

2 T. C. Jordan and M. T. Shaw, *IEEE Transactions on Electrical Insulation* 24 (1989) 849.

3 A. P. Gast and C. F. Zukoski, *Advances in Colloid and Interface Science* 30 (1989) 153.

4 W. M. Winslow, *J. Appl. Phys.* 20 (1949) 127.

5 D. L. Klass, and T. W. Martinek, *J. Appl. Phys.* 38 (1967) 67; D. L. Klass and T. W. Martinek, *J. Appl. Phys.* 38 (1967) 75.

6 J. C. Hill and T. H. Van Steenkiste, *J. Appl. Phys.* 70(1991) 1207.

7 J. E. Stangroom in *Proceeding First International Symposium on Electrorheological Fluids*, eds. H. Conrad, A. F. Sprecher and J. D. Carlson (North Carolina State University Publications, Raleigh, 1989) 81.

8 T. C. Jordan and M. T. Shaw in *Electrorheological Fluids - Proceedings of the Second International Conference on ER Fluids,* eds. J. D. Carlson, A. F. Sprecher and H. Conrad (Technomic Publishing, Lancaster, 1990) 231.

9 D. J. Peel and W. A. Bullough in *Electrorheological Fluids - Proceedings of the Second International Conference on ER Fluids,* eds. J. D. Carlson, A. F. Sprecher and H. Conrad (Technomic Publishing, Lancaster, 1990) 141.

10 H. Vejima, *Japanese J. Appl. Phys.* 11 (3) (1972) 319.

11 F. B. Growcock, *Chemtech* September (1989) 564.

12 J. R. MacDonald, *Impedance Spectroscopy* (J. Wiley & sons, New York, 1987).

13 P. Debye, *Polar Molecules* (Chemical Catalog Co., New York, 1929).

14 J. B. Hasted, *Aqueous Dielectrics* (Chapman & Hall, London, 1989).

15 L. I. Hench and J. K. West, *Principles of Electronic Ceramics* (J. Wiley & Sons, New York, 1990) Chapter 5

16 C. P. Smyth, *Dielectric Behavior and Structure* (McGraw-Hill, New York, 1955).

17 P. Hedvig, *Dielectric Spectroscopy of Polymers* (John Wiley & Sons, New York, 1977).

18 K. W. Wagner, *Arch. Elektrotech.* 2 (1914) 378.

19 R. J. Meakins, *Prog. Diel.* 3 (1961) 151.

20 B. V. Hamon, *Aust. J. Phys.* 6 (1953) 305.

21 T. Hanai, *Bull, Inst. Chem. Res., Kyoto Univ.* 39 (1961) 341.

22 K. S. Cole and R. H. Cole, *J. Chem. Phys.* 9 (1941) 341.

23 D. W. Davidson and R. H. Cole, *J. Chem. Phys.* 18 (1951) 1417.

24 J. E. Bares and J. D. Carlson in *Electrorheological Fluids - Proceeding of the Second International Conference on ER Fluids*, eds. J. D. Carlson, A. F. Sprecher and H. Conrad (Technomic Publishing, Lancaster, 1990) 93-114.

25 R. W. Phillips, *"Engineering Applications of Fluids with a Variable Yield Stress"*, Thesis, University of California, Berkeley (1969).

26 M. J. Chrzan, *Final Report Contract #N60530-90-C-0218* (Naval Weapons Center, China Lake, CA, 1991).

EXPERIMENTAL AND THEORETICAL RESPONSE TIMES OF ELECTRORHEOLOGICAL FLUIDS

JOHN C. HILL[(a)], NUNO A. VAZ
Physics Department
General Motors Research Laboratories
Warren, Michigan 48090-9055 U.S.A.

and

DANIEL J. KLINGENBERG[(b)] and CHARLES F. ZUKOSKI
University of Illinois
Urbana, Illinois 61801 U.S.A.

ABSTRACT

To date there has not been any universally accepted model of ER fluid behavior. A comparison of experimentally determined response times and those predicted by an electric dipole-dipole interaction theory indicates that the latter describes the basic operation of ER fluids.

1. Introduction

The theory of electrorheological (ER) fluids has been in an unsettled state. Two primary mechanisms, dipole-dipole interaction and double layer interaction, have been advanced to explain the formation of particle-containing fibriles which bridge a gap between electrodes upon the application of an electric field. This gap bridging is the source of the large viscosity increases that are typically observed.

One proposed source of the fibrile morphology is the electric dipole-dipole interaction.[1] The applied electric field induces polarization of the particles, which then interact under the influence of the field to form the fibriles. Usually some form of interaction with the electrode walls is also included. A mismatch of the dielectric constants of the particles and the fluid in which they are immersed is required. In addition, some form of hard-sphere repulsion is introduced to prevent overlap of particles.

The above theory is not without its detractors. Klass and Martinek[2] observed that ac fields at frequencies as high as 10 kHz could still produce some ER response. They concluded that the particles themselves could not move fast enough to follow such high frequencies. They proposed an alternate theory by which the applied field induces polarization in a "double layer" surrounding each particle. The resulting electrostatic interactions of the then distorted double layers "require the dissipation of additional energy on application of a shear stress normal to the direction of the field-induced distortion." This increased energy dissipation is cited as the source of the ER effect. Since only relatively light and mobile ions are required to move, the high frequency response is explained.

When water is the source of the double layer, field-induced contact leads to the possibility of bridging. Surface tension forces are then postulated[3] to enhance the binding between particles.

In this paper we compare values for the response times (defined below) measured as a function of the concentration of particles in the fluid with the predictions of the dipole-dipole interaction model. The observed good agreement is interpreted as supporting evidence for the validity of this model. The concentration of particles is defined in terms of the area density (ρ) of particles which can be determined by counting the number of particles in an image, calculating the total area of the particles, and then dividing by the total area of the image.[4]

The response time (τ) will be defined as the time interval between application of the field and the formation of the first bridging fibrile.[4] This is somewhat arbitrary, but there are reasons behind this choice. For one thing this event is well-defined. If, for example, the definition were in terms of formation of the last fibrile, large scatter would be introduced into the data, especially at high particle densities where the gap is crowded and it is difficult to observe the formation of the last fibrile. Finally, other researchers in this field have adopted the same definition; thus we are consistent with prior practice.

The interest in studying the response times lies on the fact that they are important to the operation of many devices, such as clutches, shock absorbers, and engine mounts. The successful operation of these devices depends on the time it takes the fluid to respond to a driving force and switch state. Typically, the millisecond range is the preferred range of values for τ. However, only recently has τ been measured for an ER fluid. Hill and Van Steenkiste[4] measured response times of an ER fluid consisting of 47 μm diameter glass spheres dispersed in silicon oil. They measured τ values in the 1-1,000 ms range after switching on a 2 kV/mm electric field.

Figure 1 shows an example of fibrile formation in one such experiment.[4] In Fig. 1(a) there is a low concentration (ca. 10%) of 47 μm diameter glass spheres suspended in silicone oil. The gap between the electrodes (black bars at top and bottom) is 600 μm. The electric field is zero, therefore the particles have a random distribution. Under this condition the viscosity is typically low, usually under 100 cP. In Fig. 1(b) an electric field of 2 kV/mm has been applied. The presence of fibriles is obvious, although they are not always linear and even have double loops in some cases. However, the viscosity has increased. This structure is similar

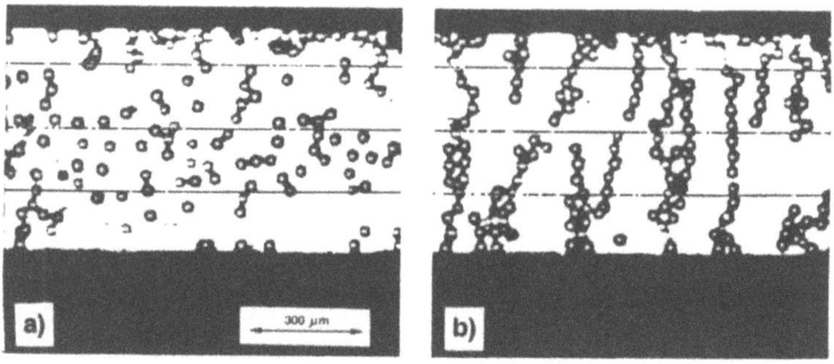

FIG. 1. (a) ER fluid before the application of an electric field; (b) after applying a field of 2 kV/mm.

to that reported by Conrad, Fisher, and Sprecher[5] for fibriles formed by 27-μm-diameter glass spheres in silicone oil at applied fields of 0-3 kV/mm. They found that the structure saturated at E ~1 kV/mm; therefore the 2-kV/mm field employed in this work can be expected to produce a fully developed fibrile morphology.

We have redone such measurements. In spite of some considerable scatter in the data, we found good agreement between the new data and the preliminary data presented in Ref. 4. We shall use the new data to make comparisons with the predictions of the theoretical model. We point out that, as in Ref. 4, sufficient care was taken to keep the particles confined to one plane. This was done to simplify the theoretical model to a two-dimensional problem.

The theoretical predictions were carried out using the leading term in the electric polarization interaction between a pair of spheres. The interaction of the spheres with the walls which confine the fluid were modeled using standard mirror image techniques. In addition, the hydrodynamic viscous resistance to particle motion was modeled by Stoke's law. Finally, the problem was solved only in two dimensions, in accord with the experimental conditions (see above).

As we will show, the dipole-dipole model successfully predicts both the magnitude of the response times and their dependence on the (two-dimensional) particle density.

2. Experiment

2.1 Fluid Properties

We dispersed 47-μm-diameter glass spheres in a silicone oil (Dow Corning No. 561). The glass spheres were treated to have a waterlayer which we found essential for obtaining an ER effect.[4] In order to obtain a reproducible water layer, the glass spheres were placed in a dessicator with a pan of water at room temperature. The water in the dessicator contained a 10 mol solution of lithium chloride, which approximately halved the humidity. An attached hygrometer indicated an interior relative humidity of 39-42%, down from 85-90% without the lithium chloride. This treatment resulted in good ER response and reproducible behavior.[4] Samples of ER fluid were prepared with a volume density on the order of 3-23%.

2.2 Set Up and Procedure

The type of sample cell employed to hold the fluids and apply an electric field is shown schematically in Fig. 2. The area density (ρ) of particles was defined as follows: (1) a known gap width was used to set the image scale, and the total area imaged between the electrodes was then determined; (2) the number of particles was multiplied by their cross-sectional area; and (3) the ratio of (2) to (1) was calculated.

The electrodes were 75-μm-thick copper shim stock simply glued to glass microscope slides. With 47-μm-diameter spheres, we had a reasonable approximation to a two-dimensional system. This reduced overlapping of fibriles and helped identify the first bridging fibrile. The electrode spacing was 600-700 μm.

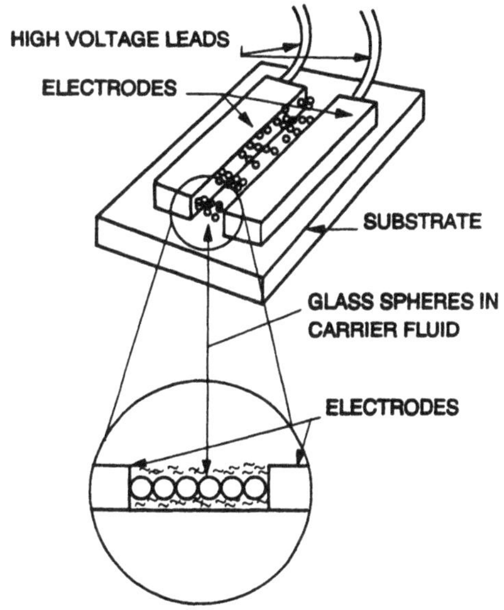

FIG. 2. Schematic drawing of sample cell.

The experimental setup[4] is shown schematically in Fig. 3. The sample cell was placed on the stage of a Reichert polarizing microscope whose eyepiece was replaced by the camera of a Spin Physics System. The overall magnification was either 156X or 387X. The high-speed image-recording capability of the Spin Physics System was essential to measure times for fibrile formation in the millisecond range. For this work a standard rate of frame grabbing of 500 s^{-1} was adopted. Spatial resolution was determined by the 192 x 240 pixels in the Spin Physics imaging array. By normalizing to a known electrode gap at a total magnification of 156X, the pixels were determined to be approximately 6 μm on a side. Therefore the 47-μm-diameter particles employed here were adequately resolved.

Since response times were on the order of milliseconds, arrangements had to be made to switch the power supplies faster than this time scale. DC excitation was with a Hipotronics HV DC power supply, which was set at a constant voltage. Power was switched to the sample cell with a mercury wetted relay. The rise time was about 20 μs and therefore fast enough for our purposes. For all the data reported here, the electric field was maintained at 2 kV/mm.

The onset of particle motion was taken as time zero, corresponding to application of the electric field. The particle motion in subsequent frames was carefully monitored up to a time at which the first bridging fibrile could be detected. The difference in frame count was converted to elapsed time, which was designated the response time (τ).

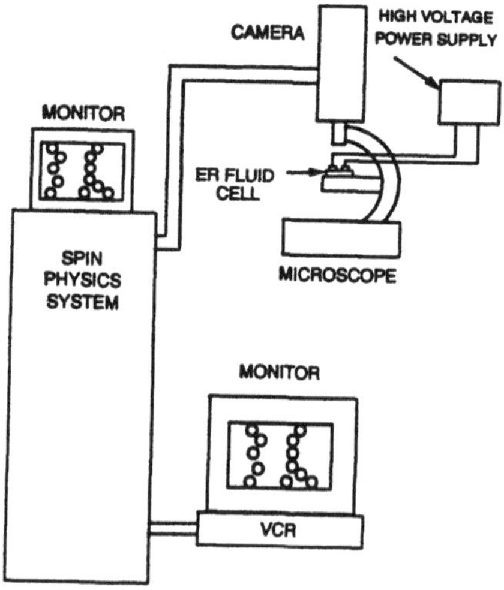

FIG. 3. Overall experimental setup.

3. Results

A plot of response time vs. area density is shown by the solid circles in Fig. 4. The area density ranges from about 0.2 to a little above 0.5, corresponding to a volume density range of 3-23%. The rapid shortening of the response time as ρ increases is typical ER fluid behavior. In spite of the somewhat limited range of ρ, the response time (τ) varies over three orders of magnitude. A regression analysis yielded a fit to an exponential dependence on ρ (Fig. 4, dashed line) characterized by a R-squared value of 0.65. This rather low R-squared value results from the appreciable scatter in the data. However, that is reasonably consistent with earlier results[4], where the R-squared values were 0.62 and 0.86.

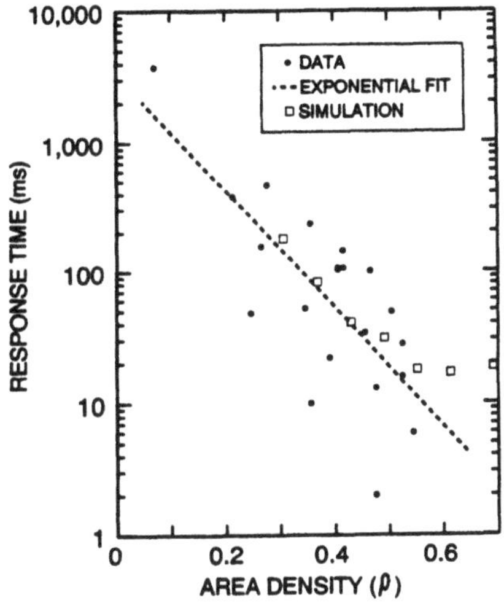

FIG. 4. Response time versus area density: (o) experimental data, (----) exponential fit, ( ) computer simulation.

4. Theory

The electric dipole-dipole interaction model employed here [1] starts with the applied electric field inducing polarization forces between the spheres, and between the spheres and the electrode walls, due to the mismatch of the dielectric constants of the disperse and continuous phases. These forces are essentially dipolar and give rise to the anisotropic, fibrile structures observed in ER fluids. The electric polarization interaction between spheres is treated in a pairwise fashion, ignoring multibody interactions. Only the leading order term in the pair electric interaction, which is that of isolated point dipoles aligned with the external electric field, is retained. The neglect of multibody interactions and higher-order corrections is not expected to produce any qualitative change in the anisotropic electric interaction. The electric interaction between a sphere and a conducting electrode wall is treated as that between a point dipole at the sphere center and the point-dipole images of all the spheres in the suspension reflected (orientation intact) about the electrode surface. This interaction satisfies the condition of the electrode wall being a constant potential surface.

The spheres are subject to hydrodynamic resistance due to their motion through the continuous phase. As an initial approximation, the hydrodynamic resistance is given by Stoke's law. While not exact, this approximation is expected to capture the essential underlying mechanical effect of the continuous phase, e.g., to produce a drag on the sphere surface proportional to its velocity.

Fibrile formation in two dimensions was simulated by randomly placing N spheres or particles in a monolayer in a simulation "box". Electrodes were placed at the 580 μm gap, and periodic boundary conditions were imposed at each end of the sample channel. The pair forces were evaluated within a cutoff radius of five particle diameters. The fibrile structure was allowed to evolve by integrating the equations of motion using a time step of 0.03 ms. All results reported here were obtained with no applied shear.

During fibrile formation, particle cluster statistics were evaluated using the "connectivity matrix" method.[6] Particles are considered connected when they are within a fraction, δ, of a particle radius from each other. By following the length (normal to the electrodes) of each fibrile, it was possible to determine the time of occurrence of the first bridging fibrile (by definition the response time). Cluster statistics were evaluated using δ = 0.01, 0.02, 0.03, 0.04, 0.05, and then extrapolating the results to δ = 0. The calculation using δ = 0.05 was not used in the extrapolation procedure for ρ = 0.6974, because the cluster statistics were considerably different than those obtained using smaller δ. This may be due to the proximity of the hard-disk phase transition, in which freezing leads to the formation of localized clusters that are less likely to form fibriles.

Simulations were performed with N_{runs} (= 20) different initial random configurations for each concentration studied. An average response time was obtained by averaging over the separate runs.

5. Discussion

The results of the theoretical calculations are plotted as the open squares in Fig. 4. Up to an area density of ca. 0.55 the agreement between experiment and theory is very good. For $\rho > 0.55$ the calculated response times deviate significantly from the experimental results. The initial configurations in this range of ρ contained regions possessing a large degree of order in the form of triangularly packed crystallites. This is not surprising, as a two-dimensional system of hard disks (equivalent to a monolayer of hard spheres) undergoes a freezing transition at an area density approaching 0.7. It is known that the presence of the electrodes will lower this freezing concentration, but a quantitative prediction is not available. As the close-packed crystalline regions represent a low energy configuration, even in the presence of an electric field, they are not as likely to rearrange to form fibrile structure. This frustration is assumed to cause the deviation in the response times from the exponential behavior observed at small ρ.

The agreement between theory and experiment supports the electric dipole-dipole interaction as the basic mechanism governing ER fluid behavior, at least up to an area density of ca. 0.55. For our sample cell geometry this corresponds to a volume density of approximately 25%. Above this density it appears that the hard-disk phase transition begins to occur, with the result that the response time saturates at about 10 ms. This is an important result, because it suggests that there will be a lower limit on the response times that can be realized in practical devices. Clearly, other examples of ER fluids should be tested to determine if this is a universal phenomenon.

6. Acknowledgments

The authors thank George W. Smith for useful discussions and Thomas H. Van Steenkiste for his assistance with the experimental work.

(a) Present address: 33 Devonshire, Pleasant Ridge, MI 48069.

(b) Present address: Department of Pathology, University of British Columbia, Vancouver, British Columbia, Canada V6T 2B5

7. References

1. D. J. Klingenberg, F. van Swol and C. F. Zukoski, *J. Chem. Phys.* 91 (1989) 7888.

2. D. L. Klass and T. W. Martinek, *J. Appl. Phys.* 38 (1967) 67.

3. J. E. Stangroom, *Phys. Technol.* 14 (1983) 290.

4. J. C. Hill and T. H. VanSteenkiste, *J. Appl. Phys.* 70 (1991) 1210.

5. H. Conrad, M. Fisher, and A. F. Sprecher, in *Proceedings of the Second International Conference on ER Fluids*, edited by J. D. Carlson, A. F. Sprecher, and H. Conrad, Technomic, Lancaster, PA (1990).

6. E. M. Sevick, P. A. Monson and J. M. Ottino, *J. Chem. Phys.* 88 (1988) 1198.

TRANSIENT RESPONSE OF ER FLUIDS

KATSUFUMI TANAKA, TAMOTSU YOSHIDA and KIYOHITO KOYAMA
Department of Materials Science and Engineering, Yamagata University, Yonezawa 992, Japan

ABSTRACT

Stress response in a steady shear flow to a step electric field and the stress response to sinusoidal strain stimulus with various amplitudes under a DC electric field were measured. It was found that the stress response to a step electric field composed of plural modes, fast and slow modes which did not take place at the same time but in succession. Also, critical strains which divided strain dependence of the storage modulus and loss tangent into three regions were found by measuring the stress response to the sinusoidal strain stimulus. The three regions were closely related to pre-yield region, post-yield region, and yield point.

1. Introduction

Mechanism of ER effects of the suspension system is considered at the present stage that polarized particles in an ER fluid under an electric field, which are dispersed at random under no electric field, begin to collide one another and form associates and bridges of particles (clusters) connecting opposite electrodes, and the clusters cause some resistance (yield stress) against the shear deformation parallel to the electrodes directions. The clusters in the ER fluids were observed under electric fields.[1,2,3] Also, the yield stress was measured experimentally[4,5] and calculated from the models of polarized particles or clusters at conditions of quasi-static shear deformations.[3,6]

The behavior of the particles at a transient state immediately after application of the electric fields can be divided into two processes: The structuring process of the clusters connecting between electrodes and the yielding process of structured clusters by shear deformation. After the transient state, the clusters yield and rearrange repeatedly. Then, a steady state can be attained.

The authors are interested in the experimental study on the structuring and yielding processes in the transient state as mentioned above. The information about the structuring process can be obtained by the measurements of the response time of the ER effects.[7] That of the yielding process can be obtained by the stress-strain relation at a quasi-static shear deformation under static electric fields.[8] In the present paper, these two processes are discussed by the measurements of response time of stress in a steady shear flow after application of a step electric field (for structuring process) and the measurements of the stress response to sinusoidal strain stimulus with a low frequency and various amplitudes under a DC electric field (for yielding process).

2. Experimental

2.1 ER Fluid

The cation exchange resin particles with diameter of about 10 μm were used for the present experiments, which were milled from the particles of Amberlite IR-124. The resin particles were dispersed in a 30 cSt silicone oil after drying and moisturizing for a proper time.[1] Concentration of the particles was 30 wt%

2.2 Stress Response to a Step Electric Field in a Steady Shear

Stress response under a step electric field was investigated by a laboratory developed sliding rheometer with a gap between two plates of 0.5 mm. The frequency sensitivity of the rheometer was up to 2 kHz.

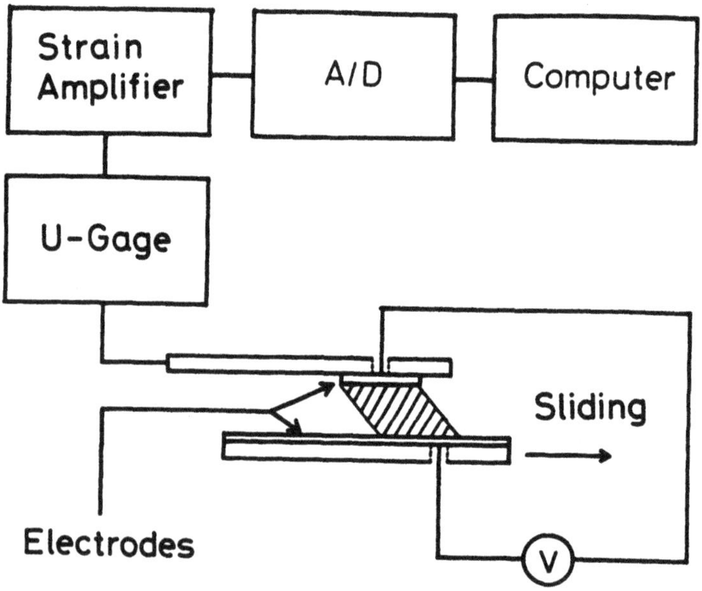

Fig.1 Schematic diagram of the system for measurements by sliding rheometer.

Figure 1 shows a block diagram of our system for measurements by the sliding rheometer. Bottom plate of the rheometer was slid to one direction for measurements. The displacement of upper plate was detected. The detected signal was digitalized by an A/D converter, and stored by a computer. The rate of the digital sampling was varied from 1μs to 20 ms. The electronics for the digital sampling was used commonly for the sliding rheometer and rotating rheometer as mentioned in the next section (2.3).

The fluid samples were shear deformed at constant strain rate about 2.3 s^{-1}. A step signal of an electric field containing frequency components up to 30 kHz was applied to the fluids with amplitude of 1.38 kV/mm. Then, stress response to the step electric field was analyzed.

2.3 Stress Response to Sinusoidal Strain Stimulus under a DC Electric Field

Stress response to sinusoidal strain stimulus was investigated by the rotational rheometer (Rheology Co., Ltd., MR-300) under a DC electric field. The diameter of the inner cylinder was 34 mm and that of the outer cylinder was 35 mm. The height of the inner cylinder was 14 mm. The inner cylinder was the detecting part of the stress and the outer cylinder was driving part of the shear deformation. The amplitude of strain was varied from 2.7×10^{-3} to 1.2 and frequency of sinusoidal strain was set to 0.5 Hz. The strength of DC electric field was set to 0 or 2 kV/mm.

3. Results and Discussion

3.1 Stress Response to a Step Electric Field in a Steady Shear

Figure 2 shows stress response to a step electric field in a steady shear deformation. Shear rate was 2.3 s^{-1}, and amplitude of the electric field was 1.38 kV/mm. At the time zero, the step electric field was applied. The stress as shown in Fig.2 is the difference between the total stress under the electric field and the stress under no electric field. The stress increases and it becomes to saturate.

The stress response is assumed to be expressed by the following equation with the plural modes.

$$\sigma = \sum_{1}^{n} \sigma_i \left[1-\exp\{-(t-t_i)/\tau_i\}\right] \tag{1}$$

$$\sigma_i = 0 \quad \text{for} \quad t-t_i<0$$

where t_i is the time that the i-mode response begins, σ_i is the maximum of the stress response of i-mode, and τ_i is response time of i-mode.

The increasing behavior of the stress response could not be expressed by single mode, but it could be expressed by the plural modes. In Fig.2, solid curve was calculated from Eq.1 with plural modes and broken curves were i-mode components of Eq.1. Stress response can be separated into two modes in Fig.2. One is fast mode at initial stage of stress response with $\tau_1 = 7$ ms. The other is slow mode with $\tau_2 = 100$ ms following the fast mode after $t_2 = 22$ ms. It is found that the stress contribution of the fast mode is about fifth times larger than that of the slow mode.

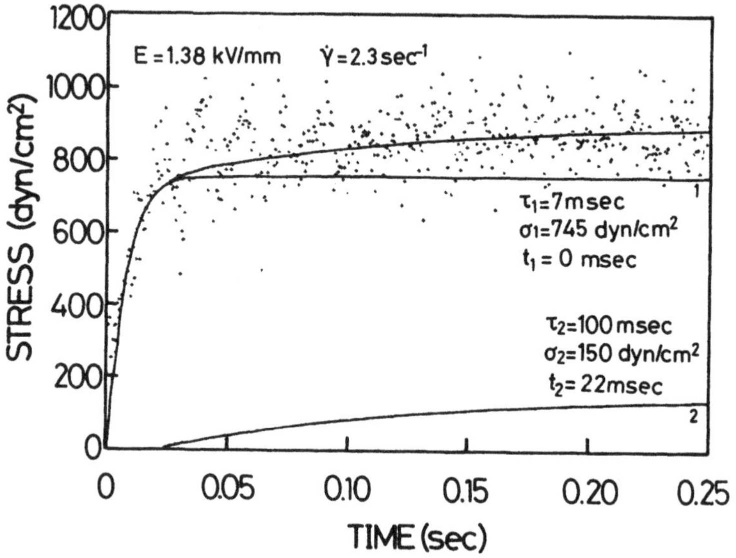

Fig.2 Stress response to a step electric field with amplitude of 1.38 kV/mm.

However, the stress response was very fast. It is difficult to evaluate the fast mode quantitatively, considering the limit of the present system for measurements. It can be thought that the τ_1 mode is synthesized from much faster modes than $\tau_1 = 7$ ms.

In any case, plural modes can be measured in the present experiments. These modes respond one after another from the fast mode to the slow mode. These results can be closely related to the structuring behavior of the polarized particles in the ER fluid.

Figure 3 shows a picture of particles behavior in the transient process.

1.When time is much shorter than $\dot{\gamma}^{-1}$ (= 435 ms), the electric force among the polarized particles dominates the structuring processes. The particles of nearest neighbors aggregate to form the short clusters.

2.Then, the short clusters aggregate to form the long clusters chaining between two electrodes while the electric force dominates in the process.

3. When the time becomes comparable to $\dot{\gamma}^{-1}$, both of the electric force and the hydrodynamic force should be taken accounted. It depends upon the contribution of the electric force and hydrodynamic force whether the long clusters yield or not.

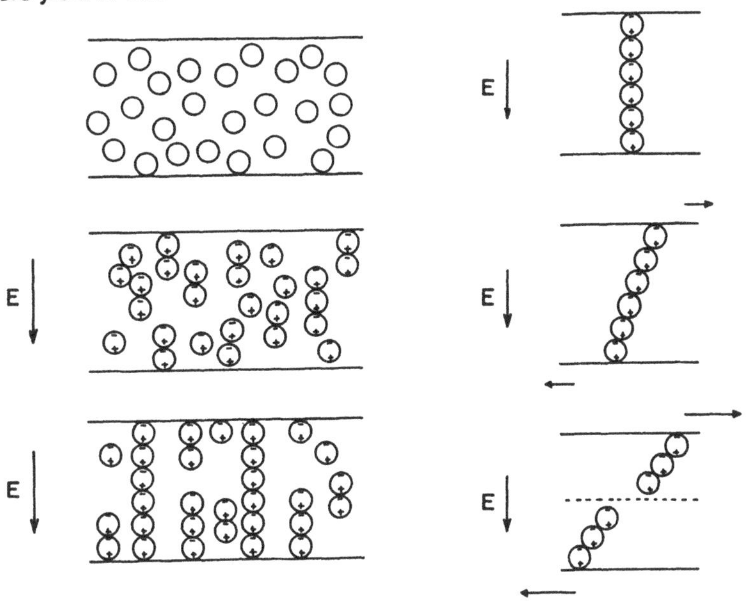

Fig.3 A picture of particles behavior in the transient process.

In the present case, it can be thought that the first and the second processes was fast and took place at the same time before τ_1. The strain calculated from the deformation time of τ_1 is 0.016. It seems to be too small to disturb the structuring process. After the time of τ_1, the third process took place corresponding the τ_2 mode. Then, the long clusters would yield and realign to reach a steady state since τ_2 is comparable to $\dot{\gamma}^{-1}$ and the strain at τ_2 of 0.23 is sufficiently large, as can be seen in the next section (3.2).

3.2 Stress Response to Sinusoidal Strain Stimulus under a DC electric field

Figure 4 shows the examples for sinusoidal wave forms of strain stimulus with frequency of 0.5 Hz and various amplitudes (γ_0). The deformation of the strain stimulus can be assumed to the quasi-static deformation since the frequency is sufficiently low.

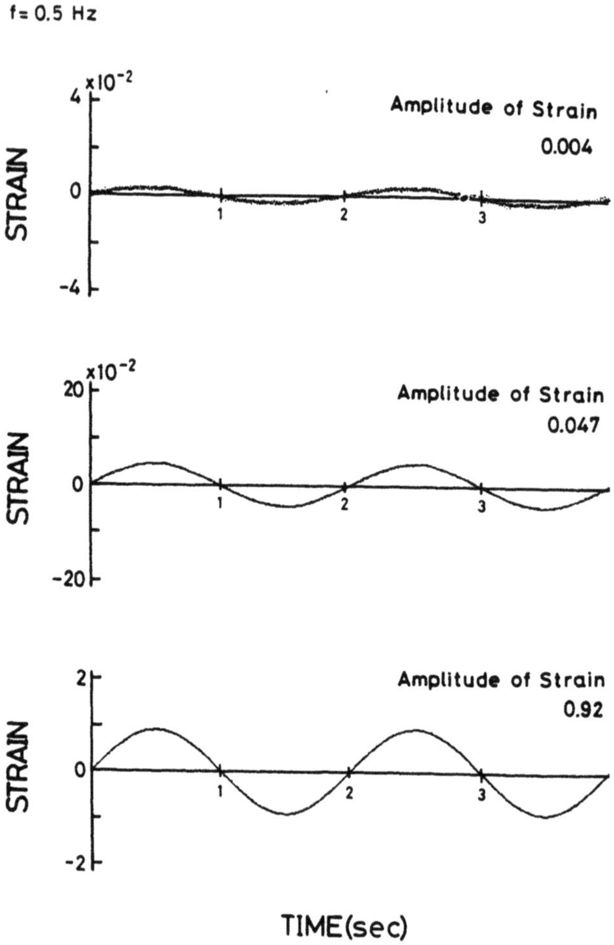

Fig.4 Examples for wave forms of strain stimulus.

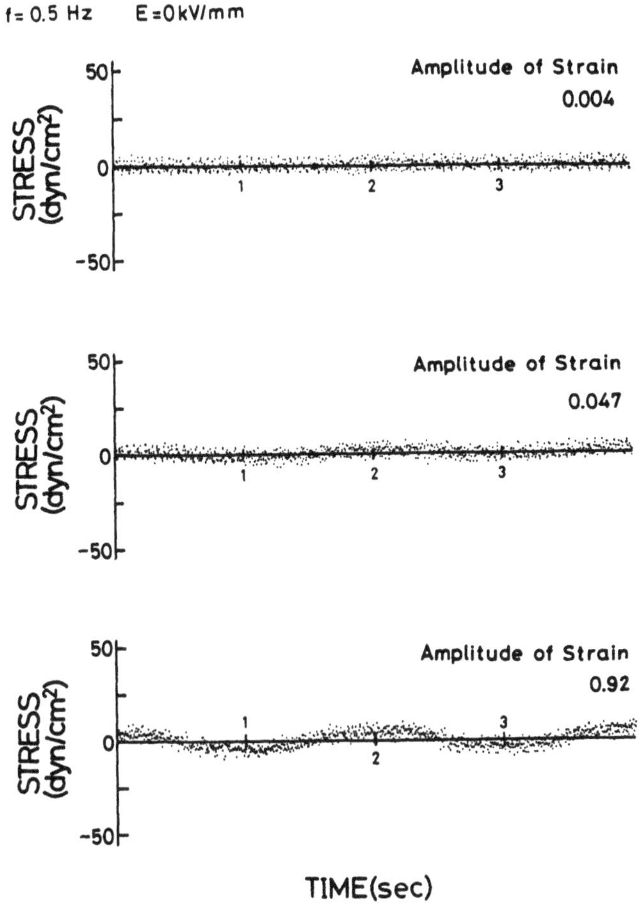

Fig.5 Stress response at 0 kV/mm to the strain stimulus.

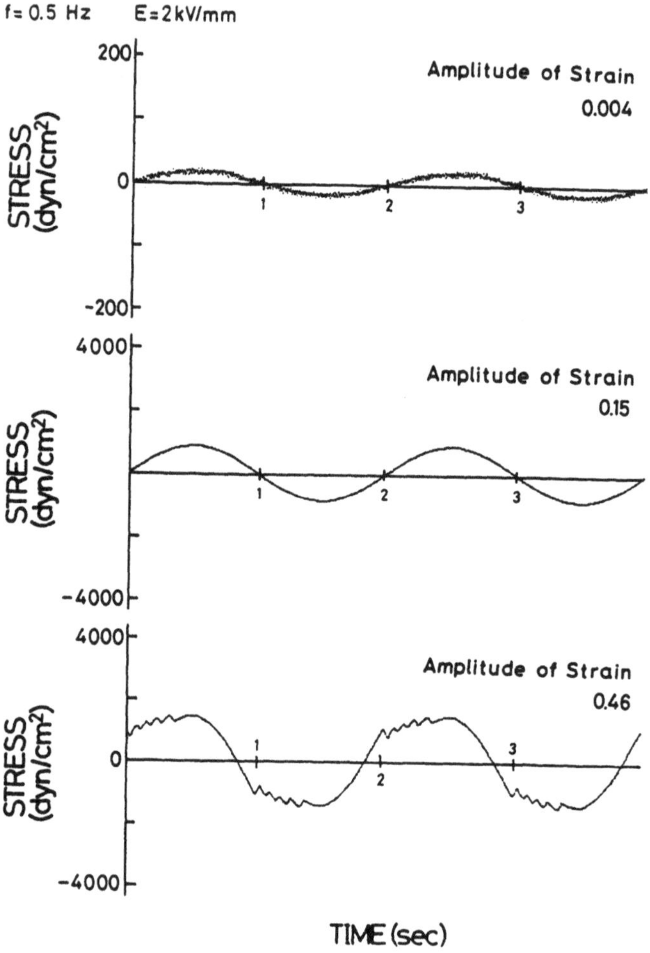

Fig.6 Stress response at 2 kV/mm to the strain stimulus.

Figure 5 shows the wave forms of stress response under no electric field to the sinusoidal strains as shown in Fig.4. The stress responses of γ_0 = 0.004 and 0.047 in Fig.5 are weak. However, it was found by digital filtering technique that the stress responses were linear and the angles of the phase difference between stress and strain (δ) increased and approached to 90° with increasing the strain amplitude.

Under a DC electric field of 2 kV/mm, the amplitudes of the stress waves become much larger than those under no electric field and the wave forms change depending on the amplitude of the strain stimulus as shown in Fig.6. The stress responses of γ_0 = 0.004 and 0.15 are linear and the stress response of γ_0 = 0.46 is non-linear. The stress wave of the non-linear response shows a "cut off" value around the peaks, which can correspond to the yield stress of the long clusters. Also, the amplitude of the stress wave increases with increasing the amplitude of the strain.

From these wave forms of linear and non-linear stress response, the storage modulus (G_1') and loss tangent (tanδ) were calculated by extracting the first harmonic components.[9] Figures 7 and 8 show the logarithmic dependence of G_1' and tanδ upon various amplitudes of the strain at 0 and 2 kV/mm. At 0 kV/mm, log G_1' decreases and log tanδ increases with increasing log γ_0.

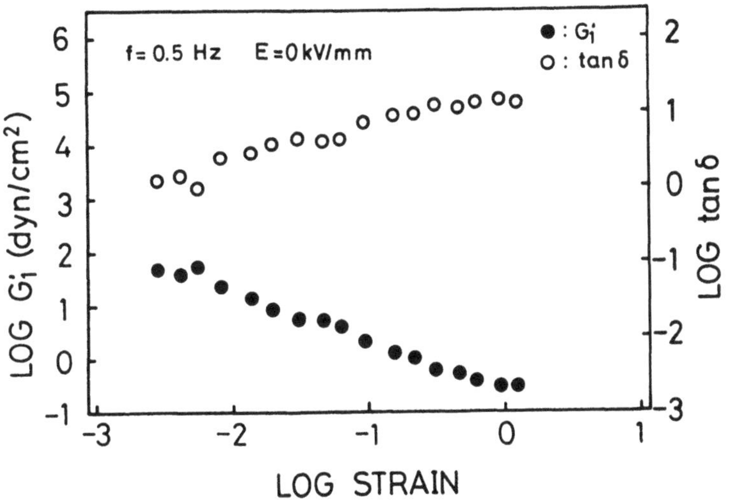

Fig.7 Logarithmic plots of storage modulus and loss tangent at 0 kV/mm as a function of strain amplitude.

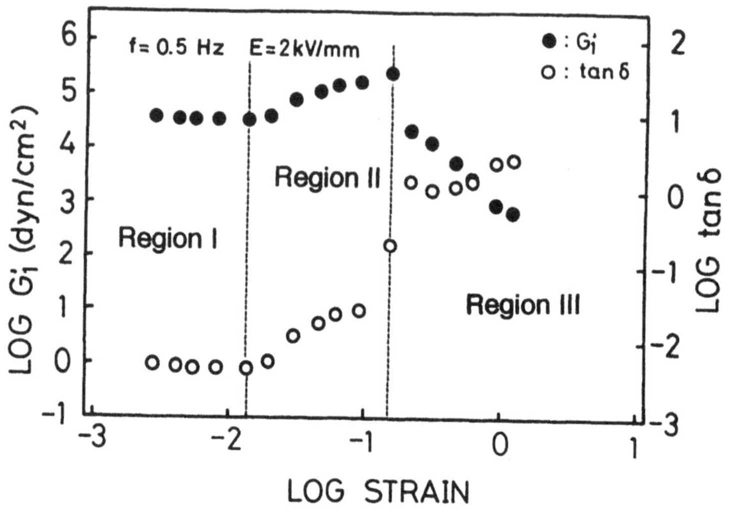

Fig.8 Logarithmic plots of storage modulus and loss tangent at 2 kV/mm as a function of strain amplitude.

The deformation of $\gamma_0 = 1.2$ as shown in Fig.7 is viscous since δ at the strain amplitude is 86°. On the other hand, the strain dependence of log G_1' and log tanδ at 2 kV/mm as shown in Fig.8 is more complicate than that at 0 kV/mm. It can be seen three regions, Region I, Region II, and Region III as shown in Fig.8. In the Region I ($\gamma_0 < 0.013$), both of log G_1' and log tanδ are almost constant and log tanδ is less than -2 ($\delta < 0.57°$). Thus, it can be thought the mechanical properties in the Region I is pseudo-elastic. In the Region II ($0.013 < \gamma_0 < 0.16$), both of the log G_1' and log tanδ increase with increasing log γ_0. In the Region III ($0.16 < \gamma_0$), log G_1' decreases and log tanδ maintains large value above 0 ($\delta > 45°$). Other researchers found only the Region III, which corresponds to relatively large strain amplitude, in their dynamic viscoelastic measurements.[10,11] These three regions in Fig.8 can be associate with the three rheological regions, pre-yield region, yield region (or yield point), and post-yield region.[11]

From the observation of the wave forms, the stress response in the Region I and II was linear. While, the stress response in the Region III was non-linear and yield stress could be seen. Therefore, it can be thought that the long clusters yield at the boundary between the Region II and III. It can be also considered that the Region I and II correspond to the pre-yield region with the viscoelastic deformation of the long clusters, and the Region III corresponds to the post-yield region with the pseudo-plastic deformation of the long clusters. The boundary between the Region II and III corresponds to the yield point.

4. Conclusions

Stress response of an ER fluid was measured for a step electric field. It was found that the response composed of plural modes, fast and slow modes which did not take place at the same time but in succession. The fast and slow modes can be closely related to the structuring and yielding process of the long clusters in the ER fluid.

Three rheological regions were found by measuring dynamic viscoelastic properties with various strain amplitudes. Region I and II in the present study were assigned to the pre-yield region since the transient responses of stress to sinusoidal strains were linear, and the fluid showed viscoelasticity. Region III was assigned to the post-yield region since the responses of the stress were non-linear, and the deformation in the region was pseudo-plastic. The boundary between the Region II and III was also assigned to the yield point since the linear responses changed to the non-linear ones at the point.

5. Acknowledgment

The authors thank to Mr. Y.Seino for the help of some experiments.

6. References

1. N.Sugimoto, *Bull. of JSME*, **20** (1977) 1476.
2. H.Conard, M.Fisher and A.F.Sprecher, *Proc. 2nd. Int. Conf. ER Fluids*, eds. J.D.Carlson, A.F.Sprecher and H.Conard, (Technomic, Lancaster, PA, 1990) p.63.
3. D.J.Klingenberg and C.F.Zukoski, *Langmuir*, **6** (1990) 15.
4. D.L.Klass and T.W.Martinek, *J.Appl.Phys.*, **38** (1967) 75.
5. H.Uejima, *Jpn.J.Appl.Phys.*, **11** (1972) 319.
6. R.T.Bonnecaze and J.F.Brady, personal communication.
7. D.J.Peel and W.A.Bullough, *Proc. 2nd Int. Conf. ER Fluids,*, eds. J.D.Carlson, A.F.Sprecher and H.Conard, (Technomic, Lancaster, PA, 1990) p.141.
8. A.F.Sprecher, J.D.Carlson and H.Conrad, *Mat. Sci. Eng.*, **95** (1987)187.
9. S.Onogi, T.Masuda, T.Matsumoto, *Trans. Soc. Rheol.*, **14** (1970) 275.
10. Z.P.Shulman, E.V.Korobko and Yu.G.Yanovskii, *J.Non-Newton. Fluid Mech.*, **33** (1989) 181.
11. D.R.Gamota and F.E.Filisko, *J.Rheol.*, **35** (1991) 399.

INSTITUTE OF HEAT AND MASS TRANSFER OF ACADEMY OF SCIENCE OF THE BSSR, 2200723, Minsk, USSR

THE INFLUENCE OF THE ELECTRIC FIELD ON ACOUSTIC PROPERTIES OF THE ELECTRORHEOLOGIC SUSPENSIONS

KOROBKO E.V

and

ZAVTRAK S.T.

ABSTRACT

The experiment has found the space anisotropy of the sound velocity in ERF placed in the external electric field. It is caused by internal structural reconstructions of the solid phase which generate a bridge oriented along the vector of the electric field strength in clearance of the capacitor.

It has been showed that the sound velocity increases with the field intensity growth in the case of the wave vector and electric field strength co- directivity.

No US velocity change has been detected at their normal location within the limits of the experimental error. Also, it has been performed the measurement of the US signal attenuation, the value of which increases up to 150%. at the expense of the field influence.

The question of the electric field influence on the ERS parameters has been theoretically considered.

The potential energy and drag coefficient values have been calculated. The calculation of the sound velocity increment has showed the satisfactory coincidence with experimental data. The US velocity depends linearly on the field strength and also, determines by the disperse particles volume fraction in suspension.

1. Introduction

The question of the ultrasonic waves amplitude or phase change passing through the electric field has been experimentally and theoretically examined in a range of works / 1 – 5/.

So, A.Baron and A.Diacomini have studied the electric field influence on orientation of the constant molecular dipoles as well as the change of viscosity and compressibility of the "pure" polar field. It has been disclosed a small change of the oscillations frequency at impressing 4kV potential. At this, the electric field strength vector has been parallel to ultrasonic wave propagation.

A.Nolle /4 / going from assumption on the sound velocity change in fluids and experimental results of the preceding authors has checked the hypothesis about the ultrasonic phase modulation by electric field at their perpendicular orientation.

It turned out that the amplitude modulation in all fluids is absent, i.e. the electric field doesn't influence on the acoustic attenuation. A.Bonetti / 5 / has confirmed this conclusion in the experiments with nitrobenzene having showed that the US velocity change is absent either.

The situation changes in the case of the imperfect dielectricfluids: for conducting fluids the phase modulation has been detected and its square-law variation from the electric field strength has been defined. When locating the particles of adsorption material (glass wool) in the field, A.Nolle also noted a considerable amplitude and phase modulation explained by him as mechanical motion of the glass wool filaments. Kindred results have been obtained for magnetic fluids and suspensions at magnetic field on influence on them /6,7/.

This study seeks to investigate the particularities of the US propagation in electrosensitive fine disperse fluid flow suspensions (ERS) showed electrorheologic effect at different intensities of electric action. This work is of scientific interest, first, as a new lead in electrorheology – the wave processes in structurizing fluids, second, in contrast with methods of vibration viscometry, the US gives the possibility of more exact estimation of real values of the ERS elasticity properties, obtained without the ERS destruction. The practical expediency of these studies is associated with perspective to create controlled acoustic devices with retuning parameters of the type of echo- ranging sonars, delay lines, filters and matching interlayers in flow detectors as well.

1.1 General notions. Particular estimates

The US propagation in a conditionally infinite medium having bulk elasticity and viscosity describes by wave equation $\Delta\Psi = \frac{1}{C^2}\frac{\partial^2\Psi}{\partial t^2}$, where $C^2 = \frac{\partial P}{\partial \rho}$ – propagation velocity of the elastic deformation, Ψ – velocity potential. The Hooke's law gives the dependence between deformation velocity and bulk modulus K in the form $C = \sqrt{\frac{k}{\rho}}$ (1) /8/.

Passing through the medium under review, the harmonic acoustic oscillations amplitude changes in an exponential law with distance: $A = A_0 \exp(-\alpha x)$ (2). where α – absorption coefficient of acoustic oscillations in the medium. The absorption coefficient in ERS as in other micro-inhomogeneous media is written as follows: $\alpha = \alpha_0 + \alpha_s + \alpha_\eta + \alpha_T$ (3) where α_s – scattering factor, α_0 – absorption coefficient in disperse medium, α_η – absorption coefficient caused by friction between suspended particles of solid phase and disperse medium, α_T – absorption coefficient defined by heat exchange between particles and medium.

The suspensions of the fine-disperse variety of the silica the diatomite in transformer oil with appropriate SAS and activators additives have been used in our experiments /9/. The losses at the cost of the scattering at particles with the size of 1-10 mcm in the range of the frequencies < 10MHz can be estimated taking into account the expression for scattering section $G = \frac{4}{a}\pi a^2 (ka)^4 (\frac{1-\beta'}{\beta})^2$ at considerable wave length λ >> a in the form $\alpha_s = \frac{1}{2} nG$ /10/.

So, let as take a = 5mcm, $\beta' = 0$ for f = 3,2MHz we get $G = 7,3 \cdot 10^{-19} m^2$, a = particle radius, $k = 2\pi/\lambda$, λ - wave length, β' - medium compressibility, β - particle compressibility.

At diatomite concentration φ = 0,2 (20%) from expression $n = \frac{3}{4} \cdot \frac{\varphi}{\pi a^3}$ we evaluate that $n = 4 \cdot 10^{14} m^{-3}$. Then $\alpha_s = 1,5 \cdot 10^{-3} m^{-1}$. If correlating this result with the value of the dispersion media absorption coefficient $\alpha \approx 100 m^{-1}$ well known from literature, then it can be considered with any assurance that acoustic losses in ERS are not determined by distance.

As a rule the viscosity component brings the main contribution in α for fluids. If considering to a first approximation the viscous strength to be proportional to the deformation velocity, then a wave equation for particles displacement can be obtained:

122

$$k \frac{\partial^2 x}{\partial x^2} + \eta \frac{\partial^2 x}{\partial x \partial t} = \rho_o \frac{\partial^2 x}{\partial t^2}, \quad \eta = \eta_o + \frac{4}{3}\eta_\varepsilon \qquad (4)$$

where K - characterizes the medium elasticity relative to its space compression, η_o, η_ε volume and shift viscosities.

The solution Eg.(4) is found in the form of the exponent Eg.(2) where α is given by expression Eg.(5):

$$\alpha = \frac{\omega^2}{2\rho_e C^3} [\frac{4}{3}\eta_\varepsilon + \eta_o + \varkappa (\frac{1}{C_v} - \frac{1}{C_p})] \quad (5)$$

Here ω - angular frequency of the acoustic oscillations, ρ_e - fluid density, $\varkappa$ - thermal diffusivity, C_v, C_p - heat capacities of the medium at constant volume pressure correspondingly.

It is known that the losses through heat conduction turn out to be significant only in molten metals and liquid helium. As for majority of fluids they don't exceed several percents.

As it has been established in /11/, the ERS thermal conductivity and thermal diffusivity coefficients increase in several times anisotropically along the vector of the field strength. Consequently, different values at α_T can be expected depending on co-directivity $\bar{E}$ and $\bar{C}$ and the last summand can't be discounted. By this means, having estimated the absorption coefficient and US propagation velocity with the help of the acoustic measurements, it is possible to calculate the ERS viscosity and elasticity with formulas (1) and (5).

3. The experimental studies

The measurement of the ERS acoustic characteristics in electric field has been performed by two methods:

1. pulse (direct)
2. interferometric (indirect)

The both of them are described in an appropriate literature /10 ,12/. In the first case, the US propagation velocity has been detected by the pulse autocirculation by frequency and distance change between source and receiver of oscillations on scheme depicted in fig.1 .

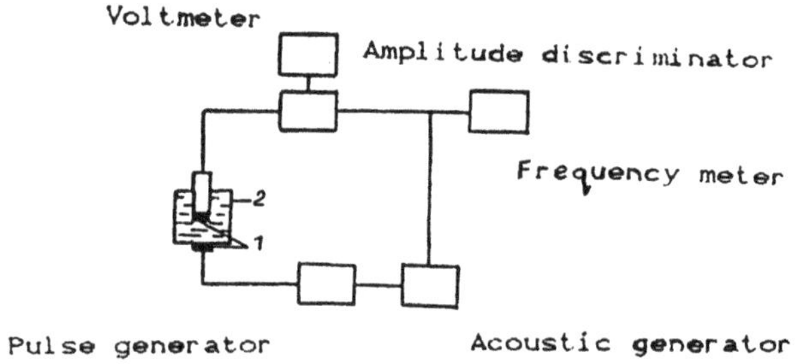

Fig 1. Scheme of the set for US velocity and attenuation coefficient measurment. 1 - receiving and emitting piezoelectric ceramics, 2 - dish.

The master oscillator generates the rectangular pulses with amplitude of 10V and duration of 3 - 5 mcs with recurrence rate of 10 kHz. The pulses generator is controlled with their help. The pulses generator generated radiofrequency pulses with maximal amplitude of 40V with duration of 2mcs of the required r.f.carrier. Under the action of the r.f.pulse the emitting piezoelectric cera mics transform the energy of these waves in energy of electric oscillations. The are analyzed in an amplitude discriminator. If maximal amplitude of r.f.pulse ís more or equal to the voltage reference V_c, then the amplitude discriminator produces a pulse for controlling the master generator which goes on to autosynchronization mode. In this case the frequency of the master oscillator change of state is determined by the time of the ultrasonic pulse propagation from radiator to receiver and delay time in electronic lay-out. The master oscillator change of state time and hence, the propagation time of the US pulse are measured by frequency meter. The voltage reference value is proportional to the pulse maximal amplitude and is measured with the help of the digital voltmeter.

The US propagation velocity has been computed on the measured autocirculation period from the formula $C = \frac{h^* + x_o}{t}$,where h^* - index of distances indicator, x_o - parameter proportional to the delay time in electronic lay-out, t - autocirculation period. The presence of the modified acoustic base affords to calculate C and X without preliminary calibration of the set. The ulatrasound damping factor is calculated on the bases of assumption that maximal amplitude is proportional to the comparation strength $A = V_c \cdot b$, where b - proportionality factor. Measu- ring h and corresponding to it V_c ,it is possible to calculate absor- ption coefficient with formula $\alpha = \frac{1}{h^*} V_c/V$.

The interferometric method as more perfect in selected frequencies range has been additionally used to define α values. In conformity with it, the relation between increment in velocity and phase change of the received acoustic signal φ^* with frequency f passed the distance h in the medium is described by expression:

$$\Delta C = \frac{C^2}{2\pi f h}\,\Delta\varphi^*$$

The measurements have been executed in following manner (fig.2).

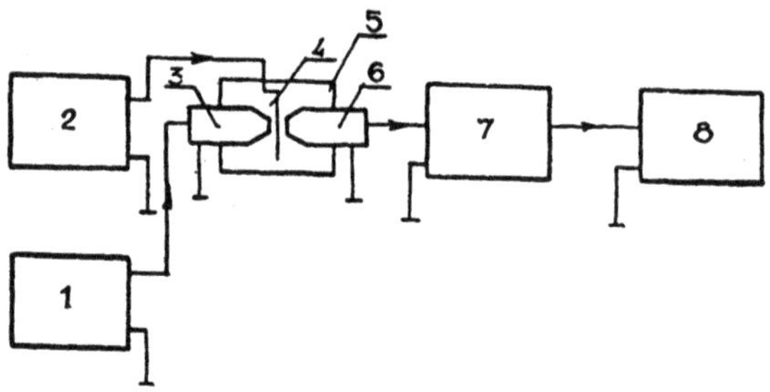

Fig. 2 Scheme of the measuring set for study of the longitudinal electric field influence on the acoustic waves propagation velocity and absorption coefficient in ERF. 1 - high frequency generator, 2 - high voltage source, 3,6, - emitting and receiving electroacoustic converters, 4 - metallic electrode, 5 - measuring chamber, 7 - amplifier, 8 - amplitude detector.

The frequency of the sounding signal selects by tuning the high frequency generator. The electric voltage of high frequency is fed from generator's output 1 on the radiator 3 which excited longitudinal acous tic oscillations of the same frequency. Having gone through the ERF layers under investigation and electrode 4 made from thin metallic foil with thickness 0,1 mm what << X the acoustic oscillations excite elec tric voltage on the receiver's output which feds on the amplitude detector's input 8 through amplifier 7, the oscillograph being used as amplitude detector. The value of the absorption coefficient has been determined from formula by voltage amplitude U_E and U_o measurement.

$$\Delta\alpha = \frac{1}{h} \ln \frac{A(x)}{A_o} = \frac{1}{h} \ln \frac{U_E}{U_o}$$

where U_E , U_o - signal amplitudes on the transformer's output in the field and without it.

In the case of the longitudinal and transversal electric field different acoustic cells have been used. For studying transient processes at the field switch-on and switch-off, the signal from measuring chamber has been fed on amplitude detector through amplifier and the voltage from output of the latter - on the storing oscillograph.

The selected procedures of study of the electric field influence on the ERF acoustic parameters embody a considerable merit residing in the fact that the application of the continuous low amplitudes acoustic waves rules out their destructive influence on the medium structure enabling proper measuring of its viscosity and elasticity parameters.

3. Discussing the obtained results

1. The ERF absorption coefficient

From the obtained results given on fig.3-6 the most interesting is the experimentally determinated difference in degree and direction of the acoustic oscillations absorption coefficient in the studied ERF according to their propagation direction with respect to the vector of the electric field strength.

Thus, at the longitudinal field action, the acoustic losses growth have been observed in frequencies range of 2 ÷ 10MHz at field strength up to 30kV/sm regardless of the solid phase concentration (φ = 5 - 40%, weight). At this the acoustic losses increase with the longitudinal field strength growth, but $\frac{d(\Delta\alpha)}{dE}$ monotonically decrease.

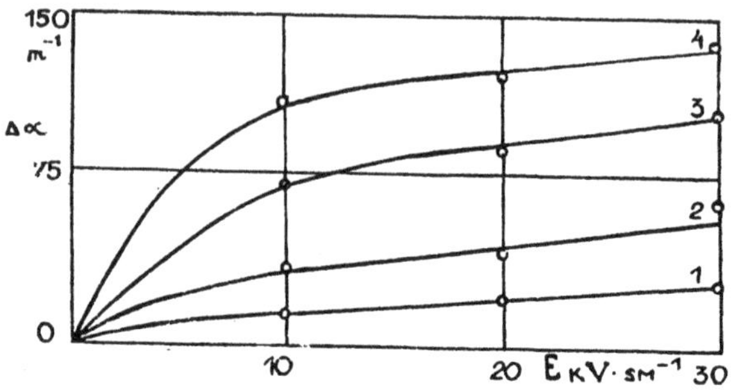

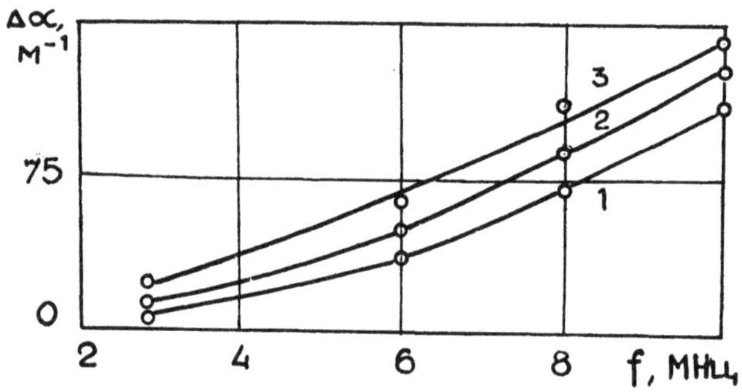

Fig. 3 Dependence of the acoustic oscillations absorption coefficient α increment from the logitudinal electric field strength at solid phase consentration φ = 10%a at frequencies f : 1 - 2,6 MHz, 2 - 6, 3 - 8, 4 - 10. Fig. 4 Dependence of the acoustic oscillations absorption coefficient Δα increment from frequency for ERF on the basis of the transformer oil at φ = 20% and electric field strength E = 1 - 10, 2 - 20, 3 - 30kV/cm

It is likely to be connected with increase of the degree and quantity of particles oriented by the field.

As it follows from fig.4 the dependence $\Delta\alpha = \Delta c(f)$ is nonlinear either. In compliance with /13/ the dependencies of absorption in solid and liquid phases from f are described by laws: $\alpha_s = af + bf^4$, $\alpha_\ell = kf^2$.
Our experimental data show that the dependence $\alpha = \alpha(f)$ behavior turns out to be more complicated. The statistical analysis gives that for each ERF concentration the dependence of the form is $\alpha = \alpha(f, f^2, f^3, f^4, E)$. In frequencies range we studied it is possible to use the formula $\Delta\alpha_1/\Delta\alpha_2 = (f_1/f_2)^n$ where n varies within the confines from 1,3 to 1,65 depending upon concentration and field intensity.

The influence of the transversal electric field at low

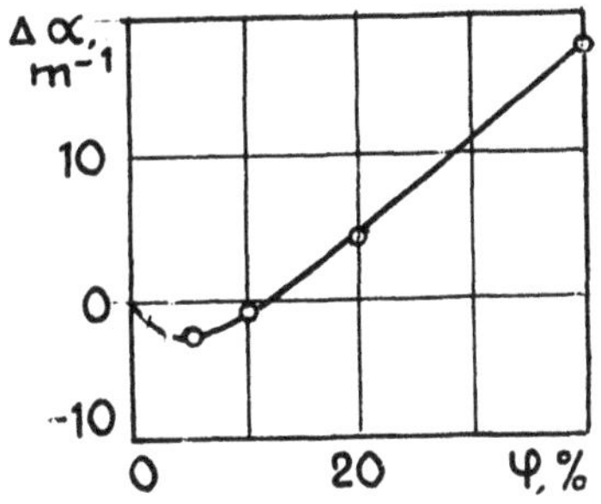

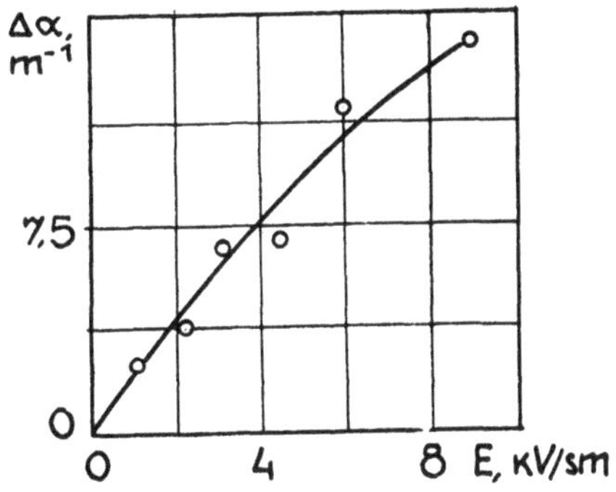

Fig. 5 Dependence of the acoustic oscillations absorption coefficient $\Delta\alpha$ increment from the from solid phase φ concentration in electric field with strangth E = 10kV/cm and acoustic oscillations frequency f = 3,8MHz.
Fig. 6 Dependence of the acoustic oscillations absorption coefficient $\Delta\alpha$ increment from the transversal electric field strength φ = 40%, f = 3,8MHz

concentrations ($\varphi = 5 \div 43\%$) boils down to the absorption coefficient curtailment and at string ones ($\varphi = 13 \div 40\%$) to its increase (fig. 5,6). The transient processes in the ERF of different concentration follow in different manner. So, for instance, at diatomite content $\varphi = 10\%$ the value of the characteristic time of structural reconstructions termination at field switch-on will be in 0,3 - 0,5 area. At $\varphi = 20\%$ it fluctuates from 0,1 (in the case of longitudinal field) up to 2 - 4s (in transversal field) fig.7. Depending upon concentration, the characteristic time varies as well at the field switch-on: from 0,3s at $\varphi = 40\%$ in transversal field up to 7,0 at $\varphi = 20\%$ in longitudinal one.

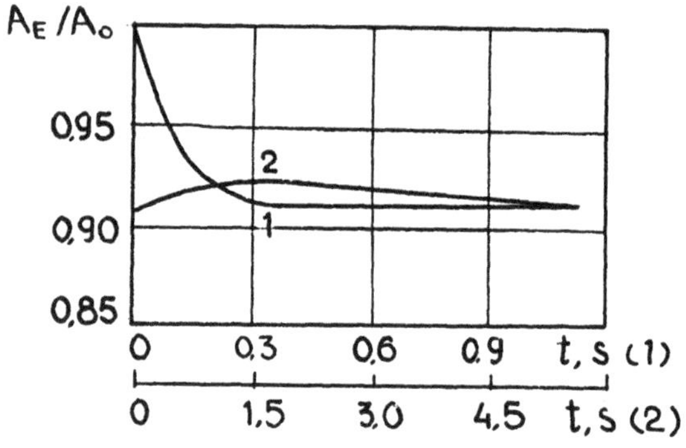

Fig. 7 Time dependence of relative change from amplitude of the received acoustic oscillations with frequency 3,8MHz. $\varphi = 20\%$. 1 - at switch-on of the lonitudinal electric field with strength E = 12,5kV/cm, 2 - at switch-off.

The noted above enables to analyze the obtained results proceeding from classic knowledge which describes the acoustic oscillations absorption coefficient by expression (5).

It is evident that the electrostatic field influence on the ERF is connected with the solid phase particles polarization in direction of the vector of strength $\bar{E}$. Taking into account the ir

regular shape of the diatomite particles, it can be suggested that along the polarization of particles their space reorientation takes place. This motivated the allocation of the semi-major axes of ellipsoids also along the vector $\bar{E}$ (this assumption steams from any system's tendency in potential field - in the given case electrostatic one - to the minimum of its potential energy). In this manner something of the ordering relating to the vector of strength E of the ERF structure occurs which might lead to the absorption coefficient change. At this it should be taken into account that the solid phase particles concentration increase, the distance between the latter reduces and the role of these particles interact--- with each other increases. If the presented arguments ho-- true, then it is evident that the indicated ordering of the solid phase - particles should lead to the ERF properties anisotropy owing to which the shift viscosity coefficient η_c in longitudinal field should be lesser and in transversal - bigger. The direct measurement of the shift velocity carried out by numerous researchers /9/ bears out this assumption. But then, acoustic losses motivated by the shift velocity should be lesser at the acoustic oscillations propagation along vector $\bar{E}$. At the same time, as it follows from plots on fig. , the overall absorption coefficient change in longitudinal field is positive and in transversal - negative (at φ = 13%), i.e., opposite to results of the direct viscosity measurement. From this it is inferred:

the electric field action really leads to the ERF properties anisotropy, specifically, the shift viscosity anisotropy but not only of it. It is necessary to take into consideration the possibility of existence of different values of the thermal diffusivity for direction coinciding with vector of electric field strength E and direction perpendicular to it [14]. Furthermore, the electrostatic field is able, depending upon its direction, to influence differently on the quasiequilibrum energy exchange between progressive and internal degree of freedom of each ERF particles at acoustic oscillation propagating in it /10/ i.e., the dilatational visco sity coefficient η_0 as well as the shift viscosity coefficient η_c, can depend upon direction of the vector $\bar{E}$. The most likely is, also, taking into account the data of the work [7] it should be presupposed that the noticed above the increment of the acoustic oscillations absorption coefficient in transformer oil in longitudinal and transversal electrostatic fields is connected precisely with difference in values of the elasticity.

It is probable that at low concentrations of the diatomite the dilatational coefficient in direction perpendicular to the electric field strength vector, decreases at its switch-on. Its contribution to the acoustic oscillations absorption coefficient turns out to be bigger than the shift viscosity coefficient growth in this direction what leads to the decrease of α. If the dilatational coefficient decreases in longitudinal field then its contribution in the acoustic losses coefficient is smaller than the shift viscosity one, that is why the growth of α observes.

At high concentration of the diatomite the main reason of the acoustic oscillations absorption coefficient increase in ERF under action of the electric field (longitudinal and transversal one as well might be a considerable increase of the shift viscosity coefficient. The rheologic effect builts up. The pulse delay time measurement at perpendicular orientation of vectors C and B has not detected any changes of US velocity in ERF.

3.1. The US velocity in ERF

Preliminarly, it has been performed the experiments for transformer oil (in delivery condition C = 1420 m/s) in which the ultrasonic vawe attenuation and velocity change is not detected. The experiments results for ER-fluid are presented on fig.8 in the form of dependence of the ultrasound velocity relative change from electric field strength at converters' resonance frequency 1 MHz, as well as in the table:

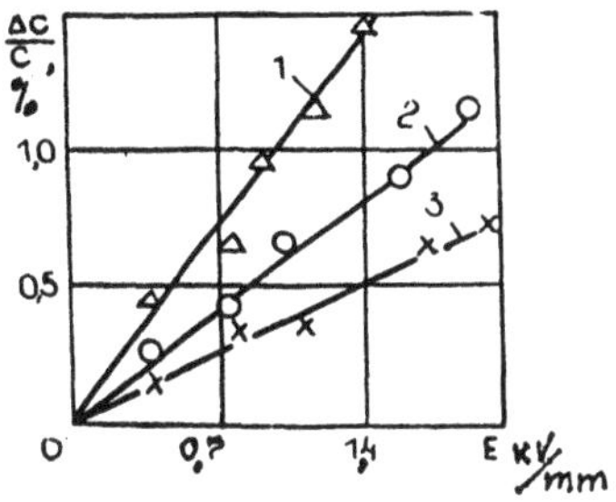

Fig.8 - Dependence of the US velocity relative change from electric field strength at solid phase content 1 - φ = 10%, 2 - 5, 3 - 2.

The average values of velocity and attenuation coefficient in fluids of different concentration

Characteristic	φ,%				
	0	5	10	20	30
	1428.2	1438.6	1433.0	1429.4	1436.2
δ_C	6.001	0.0005	0.001	0.0006	0.001
C_E	1431.3	1440.1	1434.8	1430.7	1438.4
α	0.035	0.196	0,336	0,665	0,917
δ_α	0,08	0,06	0,01	0,02	0,04

Note: C_0 – the ultrasound velocity at E = 0; δ_C – relative mean-square velocity measurement error; C_E – velocity at E = 1,5 kV/mm; α – ultrasound attenuation coefficient; δ_α – α measurement error.

The ultrasound velocity growth has been observed at electric potential increase. At this it has been ascertained the unique φ solid phase concentration influence on dependence $\Delta c/c \sim (E)$.

The observed effect strenghtens with the growth of φ. The impulse delay time measurement at perpendicular orientation of vectors $\bar{c}$ and $\bar{E}$ hasn't revealed changes in US velocity in ERF.

The US velocity decrease at the solid phase concentration growth in absence of the field can be explained on Lyakhov's theory /15/. He has showed that the sound velocity in the two- component medium liquid – solid is determined by a sufficiently complicated dependence from volume content of the solid component:

$$C^2 = \frac{1}{\rho_o}\left(\frac{a_e}{\rho_e C_e} + \frac{a_s}{\rho_s C_s}\right)^{-1}$$

Here $\rho_o = a_e\rho_e + a_s\rho_s$, where $a_e\rho_e$ – volume content and density of the dispersion medium, C_e – sound velocity in it, a_s, ρ_s, C_s – an appropriate parameters for disperse phase.

It is possible to interpret the obtained data in following manner. The visual observations and photographing have showed that under electric field influence the particles of the solid phase surrounded by activator's shells, create firm aggregates, den- dritic generations and bridge structures oriented along the strength vector. With growth of the electric potential the forces between solid phase particles and between plates and bridges' ends increases. The additional particles pull in the structural framework, the bridges become more stronger. If the medium has been quasihomogenous, then now it shows a pronounced lamination. Something like additional rigid channels have been created for ultrasonic, the crossing velocity through which is higher than it has been in liquid suspension.

The shift forces $\bar{F}$ estimate in rheologic experiments, the forces leading to the bridges destruction has showed that the particles interaction increases with the filed strength growth $\sim E^2$ and at vectors $\bar{E}$ and $\bar{F}$ co-directivity it is one order stronger than in the case of $\bar{E} \perp \bar{F}$. Beside that, the deformation of the disperse particles volume should occur in electric field, i.e., the so called electrostriction factor shows. Let consider it in more thoroughly. It is known that the electric field influence freely changes the dielectric energy /disperse particle/, located in another liquid dielectric. This change is defined by a well known expression:

$$F - F_o = -\frac{1}{8\pi} \int \bar{E}_o \, (\bar{D}_{in} - \varepsilon_o \bar{E}_{in}) dV \qquad (6)$$

Here E_o - electric field strength in pure liquid in the absence of the dispersion particles.

D_{in}, E - vectors of induction and field strength inside of the solid dielectric,

ε_e - dielectric permitivity of liquid

ε_p - dielectric permitivity of solid particle

The calculation of (Eq.6) gives /16/:

$$F - F_o = -\frac{E_o^2 \; R_o^3 \; \varepsilon_e(\varepsilon_s - \varepsilon_e)}{2(2\varepsilon_e + \varepsilon_s)} \qquad (7)$$

R_o - particle's radius.

Taking a derivative of expression (7) on the particle volume $V_o = \frac{4}{3}\pi R_o^3$ we find the volume of effective pressure deforming the particle as:

$$\Delta p = \frac{\partial\ (\Delta F)}{\partial\ V_o} - \frac{3}{8\pi}\ \frac{\varepsilon_e E_o^2 (\varepsilon_s - \varepsilon_e)}{(2\varepsilon_e + \varepsilon_s)} \qquad (8)$$

At E = 20 kV/cm the value Δp crosswise several Pa. At this the particle volume change /17/ makes up:

$$\frac{\Delta V}{V_0} \approx - \frac{\Delta p}{\rho_s C_s^2} \sim 10^{-9} \qquad (9)$$

(here ρ_s and C_s - correspondingly, the particle density and sound velocity in it).

By this means, the volume static deformation at the expense of the electric field is extremely small for solid particles.

If calculating gas bubbles instead of solid particles, for in- stance, in the form /18/ the relative change of their volume can reach very considerable values owing to the great compressivity of the bubbles.

As it showed in the work /18/ the sound velocity change in liquid with disperse inclusions is determined by minor parameter $\varepsilon^* = \Delta V/V_\alpha$.

Inasmuch as the relative sound velocity change has reached values of the order of 1% at E ~ 20kV/sm in experiment with electrorheologic suspensions, then this change can not be explained be electrostriction, i.e. particles volume deformation at the expense of the external electric field.

By this means, the explanation of the experiment's results should be searched in manifestation of the electric type interaction between particles.

Thus, let consider a sound wave propagating in electrorheologic suspension at presence of the external electric field.

Let write equation for effective pressure action on i - order disperse particle /17/

$$P(\bar{r}_i) = P_{ext}(\bar{r}_i) + P_s^m(\bar{r}_i) + P_s^d(\bar{r}_i) \qquad (10)$$

Here $\bar{r}_i$ - radius - of the i - order center vector in liquid,
$P_{ext}(\bar{r}_i)$ - pressure of the external sound wave;
$P_s^m(\bar{r}_i)$ - pressure created by oscillations
(at the expense of the other particles volumes change);
$P_s^d(\bar{r}_i)$ - pressure created by dipole oscillations of the other particles centers.

We suppose in formula (10) that the dependence from the time is determined dy exponential factor $\exp(-i\omega t)$. ω- frequency of the external wave.

The pressure $P_s^m(\bar{r}_i)$ can be easily expressed through the sound scattering amplitudes f_i on disperse particles in monopole approximation /17/.

$$P_s^m = \sum_{j\neq i} f_j \frac{e^{ikr_{ij}}}{r_{ij}} P(\bar{r}_j) \qquad (11)$$

where $k = \omega/C_e$, $\bar{r}_{ij}$ - relative distance between particles, $P(\bar{r}_j)$ - effective pressure in the j - order particle position.

In an analogous way one can express the pressures conditioned by dipole oscillations /17/, /19/.

$$P_s^d(\bar{r}_i) = \sum_{j\neq i} \frac{\rho_e \omega k^2 R_j^3}{2} (\bar{U}_j \bar{n}_{ij}) h_1(kr_{ij}) \qquad (12)$$

Here R_j radius of the j - order particle, $\bar{U}_j$ - amplitude of oscillations velocity of the j - order particle's center, $\bar{n}_{ij}$

$= (\bar{r}_i - \bar{r}_j)/\bar{r}_j$ unit vector, $h_1(z)$ - spherical Bessel's function of the 1st type /20/.

For set of equations (10) - (12) be closed it is necessary to link U_j with effective pressure $P(\bar{r}_j)$.

It is known /19/ that the amplitude of the solid particles oscillations velocity U_j is connected with the surrounding liquid velocity amplitude V_j in linear way:

$$\bar{U}_j = \alpha_j V_j = \alpha_j \frac{\bar{\nabla}_j \; P(\bar{r}_j)}{i\omega\rho_e} \tag{13}$$

The coefficient α_j will be calculated somewhat below. At this its dependence from electric field strength will be taken into ac- count. Also, a well known relation from linear acoustics /17/ is used in the last formula

$$V = \bar{\nabla} \; P(i\omega\rho_e)$$

Substituting Eq.(11) - (13) into equation (10) we have

$$P(\bar{r}_i) = P_{ext}(\bar{r}_i) + \sum_{j\neq i} f_j \frac{e^{ikr_{ij}}}{r_{ij}} P(\bar{r}_j) + \sum_{j=1} \frac{\alpha_{ji} \, kR^3_j}{2} \bar{\nabla}_j \; P(\bar{r}_j) \cdot \bar{\nabla}_i h_0 \, (kr_{ij}) \tag{14}$$

The link between Bessel's spherical functions /20/ has baen incor- porated in formula (14)

$$h_1(z) = - h_0'(z) = (\frac{1}{z} e^{iz})' z$$

We now turn to the continuous particles distribution in liquid volume, i.e. derivation of the function $N(\bar{r})$ of the particles proportion in volume unit of liquid in neighborhood of the point with radius vector $\bar{r}$, it can be done if the sound wave effective length is vastly superior to characteristic distance between particles /17/. Affecting the both parts of equation (14) by operator $/\Delta + k^2/$ and integrating the last term in the right side (14) by parts we get Dyson's equation for a self-adjusted field with regard to dipole oscillations (for simplicity's sake the particles were believed to be identical);

$$(\Delta + k^2)P\vec{r} = -\ 4\pi N(F)f - 4\pi N(F)\ \frac{\alpha' R^3}{2}\ \Delta P \qquad (15)$$

Supposing that inside of the fluid N(r) = const we get from (14) the expression for effective wave number

$$K^2_{eff} = \frac{k^2 + 4\pi Nf}{1 + 2\pi N\alpha' R^3} = \frac{k^2 + 4f\pi N}{1 + \frac{3}{2}\alpha'\varphi} \qquad (16)$$

Here $\varphi = \frac{4}{3}\pi R^3 N$ – particles volume percent concentration.

The expression (16) incorporates the dipole oscillations of particles and differs from a well-known /17/ (obtained in monopole approximation) by multiplier $(1 + \frac{3}{2}\alpha'\varphi)^{-1}$
Since at external field switch-on the change of a particle's radius conditioned by electrostriction effect is low, then, consequently, the sound velocity anisotropy which observes in experiment should ne searched in parameter α', this parameter shows the degree of the particles drag by liquid (16). It is clear that the electric field switch-on and particles alignment in chains should change α'. Let calculate this parameter with regard to electric forces.

Let consider a chain of particles located in electric field with intensity E affected by sound wave. The liquid drags the particles making them to execute the dipole oscillations (the monopole pulsations are incorporate in expressions (11), (14)÷(16) through scattering amplitude f). The equation of "i" particles motion has the following appearance /19/

$$\frac{d}{dt}(M\bar{U}_i) = \rho_e V_o \frac{d\bar{V}_i}{dt} - m\frac{d}{dt}(\bar{U}_i - \bar{V}_i) + \bar{F}_i^{el} \qquad (17)$$

Here M - particle's mass; V_o - its volume; U_i - its velocity in dipole oscillations; V_i - velocity of liquid dragging the particle; $M = \frac{2\pi}{3}\rho_e R_o^3$ - added liquid mass, $\bar{F}_i^{el}$ - the force of electric type acting on particle from the side of its neighbors. Inasmuch as the latter fastly diminishes with the growth of dis- tance between particles, then it is possible to restrict ourselves to interaction of the 1st particle only with its closest neighbors in the chain, i.e. ones with numbers "i + 1" and "i - 1".

The interaction force of electric type in correspondence with the data /21/ for two particles has the following appearance:

$$F_o = -k_1 \frac{\xi}{d} \ell^{-\xi^2/d^2} \quad (18), \qquad K = \frac{P_A^2 N_o S_o E_m^2}{4k_B T} \cdot g(1-\ell^{-E/E_o})E/E_o \quad (19)$$

Here $P_A = 8 \cdot 10^{-30}$ Kl·m, $N_o = 10^{28}$ m^{-3}
$S_o \approx 10^{-10}$m^2 - surface of particles contact;
$E_m = 10^7$ V/m - intensity of the polarization saturating at the expense of the carriers adsorption (Johnson - Rahbeck effect)
K_B - Boltzmann's constant; T = 300 ^{0} K - temperature;
d ≈ 10^{-7}m - critical distance of the particles break in the bridge; ξ - distance between particles' surfaces.
Let us assume that the electric field influence results in crea- tion of particles' chain with the same distances a_o between their surfaces. The particles start oscillating under influence of the sound wave. Let us denote particles shifting by $\xi_i(t)$. The equation (17) acquires the following form:

$$(\rho_s + \frac{\rho_e}{2})\dot{U}_i = \frac{3}{2}\rho_e \dot{V}_i - \frac{k_1}{V_o}\left\{ \frac{\bar{\xi}_{i+1} - \bar{\xi}_i + a_o}{d}\, \ell^{-\frac{(a_o + \bar{\xi}_{i+1} - \bar{\xi}_i)^2}{d^2}} - \frac{\bar{\xi}_{i-1} - \bar{\xi}_i + a}{d}\, \ell^{-\frac{(a_o + \bar{\xi}_i - \bar{\xi}_{i-1})^2}{d}} \right\} \cdot \bar{n}_E \tag{20}$$

n_E - unit vector in direction of the electric field,

ρ_s - density of disperse particles.

At this ξ_i - shiftings of particles only in direction n_E .

If the wave vector of the incident wave is directed perpendi- cularly to $\bar{n}_E$, then ξ_i = 0 . As a result, the last term in the right side /20/ converts into fields and

$$U_i = \frac{3\rho_e V_i}{2\rho_s + \rho_e} \tag{21}$$

This result coincide with the well - known /19/. That is in this case the coefficient $\alpha = \frac{3\rho_e \bar{V}_i}{2\rho_s + \rho_e}$ and electric field doesn't exhibits an inferior influence on the sound wave velocity (what corresponds to the experiment).

Now let consider the case $\bar{C} \parallel \bar{E}$ or $\bar{K} \parallel \bar{n}_E$.

Considering the particles oscillations to be small and turning to continuous dependence $\xi_i = \xi$ (z,t) (the axis z - direction along $\bar{n}_E$), we have from Eq.(20)

$$(\rho_s + \frac{\rho_e}{2})\ddot{\xi}(z,t) = \frac{3}{2}\rho_e \dot{V}(z,t) - \frac{2k_1 h^2}{V_o d}\xi''(z,t)\,(1 - \frac{2a_o}{d})\,\ell^{-\frac{a_o^2}{d^2}} \tag{22}$$

h - distance between particles' centers. When derivating (22) it has been incorporated that in trial of the three-point approximation

$$\frac{\xi_{i+1} + \xi_i - 2\xi_i}{2h^2} \approx \frac{d^2\xi}{dZ^2} = \xi''(z,t) \tag{23}$$

The particles contact with one other in the chain, that is why $h = 2R_o$.

Inasmuch as the oscillations exciting comes about with the help of the sound wave, then $\xi'' = -k^2_{eff}$.

k_{eff} - wave number in microoheterogeneous medium.

From here from Eq.(22) for drag coefficient α we have

$$\alpha' = \frac{3\rho_e}{2\rho_s + \rho_e - \frac{4k_1 h^3 k^2_{eff}}{\bar{V}_o \omega^2 d} \ell - \frac{Q_0^2}{d^2}\left(1 - \frac{2a_o}{d}\right)} \tag{24}$$

By this means, the external electric field switch-on changes the drag coefficient of disperse particles at the expense of the electric links creation between them.

Usually $\frac{ao}{d} \ll 1$ in the chain. From Eq.(24) derives

$$\alpha' = \alpha'_o + \delta\alpha'$$

$\alpha_o = \dfrac{3\rho_e}{2\rho_s + \rho_e}$ - drag coefficient in the absence of the field

$$\delta\alpha' \approx \alpha_o' \frac{4k_1 h^3 K_{eff}}{dV_o \omega^2 (2\rho_s + \rho_e)} = \alpha_o \frac{4k_1 h^2}{dV_o C_{eff}^2 (2\rho_s + \rho_e)} \tag{25}$$

drag coefficient change as a result of the electric field switch-on.

For sound velocity change from Eq.(16) we have

$$\frac{\Delta C_{eff}}{C_{eff}} \approx \frac{3}{4} \varphi\delta\alpha' \tag{26}$$

Let calculate it for $\varphi \approx$ 10% at $E = 10^6$ V/m

From Eq.(14) derives

$$k_1 = \frac{0,64 \cdot 10^{-58} \cdot 10^{28} \cdot 10^{10} \cdot 10^{11}}{4 \cdot 1,4 \cdot 10^{-29} \cdot 3 \cdot 10^{2}} \cdot 0,1 \cdot 0,63 = 0,24 \cdot 10^{-9}\ H$$

Let $\rho_e = 10^3\ kg/m^3$, $\rho_s = 2 \cdot 10^3\ kg/m^3$, $C_{eff} = 1,5 \cdot 10^3$ m/c

Then $\alpha_o' = 3/5$, $R_o = 0,5 \cdot 10^{-5}$m, $V_o = \frac{4}{3}\pi R_o^{\,3} \approx 0,6 \cdot 10^{-15}m^3$

$$\delta\alpha' = \frac{3}{5}\ \frac{4 \cdot 0,24 \cdot 10^{-9} \cdot 10^{-10}}{10^{7} \cdot 0,6 \cdot 10^{-15} \cdot 2,25 \cdot 10^{6} \cdot 5 \cdot 10^{3}} \approx 0,09$$

From here $\frac{\Delta C_{eff}}{C_{eff}} \approx \frac{3}{4} \cdot 0,1 \cdot 0,09 \approx 0,7\%$

what satisfactory correlates with the experiment.

Let note that from Eq.(25) – (26) follows that at $E > E_o$ for sound velocity change it derives

$$\frac{\Delta C_{eff}}{C_{eff}} \approx \frac{3}{4}\,\varphi\,\alpha_o' \frac{4h^2}{dV_o C_{eff}^{\,2}\ (2\rho_s + \rho_e)} \cdot \frac{P_A^{\,2}\ N \rho_\delta E_m^{\,2}}{4k_B T} \cdot g\,\frac{E}{E_0} \qquad (27)$$

i.e., the dependence from E is almost linear what correlates also well with the experiment. Beside that, $\frac{\Delta C_{eff}}{C_{eff}}$ is proportional to the volume fraction of disperse particles what also confirmes by the tests' results.

Bibliography

1. A. W. Smith and L. W. Ering, I.Chem. Phys. v 7 (1939), 632.
2. R. M. Bolt and A. Gacomini, I. Acoust. Soc. Am. v 20 (1948), 341.
3. A. Barone and A. Gacomini, Ric. Sci. v 15 (1945), 264.

4. A. W. Nolle, v 20 (1949), 589.

5. A. Bonetti, Ric. Sci. v. 18, (1949), 777.

6. A. I. Lipkin, *Magnetohydrodynamics*, N3 (1985), 25.

7. V. M. Polunin and A. A. Chernyshova, *Magnetohydrodynamics*, N3 (1984), 23.

8. V. A. Zhytikov, *Fundementals of the ultrasound physics*, (L. State University, 1980) p.318.

9. Z. P. Shulman, Yu. G. Yanovsky and E. V. Korobko, I. of *Non - Newtonian Fluids*, v.33, N1 (1989), 181.

10. A. V. Kolesnikov, *Ultrasound measurments* (M. Ed. Standards, 1982) p.248.

11. S. A. Demchuk and E. V. Korobko, *Rheophysical studies*, (Mn. IHME of the BSSR AS, 1974) p.62.

12. N. N. Brazhnikov, *Ultrasonic phasemetry* (M. Energy, 1968) p. 356.

13. N. L. Bergman, *The ultrasound and its applications in science and technique* (M. FL, 1957) p.726.

14. G. Korslow and D. Eger, *The heat conduction of solids* (M. Nauka 1964) p. 488.

15. G. M. Lyakhov, *Bul. of the USSR AS Mechanics and mechanical engineering* (1959) N1, 42.

16. L. D. Landau and E. M. Livshits, *Electrodynamics of the continuous media* (M. Nauka, 1982) p. 620.

17. V. A. Krasil'nikov and V. V. Krylov, *Introduction in physical acoustics* (M. Nauka 1984) p. 400.

18. S. T. Zavtrak and E. V. Korobko, *Journal of theoretic physics* v. 61, N3 (1991), 177.

19. L. D. Landau and E. M. Livshits, *Hydrodynamics*, (M. Nauka 1986), p. 736.

20. M. Abramovets and I. Stigan, *Reference book on special functions* (M. Nauka 1979), p. 852.

21. E. V. Korobko and A. A. Mokeev, *Journal of Intelligent Material Systems and Structures*, v. 2, N1 (1991), 25.

SOME PROPERTIES OF A DISPERSED PHASE OF ELECTRORHEOLOGICAL FLUIDS

A.D.Matsepuro, V.I.Kordonsky
Luikov Heat & Mass Transfer Institute of Academy of Sciences Republic of Belorus, Minsk, USSR

Visualization has shown that the field intensity distortion due to coarse platinum and silica gel particles with the adsorbed water content is the same. Measurement is made of a charge of tablets preliminarily to be pressed from a dispersed phase of electrorheological fluids (ERF) after their contact with a charged electrode. Using the developed method of measuring voltage amplitude-frequency characteristics has enabled one to determine limiting charge-carrier velocities on tablets of aerosyl containing 0 to 10 wt% water.

For a long period of time we tried to choose a dispersed phase of electrorheological fluids (ERF) in an empirical manner: we prepared a suspension composed of particles and checked its apparent viscosity in an electric field. Attempts to take as a basis the nature of substance, e.g. the presence of hydrogen bonds or semiconductor properties, as well as the polarizability, i.e. dielectric properties or electrical conductivity or a combination of dielectric properties and electrical conductivity did not give equivalent results. The existing approaches enable one to explain physics of the electrorheological effect (ERE) but using the known parameters we had no success in choosing, among a variety of dispersed materials, the one suitable for preparing a suspension. Moreover, a sensation was present all the time that the reason for such a powerful phenomenon as the ERE must manifest itself in some pronounced property not discovered by us.

Based on the results of the previous works dealing with model particles, or silica gel spheres with the adsorbed water or diethylamine content, the following conclusions were made: (1) a separate particle of ERF acquires, from an electrode, a charge having the same value as platinum one, i.e.,a dispersed ERF phase is charged, in magnitude, like a conductor; (2) an interelectrode bridge composed of these particles obligatorily conducts the current but like a bad dielectric or semiconductor; (3) if in an interelectrode bridge two adjacent particles are somewhat moved apart, then between the particle surfaces exposed to each other there appears a high potential difference close to, in magnitude, to the one between the electrodes [1].

However, up to now it is not clear how much the polarization, electrical conductivity and the ability of particles to be charged contribute to the ERE. There is sufficient information about the ERF particle polarization [2,3], nevertheless, the ability of particles to be charged has not been almost studied.

Thus, the present work is aimed at studying the ability of ERF particles to be charged and at obtaining the parameters of this process. The following experiments were made: (1)visualization of a field intensity distribution in an interelectrode gap; (2) measurement of a charge to be transferred from an electrode to a dispersed phase particle, and (3) measurement of a charge-carrier velocities on particles.

Equipment and Method

To make visualization a plexiglas cell 35 mm long and 10 mm high was used; a gap between vertical stainless steel electrodes amounted to 10 mm. About 6 wt % water-containing silica gel and platinum spheres were placed at the cell bottom separately, together, or in different combinations. Then the cell was filled with a 1 % suspension of polymer particles (dispersed ERF phase). The cell was located on a special device and then photographed: the light flow was directed via the cell bottom downward; heating of an object under study was not allowed. Photographing was made after the voltage had been supplied to electrodes and the particle motion had ceased.

To measure a charge, samples were pressed from dispersed materials in a special mould in the form of tablets 10 mm in dia and 2 - 4 mm in thickness. Preliminarily, transformer oil-based 5 wt % suspensions were prepared from these powders. The apparent viscosity of suspension was measured in ad.c. and a.c. (50 Hz) fields with a strength up to 3 kV/mm (viscometer Reotest 2).

A device for measuring the tablet charge was composed of a high direct current source, with a steel 10 mm dia electrode being fixed on its high-voltage lead. A terminal of the electrometric amplifier U5-6 was located at a distance of 150 mm from this electrode. A tablet of the test material was weighed and fixed via a thin plexiglas holder 160 mm long. Then a voltage was supplied to the high-voltage electrode, a tablet was set into contact with the electrode, was moved and touched the electrometer terminal. The amplified electrometer signal was supplied to the screen of the storage electron oscillograph. A value of the charge Q was calculated in terms of charging signal area and was divided by the tablet mass.

Charge-carrier velocities on particles were measured in terms of measured voltage amplitude-frequency characteristics. Figure 1 displays a schematic of a device.

Test tablet 2 was placed between two steel electrodes 1 and 3. A voltage U_o with a frequency 0 to 1000 Hz was supplied from power source 5 to input electrode 1; at each frequency the voltage value was varied step by step: 500, 1000, 1500 and 2000 Volts. The input voltage U_o was measured by an electrostatic kilovoltmeter 4 with an input capacity of 3.4 pF and with an internal resistance of $3.8 \cdot 10^9$ Ohm. Output electrode 3 was connected with the electrostatic kilovoltmeter 6 where the voltage U was measured. The earthed circuit of the output kilovoltmeter 6 incorporated R_1=5200 Ohm resistance; in terms of a voltage drop on this resistance measured by voltmeter 7 the current in the source- tablet- kilovoltmeter- earth circuit was calculated. To measure the conduction current through the tablet to output electrode 3, R2=1400 Ohm resistance was incorporated instead of kilovoltmeter 6, and a voltage drop on it at a direct current and alternating current (50 Hz) voltage was determined by voltmeter 8.

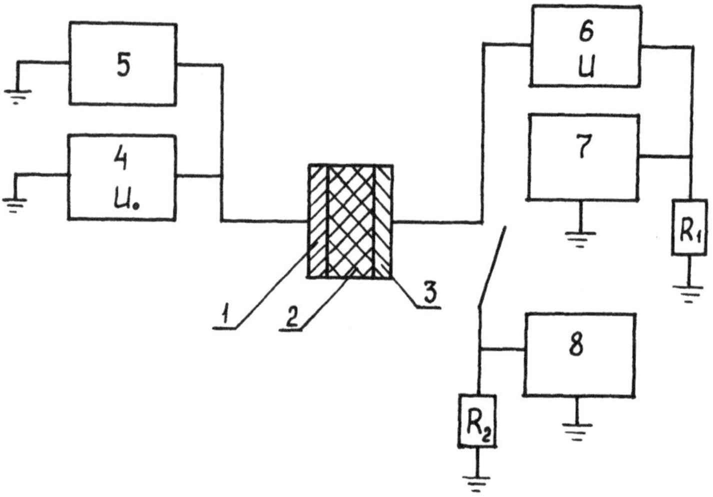

Fig.1. Measuring scheme of voltage amlitude- frequency characteristics of a disperse ERF phase.

Study was made of aerosyl tablets with a specific surface of 380 m^2/g and with different content of adsorbed water: from 0 to 10 wt %. Measurements were performed on tablets having no less than two thicknesses at each value of a moisture content; 2 or 3 tablets were put together to get a required thickness. Electrodes 1 and 3 were placed into a plexiglas casing, and in experiment a tablet was protected to avoid a contact with atmosphere.

Results and Discussion

Current conducting particles distort electric field force lines in a gap between electrodes (Fig. 2a). The smaller the distance between a particle and electrode, the higher is the local field strength and the greater amount of particles sensitive to a field moves into this region. The same picture is observed, with activated silica gel spheres (Fig.2b) being placed into the gap. Here it is also seen that the force line direction changes, with coarse particles being placed into the gap: force lines are along the shortest path between them. In a bridge composed of silica gel and platinum spheres, the force lines are concentrated in particle contact regions: when particles touch each other or when particles are separated by medium sublayers (Fig.2c). The closed bridge between electrodes draws a current, nevertheless, the elevated field strength remains in the contact regions.

Analysis of the photographs in Fig.2 shows that: (a) there is no difference in the behaviour of activated silica gel and platinum spheres in an electric field both for separate particles and for bridges; (b) distortion of force lines due to all coarse particles is very strong; (c) particle surface in the bridge, via which the current flows, remains charged.

Thus, it has been supported once more that ERF particles are charged, in magnitude, in an electric field like conductors. But it is not clear why a strong electric field remains in the ERF-filled gap while a short circuit immediately originates in the metal particle suspension. Earlier, it was shown that equal-size platinum and silica gel spheres were, in magnitude, charged identically: a charge depends on a field strength and particle radius [4]. We made an attempt to find a relationship between a charge of tablets of a polymer water-free dispersed phase and a relative viscosity, $\eta app/\eta o$, of suspensions of these particles in d.c. and a.c. fields. The electrical suspension conductivity σ was measured in the d.c.

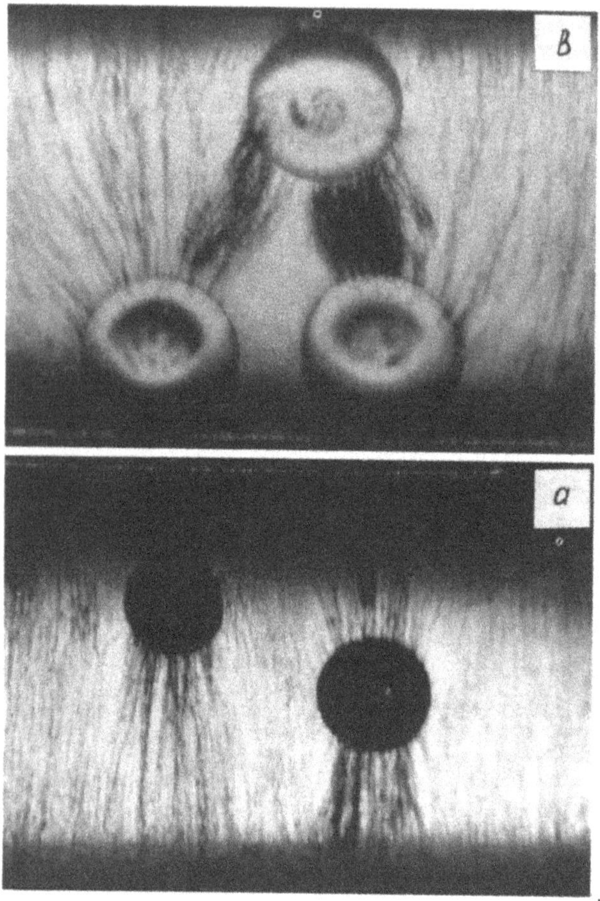

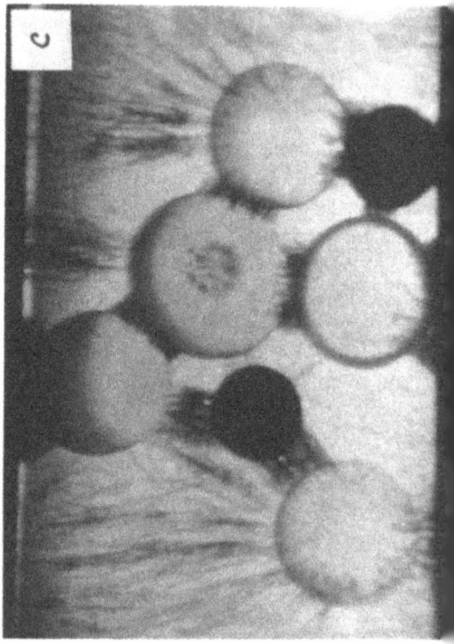

Fig.2. Photoes of platinum (a), moistened silica gel (b) and platinum - silica gel (c) spheres placed into polymer particle suspension in hexadecane. Interelectrode voltage is 6 kV (a,b) and 3 kV (c). Dark spheres, platinum.

field, and ε and tgδ of suspensions and tablets were determined at a frequency of 1 kHz by using the bridge R5010. Table 1 comprises 4 examples of 40 test samples.

Table 1 - Experimental data

N sample	U, V	Suspensions 5 wt % ηapp/ηo, f=50Hz	f=0	$\sigma \cdot 10^{-8}$ Ohm$^{-1} \cdot$ m^{-1}	ε	tgδ	Tablets Q$\cdot 10^{-10}$ C/g	ε	tgδ
1	500	1.0	1.0	0.004	2.2	0.04	3.36	1.25	0.17
	1000	1.0	1.1	0.045			8.58		
	1500	1.1	1.2	0.044			19.78		
2	500	2.2	2.3	4.5	3.8	0.11	5.67	12.3	0.2
	1000	5.8	6.9	16.6			12.4		
	1500	11.9	13.5	32.6			18.9		
3	500	1.6	2.0	0.52	3.0	0.07	5.6	500	>1
	1000	3.9	4.0	1.13			11.7		
	1500	7.2	7.2	2.15			21.6		
4	500	1.0	1.0	0.05	2.3	0.01	3.7	25	0.03
	1000	1.2	1.2	0.06			4.0		
	1500	1.5	1.2	0.064			5.4		

In all cases, the tablet charge Q grows with increasing electrode voltage. A very small electrical conductivity σ (examples 1 and 4) corresponds to small values of the relative viscosity. But the main goal, to establish a direct relationship between a dispersed- phase particle tablet charge and relative viscosity of suspensions of these particles, has not been achieved: no correlation has been established. However, it has been found that a contact time of a tablet with a charging electrode sometimes affects a charge quantity, which is, apparently, attributed to a charge-carrier velocity on particles.

Measuring charge-carrier velocity on particles must be made under the conditions, eliminating a time-variation of parameters to be found (e.g., current growth or drop). As we do not know types of charge-carriers and their number, an attempt has been made to measure a velocity of the most slow charge-carrier.

Use was made of the method of measuring voltage amplitude-frequency characteristics of tablets pressed from a dispersed phase (Fig.1). A supply source voltage Uo on the one side a tablet served as an input parameter while a voltage U on the other side, as an output parameter. Values of Uo and U were measured by electrostatic kilovoltmeters, time variations of these parameters were not observed.

With a small water content, the output voltage U on aerosyl is less than Uo even at a constant voltage; increasing a field frequency reduces a value of U up to zero (Table 2). Increasing the content of the adsorbed activator (water, for example) yields values of U (at constant Uo) over some frequency range.

Table 2 - Values of U

Moisture content %	Tablet thickness m	Values of Uo at f=0			Values of Uo at f=50Hz		
		1000 V	1500 V	2000 V	1000 V	1500 V	2000 V
0	0.0067	0	0	0	0	0	0
0	0.003	0	0	0	0	0	0
1.27	0.0069	400	750	1000	0	0	0
1.27	0.0031	950	1400	1920	0	0	0
1.75	0.0073	950	1440	1930	0	0	0
1.75	0.0035	980	1480	1970	0	0	0

Figures 3 and 4 plot an output voltage U vs frequency f (Hz) at values of Uo equal to 500,1000,1500 and 2000 V for aerosyl tablets with the moisture content of 4.9 and 7.8 wt % for two different thicknesses. At other conditions being equal, values of U decrease with increasing a frequency more intensively: the smaller is the amount of adsorbed activator the more thick is the tablet. This means that charge-carriers cannot reach wall 3, in changing field, to cover a distance between electrodes 1 and 3, i.e. selecting a tablet thickness

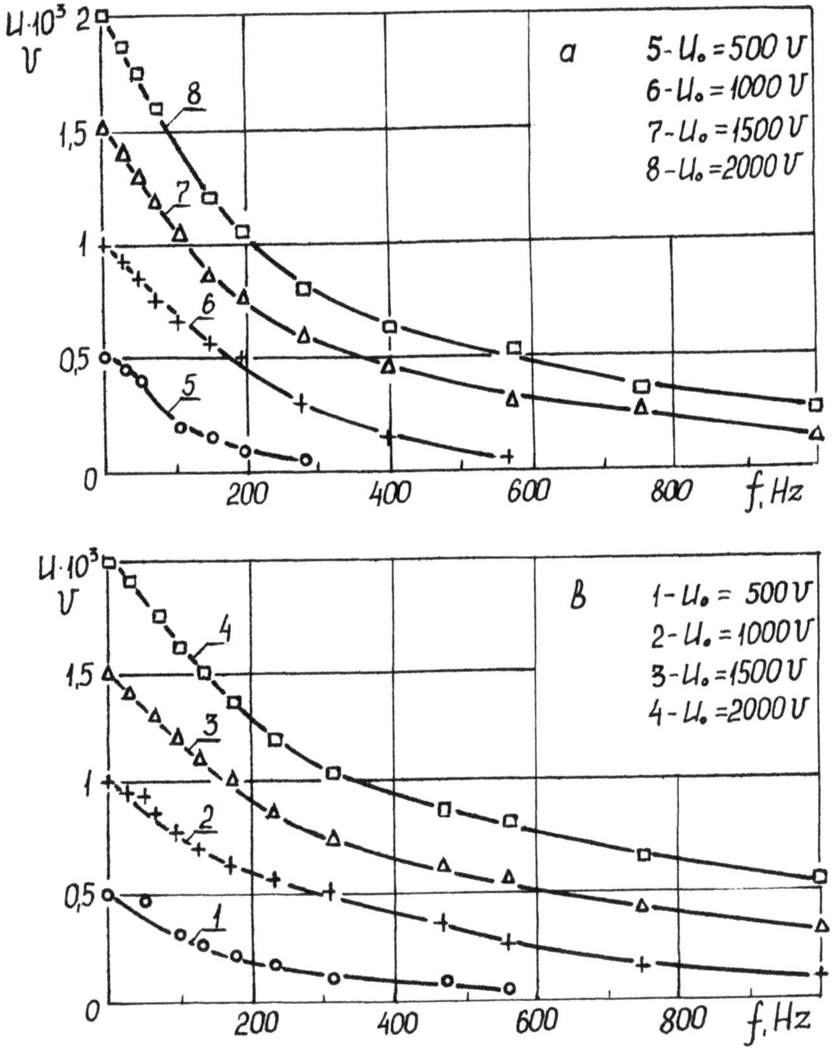

Fig.3. Output voltage U vs field frequency f for aerosyl tablets with a moisture content of 4.9% and thickness of 0.0075 m (a) and 0.0042 m (b) at different input voltages Uo.

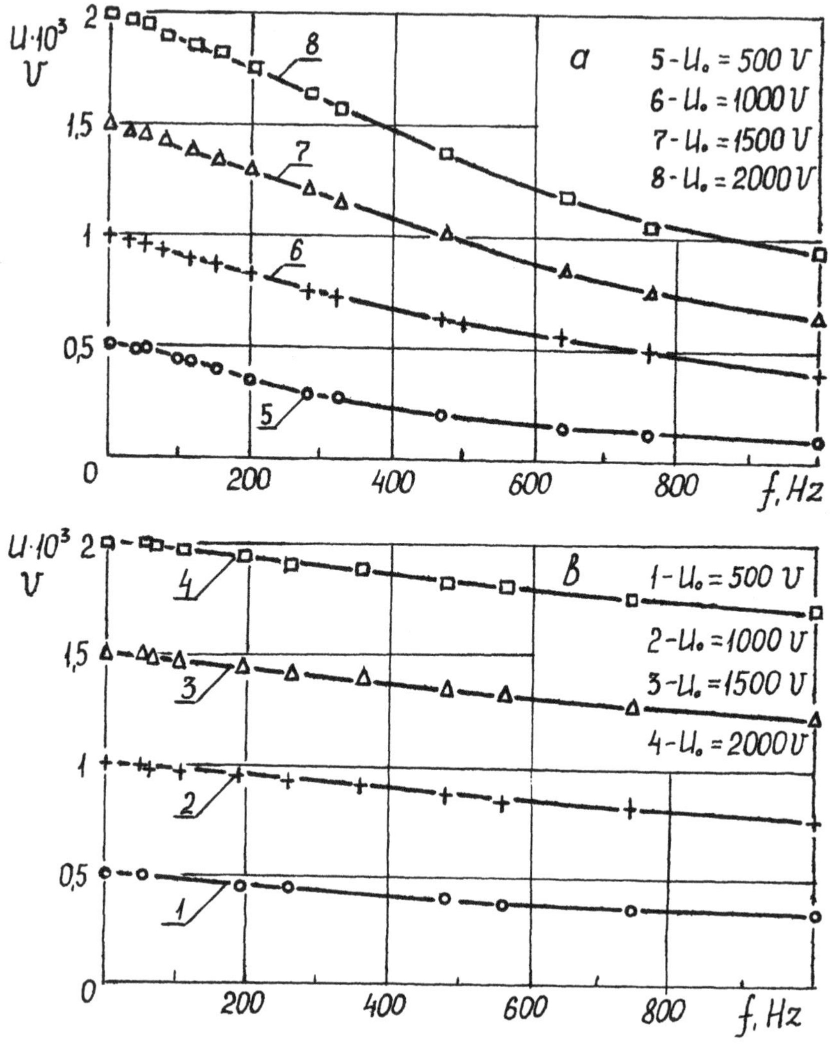

Fig.4. Output voltage U vs field frequency f for aerosyl tablets with a moisture content of 7.8% and thickness of 0.0072 m (a) and 0.0032 m (b) at different input voltages Uo.

or a frequency range, conditions may be found, under which a charge-carrier velocity will be insufficient to transfer a total voltage U_o from electrode 1 to electrode 3.

For convenience of subsequent processing, the data in Figs. 3 and 4,are plotted as U/U_o vs. an angular frequency ω, i.e., as ordinary amplitude-frequency characteristics (Fig.5a). Additional plots of U/U_o vs. $\lg\omega$ (Fig.5b) are close to linear over a frequency range of interest to us. This enables one to use the known methods of the linear control theory for determining a time constant, especially when $U/U_o=0.71$, T=1 or at $U/U_o=0.45$ T=2.

Results of such processing of amplitude- frequency characteristics have shown that a time constant T is the thousandth fractions of a second for aerosyl moisture content of approximately 5% and ten thousandth, for moisture contents of about 10% . It depends on a tablet thickness but almost does not depend on the applied voltage.

If the charge transfer is assumed to occur uniformly within the entire tablet volume, then a charge carrier velocity may be calculated. $V=l/T$ where l is the tablet thickness.

Values of the characteristic charge transfer rate as a function of moisture content on aerosyl are plotted in Fig.6. The charge transfer rate is weakly increased with growing aerosyl moisture content up to 4 wt % . After this, the rate sharply increases.

The ability of dispersed phase to be charged and the rate of this process may serve as one of the parameters to choose particles: the more fastly a maximum charge leaks-in onto a particle, the better is it. An electrical conductivity may be another parameter. The useful part of the electrical conduction must provide a desirable charge leak-in velocity; increasing the conduction above this necessary value gives rise to degenerating ERE. Fig.6 shows that the aerosyl tablet resistance sharply drops at a moisture content up 5% . Earlier, it has been found that an optimum of the effective ERF viscosity in an electric field as a function of aerosyl

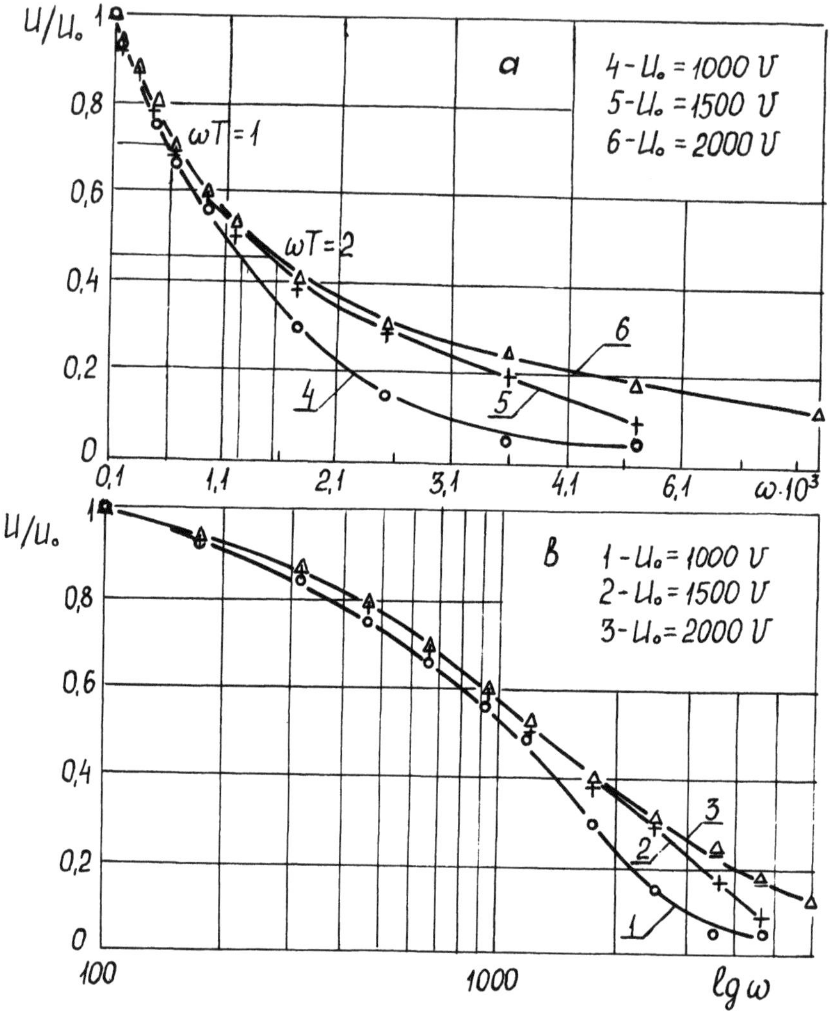

Fig.5. Output (U) - to - input (Uo) voltage ratio vs angular frequency ω (a) and lgω (b) for an aerosyl tablet with a moisture content of 4.9% and thickness of 0.0075 m at input voltages Uo equal to 500, 1000, 1500 and 2000 V.

moisture content is about a moisture content range 5-6% . This means that a high charge transfer rate and moderate electrical conductivity not being avalanche in nature correspond to the optimum content of the activator.

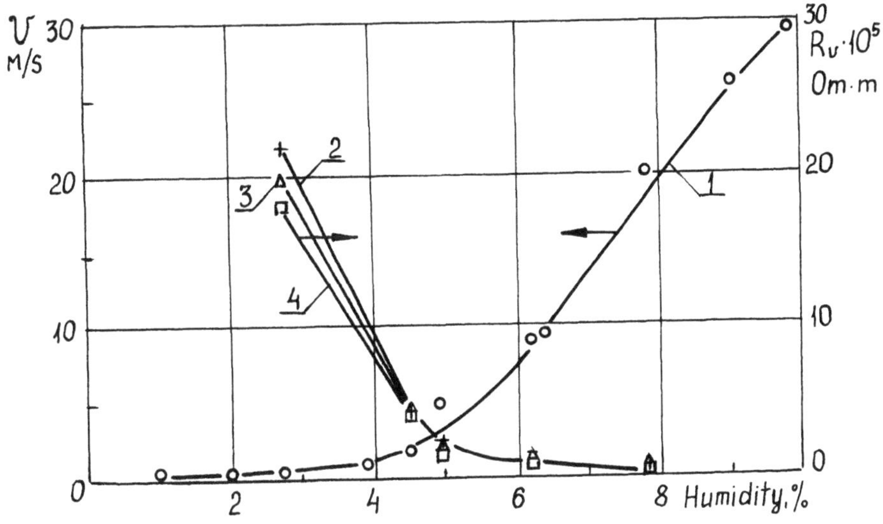

Fig.6. Charge-carrier velocity V (1) and specific volume electricresistance Rv (2-4) as a function of aerosyl tablet moisturecontent at input voltages Uo equal to 1000, 1500 and 2000 V.

Voltage amplitude-frequency characteristics of pressed tablets can characterize the dispersed phase material as well as to formulate requirements for the results on synthesizing new materials. The first requirement is: particles must be capable of being charged with a high charge density, i.e. they must have vacancies for charge-carriers.

Explain this by way of aerosyl example. It has appeared useful to allow for the information on static powder electrization under friction: in this case, charges are also accumulated on particles. So perfectly dry aerosyl has a negligibly small electrical conductivity and is not electrized under friction. Moistening a powder up to 2-3 wt% results in its strong electrizability, i.e., in the adsorbed water layer on a silica surface there are conditions for charge accumulation and separation but no conditions for their drain off. At a further increase in the aerosyl moisture content there appear means to charge transfer by its particle: the higher the electric field strength, at the smaller water content this occurs. Under these very conditions, in aerosyl suspensions the ERE start manifesting itself, at which particles are being charged, already while contacting an electrode or another particle.

The second requirement is that a charge-carrier velocity on particles must be high, for a particle might take an ultimately high charge during a short contact with an electrode or another particle. The problem of meeting this requirement at not a high electrical conductivity is extremely difficult.

Voltage amplitude-frequency characteristics as a method to compare different powders enables one to determine a charge-carrier velocity V in a strong electric field. Naturally, the thus measured values of V are also valid for particle and aggregate polarization and are responsible for the most slow polarization process. Experiments with tablets were made to study properties of disperse-phase substances. This method makes it possible to measure a charge transfer rate in disperse systems, too and to determine, how this process is affected by such complex phenomena as interparticle interactions and how the latter are, in their turn, affected by a carrier medium and surfactants.

There is a point to emphasize that the mentioned information not only supports that the dispersed ERF phase is able to be charged at a contact with an electrode, in magnitude, as metal but also with a smaller velocity and also that under certain conditions hydrated systems, e.g., silica, alumina are charged in the same manner. This is of importance for safety engineering, thunder protection and electricity control.

1. Z.P.Shulman, A.D.Matsepuro, in *Electrorheological fluids*,Proc. 2nd Int. Conf. on ERF (Technomic Publ. CO INC), 1989, p.207.

2. H.Block, J.P.Kelly, *J. Phys.* **D**, 21 (1988) 1661.

3. T.B Jones, in *Electrorheological fluids*, Proc. 2nd Int. Conf. on ERF (Technomic Publ. CO INC), 1989, p.14.

4. Z.P.Shulman, A.D.Matsepuro, B.M.Khusid, *Vesti AN BSSR, ser. fiz.-energ. nauk*, 4 (1974) 62.

IV. APPLICATIONS OF ER FLUIDS

A Model For The Electrical Characteristics of an ER valve.

M.Whittle
Department of Mechanical and Process Engineering
University of Sheffield
Mappin Street
Sheffield S1 3JD. United Kingdom.
Tel: (0742) 768555 Etxn. 5435
Fax: (0742) 753671

R. Firoozian, D.J.Peel and W.A.Bullough.
Department of Mechanical and Process Engineering
University of Sheffield

ABSTRACT

We present a simple model which can account for the electrical response characteristics of an Electro-Rheological (ER) near-parallel plate valve restrictor. We find that a self-consistent electrical model which can describe the current response to transient and biased sine wave voltage exitations must incorporate the non-linear dependence of current on voltage found at steady state. In addition, the steady state current characteristics are a function of the fluid flow rate. This can be described analytically and included in the model.

Analytical solutions show that non-linearity affects the detailed shape of the current response to large voltage inputs. The initial current response is well described by a numerical solution to the model differential equation using the experimental voltage as input. In the case of the relatively small fluctuations involved in the sine wave experiments, a linearised form of this model is found to be an adequate approximation.

Parameters are obtained for the model which are, within a generous margin of error, independent of valve dimension.

1. Introduction

Electro-Rheological (ER) devices promise rapid response times commensurate with the needs of modern high speed equipment operated by computer. The control of these devices requires a detailed understanding of the response characteristics. This paper describes a model which can account for the sinusoidal, steady-state and transient electrical behaviour of an Electro-Rheological (ER) parallel plate valve restrictor. The experimental setup has been described in detail previously[1] and a flow diagram of the apparatus is shown in Fig. 1.

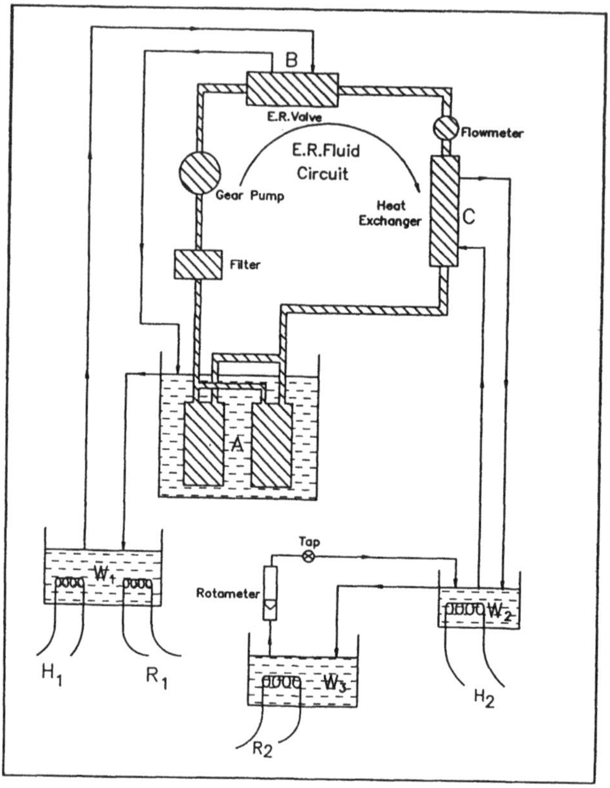

Fig. 1. Diagram of experimental system as described in the text.

In summary ER fluid is passed through an experimental valve by means of a gear pump set in a closed circuit with a flow meter. The flowrate is maintained effectively constant by means of a DC tacho/thyristor controlled drive and the fluid temperature is kept to within $\pm 0.1^{o}C$ of $30^{o}C$ at the inlet. With these parameters constrained, the electrorheological changes in the fluid induced by a voltage input appear only as the observed time dependent current and pressure responses. Although there is some interaction between these responses it is of second order importance and will be the subject of a future publication. In this work we examine the voltage-current relationships in isolation and develop a model independent of the pressure. Data is presented on valves of different lengths and gap size and the results are compared and related. The valve dimensions are given in table 1.

Table 1. Valve dimensions: length l, gap h, surface area A.

Valve	l/mm	h/mm	A/m^2
A	100	0.5	.0362
B	50	0.5	.0181
C	100	1.0	.0362
D	140	0.5	.0507
E	100	0.75	.0362

Previous work[1] has shown that the ER valve current response to a sinusoidal voltage input can be modelled by a system comprising a resistor R and capacitor C in parallel. Such a model, however, cannot fully describe the valve step-function current response which requires the inclusion of a delay time constant. This has involved the incorporation of a new resistance r into the original model which can be either in parallel (Fig. 2) or in series with the previous circuit. It turns out that the parallel circuit yields the more convenient equations and we limit our discussion here to this model.

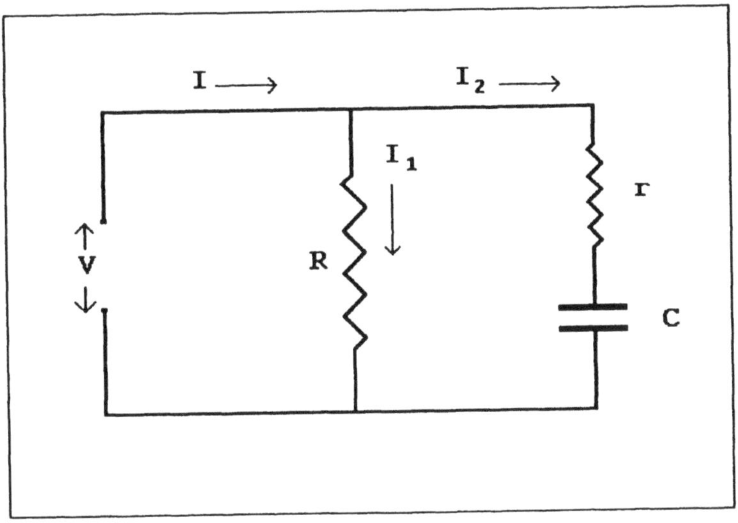

Fig. 2. Equivalent electrical circuit for the valve system. In the non-linear analysis R is replaced by a non linear conductivity coefficient Ξ and r is replaced by ξ (see eqs 3 and 4).

A self consistent electrical model which can account for the current response to a large voltage step must include the non-linear dependence of current on voltage found at steady state. Hence the resistances R and r in Fig. 2 are in fact non-linear in the field as we describe in the next section. Nevertheless, for the relatively small voltage input fluctuations involved for the biased sinusoidal response we find that a linearised form of this model is an adequate description. Perhaps a more refined model would also need to include non-linear characteristics of the capacitance, but we have not found evidence of this.

2. Experimental control

The nominal variation in bulk (zero volts) viscosity and conductivity of an ER fluid 5% and 15% per °C respectively requires close temperature control for any series of tests on which correlation or identification studies are to be carried out. In practice this places a 0.1°C limit on detection, control and accuracy. All measurements quoted here are for 30°C. Similar precision schemes were applied to the other variables involved and are described in references 1-3. Although the time and frequency response in a flow mode valve is an order easier to measure than in a shear clutch (most sensitive torque meters have a low natural frequency) the former mode of operation was chosen principally on temperature control grounds. Even so, an elaborate cooling mechanism had to be used to achieve the required limits. It has the disadvantage that the shear rate is not constant over the electrode gap.

Temperature monitoring was by BS1904 P.T. grade platinum resistance thermometers which were accompanied by digital readout equipment with 0.1°C resolution. The elements were set in the wall stainless steel tube 6mm diameter x 6mm (approx.) length. These units were calibrated against a class 1 BSI certificated thermometer which had itself been checked at the ice point.

The ER fluid bulk temperature was measured close to the pressure transducer locations immediately upstream and downstream of the valve. Sealing of the probes was by compression fittings. A third temperature probe and indicator was used for water temperature measurement. The volume of ERF in the circuit was made sufficiently large to ensure a mean circulation time of approximately sixty seconds at maximum flow rate, increasing to some 20 minutes at minimum flow. ER fluid temperature control was effected by means of :

(i) immersion of the storage tanks containing the bulk of the fluid under test in a constant temperature water bath (A) maintained at the fluid test temperature. (see Fig. 1).
(ii) Diversion of water from the water bath (A) to the water jacket (B) around the earth potential electrodes of the ER valve.
(iii) A water cooled heat exchanger (C) downstream of the ER valve, in the return line to the storage tanks.
(iv) Insulation of all water baths with high density polystyrene sheet, and of all pipes with cotton wool wrapping.

The effect of (i) and (ii) above is to damp out fluctuations in the ER fluid operating temperature at the ER valve. For steady state ER fluid tests, adequate settling times were allowed and a tolerance of ± 0.25°C allowed on ER fluid temperature at the valve inlet and outlet. The chief means of heat transfer to and from the ER fluid is (iii) above.

In more detail, water is pumped from water bath W_1 via the ER valve water jacket B to water bath A. Temperature is regulated by means of the constantly running refrigerator R1 which provides up to 0.75kW cooling in opposition to the thyristor controlled heating coil H_1 rated at 1.5kW. Thermostat control is by means of a mercury contact thermometer giving control to approximately 0.1°C. The water circulation rate is of the order of 50 lit min^{-1}, and the water bath capacities of W_1 and W_2 are 50 litres and 150 litres respectively.

Water is pumped to the heat exchanger C from the water bath W_2, where the temperature is regulated by means of a regulated heater as in W_1 working in opposition to a cold water supply. The cold water supply is maintained in a water bath W_3; a refrigerator provides up to 2kW cooling, operating constantly with a thermostatically controlled by-pass valve in the refrigerator circuit. The water is pumped at 5°C via the tap and a Rotameter flowmeter. It was found most convenient to monitor the temperature of water bath W_2 and after calibration to regulate it by manual control of the cold water tap. The cold water tank W_3 and water bath W_2 each have approximately 25 litres capacity.

3. Steady state current characteristics and ramp response

The short time (0-10ms) response to a step function voltage input will be described in the next section. At longer times the current and pressure response to this input normally reach a steady state. However, especially in the case of the electrical response, the approach to this state may contain some very slow relaxations indeed (~100s). These timescales are beyond the expected region of operational interest and we have therefore defined "steady state" for our purposes to be at 0.75s. At this point the pressure response has normally reached its limiting value.

Plotting the steady state pressure or current against the input voltage for each valve reveals markedly non-linear characteristics. This seems to be a general feature of electrorheological properties. Here we concentrate on the electrical characteristics and examine the dependence of steady state current on voltage. These results all display near parabolic curves with a weak dependence on flow rate[1]. Attempts have been made to fit this data by polynomial expressions with only moderate success. Up to four terms are required and the best fits are obtained when one or more of the coefficients are negative.

Other descriptions that have been favoured by some workers involving a hyperbolic sine are a rather poor fit to our data.

Plotting steady state current against the square of the voltage shows that the major deviation from simple parabolic behaviour occurs at low voltages. Some response data for a ramp voltage input is available for valve A which shows the behaviour in this region in more detail. We have found that a good empirical fit for this data involves a threshold voltage V_0 and has the form

$$I = \Xi(V - V_0)^2 \quad ; \quad V > V_0$$

$$I = 0 \quad ; \quad V < V_0 \tag{1}$$

where Ξ is a nonlinear conductivity coefficient, I is the total current and V the measured potential difference. Since we are interested in comparing valves of different dimensions we also define a coefficient Ξ^* in terms of the current density $i = I/A$ and the electric field $E = V/h$ where A is the surface area of the valve electrode and h the electrode gap

$$i = \Xi^*(E - E_0)^2 \quad ; \quad E > E_0 \tag{2}$$

This differs only in a numerical scaling factor from Eq. 1 , i.e $\Xi^* = \Xi h^2/A$. It is also appropriate to compare data with respect to a mean fluid velocity U rather than the volume flow rate Q , where, for a valve of length l

$$U = Q\, l / Ah \tag{3}$$

The residence time for fluid between the electrodes can then be esimated as

$$\tau_R = l / U \tag{4}$$

Data for a ramp speed of 50 Vs^{-1} (corresponding to a total measurement time from V = 0V of ~30s) is shown in Fig. 3 and demonstrates that in this case the fit is extremely good over the whole range.

Parameters obtained by fitting Eq. 2 to this data are shown in Table 2. For the faster ramp speed of 500 Vs^{-1} (corresponding to a total measurement time of ~3s) the fit of Eq. 2 was also quite good but comparatively lower values of E_0 and Ξ are obtained. At the faster ramp speed the shape of the curve is different because the current

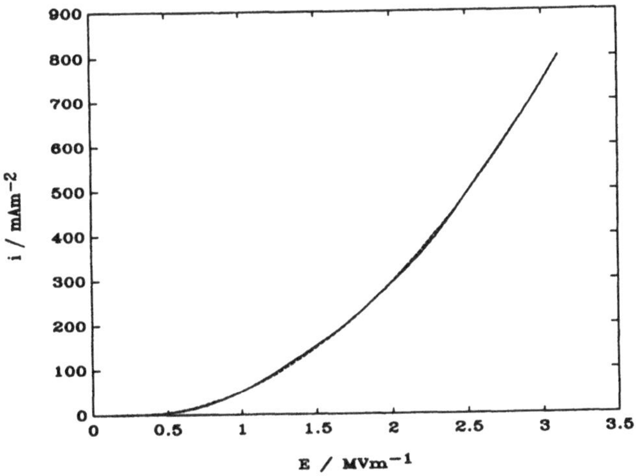

Fig. 3. Fit of Eq. 2 to ramp response. Valve A, ramp speed dE/dt = 0.1 $MVm^{-1}s^{-1}$.

response is still subject to the very long relaxations present in this system and has not reached true dynamic equilibrium by the end of the measurement. Nevertheless, we can expect the data obtained at 500 Vs^{-1} to be closer to the steady state, as we have defined it, since the measurement times in each case are comparable .

Table 2 Parameters for the valve current/voltage characteristics derived from ramp input for valve A at flow rate Q = 9 lit min-1, mean fluid velocity = 0.828 ms^{-1}.

Ramp Speed /Vs^{-1}	Ξ^*/ $mAMV^{-2}$	E_o/ MVm^{-1}
50	100.2	0.284
500	78.1	0.131

The expression given above has also been applied very successfully to the steady state data as shown in Fig. 4. The parameters obtained are collected in Table 3.

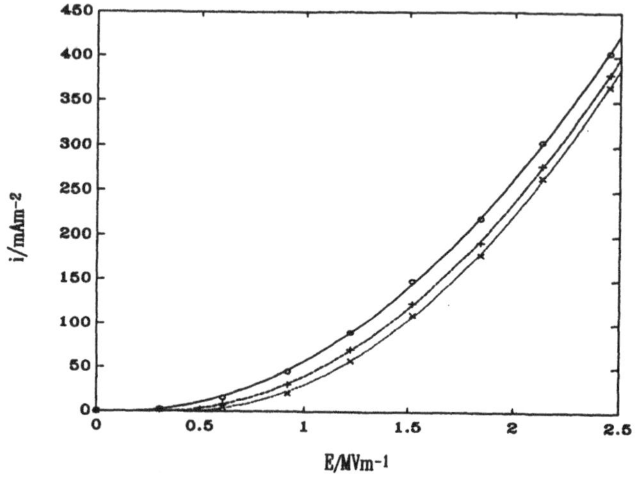

Fig. 4. Fit of Eq. 2 to steady state current at three flow rates. - o - Q = 3 Lit min^{-1}; - + - Q = 9 Lit min^{-1}; - x - Q = 15 Lit min^{-1}.

The the coefficient Ξ^* obtained at a flow rate of 9 lit/min compares well with the ramp result at 500 Vs^{-1} however, the threshold voltage compares more favourably with the results obtained from the ramp experiment at 50 Vs^{-1}. The steady state threshold field significantly increases with flowrate and a plot of E_0 against Q (Fig. 5a) shows that it may have a small positive intercept at zero flowrate, however, it is also not unreasonable to suppose that it vanishes.

For individual valves Ξ^* apparently increases or decreases with flow rate and there is no clear trend in the collected results (Fig. 5b). However, for valves of constant gap (B,A and D), Ξ^* increases with length of valve and for the longer valves (A and D), Ξ^* increases with flow rate. For the valves of wider gap (C and E), Ξ^* decreases with flowrate. Results given in later sections provide further estimates of this parameter and suggest that the accuracy is about ± 10% (see Fig. 10) putting the variations discussed here just outside the error bars. The mean and standard deviation of this parameter for all valves is 78.8 ± 12.3 mAMV^{-2}, and for many practical purposes it could be considered a constant.

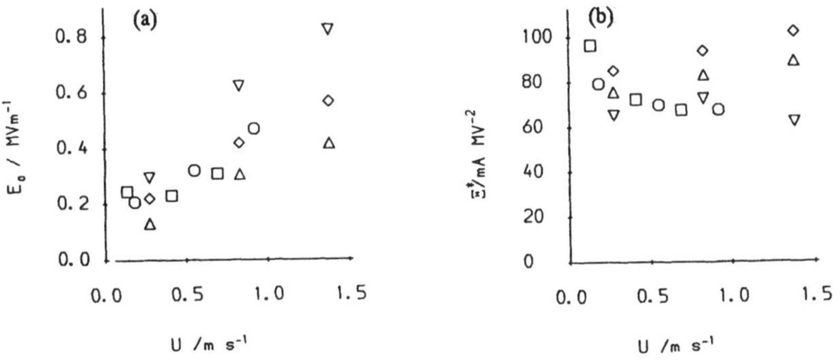

Fig. 5. Comparison of steady state parameters for different valves as a function of mean flow velocity U. - △ - Valve A; - ▽ - Valve B; - □ - Valve C; - ◇ - Valve D; - ○ - Valve E.
(a)Threshold field strength E_0. (b)Non linear conductivity coefficient Ξ^*.

Table 3. Parameters obtained from "steady state" results fitted by Eq. 1.

Valve	Q / Lit min^{-1}	U / m s^{-1}	Ξ^*/ mAMV^{-2}	E_0/ MVm^{-1}	τ_R / ms
A	3.0	0.276	75.7	0.132	362
	9.0	0.828	83.1	0.307	121
	15.0	1.38	89.3	0.418	72.4
B	3.0	0.276	65.3	0.295	181
	9.0	0.828	72.4	0.624	60.3
	15.0	1.38	62.1	0.822	36.2
C	3.0	0.138	96.5	0.246	725
	9.0	0.414	72.3	0.231	242
	15.0	0.69	67.5	0.310	145
D	3.0	0.276	85.2	0.220	507
	9.0	0.828	93.7	0.420	169
	15.0	1.38	102	0.569	101
E	3.0	0.184	79.3	0.208	543
	9.0	0.552	69.8	0.321	181
	15.0	0.92	67.5	0.472	108

4. Current response to voltage step input.

4.1 Overview

The short time (0-10ms) current response to a voltage step is characterised by an initial surge which dies away within ~1ms followed by a slow increase to the steady state value. A typical example is shown in Fig. 6 where it will be noted that the data has been truncated to observe details of the decay. When a step input is applied to the circuit of Fig. 2 current is drawn immediately through the resistor R and to charge the capacitor C. This current, limited by the resistor r decays with a time constant rC to a steady state determined only by the resistor R. This electrical network thus provides a good starting point for a discussion of the experimental results and we now describe the modifications that must be made to the model to account for detailed features of the valve response.

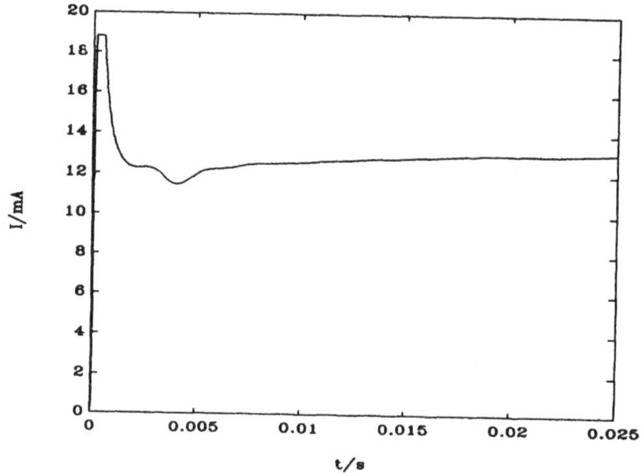

Fig. 6. Typical current response to a voltage step input. Valve A, Q = 3 Lit min^{-1} , V = 1.2 kV : step input.

4.2 Incorporation of Non - linear elements in the electrical model.

Experimental voltage steps corresponding to fields of around 2.4 MVm^{-1} and 1.2 MVm^{-1} were used as input and reference to the steady state data (Fig. 4) shows that the current response over this range is markedly nonlinear. For this reason we need to replace R and r in the circuit with resistances that vary according to the characteristics derived from the steady state data. The capacitance may also be non-linear but at this stage we have no information about this and it is treated as a linear element.

We suppose that both resistances in the circuit have non-linear characteristics. At the steady state the capacitive arm of the circuit can be neglected hence R is replaced by the element Ξ which we have discussed in the last section. The current I_1 is then

$$I_1 = \Xi(V - V_0)^2 \tag{5}$$

we assume that r behaves in the same way and replace it with ξ where

$$I_2 = \xi(V_r - V_0)^2 \tag{6}$$

and V_r is the voltage across r (or in this case ξ) in Fig. 2. By applying Kirchoffs laws we can now obtain a separable differential equation for i_2 with a step function input

$$\frac{dI_2}{dt} = -\frac{2\sqrt{\xi}}{C} I_2^{3/2} \tag{7}$$

giving for $I(t) = I_1 + I_2(t)$

$$I(t) = \left[\Xi(V - V_0)^2 + \frac{\xi C^2(V - V_0)^2}{(\xi(V - V_0)t + C)^2}\right]H(0) \tag{8}$$

where H(t) signifies a step at time t. In terms of the current density

$$i(t) = \left[\Xi^*(E - E_0)^2 + \frac{\xi^* C^{*2}(E - E_0)^2}{(\xi^*(E - E_0)t + C^*)^2}\right]H(0) \tag{9}$$

where ξ^* is defined in the same way as Ξ^* and C^* is a reduced quantity calculated from the capacitance C and the valve dimensions assuming that the system behaves as a parallel plate capacitor.

$$C^* = C\frac{h}{A} \tag{10}$$

C^* can be thought of as an effective permittivity.

For comparison, the corresponding linear circuit gives a more familiar exponential decay:

$$I(t) = \left[\frac{V}{R} + \frac{V}{r} \exp(-t/\tau_p) \right] H(0) \tag{11}$$

where $\tau_p = rC$ is a relaxation time.

The effective resistance at any voltage can be related via a generalised ohms law to the slope of the current voltage characteristic

$$r_V = \frac{dV}{dI} \tag{12}$$

Thus a resistance r_V can be related to the to the non-linear parameter ξ

$$r_V = \frac{1}{2\xi(V - V_o)} \tag{13}$$

and we can define a relaxation time τ_1 by comparison with (11)

$$\tau_1 = \frac{C^*}{2\xi^*(E - E_o)} \tag{14}$$

Equations 9 and 11 have been fitted to the observed results using a Nelder-Mead simplex algorithm[4]. The resulting exponential (Fig. 7a) and non-linear fits (Fig. 7b) to the data demonstrate that the non-linear version more closely describes the data.

4.3 Initial response.

In solving Eq. 7 we used an idealised step function input. For any real system the input must have a finite rise time and this will have some effect on the output. For linear systems it is possible to include the experimental voltage input shape by convolution with the theoretical response to a delta function input[5]. However, this method is not viable for the non-linear case and we have to resort to numerical methods.

For an input which is not a pure step Eq. 7 becomes

$$\frac{dI_2}{dt} = 2\sqrt{\xi I_2}\left(\frac{dV}{dt} - \frac{I_2}{C}\right) \tag{15}$$

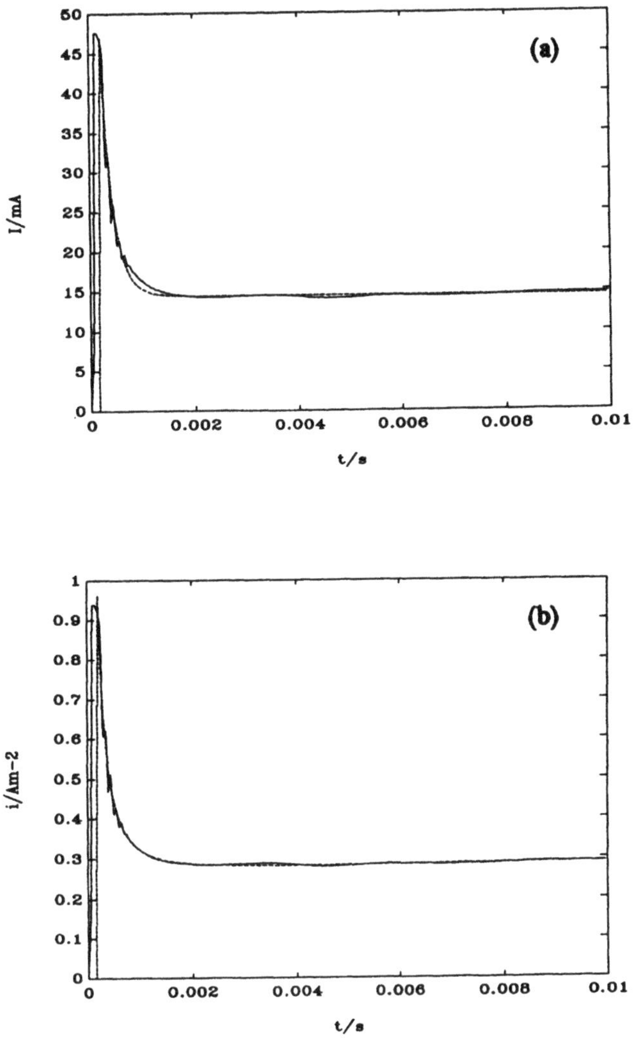

Fig. 7. Fits to current response for valve D, Q = 3 Lit min^{-1} , V = 1.2 kV : step input. (a) Linear prameters; eq (9). (b) Non-linear parameters; eq(15).

which must be solved numerically. However, with an initial condition of $I_2 = 0$, the square root is zero and an iterative solution can never begin. This problem can be avoided by substituting $J = \sqrt{I_2}$ when we get instead

$$\frac{dJ}{dt} = \sqrt{\xi}\left(\frac{dV}{dt} - \frac{J}{C}\right) \qquad (16)$$

which has a non-zero value of dJ/dt when dV/dt is non-zero. This can now be evaluated using a standard Runge-Kutta procedure[4] and the solution incorporated into the Nelder-Mead fitting routine. An example of this fit is shown in Fig. 8 (carried out by zero weighting the points over the truncation range) and shows that the short initial rise time of the current output can be adequately accounted for by the voltage input rise time; there is no need to include an inductive term. Ringing of the voltage input signal is responsible for the oscillations that appear in the experimental current and fit during the first millisecond. Slightly different parameters are obtained by this method but we have ignored this in carrying out the main programme of results.

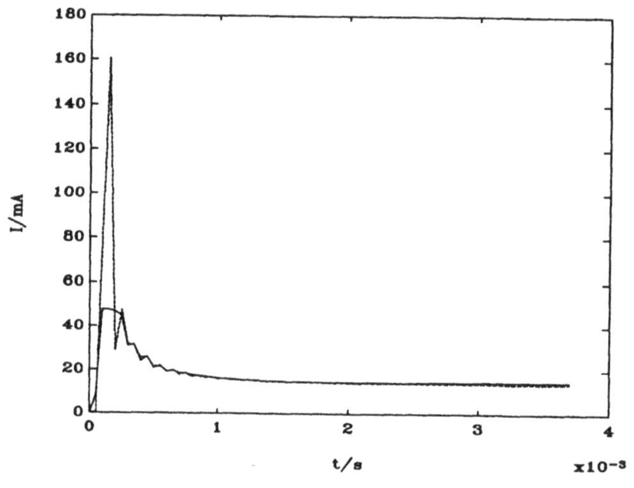

Fig. 8. Non-linear fit to the early part of the response using the experimental voltage data as input (see section 3.2). Valve D, Q = 3 Lit min^{-1} , V = 1.2 kV : step input.

4.4 Approach to the plateau.

Equations 9 and 11 both describe an instantaneous increase in current followed by decay to a constant plateau. The experimental current response contains a slow relaxation associated with a gradual increase in the current following the initial decay. In an attempt to improve the comparison with experimental data we have included an exponential term with a single relaxation time τ_2 replacing the constant term in Eq. 9 .

$$i(t) = \left[(\Xi^* - \Xi_o^* \exp(-t/\tau_2))(E - E_o)^2 + \frac{\xi^* C^{*2}(E - E_o)^2}{(\xi^*(E - E_o)t + C^*)^2} \right] H(0) \qquad (17)$$

Table4 Parameters derived from step function analysis

(a) Voltage step 600 V.

Valve	Q / Lit min^{-1}	Ξ^*_∞ / mAMV^{-2}	Ξ^*_o / mAMV^{-2}	ξ^*_∞ / mAMV^{-2}	C^* / nFm^{-1}	τ_1 / ms	τ_2 / ms
A	0.0	34.7	6.75	1492	0.183	0.051	18.7
	3.0	67.6	20.0	466	0.223	0.223	5.69
	9.0	81.7	18.7	1784	0.237	0.074	5.33
	15.0	81.4	73.9	1132	0.257	0.148	2.07
B	3.0	67.4	75.3	320	0.388	0.682	4.04
	9.0	66.6	43.0	328	0.223	0.575	6.00
	15.0	99.6	33.1	1836	0.266	0.201	19.2
C	3.0	89.5	14.1	513	0.200	0.213	4.07
	9.0	63.0	23.7	906	0.276	0.163	4.78
	15.0	63.1	15.2	438	0.127	0.169	14.3
D	3.0	64.8	12.1	497	0.203	0.214	8.15
	9.0	57.0	25.7	1558	0.252	0.107	4.85
	15.0	48.9	18.2	1133	0.178	0.125	8.69
E	3.0	71.2	25.5	290	0.229	0.405	6.00
	9.0	59.6	31.8	508	0.235	0.266	5.35
	15.0	55.8	30.4	275	0.184	0.466	7.00
Mean		67.0	29.2	842	0.229	0.255	7.76
S.D.		15.7	19.9	565	0.057	0.182	5.12

(b) Voltage step 1200 V.

Valve	Q / Lit min^{-1}	Ξ^*_∞ / mAMV^{-2}	Ξ^*_0 / mAMV^{-2}	ξ^*_∞ / mAMV^{-2}	C* / nFm^{-1}	τ_1 / ms	τ_2 / ms
A	0.0	63.4	15.5	1113	0.298	0.057	5.76
	3.0	72.2	14.7	338	0.225	0.149	6.69
	9.0	92.3	16.6	1769	0.332	0.045	12.6
	15.0	86.3	24.5	753	0.258	0.087	4.93
B	3.0	68.8	8.81	175	0.221	0.305	10.2
	9.0	57.8	16.0	427	0.234	0.154	3.76
	15.0	47.4	34.7	742	0.342	0.146	2.73
C	3.0	72.6	0.1	469	0.196	0.092	6.61
	9.0	65.1	5.89	88.8	0.168	0.412	2.99
	15.0	59.4	105	200	0.878	0.925	2.33
D	3.0	63.3	11.5	446	0.224	0.115	5.99
	9.0	62.8	49.1	1989	0.37	0.409	3.03
	15.0	68.1	25.2	484	0.219	0.123	3.82
E	3.0	70.5	10.2	496	0.219	0.101	7.24
	9.0	61.6	47.2	253	0.384	0.370	2.80
	15.0	58.2	35.6	322	0.274	0.226	2.73
Mean		66.9	26.6	637	0.303	0.237	5.20
S.D.		10.9	25.2	547	0.166	0.220	2.91

Inserting the appropriate values for E_0 obtained from the steady state results (table 2), this expression has been fitted to the voltage step response data using the simplex algorithm and the parameters obtained are given in table 4.

Three factors have detracted from the accuracy of the parameters obtained here. At short timescales (<0.2ms) the current response was normally truncated to obtain a good scale for the rest of the signal. The first few points were therefore neglected in the fitting process. In addition, at timescales between ~0.2 - 0.5 ms an oscillation often appears in the current response (e.g. Fig 6) which appears to be related to a simultaneous feature in the pressure response. This is beyond the scope of the present model and may distort some of the fits we have obtained here. Also superimposed in all results for this system is a sinusoidal signal related to the gear pump frequency. This is more troublesome at high flow rates but can be removed by fitting a sine wave to the steady state portion of the result before further analysis.

5. Response to biased sinusoidal input

The current response to a biased sinusoidal voltage input has been obtained for valve A only. A bias of 600V and 1200V d.c. was used in conjunction with sinusoidal voltages of amplitude around 50 V. After removal of the mean bias current the resulting current fluctuations ΔI can be described by a superposition of a phase shifted sinusoidal component ϕ_I at the voltage input frequency ω_v and an additional component of arbitrary phase ϕ_q at the known pump frequency ω_q.

$$\Delta I = \frac{V}{|Z|}\sin(\omega_v t + \phi_I) + \frac{\delta I}{\delta Q} Q_o \sin(\omega_q t + \phi_q) \tag{18}$$

With only two frequencies present a time domain analysis was viable and the Nelder-Mead simplex algorithm[4] has again been used to fit this form to the current output. The resulting fit is extremely good over all frequencies studied (1 - 200 Hz). Knowing the phase and amplitude of the input signal the electrical impedence |Z| and phase change tanϕ can be found at each frequency. A typical example is shown in Fig. 9. and the derived parameters for the full range of frequencies are given in Table 5.

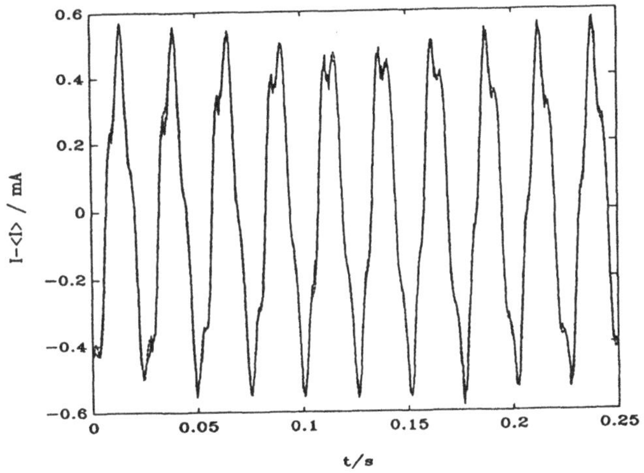

Fig. 9. Typical time domain fit to the sinusoidal response using eq(18). Valve A, Q = 9 Lit min -1 , d.c. voltage bias = 1.2 kV , ω_v = 40Hz. ——— experimental data. - - - - fit.

Table 5 Impedence data derived from sinusoidal response.

(a) With D.C. bias 600 V.

Flow Rate /lit min^{-1}	3.0		9.0		15.0	
Freq./ Hz	Z / kΩ	tanφ	Z / kΩ	tanφ	Z / kΩ	tanφ
1	88.51	-0.0104	90.66	-0.03	91.18	-0.065
2	90.32	0.0227	97.68	0.015	94.32	-0.144
5	92.97	0.0968	102.7	0.080	102.54	-0.051
10	94.20	0.1545	105.5	0.142	111.6	0.1097
20	93.06	0.286	105.9	0.277	112.4	0.2478
40	85.33	0.532	97.67	0.586	102.1	0.550
60	75.10	0.813	85.08	0.852	88.58	0.871
80	65.51	0.947	73.1	0.984	77.0	1.019
100	56.73	1.140	63.41	1.202	65.27	1.148
200	34.70	1.278	38.3	1.509	38.71	1.582

(b) With D.C. bias 1200V.

Flow Rate /lit min^{-1}	3.0		9.0		15.0	
Freq./ Hz	Z / kΩ	tanφ	Z / kΩ	tanφ	Z / kΩ	tanφ
1	44.13	0.0125	90.66	-0.053	42.78	-0.074
2	44.53	0.0156	97.68	-0.0159	44.50	-0.014
5	45.70	0.0815	102.7	0.070	45.61	0.057
10	45.66	0.0943	105.5	0.071	46.53	0.0787
20	46.30	0.149	105.9	0.132	47.72	0.131
40	44.81	0.247	97.67	0.244	46.60	0.239
60	43.52	0.378	85.08	0.376	45.44	0.374
80	41.46	0.412	73.1	0.398	43.18	0.415
100	39.39	0.507	63.41	0.511	41.03	0.508
200	30.49	0.761	38.3	0.797	31.25	0.795

The applied sinusoidal voltage is small compared to the bias and in the following analysis we approximate the current-voltage characteristics by linear terms over the range of interest. Thus we replace Ξ and ξ with R_v and r_v calculated according to Eqs. 5 and 6 . In this section we drop the subscript v and understand the resistances to refer to the bias voltage. For the linearised circuit the complex impedence is then

$$Z = \frac{R(\omega^2 r^2 C^2 + 1)}{(\omega^2 C^2 r^2 (1 + R/r) + 1) + j\omega RC} \tag{19}$$

Plotted against frequency the modulus of this function is monotonically decreasing and compares well with the experimental results above 20Hz. However, using frequency independent parameters eq 19 does not account for the maximum found in the experimental results around 10Hz (Table 5) . The phase angle ϕ calculated in the usual way

$$\tan\phi = \frac{\mathrm{Im}(Z)}{\mathrm{Re}(Z)} \tag{20}$$

is monotonically increasing with frequency and similarly fails to account for the negative values of ϕ observed at low frequency. In section 4.4 we wrote down the non-linear response to a step function including a slow relaxation τ_2. If this relaxation is included in the step function response for the linear circuit (Eq. 11) we obtain

$$i_H(t) = (E - E_o)^2 [(\Xi^* - \Xi_o^* \exp(-t/\tau_2) + \frac{1}{r}\exp(-t/\tau_1)] H(0) \tag{21}$$

Differentiating gives the response to an impulse; i.e. the transfer function W(t)

$$W(t) = (\Xi^* - \Xi_o^* + r^{-1})\delta(0) + [\frac{\Xi_o^*}{\tau_1}\exp(-t/\tau_1) - \frac{1}{r\tau_2}\exp(-t/\tau_2)] H(0) \tag{22}$$

Where $\delta(t)$ is the dirac delta function. Taking the one sided fourier transform we can obtain the expected frequency response[5]

$$W(j\omega) = \int_0^\infty W(t) \exp(-j\omega t)dt \tag{23}$$

from which we obtain

$$W(j\omega) = (G_\infty - G_o + \frac{1}{r}) + \frac{G_o}{(1 + \omega^2\tau_2^2)} - \frac{1}{r(1 + \omega^2\tau_1^2)} + j\,[- \frac{G_o\omega\tau_2}{(1 + \omega^2\tau_2^2)} + \frac{\omega\tau_1}{r(1 + \omega^2\tau_1^2)}] \tag{24}$$

Identifying the effective limiting long time resistance by

$$G_\infty = 1/R_\infty \tag{25}$$

and R_o the value of R at short time with

$$G_\infty - G_o = 1/R_o \tag{26}$$

we obtain

$$W(j\omega) = (\frac{1}{R_o} + \frac{1}{r}) + (\frac{R_o - R_\infty}{R_oR_\infty})\frac{1}{(1 + \omega^2\tau_2^2)} - \frac{1}{r(1 + \omega^2t_1^2)} + j[- (\frac{R_o - R_\infty}{R_oR_\infty})\frac{\omega\tau_2}{(1 + \omega^2\tau_2^2)} + \frac{\omega\tau_1}{r(1 + \omega^2\tau_1^2)}] \tag{27}$$

The impedence and phase angle are then given by

$$Z(\omega) = V / |W(j\omega)| \tag{28}$$

and

$$\tan\phi = Im[W(j\omega)] / Re[W(j\omega)] \tag{29}$$

In the limit $\tau_2 \to \infty$ this is identical to Eq.18 , but for finite values Eq. 27 introduces a narrow negative contribution to the impedence at low frequency as seen in the results. It can also account for the negative values of phase angle observed for frequencies below 5Hz.

The real and imaginary parts of this function $Z' = Z\cos\phi$ and $Z'' = Z\sin\phi$ have been fitted to the experimental counterparts by minimising the separation in the complex plane. An example is shown in Fig. (9) and parameters obtained are given in Table 5. These values represent a parameterisation of the data by frequency independent variables.

The estimate of ξ^* by this means is likely to be more accurate than the step function method and the mean is somewhat lower than the value given in the last section.

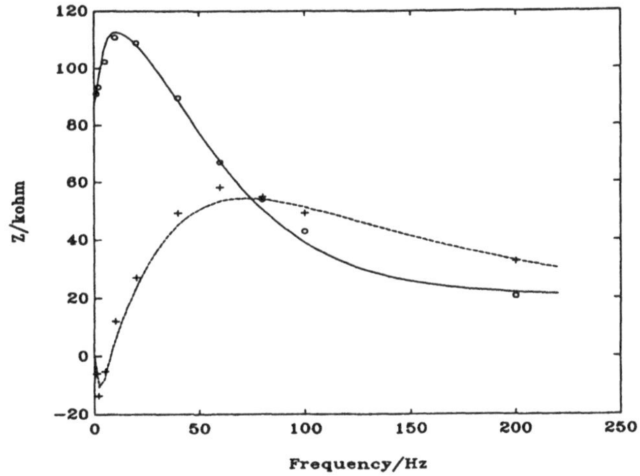

Fig. 10. Typical fit to the complex impedence using Eq. 27. Valve A, Q = 9 lit min^{-1}, d.c. voltage bias = 600 V. -- o -- Re(Z). -- + -- Im(Z).

Table 6 Linearised parameters derived from sinusoidal response. Obtained by simultaneously fitting Re(Z) and Im(Z) calculated from eq.(26).

(a) With D.C. bias 600 V.

Q/ lit min^{-1}	R_o/ kΩ	R_∞ / kΩ	r / kΩ	C / nF	τ_1 / ms	τ_2 / ms
3.0	95.8	91.0	21.0	20.42	0.430	68.8
9.0	107.9	95.9	21.28	18.48	0.393	59.7
15.0	115.8	87.6	23.2	17.97	0.418	62.0

(b) With D.C. bias 600 V.

Q/ lit min^{-1}	R_o/ kΩ	R_∞ / kΩ	r / kΩ	C / nF	τ_1 / ms	τ_2 / ms
3.0	46.26	45.7	16.2	18.16	0.294	69.3
9.0	46.80	41.88	13.88	17.29	0.24	98.7
15.0	47.54	42.2	14.97	17.39	0.26	129.9

Table 7 Nonlinear parameters for the sinusiodal response derived from the data of table 6 .

(a) With D.C. bias 600V.

Q / Lit min^{-1}	Ξ^*_∞ / $mAMV^{-2}$	Ξ^*_o / $mAMV^{-2}$	ξ^*_∞ / $mAMV^{-2}$	C^* / nFm^{-1}
3.0	72.5	3.63	314.2	0.282
9.0	79.8	8.87	359.4	0.255
15.0	102.4	24.9	386.7	0.248

(b) With D.C. bias 1200V.

Q / Lit min^{-1}	Ξ^*_∞ / $mAMV^{-2}$	Ξ^*_o / $mAMV^{-2}$	ξ^*_∞ / $mAMV^{-2}$	C^* / nFm^{-1}
3.0	67.3	0.814	189.7	0.251
9.0	78.4	8.24	236.6	0.238
15.0	83.1	9.33	234.2	0.240

6. Conclusions

The model we have presented can account for the electrical behaviour of our valve systems in response to step and sinusoidal responses over a wide range of timescales. At long times the steady state results are fit by a quadratic form (Eq. 2) which includes a threshold field strength E_o. At shorter times (~0.2 - 2 ms) and for frequencies in the range 20-200Hz the circuit of Fig. 2 with non-linear resistances based on the steady state results can account for the experimental responses. At very short times (< 0.1ms) the numerical solution of the model equation using the experimental voltage step as an input parameter can account for the initial rise of the current response. This is equivalent to a convolution of the system transfer function with the input for a linear system. To describe the response below ~20Hz and for times > 5ms an extra exponential term must be included. While this may not accurately describe the response in this region it is analytically convenient and sufficient to link the general form of the responses in the time and frequency domains.

In addition to E_o, five parameters Ξ, Ξ_o, τ_2, ξ and C, are required for the analysis. When normalised by the valve dimensions the values of these parameters are reasonably consistent under both modes of exitation and apply to valves of different size. Some variations do occur however, which we now summarise.

Of these parameters Ξ_o^* and the associated timeconstant τ_2 are probably the least reliable. These are derived from the slow background current drift in the time domain or equivalently from a few points at low frequency for the biased sine results. This background relaxation is sometimes very long lived and an attempt to fit with a single exponential is at best a first approximation.

Measurements of ξ^* from the transient response (Table 4) are probably not very reliable because the data has been truncated. The form of those results nevertheless suggests that a nonlinear expression for this term is justified. The apparently more accurate estimates of this parameter from the sinusoidal response (Table 7) suggest that ξ^* may be dependent upon the bias voltage. In deriving this parameter we assumed that the resistance r in Fig. 2. is subject to the same non linearity seen in the steady state results for R. This may be incorrect, but equally the variation of r with bias voltage shown in Table 6 again indicates that a linear model for this term is inadequate.

Values of Ξ^* for valve A derived from all methods (Table 8) and all valves (tables 3 & 4) are generally within 20% of the mean and show a small variation with flow rate, increasing for small gap size and decreasing for wider gaps. Fig. 11 shows that for valve A consistent values of this parameter are obtained by all methods and that Ξ^* increases with flow rate by some 20% over the range studied.

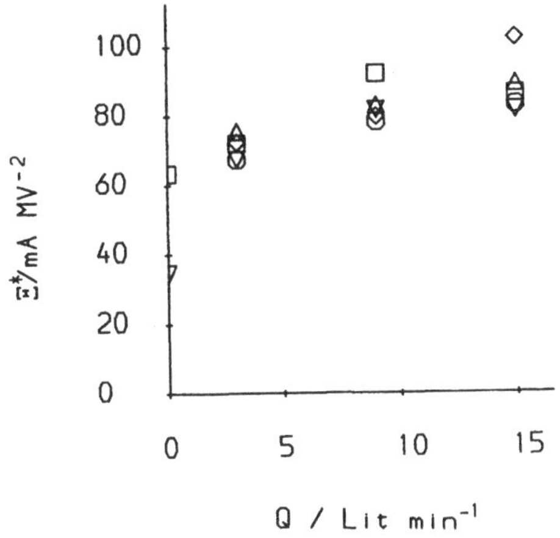

Fig. 11. Variation of Ξ^* with mean flow rate Q by different methods.
- △- Steady state; - ▽- Step input 600 V; - □- Step input 1200 V; - ◇- Sine input 600 V; -○- Sine input 1200 V.

The scaled capacitance derived from both time and frequency domain responses is the most consistent parameter and there appear to be no significant trends in the values for this quantity.

The other parameter involved is the threshold field E_0 and is obtained only from steady state data. It varies with flow rate and valve size and underpins the analysis. Inaccuracies in this parameter can be expected to lead to errors in the other derived parameters.

Equation (1) is an empirical fit and in fact, of course, a small current does flow below V_0. Nevertheless, the occurrence of a threshold may be linked to a physical effect. It is well known that ER fluids form chains of particles along the field direction[6], but it is not clear to what extent this affects the flow of current. In a flowing ER fluid we may expect viscous forces to disrupt the formation of particle chains and it is reasonable to suppose that higher field strengths are required before related dynamic structures can form. Our observations therefore seem compatible with an argument that current flow is in some way connected to the formation of chains. However, the appearance of plug flow in this system complicates such an analysis.

Acknowledgments

The authors thank the Science and Engineering Research Council for support of this work.

References

(1) R. Firoozian, W.A. Bullough, and D.J. Peel , ASME Publication DE-Vol. **18-4**, (1989) pp45-50, ASME.

(2) D.J. Peel, D. Brooks and W.A. Bullough. Proc. 9th Conf. Fluid Machinary. Hung. Acc. Sci. Budapest. Sept. 1991.

(3) W.A Bulluogh. I. Mech Eng. Paper C419/043, pp 171-177, (1990).

(4) W.H. Press, B.P. Flannery, S.A. Teukolsky and W.T. Vetterling, , Numerical Recipes (CUP 1986) p289.

(5) J.A. Aseltine, Transform Method in Linear System Analysis. McGraw-Hill (1958).

(6) M. Whittle, J. Non-Newtonian Fluid Mech., **37**, (1990) pp233-263, and Physics World **2**, 39, (1989).

DESIGN AND DEVELOPMENT OF FLOW BASED ELECTRO-RHEOLOGICAL DEVICES

DOUGLAS ALAN BROOKS
Advanced Fluid Systems Limited
10/14 Pensbury Industrial Estate
Pensbury Street
London SW8 4TJ
England

Abstract

The foundations for the design of electro-rheological (ER) actuating devices that use Poiseuille flow are laid. Systems of measurement to assess ER fluids are described including a high pressure flow apparatus. This was used to examine alternative valve geometries and fluids. A method is shown for the reduction of basic flow data to excess shear stress and shear rate.

Standard actuator arrangements are reviewed before the design of a flying control surface servomechanism is discussed. The evaluation of this validates the methods and indicates that the forces developed are within striking distance of those required for operational systems.

1 INTRODUCTION

The transition from fundamental rheological measurements to data applicable to the design of engineering equipment is often indistinct. It could be argued that as the complexity of the material under investigation increases, the correspondence between rheological measurements and device design decreases. The result is that one is often forced into experimental work on pre-production prototypes to generate viable designs[1]. In this respect an electro-rheological (ER) fluid must be seen as a particularly complex material and as a consequence designing from first principles is fraught with difficulties.

The initial consideration in this paper, which is also the first step in the design process, is to assess the applicability of available data and, perhaps more importantly, under what conditions such data was obtained. Discussed next is a laboratory high pressure flow evaluation apparatus that provides data directly applicable to the design and development of flow based ER devices. The data collected is subsequently described, together with some of the data reduction techniques used. The data is then applied to the evolution of the design of some actuating concepts and a flying control surface servomechanism.

2 SYSTEMS OF MEASUREMENT

Since ER fluids are complex rheological materials, measurement of their properties has generated numerous and varied measurement devices. This has led to considerable confusion in interpretation of data and to great difficulties in correlating data obtained from different rheological devices[2-5]. Unless consideration is given to the measurement system used to obtain data it is not possible to compare materials, and more importantly, use generalised data in the design of devices.

Broadly speaking, three separate measurement methods can be identified as indicated in figure 1. These are the static type measurements, rotary shear measurements, and flow based measurements. These will be briefly considered in turn.

Static measurements are assumed to take place at zero shear rate. The sample is placed in a small cell with one fixed electrode. The other electrode is pulled or pushed and the force required to fracture the sample is measured. The measurement of the force can be obtained from a simple cantilever beam strain gauge right the way through to a sophisticated proprietary materials testing machine. Whilst very simple in principle, being apparently just *go* or *no-go*, the test yields results which are complex and difficult to interpret. This is because once the sample fractures, the load being resisted is relaxed which enables the sample to restructure. The loading cycle can then be repeated. What complicates the situation is that fracturing and restructuring of the sample occurs under the electric field. Consequently the disposition of the particles and the resulting

Static or Laminar Shear Cells

- simple in concept

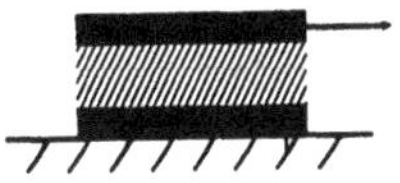

- complex in interpretation

Ranges of Equipment - in house shear cells to
Instron test machines

Rotary Shear Devices

- Couette

Constant Stress
Constant Strain
Oscillatory Stress
Oscillatory Strain

- parallel plate

Ranges of Equipment - Brookfield viscometer to
Weissenberg, Bohlin & Rheometrics

Poiseuille or Valve Flow Devices

- Steady Flow

- Oscillatory Flow

Ranges of Equipment - in-house designed.
Can be small but usually large
due to high pressures involved

Figure 1 Systems of Measurement

structural network differ from fracture to fracture. These changes usually enhance the yield strength and as a consequence a sawtooth type profile with shearing time is observed.

Rotary shear measurements are probably the most commonly used in rheological measurements. These are either of a concentric cylinder form, giving rise to Couette flow, or a cone and disc arrangement. As the latter is not particulary suitable for applying the electric field, a parallel plate geometry is preferred[6]. Either constant shear or constant stress is applied to the sample. As the stress or strain can be steady or oscillatory, this gives rise to numerous possible alternatives. Other variables that can complicate the situation are time dependent, thixotropic, or dilatant behaviour.

Despite these difficulties and complications rotary shear methods have a lot offer. Not only are sample sizes small, but considerable background understanding and expertise exist using these techniques.

The use of flow based evaluation equipment is less well covered. Whilst proprietary equipment exist for rotary shear measurements, including the development of equipments specifically for ER fluids[20,23], flow based systems are usually of in-house design. Some small syringe driven devices exist[7], but on the whole flow based methods are large, requiring considerable quantities of ER fluid, and operate at high pressures. The flow restriction is usually termed a valve, with the most common form being such that Poiseuille flow is established. This arrangement is not the only one possible, but it does have the considerable advantage of being relatively tractable to analysis. This is of significant importance in the design of flow based devices.

In a similar vein to rotary measurements, flow based measurement can be conducted, in principle, with either a steady or an unsteady flow. Steady flows can be obtained from, for example, a pumped system as described below or from a blow-down or expulsion vessel. Controlled, reproducible unsteady flows on the other hand are very difficult to obtain. The only such device known to the author was designed by Bullough[8], at Sheffield University. Unfortunately little experimental work has been conducted on the apparatus.

3 VALVE AND FLUID DEVELOPMENT APPARATUS

The Valve and Fluid Development Apparatus (VFDA) was designed for the evaluation of ER fluids and valve geometries. A secondary role was to provide a controllable ER fluid source for the evaluation of actuating devices. The apparatus, conceived to be a fully automated research apparatus, is shown in block diagram form in figure 2. It can be conveniently split into two parts, the first being the hydraulic circuit which uses, wherever possible, conventional hydraulic components. And secondly, the control and data capture circuit which, as implied by the name, handles the control and data capture under computer control.

The basic principle of this apparatus was to measure the pressure dropped through an ER valve at different conditions of flow, applied electric field and temperature. Different ER fluids and different valve geometries were evaluated. If sufficient experimental evaluations, i.e. combinations of fluid and geometries, are conducted these two aspects can be separated thereby enabling the design of other flow based devices. This apparatus will be discussed in three sections, the hydraulic circuit, the electronics, and the software.

The hydraulic circuit was straightforward and consisted of a variable speed motor driving a pump. The fluid was passed through an accumulator, then through the ER valve, and finally to the tank via a filter and cooler. The tank was triple skinned, with the two inner skins providing a flow path for the coolant and the outer skins providing a space for an insulating layer. The tank could be evacuated, or alternatively purged with dry nitrogen. It was fitted with a stirrer to keep the particles in suspension at all times. The motor, a fan cooled dc shunt wound unit, was rated at 4.1 kW with a maximum speed of 2,000 rev/min. The controller was a thyristor type which provided a fixed excitation voltage and a varying armature voltage. The control circuit was isolated from the drive circuit and thus enabled remote speed control. The armature voltage provided the feedback signal to the control circuit, giving speed control over a 20:1 range. Other features included a maximum and a minimum speed adjustment, a maximum current limit, current stability adjustment, and protection for over temperature and stalled conditions. In-house modifications included changeover switching for local or remote operation and the provision of a remote speed signal input.

The pump, a Hydracell D4 unit manufactured by Wanner Engineering, delivered a maximum flow rate of 15 litres/min at pressures of 60 Bar. It was a three cylinder axial piston oil pump in which the pumping action of the pistons was transferred through hydraulic cells to the pumped fluid. The hydraulic cells were isolated from the pumped fluid by means of three membranes. Pressure was maintained in the system by self aligning, cleaning, and seating, poppet valves. Since the hydraulic circuit was separated from the pumping circuit, virtually any fluid or suspension could be handled. ER fluids are pumped with ease and studies had shown that pumps suffer negligible wear[9]. The pump was a positive displacement device with flow rate varied by pump speed. When an increased delivery pressure was demanded, a slight fall in the flow rate was observed. This was compensated by the controller and the software to some extent however, the flow rate was evaluated by monitoring both the pump pressure and speed. Speed was monitored by a digital tachometer which derived an input signal from a 60 tooth wheel fitted to the pump shaft. A frequency to digital converter provided the analogue signal for the data acquisition section. The remainder of the hydraulic circuit was conventional hydraulics. A piston accumulator followed the pump, efficiently eliminating ripple from the pump outlet. Standard fittings connected the ER valve to the hydraulic circuit. There then followed a conventional water/oil cooler and a large mesh (250 micron) filter. The filter outlet drained into the tank which had a capacity of 5 litres. Temperature control was exercised using a cryostat which had an operating range of -40°C to 120°C with a cooling capacity of 5,000 watts at 20°C and a heating capacity of 2,000 watts.

Included in the electronics section were the high voltage source, the measuring instrument, miscellaneous circuits to drive pressure transducers, and the control computer. The power supply used was a standard high voltage laboratory unit operated in the voltage programmed mode. It provided 5 kV at 20 mA and was fitted with a remote input module that required 5 volts for 5 kV output. Current and voltage monitoring was accomplished internally and was available as a 100 μA signal. Other features included a remote inhibit of the output, polarity changing and short-circuit proof design.

The data capture section consisted of a microprocessor voltmeter equipped with an IEEE 488 communication bus and a processing interface. Accuracy was 0.002% plus 1 digit of the input scale, the unit being auto-ranging from 10 mV to 1000 volts. Input sensitivity on the bottom range was 1 μV and 10mV on the upper range. Maximum reading rate without filter was 330 per second, falling to 6 per second with the filter. A scanner complemented the voltmeter and derived both power and timing signals from the voltmeter. All control signals were provided by the voltmeter which in turn received them from the system controller, the computer. The scanner provided 16 channels, each 4 wire plus guard, and generated a reading rate of 130 per second with less than 1 μV error due to switching noise.

An 8 channel, 8 bit resolution, D to A converter provided signals to control pump speed, voltage output, voltage inhibit signals and the alarm circuits. The maximum output was 5 volts with an accuracy of 0.5%.

The software controlled the apparatus and directed the data acquisition. Data was collected on the pressure drop characteristics for a particular geometry against variables of flow rate and voltage. Other parameters monitored included temperature and current. The control software was designed to collect and store data, as well as monitoring the system in terms of its integrity, for example, possible dangerous or erroneous conditions of pressures, flow rate and/or temperature. The initial conditions were set as the first operation. These included voltage and flow rate ranges and a subroutine that measured all the system zero errors so that all subsequent measurements were compensated. The programme commenced with the first, lowest, flow rate and incremented the voltage from zero to the pre-determined maximum in steps that could be linear or a programmed function. At each voltage setting the following analogue variables were read:

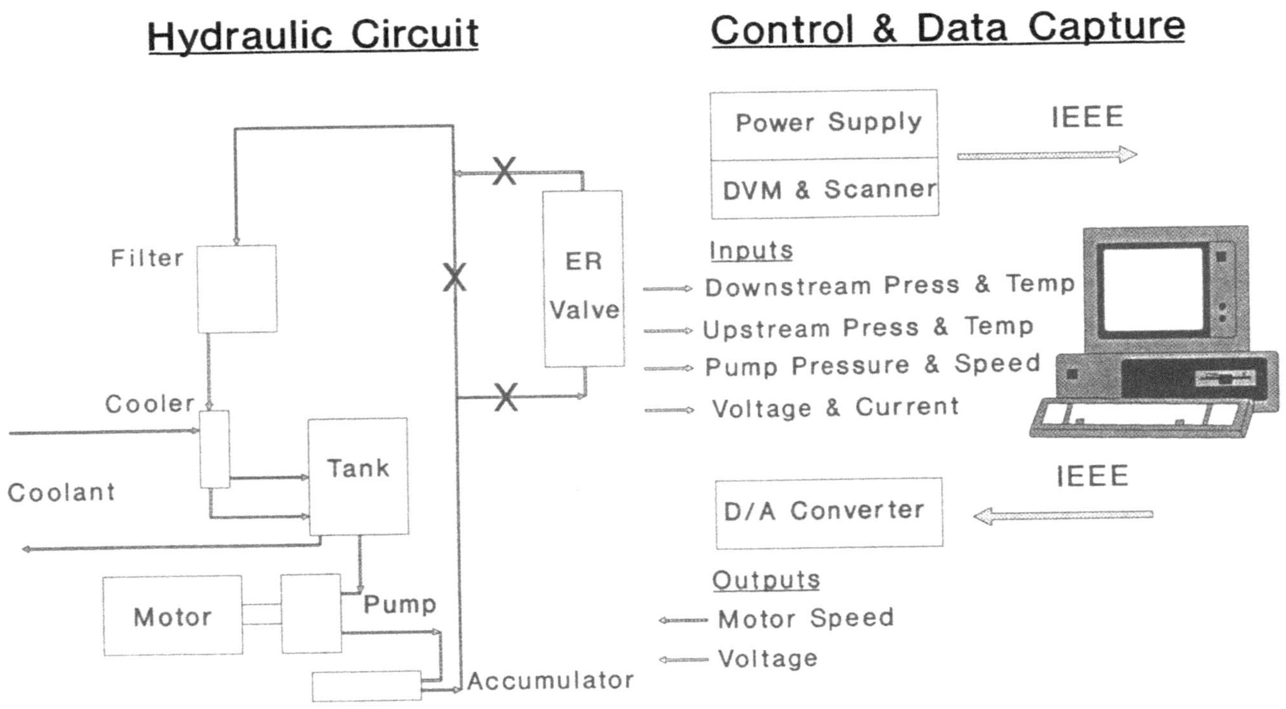

Figure 2 Valve and Fluid Development Apparatus

a) Voltage Output
b) Current Drawn
c) Upstream Pressure
d) Downstream Pressure
e) Upstream Temperature
f) Downstream Temperature
g) Pump Speed
h) Pump Pressure

Each reading was taken in sequence and 10 sequences were taken i.e. 80 readings being taken, at the unfiltered scanner rate, in 0.62 seconds. Each set was then averaged and provided eight data points. The analogue readings were converted to their engineering equivalents, e.g. pressure signals to pressure in Bar and pump speed and pump pressure to a flow rate. The voltage was incremented until the desired maximum voltage was reached, typically some 16 steps being used. At this stage the data was transferred to disc, the data arrays re-set, and the sequence repeated at an increased flow rate, typically 12 flow rates were used. Additional features included data checking and system integrity checking at each voltage step, computation of the viscosity of the fluid and the wall shear rate at each flow setting, a theoretical modelling of the experimental conductivity data at each voltage set and a flushing procedure to fully mix the fluid prior to the next flow increment. The apparatus has been described in some depth and it is now worthwhile to portray some of the valve arrangements that were evaluated.

4 TYPE OF ER VALVES

A number of alternative valve geometries were assessed and four distinct types are presented here, including a Poiseuille flow geometry which, as noted earlier, is relatively tractable to analysis[10-12]. This feature, in combination with the ease of construction and that it produced the largest pressure drop, has led to a more detailed evaluation of this geometry.

4.1 Pin Valve

The Pin valve is shown schematically in figure 3. The main parts were the body, the pin block assembly, and two end covers. All components were produced from lightweight reinforced nylon apart from the pins and electrical contacts. The pin block which consisted of a series of interspaced positive and negative pins. In total 147 pins were used, 70 negative in six rows and 77 positive in seven rows. The pins were a press fit into a mild steel contact surface and were supported by a 5 mm thick nylon disc. The pins located in the nylon disc of the opposite electrode, hence maintaining pin spacing. The pins were 1 mm diameter with a centre to centre distance of 2 mm, row to row spacing for was 4 mm. The pin block could be rotated with respect to the flow and hence the electric field could be parallel, perpendicular, or rotated to any angle relative to the flow.

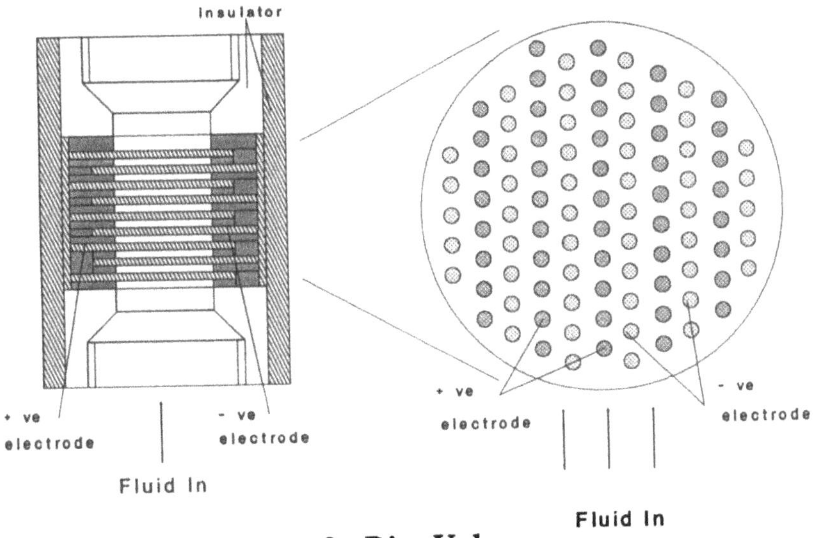

Figure 3 Pin Valve

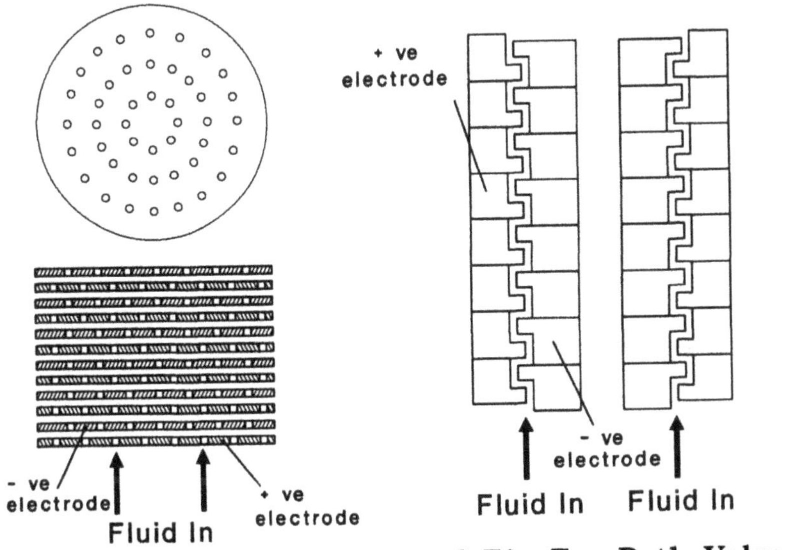

Figure 4 Perforated Plate and Zig-Zag Path Valve

4.2 Perforated Plate Valve

The perforated plate valve, shown schematically in figure 4, consisted of a number of perforated alternate positive and negative plates. 30 plates were used, 15 of each polarity. The plates were 50 mm in diameter, with a spacing of 0.75 mm. Holes in the positive plate were 2 mm diameter placed on the intersection of the pitch circle diameters (PCDs) of 0, 10, 20, 30, and 40 mm and angles of 45, 22.5, 15, and 10 degrees giving a total of 95 holes in each positive plate. The negative plate holes, also 2 mm diameter, were placed on PCDs of 5, 15, 25, 35, and 45 mm at the intersection with angles of 45, 22,5, 15, 10 and 8 degrees giving a total of 129 holes. This staggered arrangement between the PCDs on each plate forced the fluid to take a meandering path through the valve. The end caps were mild steel and the body, that supported the plates, was of reinforced nylon.

4.3 Zig-Zag Path Valve

The zig-zag path valve electrode arrangement is shown schematically in figure 4. The objective was to force the fluid through a twisted flow path so that at the changes of direction the electric field lines would be at an angle with respect to the fluid flow. By using this approach, we hoped to increase the total pressure drop generated. The valve consisted of five interlocked and stepped electrodes. The total valve length in line with the inlet and outlet was 50 mm, with 27 mm perpendicular to the flow giving an overall length of approximately 100 mm. This could be compared with a concentric cylinder arrangement of similar length.

4.4 Concentric Cylinder Valve

The ER valve is shown schematically in figure 5. This design has evolved from a number of previous designs and is an elegant, simple, and reliable one which can be cleaned and changed with particular ease. The valve consisted of an inner electrode assembly, an outer earthed electrode and an end cap. The inner electrode was a mild steel sleeved insulating rod, relieved at both ends, which provided a smooth flow entry and exit to the electrode. The steel sleeve, which was the high voltage electrode with an internal high voltage connection, was ground concentric to a location collar at one end and a counterbored centring hole at the other end. The end cap was suitably machined to locate into this hole. The cap also had a location collar similar to the inner electrode. The outer body was a square tube with a suitably finished hole through its centre; tapped holes were provided to secure the end cap and the inner electrode assembly. The concentricity was readily maintained and the attention to detail at the electrode edges[21] are such that electric field strengths in excess of 12 kV/mm can be maintained.

All four of these valves developed a field controllable pressure drop and comparative data is shown in figure 6. Different valves exhibited different characteristics, for example, the perforated plate valve had a very low no-field pressure drop. Rotating the pin block in the pin valve had little effect on the pressure drop. The zig-zag path valve was very difficult to operate, requiring great care in its assembly. The concentric cylinder valve proved to be easy to operate and exhibited considerable pressure drops. It was therefore chosen for further study in relation to the development of actuating devices.

Figure 5 Concentric cylinder Valve

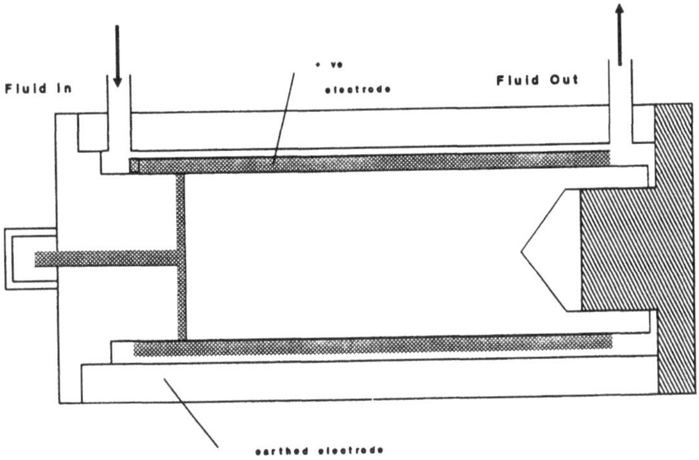

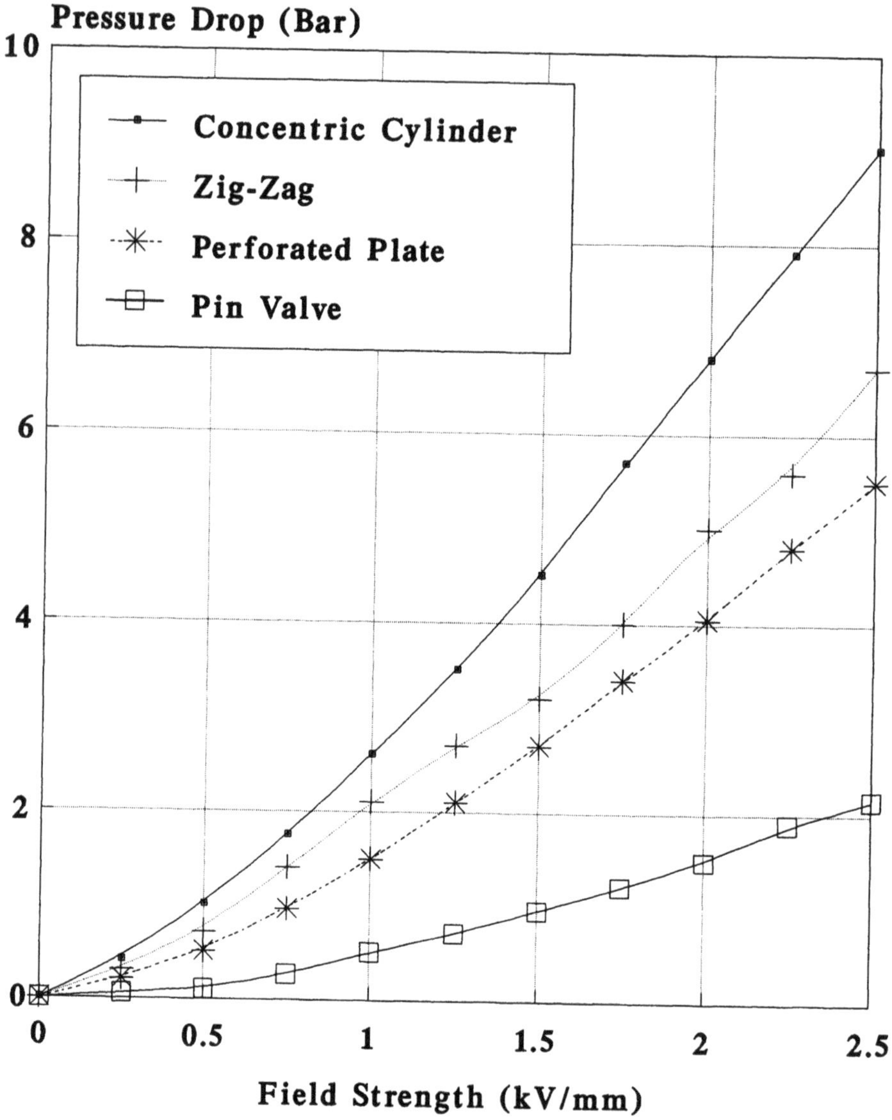

Figure 6 Pressure Drop Characteristics

5 DATA AND DATA REDUCTION

During the course of the experiment the measured voltages were converted to their engineering equivalents. In this section, the engineering equivalents will be discussed along with their reduction to a form suitable for device design. The raw data falls into two categories: first, the input variables of voltage and flow rate, at a given temperature, and secondly the output variables of pressure drop and current drawn. These give rise to the following possible comparisons:

Pressure drop versus flow rate	(no voltage)
Pressure drop versus flow rate	(fixed voltage)
Pressure drop versus voltage	(fixed flow rate)
Current drawn versus flow rate	(fixed voltage)
Current drawn versus voltage	(fixed flow rate)

However, voltage (V), flow rate (Q), pressure drop with no voltage (P_n), pressure drop with voltage (P_{tot}), and current drawn (I) are of little use in the subsequent design process. These variables need to be reduced in relation to the geometry of the valve. Some of these are relatively straightforward as follows:

Electric Field from the Voltage

$$E = V/H \tag{1}$$

In the case of a concentric cylinder valve, this assumes that the mean diameter, D_m, is large with respect to the gap, H.

Current Density from the Current

$$J = I/(\pi D_m L) \tag{2}$$

Here the mean diameter is used again, the effective electrode length is L.

Shear Rate from the Flow Rate

$$G_w = 2Q/(H^2 D_m) \tag{3}$$

The shear rate is taken as that occurring at the electrode, hence the subscript w, implying a wall shear rate[13]. Again the mean diameter is used and the assumption that the mean diameter, D_m, is large with respect to the gap, H, is taken as valid.

Reducing the pressure data to a yield stress, T_y, and a viscosity, μ, is more complicated. The total pressure drop P_{tot} can be taken as the sum of two components, one due to the viscosity, i.e. the no-field or viscous pressure drop, P_n, and that due to the field, the excess pressure drop, P_e. This implicitly assumes no interactions or cross product terms.

$$P_{tot} = P_n + P_e \quad (P_n * P_e = 0)$$

The viscous pressure drop is given by:

$$P_n = \frac{12 * \mu * Q * L}{H^3 * \pi * D_m} \tag{4}$$

where:

μ is the viscosity
Q the volume flow rate
H the electrode gap
D_m the valve mean diameter
L the valve (electrode) length.

One simple expression for the excess pressure drop is given by:

$$P_e = \frac{2 * T_y * L}{H} \tag{5}$$

where T_y is the yield stress due to the applied electric field.

Equation 4 has been shown to adequately model the viscous component,[14] but the use of Eq. 5 will lead to non-systematic and variable errors in the total pressure developed in a concentric cylinder valve[15,22]. Although this is unfortunate, it does provide a rapid and approximate assessment of the required valve geometry and as a consequence should not be lightly dismissed.

The yield stress is now redefined as **excess shear stress**, to distinguish it from that used in the simple approximation. To quantify excess shear stress we need to solve the following equation:

$$T_x = T_e + \mu \, \underline{dU}/dX \qquad (6)$$

where T_x, is the total stress in a fluid with an excess shear stress, T_e, a viscosity, μ, and a fluid velocity, U.

To solve this equation, an assumption must be made concerning the velocity profile in the geometry under consideration. The assumption is that there is a central "plug" of unsheared fluid in the valve with two sheared regions either side of it. Details of the profile are shown in figure 7. In the case of Poiseuille flow:

$$\frac{dT_x}{dX} = -G_p \qquad (7)$$

where G_p is the pressure gradient given by P_{tot}/L.

Figure 7 Flow Profile in Electrode Gap

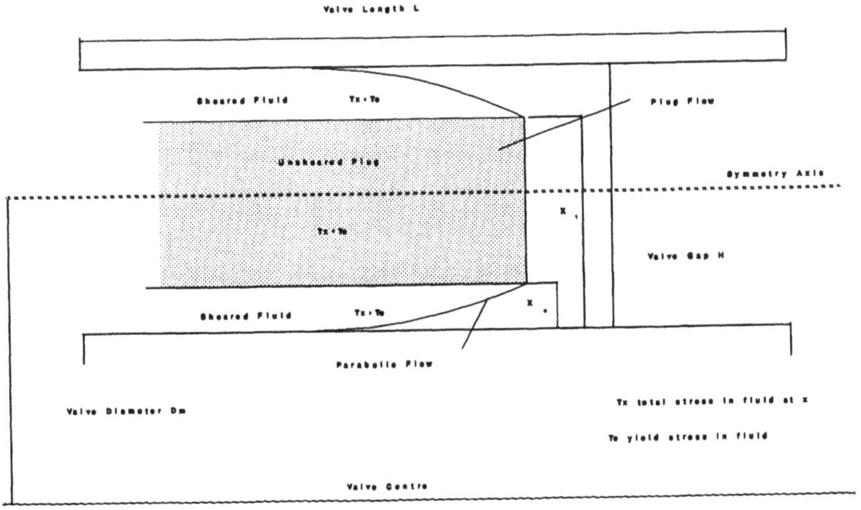

Normally Eq 6. would be substituted into Eq. 7 and the differential equation formed duly solved. However, because of the discontinuity that exists at the boundaries of the sheared and the unsheared regions, this is inadvisable. The differential equation can be integrated in two parts, the sheared and the unsheared regions. This has been carried out by a number of people yielding results of the general form $S^3 - (1-3T)^2 + 4T^3 = 0$[15-18]. This looks fairly straightforward until S and T are substituted.

$$S = \pi D_m H^3 P_{tot}/12\mu QL \text{ and } T = \pi D_m T_e H^2/12\mu Q \tag{8}$$

To extricate the excess shear stress, T_e, can be quite difficult.

An alternative is to consider the volume flow rate through the valve[15]. This results in the following expression which has been covered in some detail elsewhere[19,22]:

$$Q = \frac{G_p H^3 W}{12\mu} * \left(1 - \frac{2T_y}{G_p H}\right)^2 * \left(1 + \frac{T_e}{G_p H}\right) \tag{9}$$

In practice, it is more useful to express Eq.9 in the form of pressure drop and flow rate. If the following substitutions are made:

$$x = T_e/G_p H \text{ and } b = 12\mu Q/G_p H^3 W$$

Eq. 9 becomes:

$$b = (1 - 2x)^2 * (1 + x)$$

which expands to:

$$4x^3 - 3x = b - 1$$

Using the trigonometric identity $4\cos^3\Theta - 3\cos\Theta = \cos3\Theta$, Θ can be evaluated:

$$\Theta = \cos\left(\frac{1}{3}\cos^{-1}(b - 1) + \frac{4\pi}{3}\right) \tag{10}$$

As $G_0 = 12\mu Q/H^3W$, that is, the pressure gradient with no excess shear stress, then $b = G_0/G_p$. Additionally, since $x = T_e/G_pH$, where $G_p = P_{tot}/L$ and $G_0 = P_n/L$, where P_n is the viscous pressure drop, Eq 10. can be re-written as:

$$T_e = \frac{P_{tot}H}{L} \cos\left(\frac{1}{3}\cos^{-1}\left(\frac{P_n}{P_{tot}} - 1\right) + \frac{4\pi}{3}\right) \tag{11}$$

The attractiveness of this form is the ease by which the excess shear stress can be determined. That is, solely from measurements of pressure drop with and without the applied electric field, P_{tot} and P_n, in conjunction with details of flow rate and the valve geometry.

The above expression has been successfully assessed with a number of valve geometries, different lengths, gaps, and diameters as shown in table 1[19]. As an example of the value of this approach, the reduction of some typical data obtained for three different valve gaps will be demonstrated.

Body Internal Diameter	Diameter	Electrode Gap	Length	Shear Rate Range
50.015	47.991	1.012	150	600 - 5000
	48.489	0.763	150	1000 - 8000
	48.983	0.516	150	2500 - 21000
	48.983	0.516	150	2500 - 21000
	48.466	0.775	100	600 - 5000
	48.466	0.775	100	600 - 5000
50.013	48.489	0.762	150	600 - 5000
50.055	48.489	0.783	150	600 - 5000
50.005	48.984	0.510	96.84	2500 - 21000
	48.984	0.510	72.74	2500 - 21000
	48.984	0.510	48.46	2500 - 21000
	48.984	0.510	24.22	2500 - 21000
30.005	28.386	0.810	60	1700 - 14000
	28.594	0.706	60	2300 - 18000
	28.785	0.610	60	3000 - 25000
	28.989	0.508	60	4000 - 36000

TABLE OF VALVE DIMENSIONS

The two plots, figures 8 and 9, show the some of the raw data. Figure 8 is a plot of developed pressure plotted against field strength. Three valves are indicated, the basic geometry being similar but with different gaps. As would be expected a reduced gap results in an increased pressure drop at zero field. Also of interest is the magnitude of the developed pressure which increases more rapidly with field as the gap is reduced. Figure 9 shows developed pressure, now plotted against flow rate, for a fixed field of 2.66 kV/mm. Data for the same three valve geometries is shown. If data from these two figures is processed using Eq. 11, the developed pressure can be reduced to an excess shear stress and the flow rate to a shear rate via Eq. 3. Processing the data and re-plotting the results transforms figure 8 to figure 10 and figure 9 to figure 11.

Figure 10 demonstrates that the three curves of figure 8 can be collapsed to a single plot of T_e against E_0. Similarly figure 11 collapses to figure 9. The reduction is not as good as that seen in figure 10 but it is still acceptable. These are significant results. We are now in a position to be able to predict the flow behaviour from basic data concerning the fluid's excess shear stress and viscosity. Given such information, it is possible to accurately specify the pressures developed for any given Poiseuille geometry. Hence we are able to design devices using ER fluids in flow.

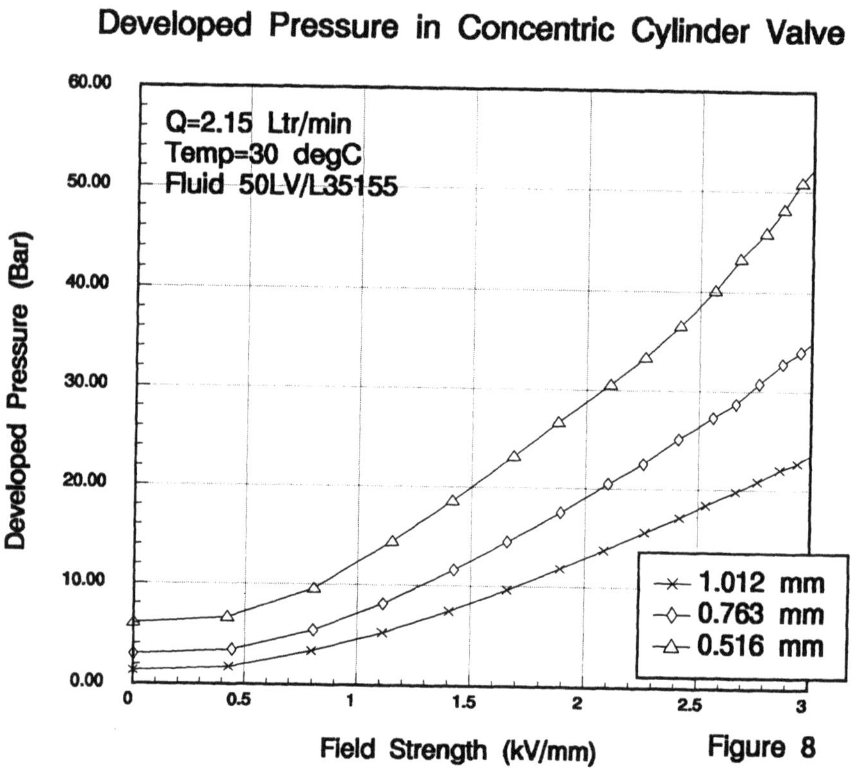
Developed Pressure in Concentric Cylinder Valve
Q=2.15 Ltr/min
Temp=30 degC
Fluid 50LV/L35155
Developed Pressure (Bar)
60.00
50.00
40.00
30.00
20.00
10.00
0.00
0
0.5
1
1.5
2
2.5
3
1.012 mm
0.763 mm
0.516 mm
Field Strength (kV/mm)

Figure 8

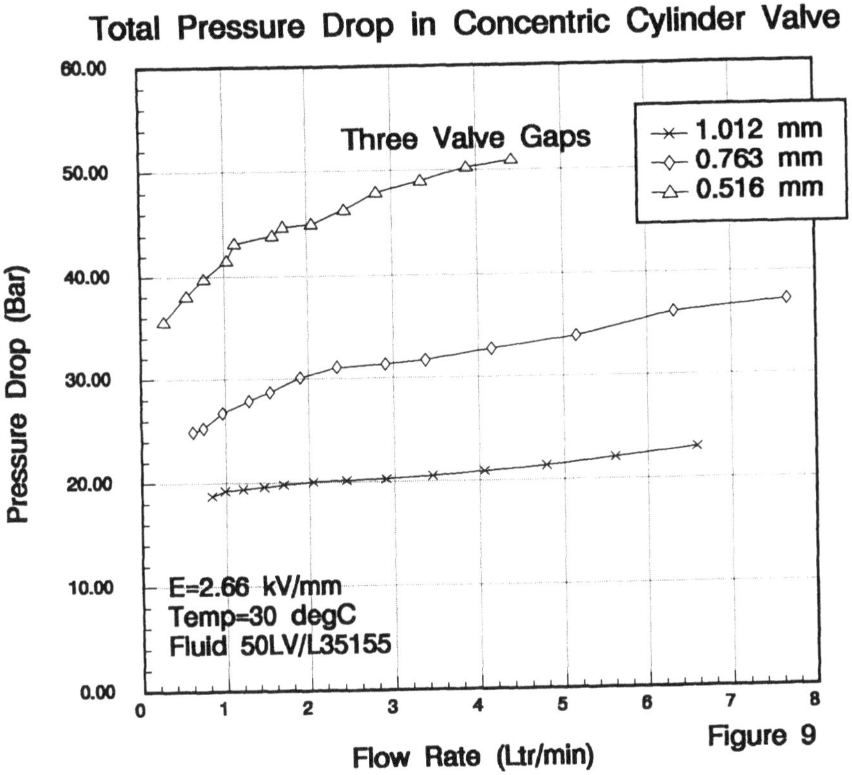

Total Pressure Drop in Concentric Cylinder Valve
Three Valve Gaps
1.012 mm
0.763 mm
0.516 mm
E=2.66 kV/mm
Temp=30 degC
Fluid 50LV/L35155
Pressure Drop (Bar)
60.00
50.00
40.00
30.00
20.00
10.00
0.00
0
1
2
3
4
5
6
7
8
Flow Rate (Ltr/min)

Figure 9

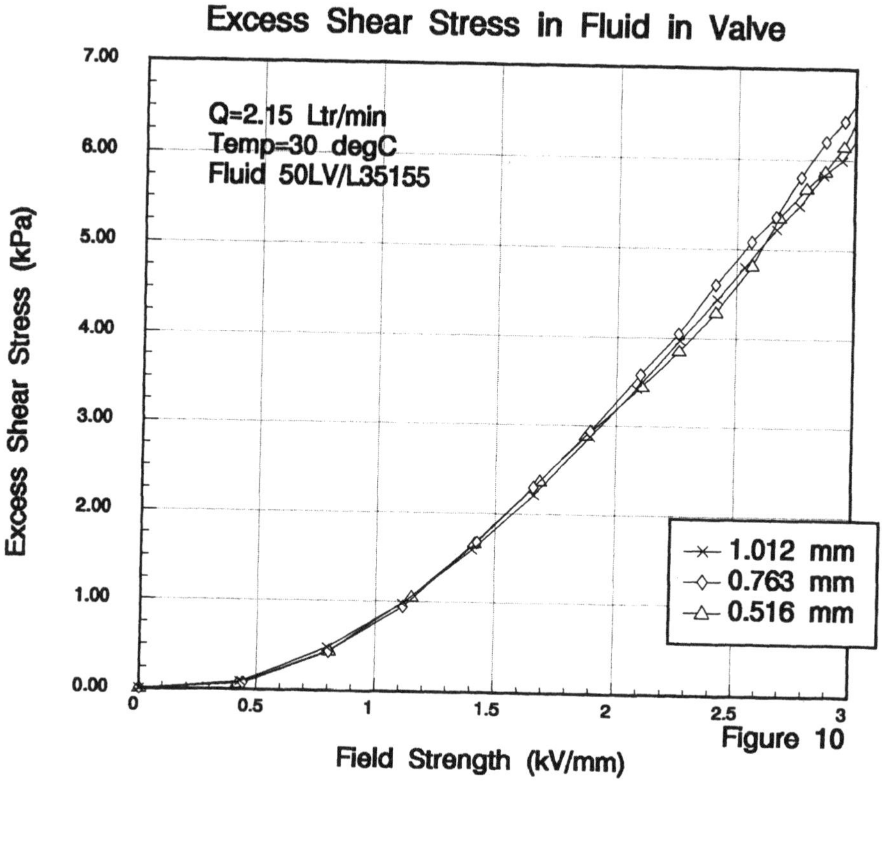
Excess Shear Stress in Fluid in Valve
Q=2.15 Ltr/min
Temp=30 degC
Fluid 50LV/L35155
Excess Shear Stress (kPa)
7.00
6.00
5.00
4.00
3.00
2.00
1.00
0.00
0
0.5
1
1.5
2
2.5
3
1.012 mm
0.763 mm
0.516 mm
Field Strength (kV/mm)

Figure 10

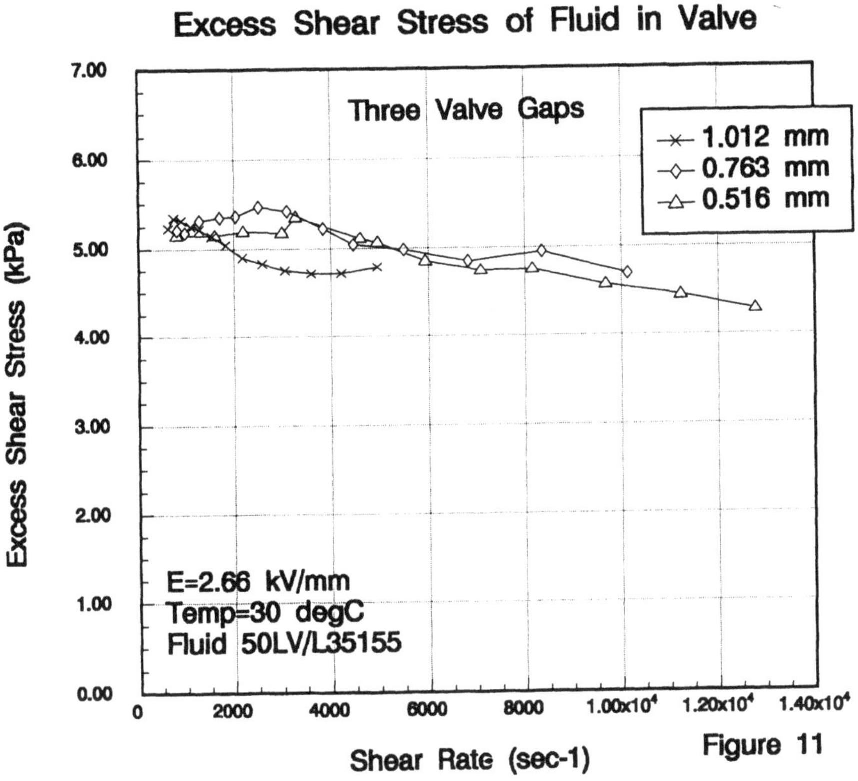
Excess Shear Stress of Fluid in Valve
Three Valve Gaps
1.012 mm
0.763 mm
0.516 mm
E=2.66 kV/mm
Temp=30 degC
Fluid 50LV/L35155
Excess Shear Stress (kPa)
7.00
6.00
5.00
4.00
3.00
2.00
1.00
0.00
0
2000
4000
6000
8000
1.00x10⁴
1.20x10⁴
1.40x10⁴
Shear Rate (sec-1)

Figure 11

6 ELECTRO-RHEOLOGICAL ACTUATING DEVICES

An ER valve will provide a restriction to the passage of the ER fluid because of the viscous pressure drop and a component induced by the electric field. An external source is required to create the flow and the restriction will result in an increased pressure upstream of a valve. This increased pressure can, in turn, be used to perform some action, for example to generate bi-directional motion as in the four valve bridge arrangement shown in figure 12.

6.1 Basic 4 Valve Bridge

The basic four valve bridge one of the simplest arrangement to explain. The bridge is balanced when equal voltages are on each of the four valves. The pressure drop down each valve is the same and no differential pressure is generated across the piston, or ram, i.e. Pa=Pb. However, if the voltage on valves "a" and "d" were to be increased the bridge balance would be upset. The increased pressure would be met by the pump, Pd. If at the same time the voltage is removed from valves "b" and "c", the fluid can readily flow out of the low pressure side of the ram to the tank. In this case Pa>Pb and the piston will be moved to the left. If the reverse voltage condition is made, "a" and "d" low with "b" and "c" high, motion is in the opposite direction as Pb>Pa. The thrust generated is the product of the piston area and the developed pressure. Since pressure drops in the region of 60 Bar can be produced down a valve of some 100 mm length, significant thrusts can be developed. Of considerable interest in this mode is the speed at which this pressure reversal can be induced. It has been demonstrated that ER fluids can be made to change their state in a few milliseconds and hence pressures of 40 to 60 Bar can be modulated at frequencies in the order of 200 to 300 Hz.

6.2 Basic 2 Valve Bridge

The two valve bridge is one of the more obvious variants in actuator design. Its operation can be appreciated in relation to the schematic, figure 12. The supply pressure, P_s, under balanced field conditions, will be the sum of the viscous pressure drop through valves x and y and the field induced pressure drop. This is applied to one side of the differential piston, side a. The pressure drop through valve y is applied to the other side of the ram, denoted 2a, indicating that the area is twice that of side a. When the pressure drop through x equals that through y there is no movement. If however, the field on valve x is raised as that on y is lowered, fluid will flow into side a and out of side 2a, allowing motion to the left. Similarly if y is raised and x lowered, fluid flows into side 2a and out of a, via valve x. In this case motion is to the right.

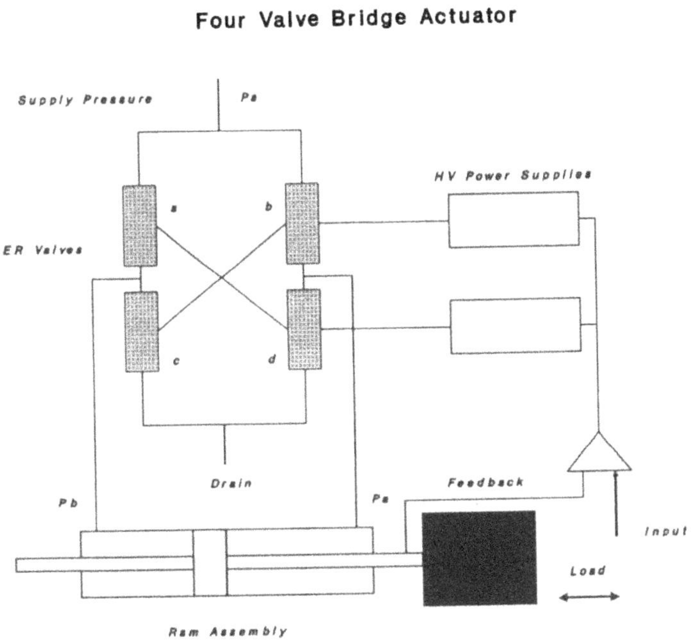

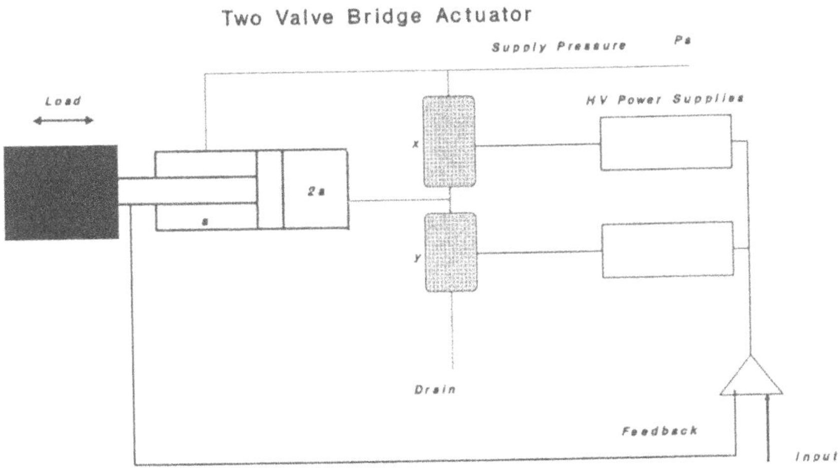

Figure 12 Actuator Arrangements

In either of the two designs noted above, the ER valves can be local to the ram or remote from it, with conventional hydraulic connections. As we are in the main considering concentric cylinders as the valves, it is feasible to wrap these around the ram thereby giving a self contained unit. As an alternative, the valves can be connected to the ram by means of conventional hydraulic hoses or pipes. The small servomechanism, to be discussed next, uses this approach and was designed to control the motion of a fin fitted to an airborne system. It is however, equally applicable to any other high speed control system.

7 ER FLUID DESIGN - A FLYING CONTROL SURFACE SERVOMECHANISM

The objective of any actuation system is to deliver controlled mechanical power in a cost effective manner. Outlined above are the details necessary to design some basic ER servomechanism. The realisation of these basic components into a prototype design for a flying control surface servomechanism will be discussed. A significant point to note is that although this design relates to airborne systems, it is equally applicable to marine surface or sub-surface applications, to nuclear, and to any general industrial application that requires a high performance servomechanism.

In the conventional hydraulic circuit used with flying control surfaces, the servomechanisms represent the majority of the system cost and mass. They are also unreliable. The anticipated potential benefits of an ER system are:

* reduced cost - via easier manufacture, uncomplicated parts, and simpler assembly
* reduced mass - less fluid, valves integral to structure
* increased reliability - less sensitive to contamination, simpler design, no moving parts
* higher bandwidth - possibly as high as 1,000 Hz compared to 50 Hz currently available
* low control power - 5 to 10 watts

Two basic flight times can be considered in airborne systems: short flights, less than 20 seconds, and longer flights, greater than 20 seconds. For the former, a blowdown system is preferable, and for the latter a recirculating system is more suitable. Blowdown systems using a conventional expulsion cylinder have been shown to be an effective means of delivering ER fluid to the servomechanism. However, in terms of development, a recirculating system enables a more detailed study of servomechanism performance. The valve and fluid development apparatus described above was modified to provide a source of high pressure ER fluid to test the servomechanism.

The following system parameters were taken as initial design targets:

System Pressure	less than 60 Bar
Flow rate	between 25 and $100 cm^3/sec$
Flight duration	20 seconds
Control surface torque	10 Nm
Control surface velocity	$300°/sec$
Control surface deflection	$\pm 15°$
Bandwidth	60 Hz
Inertial mass	$7 * 10^{-4}$ kg m^2

A two valve bridge was chosen with the valves separated from the fin actuator assembly. This enabled the valves to be designed with low cost, in terms of materials, complexity, and assembly as the objective. The design, shown in figure 13 met these objectives. It was based on a length of ½ inch outside diameter precision drawn nickel alloy tube available off the shelf. The wall thickness was such that when a length of standard, ground finish, silver steel rod was fitted inside, the required gap of 0.7 mm was achieved.

Manufacture was very simple. The tube was cut to length. The central rod was turned at each end to accommodate the insulating spacers. A port was drilled through the rod centre with sealing being achieved by a single O-ring. The ends were threaded to accept a standard pipe fitting. The high voltage electrical connection was to the tube and was insulated using standard heat shrink sleeving. The earthed return connection was via the pipework.

The actuator, shown in figure 14, was a push-push type with two separate pistons. Based on the two valve design, one piston was twice the area of the other. Fluid acted on one side only, thereby reducing both sealing and friction problems. Experience had indicated that spring energised U-section PTFE seals were suitable. The actuator consisted of two main body sections. One section was finished in a fashion suitable for the pistons. The other section supported a yoke, in suitable bearings, and the feedback sensors. The yoke was connected to the fin.

To fit in the design envelope, the maximum radius arm on the yoke was 25 mm and therefore to achieve the required torque of 10 Nm, a thrust of 400 N was required. Taking the larger piston to be 18.0 mm diameter, required that the smaller one be 12.7 mm, the respective areas being 253 and 127 mm^2. To develop a thrust of 400 N over an area of 127 mm^2 a pressure of 32 Bar was required. Using the simple expression, Eq. 5, for the developed pressure, $P_e = 2T_yL/H$, indicated an electrode length of approximately 225 mm. The fluids' yield stress

Figure 13 Schematic of Valve for Flying Control Surface Servomechanism

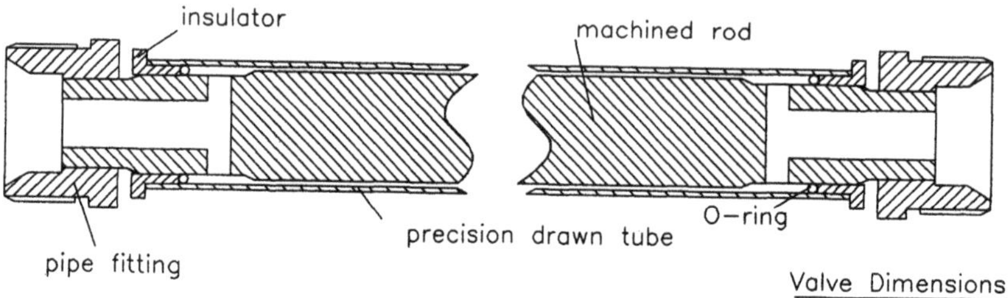

Valve Dimensions	
Length	225 mm
Gap	0.7 mm
Diameter	15.0 mm

Figure 14 Schematic of Push-push Actuator

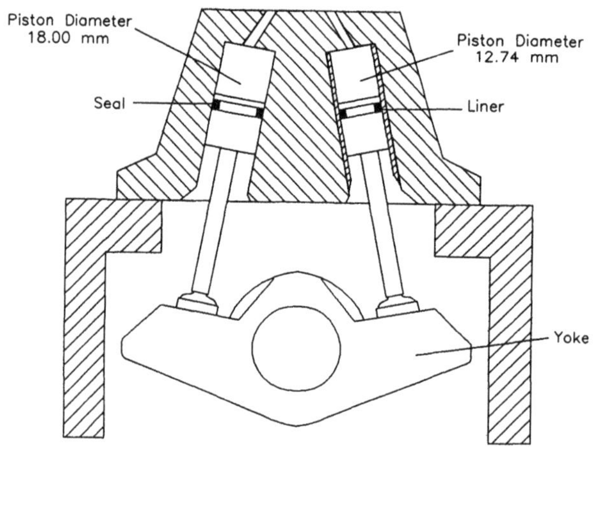

was taken as 5 kPa which could be achieved using a typical 35% polymer based ER fluid at 2.7 kV/mm. The power supply therefore needed to deliver approximately 2,000 volts. However, using the full expression, Eq. 9, indicated that the pressure drop due to the field would be almost 40 Bar for a length of 225 mm. It was decided to use this length but also to monitor the pressure drop during testing.

The flow rate required to slew at 300°/sec over the full arc of 15° into the larger piston area of 253 mm^2 was 50 cm^3/sec. For a 20 second flight the total fluid required was approximately one litre of fluid which was quite acceptable. Having established the maximum flow rate required the pressure drop with no-field using, Eq. 4, $P_n = 12\mu QL/(H^3\pi D_m)$, could be accomplished. With D_m at 15 mm and a viscosity of 0.07 Pas gave a result of 5.9 Bar. Hence the total system pressure in the order of 50 Bar ($2^*P_n + P_e$) was expected. This was within the maximum system pressure of 60 Bar. The maximum shear rate in the valve was calculated, using Eq. 3, and a value of 13,640 sec^{-1} was obtained, within the known performance capabilities of the ER fluid used.

Finally some consideration was required in relation to the maximum current demand. The total valve area was approximately 100 cm^2. Allowing a maximum current density of 50μA/cm^2 indicated the need for a 5 mA power supply, that is 10 watts per valve. Whilst this would be the maximum demanded, the precise requirement will depend upon the method of control, analogue, bang-bang, or pulse length modulation (PLM). Each method had inherent advantages and disadvantages, although PLM was known to be most suitable for digital autopilots and offered the real possibility of multiplexing the power supply. The total electrical requirement would therefore be reduced.

Once the majority of the components had been defined, the system could be assembled and figure 15 shows the finalised blowdown version. For the early development work, the expulsion cylinder, pressure cylinder, and regulator were replaced by the VFDA. When evaluated, a slew rate of 270°/sec was achieved. The developed pressure was 38 Bar at 2.5 kV/mm giving a torque of 13.1 Nm, well in excess of the required torque. It should be noted that the measured pressure drop was similar to that predicted by the full expression, Eq. 9. Using the simple equation, the developed pressure was underestimated by some 6 Bar. The pressure rise was measured at 2.3 Bar/msec for a voltage rise time of 0.5 kV/msec. With reduced seal friction and improved flow paths, the performance would be enhanced. No condensation problems were experienced below freezing point, -15°C, using aqueous based polymer fluids. However, at these temperatures the no-field viscosity was high and a reduction in the excess pressure developed was noted. High temperature tests were limited to +60°C due to excessive current demand. To a large extent, especially in short duration flights, temperature could be controlled using a decompressing gas. In this case the full expression for the developed pressure can be modified and the thermal energy input to the valve due to the shearing of the fluid calculated. This was 86 watts under full flow conditions for the valve geometry used. A decision for designers is to balance

Figure 15 General Arrangement of Flying Surface Servomechanism

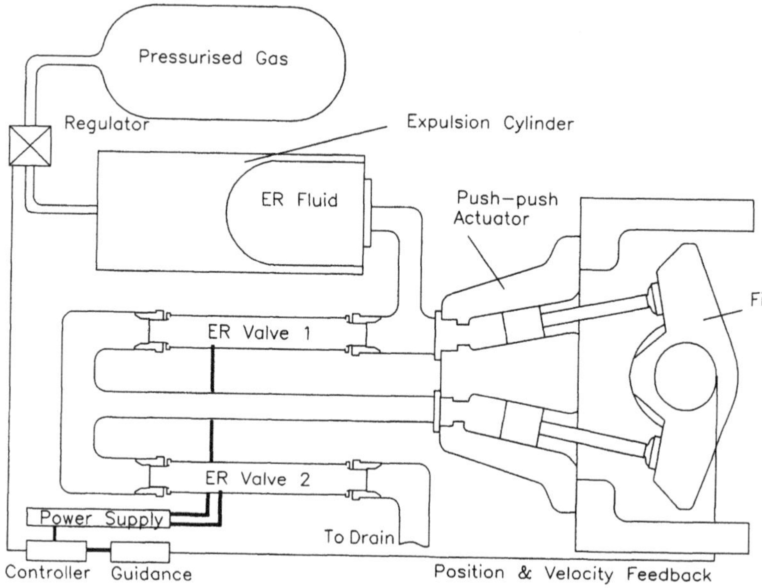

weight savings in the power supply against increased complexity of a gas cooling system. No shortcoming were experienced during vibration and acceleration tests.

This simple design and subsequent evaluation exercise has indicated that a high performance servomechanism can be designed using straightforward mathematics and known fluid properties. There are no unknown aspects of fluid behaviour to be accounted for. The shear rates, excess shear stresses, applied voltages, and valve dimensions are within known engineering constraints and fluid capabilities. Additional design work is required to assess the servomechanism performance with respect to temperature. This revolves around the increased no-field pressure drop as the temperature falls and the increased current required as the temperature rises.

References

1 Brooks D A, Devices Using Electro-Rheological Fluids *Proc 2nd Conf on ER Fluids*, Raleigh NC Aug 7-9 1989 p371-401

2 Block H, *US patent* No 4,687,589

3 Brooks D A, Goodwin J, Hjelm C, Marshall L and Zukoski C, Visco-Elastic Studies on an Electro-rheological Fluid *Colloids & Surfaces* **18** (1986) 293-312

4 Bullough W A, Miscellaneous Electro-Rheological Phenomena Part I *Proc 2nd Conf on ER Fluids*, Raleigh NC Aug 7-9 1989 p115-123

5 Block H & Kelly J P, Electro-rheology, *J Phys D Appl Phys*, Vol 21 1988 p1661-1677

6 Marshall L, Electrically Induced Charges in the Rheology of Dispersed Systems, Ph.D. Bristol University 1987

7 Bares J E and Carlson J D, Electrorheological Fluid Design: An Overview, *Proc 2nd Conf on ER Fluids*, Raleigh NC Aug 7-9 1989 p93-114

8 Peel D J, Electroviscous Control of Dampers, *Departmental Report 90A* 1977, Sheffield University

9 Peel D J, Brooks D A and Bullough W A, Experiences in Pumping Electro-Rheological Fluids, *9th Conference on Fluid Machinery*, 17-19th Sept. 1991, Budapest

10 Reiner M, *Advanced Rheology*, MacLehouse & Co. Ltd 1971

11 Van Waser J R, Lyons J W, Kim K Y and Colwell R E, *Viscosity and Flow Measurement* Wiley 1963

12 Phillips R W and Auslander D M, The Electroplastic Flow Modulator, *Meas in Sci & Ind. Symo.Proc.* Pittsburg 1971 p1243-1251

13 Peel D J, Electroviscous Control of Dampers, Sheffield University *Departmental Report DDR 89* May 1977

14 Peel D J, Electroviscous Control of Dampers, *Departmental Report 96A* 1978, Sheffield University

15 Lovett C, The Behaviour of Bingham Fluids, Laser Engineering (Development) Limited Internal Report
No R672/10/1 January 1977

16 Duclos T G, Design of Devices using Electrorheological fluids, *SAE Paper* No 881134 1988

17 Stangroom J E, Electrorheological Fluids, *Phys.Tech.* **Vol 14** 1983 290-296

18 Turnbull D E, Simplified Models of an Electroviscous Fluid and Valve and their Application to a Position Control System, Sheffield University *Departmental Report 1983*

19 Brooks D A, A Characterisation of Electro-Rheological Fluids, PhD Thesis 1988, Imperial College of Science and Technology

20 Rheometrics RMS-800 and RDS II rheometers Data Sheets

21 Miles O J, Electrical Breakdown at Electrode Edges
Laser Engineering (Development) Limited Internal Report
No R497/7/1 April 1981

22 Brooks D A, A Model for Electro-rheological Fluid in Poiseuille Flow, *63rd Annual Meeting Society of Rheology*, 1991 Rochester NY

23 Rheology in Action No 2 Electro-rheology, Carri-Med Data Sheet 1991

BEHAVIORS OF ELECTRORHEOLOGICAL VALVES AND BRIDGES

ZHENG LOU*, ROBERT D. ERVIN*, AND FRANK E. FILISKO**
**Transportation Research Institute*
*** Material Science and Engineering Department*
The University of Michigan, Ann Arbor, MI 48109

ABSTRACT

Electrorheological valves and bridges were analyzed and evaluated experimentally in its ability to control the flow and pressure conditions.

A Bingham plastic model was used for ER fluids. The model provided characterization of the ER bridge in terms of the load pressure, load flow rate, volume efficiency, power efficiency and working range. The bridge flow rate is nonlinearly related to the pressure drop at a constant level of electric field strength. The working range and efficiencies increase when fluids having higher values of Bingham yield stress are considered. Further, the efficiencies of the bridge drop as the dimensionless load pressure rises. With an ER fluid having a Bingham yield stress of 6.895 kPa and bridge arms having an internal shear length of 1.5 meters, a highly efficient ER bridge can be built for a 206.85 bar fluid power system.

An approximate limit on the effective bandwidth of an ER valve was obtained. Due to the inertial force accompanying oscillatory flow conditions, the dynamic response of an ER valve is constrained by, and linearly related to, the shear stress limit of an ER fluid. With current ER fluids, this effective bandwidth is much lower than the nominal bandwidth of the ER effect, itself. The dynamic limit was verified experimentally in an ER valve.

Experimental results from a particular ER bridge show good qualitative agreement with the analysis. Some substantial asymmetry in bridge performance at high frequencies was observed, and it could have been caused by asymmetric design in conneting passages.

1. Introduction

Electrorheological (ER) fluids have received increased attention both in industry and academia recently. One of their main advantages is their high bandwidth, which is typically at or beyond 1 kHz. Among their many potential applications, ER fluids can be used in a new generation of valves and bridges, which control hydraulic flow and pressure in fluid power and control systems[1,2].

This study is to understand the characteristics of ER valves and bridges. The Bingham plastic fluid will be used as the rheological model. Parameters of industrial concern, like efficiencies, load pressure, and load flow rate, are going to be derived. Some basic design criteria for ER valves and bridges are to be recommended.

2. Static Analysis of an ER Valve

ER valves used in this study consist of two or more parallel electrodes or plates (Fig.1) of a width of W and a total length of L. The gap between the electrodes is h.

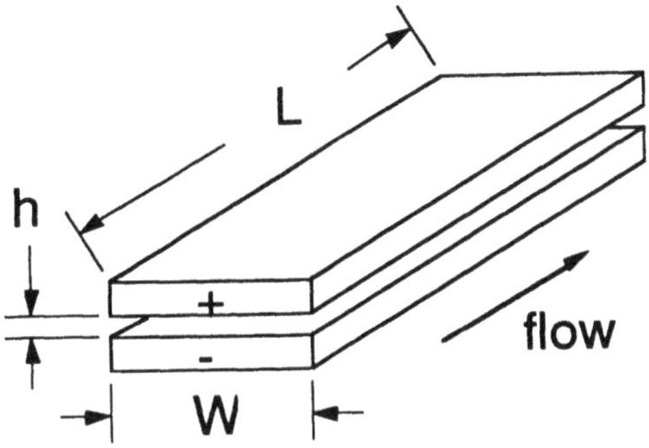

Fig. 1 Geometry of an ER valve

The rheology of an ER fluid can be approximated as that of a Bingham plastic fluid, which has the following constitutive equations:

$$\tau = \begin{cases} -\tau_y + \eta\dfrac{du}{dy} & \text{if } \dfrac{du}{dy} < 0 \\ \tau_y + \eta\dfrac{du}{dy} & \text{if } \dfrac{du}{dy} > 0 \end{cases} \tag{1}$$

or

$$\frac{du}{dy} = \begin{cases} \dfrac{\tau + \tau_y}{h} & \text{if } \tau < -\tau_y \\ 0 & \text{if } |\tau| \le \tau_y \\ \dfrac{\tau - \tau_y}{h} & \text{if } \tau > \tau_y \end{cases} \tag{2}$$

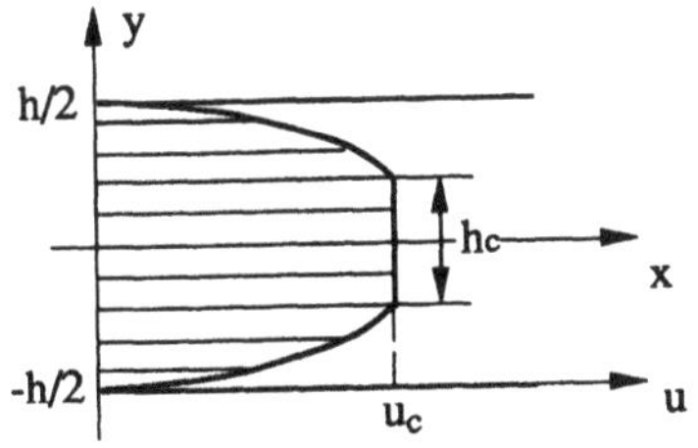

Fig. 2. A flow between two parallel plates.

For a two-dimensional fully-developed steady laminar flow between two parallel plates (Fig. 2), the governing equation is

$$\frac{dp}{dx} = \frac{d\tau}{dy} \tag{3}$$

which gives

$$\tau = \frac{dp}{dx} y \tag{4}$$

The shear stress changes linearly across the gap. Its value is equal to -/+ τ_y when y is at +/- $h_c/2$ (the edges of the blunt core of the flow), therefore

$$h_c = \frac{2\tau_y}{\frac{dp}{dx}} \tag{5}$$

where h_c increases with τ_y and is inversely proportional to (-dp/dx). The maximum value h_c can reach is h, when the pressure gradient (-dp/dx) is less than or equal to the following lock-up pressure gradient

$$(-\frac{dp}{dx})_c \equiv \frac{2\tau_y}{h} \tag{6}$$

or when the pressure drop Δp is less than or equal to the following lock-up pressure drop, which is that value needed for overcoming the Bingham stress,

$$\Delta p_c \equiv 2(\frac{L}{h})\tau_y \tag{7}$$

where L is the length of the plates. The flow is completely stalled when h_c is equal to h.

Substituting Eq. 2 into Eq. 4 and performing integration gives the following velocity profiles:

$$u = \frac{1}{2\eta}(-\frac{dp}{dx})[(\frac{h}{2})^2 - y^2] - \frac{\tau_y}{\eta}(\frac{h}{2} + y) \quad \text{when } -\frac{h}{2} \leq y \leq -\frac{h_c}{2} \tag{8a}$$

$$u = u_c = \frac{1}{8\eta(-\frac{dp}{dx})}[h(-\frac{dp}{dx}) - 2\tau_y]^2 \quad \text{when } |y| \leq \frac{h_c}{2} \tag{8b}$$

$$u = \frac{1}{2\eta}(-\frac{dp}{dx})[(\frac{h}{2})^2 - y^2] - \frac{\tau_y}{\eta}(\frac{h}{2} - y) \quad \text{when } \frac{h_c}{2} \leq y \leq \frac{h}{2} \tag{8c}$$

If the width of the flow channel between the plates is w, then the flow rate is

$$Q = \frac{w}{12\eta}(\frac{L}{\Delta p})^2[4\tau_y^3 - 3\tau_y h^2(\frac{\Delta p}{L})^2 + h^3(\frac{\Delta p}{L})^3] \tag{9}$$

When τ_y approaches zero, the flow rate becomes that of a Newtonian fluid

$$Q_N = \frac{h^3}{12\eta}\frac{w}{L}\Delta p \tag{10}$$

To simplify the analysis, the following two dimensionless parameters are defined

$$Q^* \equiv \frac{Q}{Q_N} \tag{11}$$

and

$$\Delta p^* \equiv \frac{\Delta p}{\Delta p_c} = \frac{\Delta p}{2(L/h)\tau_y} \tag{12}$$

where Δp_c is the lock-up pressure drop as defined in Eq. 7. Q^* is a dimensionless flow rate and Δp^* is a dimensionless pressure drop. When Δp^* is less than 1 or the pressure drop is less than the lock-up pressure drop, the flow rate Q or Q^* will be zero. Substitution of Eqs. 10, 11 and 12 into Eq. 9 gives

$$Q^* = 1.0 - \frac{1.5}{\Delta p^*} + \frac{0.5}{\Delta p^{*3}} \qquad \text{if } \Delta p^* > 1 \tag{13a}$$

$$Q^* = 0 \qquad \text{if } \Delta p^* \leq 1 \tag{13b}$$

The relationship between Q^* and Δp^* is also presented in Fig. 3. The maximum Q^* value is 1.0; i.e., the maximum flow rate is equal to that of a Newtonian fluid, which can be achieved only by having an infinite pressure drop Δp if the τ_y value is finite. In an ER valve, however, the region of interest is the one where a very small or zero Q^* is delivered.

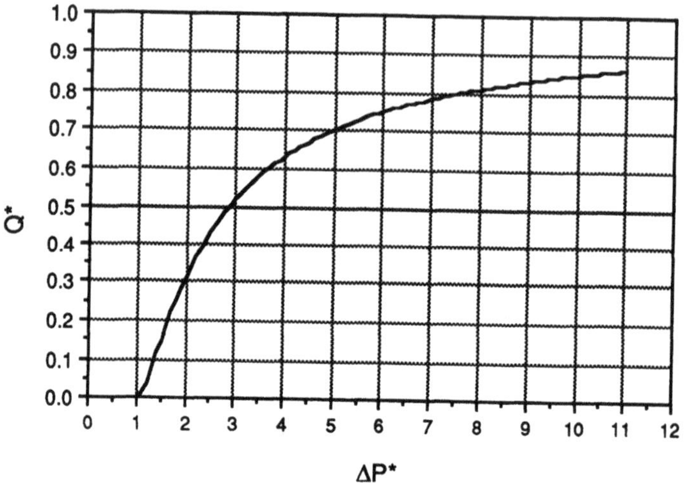

Fig. 3. Flow characteristic of an ER valve.

As shown in Eq. 13b and Fig. 3, Q* is 0 when Δp^* is less than 1. If Δp is 3,000 psi, h 1 mm, τ_y 1 psi, and Δp^* is less than 1, then, according to Eq. 12, the valve length L has to be equal to or longer than 1.5 meters. Such valves, which are able to completely stop flow against the full pressure head, may be practicably constructed from some form of labyrinth geometric designs that accumulate a long total shear length within a compact package.

3. An ER Bridge

An ER bridge, as illustrated in Fig. 4, consists of four ER valves that are arranged in analogy to the electrical Wheatstone bridge, to provide pressure amplification across a controlled load. That is, even if the valves installed in each arm are incapable of completely restraining the supply-level pressure drop, the full range of pressure control can be achieved across the diagonally-situated load, with resulting flow and power efficiencies to be defined by the analysis that is developed here. The load of the bridge can be either the "power spool" of a hydraulic valve or the hydraulic actuator itself. The flow directions and rates can be controlled by applying a combination of different electric fields to the four individual valves in the bridge.

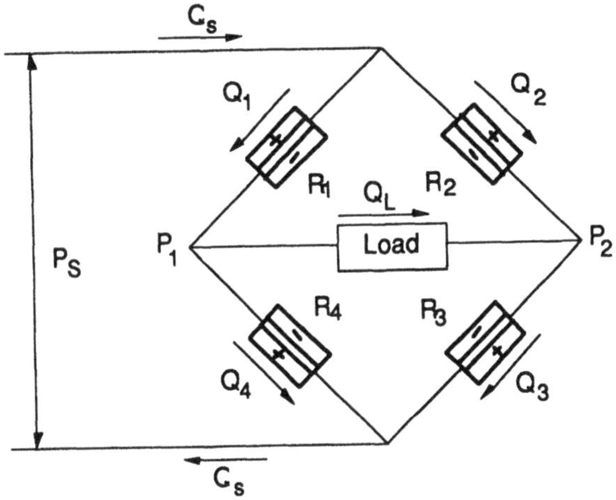

Fig. 4. An ER bridge

The load pressure P_L and load flow rate Q_L are defined as follows:

$$P_L = P_1 - P_2 \tag{14}$$

$$Q_L = Q_1 - Q_4 \tag{15}$$

or

$$Q_L = Q_3 - Q_2 \tag{16}$$

where P_1 and P_2 are the exit pressures of ER valves 1 and 2, respectively. If R_1, R_2, R_3, and R_4 are the hydraulic resistances of the four ER valves, their flow rates can be calculated as follows:

$$Q_1 = \frac{P_s - P_1}{R_1} \tag{17}$$

$$Q_2 = \frac{P_s - P_2}{R_2} \tag{18}$$

$$Q_3 = \frac{P_2}{R_3} \tag{19}$$

$$Q_4 = \frac{P_1}{R_4} \tag{20}$$

The system flow rate Q_s is calculated using the following equation:

$$Q_s = Q_1 + Q_2 = Q_3 + Q_4 \tag{21}$$

If the valves are designed and controlled in such a way that $R_3 = R_1$ and $R_4 = R_2$, then the following can be obtained:

$$P_s = P_1 + P_2 \tag{22}$$

$$P_1 = \frac{1}{2}(P_s + P_L) \tag{23}$$

$$P_2 = \frac{1}{2}(P_s - P_L) \tag{24}$$

$$Q_1 = Q_3 = \frac{\frac{1}{2}(P_s - P_L)}{R_1} \tag{25}$$

$$Q_2 = Q_4 = \frac{\frac{1}{2}(P_s + P_L)}{R_2} \tag{26}$$

$$Q_L = \frac{1}{2R_1R_2}[(R_2 - R_1)P_s - (R_2 + R_1)P_L] \tag{27}$$

$$Q_s = \frac{1}{2R_1R_2}[(R_2 + R_1)P_s - (R_2 - R_1)P_L] \tag{28}$$

The volumetric efficiency of the bridge is defined as

$$\eta_v \equiv \frac{Q_L}{Q_s} = \frac{(R_2 - R_1)P_s - (R_2 + R_1)P_L}{(R_2 + R_1)P_s - (R_2 - R_1)P_L} \tag{29}$$

The pressure efficiency of the bridge is defined as

$$\eta_P \equiv \frac{P_L}{P_s} \tag{30}$$

The power efficiency of the bridge is defined as

$$\eta_{power} \equiv \eta_v \eta_P \tag{31}$$

To simplify the analysis, the variables are normalized as follows

$$R^* \equiv \frac{R_2}{R_1} \tag{32}$$

$$P_L^* \equiv \frac{P_L}{P_s} = \eta_P \tag{33}$$

$$Q_L^* \equiv \frac{Q_L}{Q_{L\,max}} \tag{33b}$$

R^* can be called relative flow resistance, the value of which is not only a function of electrical voltages applied to the valves but also an implicit function of flow conditions inside the valves, as analyzed later. For a normal hydraulic system, P_S is constant, and it is used as a reference pressure. P_L* is identical to η_P. Q_S is not used as a reference flow rate because it is not a constant and changes with the load pressure and other system parameters. Instead, the maximum possible load flow rate, $Q_{L\,max}$, is used and it is obtained when the system is load-free ($P_L = 0$) and two of the valves (valves 1 and 3 or 2 and 4) are completely closed.

If the analysis is limited to the working conditions where all the flow rates, the directions of which are shown in Fig. 4, are positive and the work is positive, i.e.,

$$P_L \geq 0$$

$$R_2 \geq R_1$$

$$Q_L \geq 0 \tag{34}$$

then

$$Q_{L\,max} = \frac{P_s}{2R_1} \tag{35}$$

and

$$Q_L^* = \frac{(R^*-1) - (R^*+1)P_L^*}{R^*} \tag{36}$$

It appears that Eq. 36 gives the pressure-flow curve that is very important in bridge design. However, R* is also related to P_L*(see Section 3) in addition to other parameters, and it cannot be treated as an independent variable. Rearranging Eq. 36 gives

$$P_L^* = \frac{(R^*-1) - R^*Q_L^*}{R^*+1} \tag{37}$$

With the dimensionless variables defined above, the efficiencies can be expressed as:

$$\eta_v = \frac{(R^*-1) - (R^*+1)P_L^*}{(R^*+1) - (R^*-1)P_L^*} \tag{38}$$

$$\eta_P = P_L^* \tag{39}$$

$$\eta_{power} = \frac{(R^*-1)P_L^* - (R^*+1)P_L^{*2}}{(R^*+1) - (R^*-1)P_L^*} \tag{40}$$

4. The Relative Flow Resistance R^*

As shown in the previous section, the relative resistance R* is an important variable in determination of the flow curve, volume efficiency and power efficiency. The objective of this section is to lay out the derivation of R*, itself. As defined in Eq. 32, the relative flow resistance is

$$R^* = R_2/R_1 \tag{41}$$

and

$$R_2 = \Delta P_2/Q_2 \tag{42}$$

$$R_1 = \Delta P_1/Q_1 \tag{43}$$

where ΔP_1 and ΔP_2 are the pressure drops developed across valves 1 and 2, respectively. Substituting Eqs. 42 and 43 into Eq. 41 results in

$$R^* = \frac{\Delta P_2}{\Delta P_1}\frac{Q_1}{Q_2} \tag{44}$$

From Eq. 11,

$$\frac{Q_1}{Q_2} = \frac{Q_1^*}{Q_2^*}\frac{Q_{N1}}{Q_{N2}} \tag{45}$$

If the design of the bridge is symmetric, then, according to Eq. 10

$$\frac{Q_{N1}}{Q_{N2}} = \frac{\Delta P_1}{\Delta P_2} \tag{46}$$

and thus

$$\frac{Q_1}{Q_2} = \frac{Q_1^* \Delta P_1}{Q_2^* \Delta P_2} \tag{47}$$

Substituting the above equation into Eq. 44 gives

$$R^* = \frac{Q_1^*}{Q_2^*} \tag{48}$$

Using the reference case as assumed in the bridge analysis, namely that valves 2 and 4 are energized and valves 1 and 3 are not, then,

$$Q_1^* = \frac{Q_1}{Q_{N1}} = 1 \tag{49}$$

$$Q^* \equiv Q_2^* = \frac{Q_2}{Q_{N2}} \neq 1 \tag{50}$$

$$R^* = \frac{1}{Q^*} \tag{51}$$

Substituting Eq. 13 into the above equation gives

$$R^* = \frac{\Delta P^{*3}}{\Delta P^{*3} - 1.5\Delta P^{*2} + 0.5} \qquad \text{if} \quad \Delta P^* > 1 \tag{52}$$

$$R^* = \infty \qquad \text{if} \quad \Delta P^* \le 1 \tag{53}$$

where

$$\Delta P^* = \Delta P_2^* = \frac{\Delta P_2}{\Delta P_c} = \frac{\Delta P_2}{2(L/h)\tau_y} \tag{54}$$

Because $\Delta P_2 = P_s - P_2 = P_s - (P_s\text{-}P_L)/2 = (P_s\text{+}P_L)/2$, therefore

$$\Delta P^* = \frac{\frac{1}{2}(P_s + P_L)}{\Delta P_c} \tag{55}$$

ΔP^* is a function of both the load pressure and the Bingham stress, as is R^*. Let

$$\tau_y^* \equiv \frac{\Delta P_c}{P_s} = \frac{2(L/h)\tau_y}{P_s} \tag{56}$$

and

$$P_L^* \equiv \frac{P_L}{P_s} \tag{57}$$

then

$$\Delta P^* = \frac{1 + P_L^*}{2\tau_y^*} \tag{58}$$

The substitution of the above equation into Eqs. 52 and 53 gives

$$R^* = \frac{(1+P_L^*)^3}{(1+P_L^*)^3 - 3(1+P_L^*)^2\tau_y^* + 4\tau_y^{*3}} \qquad \text{if} \quad P_L^* > 2\tau_y^* - 1 \tag{59a}$$

$$R^* = \infty \qquad \text{if} \quad 0 \le P_L^* \le 2\tau_y^* - 1 \tag{59b}$$

Under the working condition assumed in this analysis, P_L^* should be always greater than or equal to zero, therefore Eq. 59b will be useful only if τ_y^* is greater than 0.5. Eq. 59a is graphically presented in Fig. 5, from which it is clear that R^* is a function of the load P_L^* on the bridge as well as the level of the Bingham stress τ_y^*. With a fixed value of τ_y^*, the relative flow resistance and, thus, the ER effect decrease with an increasing P_L^*. Overall, the ER effect is stronger with a higher Bingham stress τ_y^*. When τ_y^* exceeds 0.5, there is a certain range of P_L^*, from 0 to $(2\tau_y^*\text{-}1)$, where R^* is at infinity. That is, there is a saturation zone.

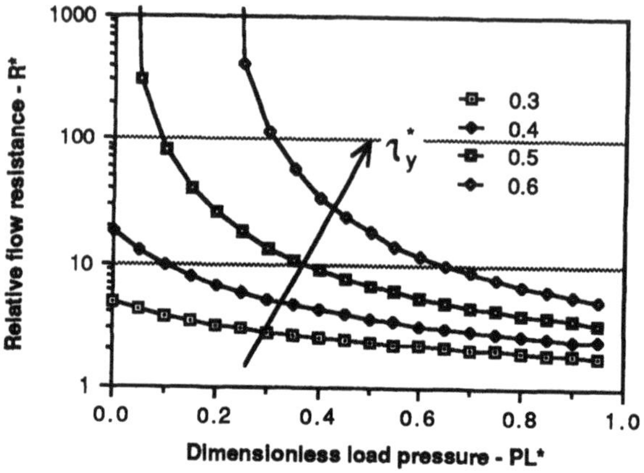

Fig. 5. The dependence of R^* on P_L^* as well as τ_y^*.

5. Static ER Bridge Performance

As shown in Eqs. 36, 38, and 40, the flow rate, volumetric efficiency, and power efficiency are all functions of R^* and P_L^*, while R^* itself is a function of P_L^* and τ_y^* (Eq. 59). If Eq. 59 is substituted into Eqs. 36, 38, and 40, the flow rate, the volume efficiency, and the power efficiency are all functions of P_L^* and τ_y^* only, and are shown in Figs. 6 to 9. Therefore, R^* has served merely as an intermediate variable, and it can be eliminated in the calculation. P_L^* and τ_y^* are two independent variables.

Looking at Fig. 6, which represents the case for which τ_y^* is 0.2, Q_L^* is linearly related to P_L^*. $Q_L^*{}_{max}$ (the maximum value for Q_L^*), which is achieved when $P_L^* = 0$ (zero load pressure), is about 0.568 or 56.8% of Q_{Lmax}. The bridge is stalled at $P_L^* = P_L^*{}_{max} = 0.29$ where the load flow rate is zero. The load pressure P_L^* can not go beyond $P_L^*{}_{max}$. In this case, the bridge can only use 29% of the available system pressure. The volume efficiency, η_v (the variable "Evolume" in Fig. 6), follows the same pattern as Q_L^*. However, it is not linearly related to P_L^*. The power efficiency, η_P ("Epower" in Fig. 6), is equal to zero at both ends of the working range (from $P_L^* = 0$ to $P_L^* = P_L^*{}_{max}$), where either Q_L^* or P_L^* is equal to zero, which makes the product of η_v and P_L^* zero. The maximum power efficiency is reached in the middle of the working range, where the product of η_v and P_L^* reaches its maximum value. With $\tau_y^* = 0.2$, the maximum power efficiency is very small, about 0.58%. $P_L^*{}_{max}$, $Q_L^*{}_{max}$, $\eta_{v\,max}$ and $\eta_{P\,max}$ increase with τ_y^*.

For the case of τ_y^* equal 0.5 (Fig. 7), i.e. the lock-up pressure drop ΔP_C is half the system pressure, the bridge can use up to 66% of the available system pressure. It can also use up to 100% of the available system flow rate, $Q_L* = 1.0$, when, with no load pressure, a pair of valves in the bridged are just locked up by the ER effect. In order to achieve a pressure drop across the load, the bridge valves must be energized below the lock-up condition. The maximum power efficiency with $\tau_y* = 0.5$ is about 24%.

As τ_y* increases further, the working condition is much improved. If $\tau_y* = 0.8$ (Fig. 8), P_L*_{max} can reach a value of .925 and the maximum power efficiency becomes about 66%. Further, the bridge is able to lock up a pair of valves completely at a load pressure up to 60% of the system pressure. When $\tau_y* = 1.0$ (Fig. 9), half of the bridge can be locked up over the entire working range, delivering up to 100% of the system pressure and 100% of the system flow rate to the load. The power efficiency is equal to P_L*, and goes up to 100%.

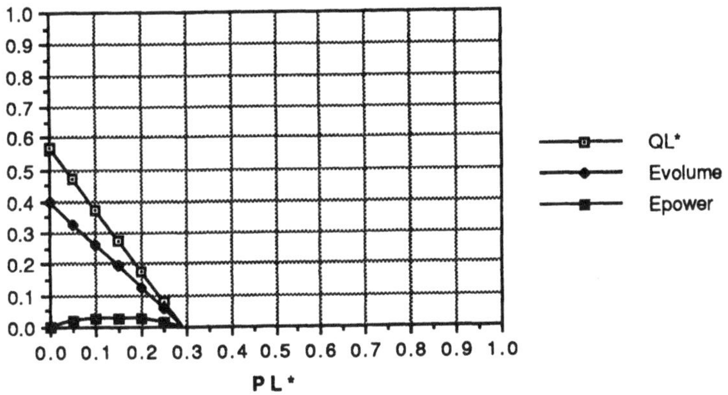

Fig. 6. Bridge characteristics at $\tau_y* = 0.2$.

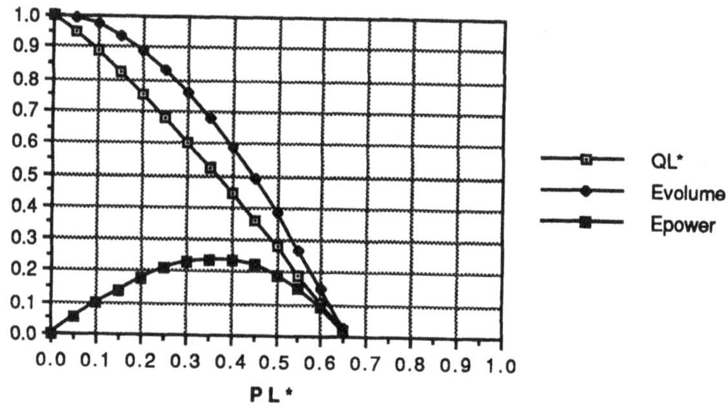

Fig. 7. Bridge characteristics at $\tau_y{}^* = 0.5$.

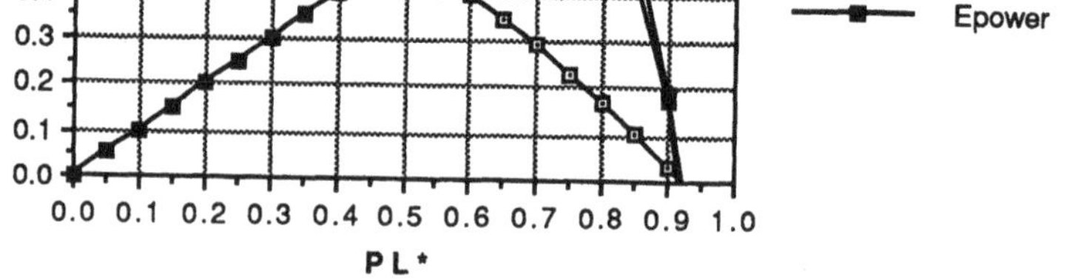

Fig. 8. Bridge characteristics at $\tau_y{}^* = 0.8$.

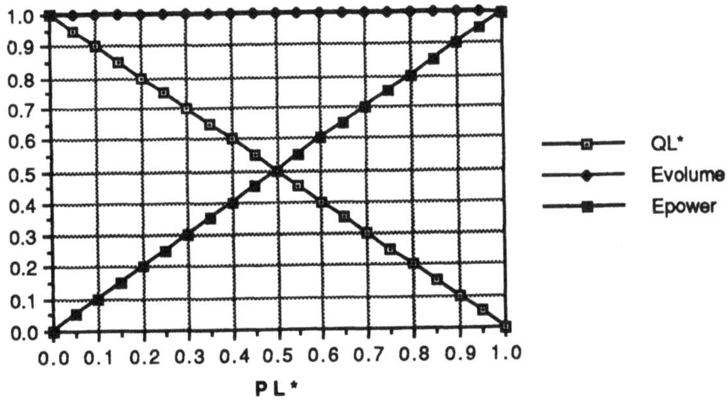

Fig. 9. Bridge characteristics at $\tau_y^* = 1.0$.

The maximum load flow rate and the maximum volumetric efficiency are obtained when the load pressure is zero. From Eqs. 36, 38 and 59

$$Q^*_{L\,max} = \begin{cases} 3\tau_y^* - 4\tau_y^{*\,3} & \text{if } \tau_y^* < 0.5 \\ 1 & \text{if } \tau_y^* \geq 0.5 \end{cases} \tag{60}$$

$$\eta_{v\,max} = \begin{cases} \dfrac{3\tau_y^* - 4\tau_y^{*\,3}}{2 - (3\tau_y^* - 4\tau_y^{*\,3})} & \text{if } \tau_y^* < 0.5 \\ 1 & \text{if } \tau_y^* \geq 0.5 \end{cases} \tag{61}$$

The maximum load pressure is obtained at zero load flow rate. From Eqs. 37 and 59,

$$P^*_{L\,max} = \frac{3(1+P^*_{L\,max})^2\tau_y^* - 4\tau_y^{*3}}{2(1+P^*_{L\,max})^3 - 3(1+P^*_{L\,max})^2\tau_y^* + 4\tau_y^{*3}} \tag{62}$$

Since there is no closed-form solution for $P^*_{L\,max}$, an interative computation is required. The maximum power efficiency is even more difficult to obtain analytically, and has not been attempted in this study. This value is simply illustrated in the numerical results that have been produced. The maximum values for volume efficiency, load flow rate, load pressure, and power efficiency are plotted with respect to the dimensionless Bingham stress in Fig. 10. All of them increase with the dimensionless Bingham stress.

The higher the dimensionless Bingham stress of an ER bridge, the broader its working range will be and the more efficient the bridge will be.

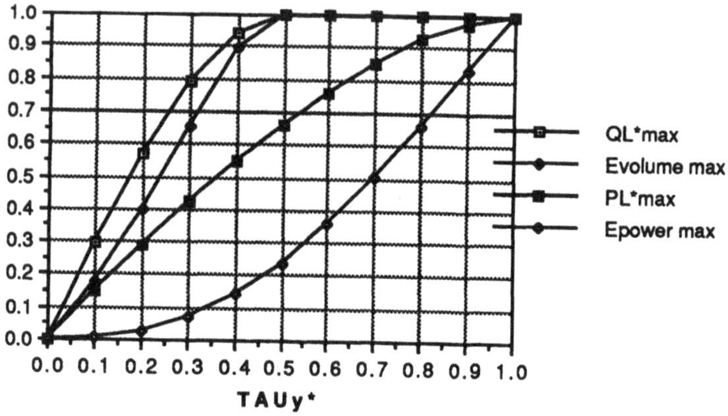

Fig. 10 The effect of τ_y*.

From the above discussion, it is very clear that the value of τ_y* has elementary importance for bridge design. Further, τ_y* is not a constant for a given system. According to Eq. 56, it is a ratio of the lock-up pressure drop over the system pressure. For a given system pressure and a given geometric parameter, (L/h), τ_y* linearly increases with τ_y that is mainly a nonlinear function of the electric field strength, E, in addition to many other parameters. In an automatic control system, the operating point of an ER bridge can be controlled by changing τ_y*, which is directly regulated by adjusting the field strength.

6. ER Bridge Design

Eq. 56 can be re-written as

$$\frac{L}{h} = \frac{P_s}{2\tau_y}\tau_y^* \tag{63}$$

For a given fluid power system, P_s is a given constant or system requirement. For a given ER fluid, τ_y is a variable; however it has an upper limit τ_{ymax}. τ_y* changes with τ_y for a given (L/h) and a given P_s, and it reaches the peak level, $\tau_y{*}_{max}$, when τ_y is equal to τ_{ymax}. Therefore,

$$\frac{L}{h} = \frac{P_s}{2\tau_{y\,max}}\tau_{y\,max}^* \tag{64}$$

indicating that (L/h) depends on the selection of $\tau_y{}^*{}_{max}$. According to Fig. 10, the best choice is $\tau_y{}^*{}_{max} = 1.0$, which gives the maximum values of flow rate, the load pressure, and the efficiencies. For example, if τ_{ymax} is 1 psi and P_s is 3,000 psi, then (L/h) will be 1,500 with $\tau_y{}^*{}_{max} = 1.0$. If the valve gap is 1.0 mm, then the total plate length for each ER valve should be 1.5 meters, which is indeed long but can be achieved in a compact package using labyrinth-plate valves. When $\tau_y{}^*{}_{max} = 1.0$

$$2\frac{L}{h}\tau_{y\,max} = P_s \tag{65}$$

and according to Eq. 7,

$$\Delta p_{c\,max} = P_s \tag{66}$$

Therefore, the design criterion is fundamentally one of selecting (L/h) such that the maximum lock-up pressure drop is equal to the system pressure. If τ_{ymax} is too small, then L may become too long to be practicably fabricated. For example, if τ_{ymax} is 0.1 psi, then L must be 15 meters to keep $\tau_y{}^*{}_{max} = 1.0$. In another application of such a fluid, packaging constraints may require that, with $\tau_y{}^*{}_{max} = 0.1$, L is limited at 1.5 meters, thereby limiting the maximum load pressure to only 15% of the system pressure and the maximum power efficiency to only 0.58% according to Fig. 10.

As for the width of the valve W, it is selected to accomodate the maximum flow rate, which is the zero-field or Newtonian flow rate Q_N. Therefore, one gets

$$W = \frac{12\eta L}{h^3 \Delta p} Q_N \tag{67}$$

7. Dynamic Limit of an ER Valve

For a parallel-plate ER fluid valve (Fig. 1), the total fluid mass enclosed is

$$m = \rho\,(WhL) \tag{68}$$

where ρ is the fluid density. The control force on the fluid due to the electrorheologically-induced Bingham shear stress, τ_y, can be approximated as,

$$F = 2\,(WL)\,\tau_y \tag{69}$$

The Bingham shear stress can be changed in an approximately sinusoidal wave with a peak stress of τ_{y0}, thus modulating the flow through the valve section. If the flow pulsating frequency (twice that of the electric field) is f and u_0 is the peak value of the mean velocity across the valve gap, the amplitude of the flow acceleration is $2\pi f u_0$. One can note that the most basic level of valve control requires that the amplitude of the ER-controlled force (due to shear stresses along the plate surfaces) should be equal to or greater than the peak inertia force, i.e., the mass times the amplitude of the acceleration prevailing at the desired bandwidth frequency. This inequality is expressed by

$$2\,(WL)\,\tau_{y0} \geq \rho\,(WhL)\,2\pi f\,u_0 \tag{70}$$

or

$$\tau_{y0} \geq \pi\rho h u_0 f \tag{71}$$

For fixed values of Bingham shear stress, fluid density, valve gap height and velocity amplitude, the valve can be controlled up to the frequency value,

$$f_n = \frac{\tau_{y0}}{\pi\rho h u_0} \tag{72}$$

which constitutes an obvious maximum bandwidth. We see that f_n is proportional to τ_{y0} and inversely related to h and u_0. The corresponding flow rate amplitude is

$$Q_0 = hW\,u_0 \tag{73}$$

thus

$$f_n = \frac{W\tau_{y0}}{\pi\rho Q_0} \tag{74}$$

Figures 11 and 12 show the relationship between the bandwidth frequency and the flow rate amplitude at various values of Bingham stress limit for valves with width of 1 cm and 10 cm, respectively. A tenfold increase in W, from 1 cm to 10 cm, raises the bandwidth by a factor of 10 for a fixed flow rate amplitude, or conversely, raises the flow rate amplitude by a factor of 10 for a given bandwidth. For a valve with W = 10 cm and a flow rate amplitude of 15.9 gpm, the bandwidth is 219.5 Hz.

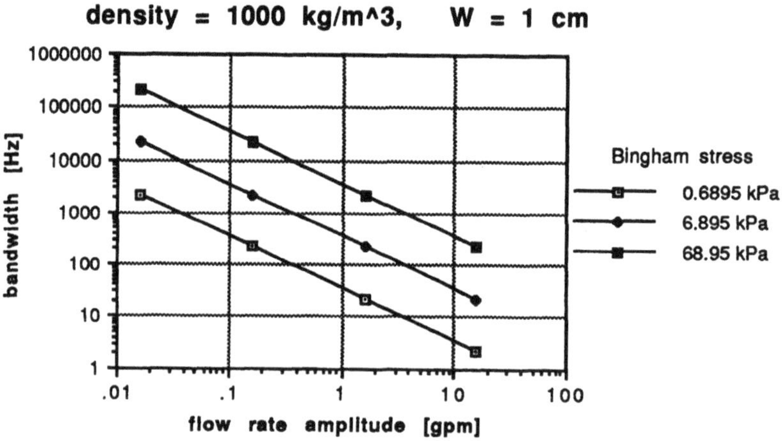

Fig. 11 Bandwidth and flow rate relationship for a valve with W = 1 cm.

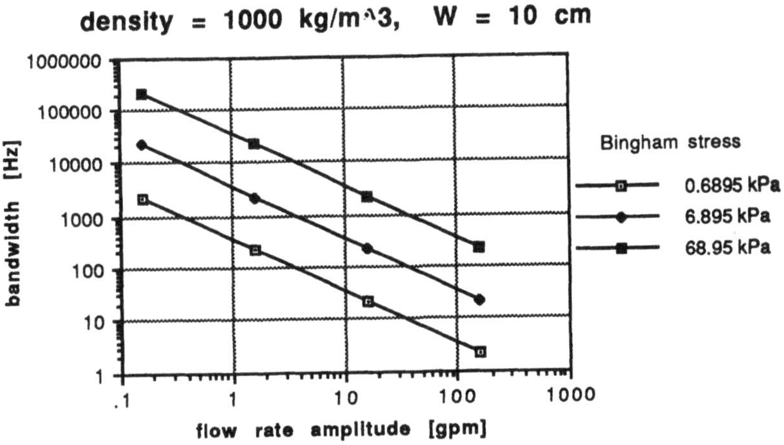

Fig. 12 Bandwidth and flow rate relationship for a valve with W = 10 cm.

Finally, the bandwidth limitation considered here pertains only to the constraint imposed when inertial fluid stresses reach the ER shear stress limit. Further, this analysis provides only a crude approximation of the bandwidth-limiting phenomenon, for purposes of illustrating numerical values. To determine the bandwidth of an actual mechanical system, one must include the dynamics of the load as well as the effect of additional fluid mass in passages between the load and the valve.

8. Experimental Characterization of an ER Valve

A parallel-plate ER valve was experimentally characterized. The valve has a length of 1 meter, a width of 1 cm and a gap of 1 mm. Figures 13 through 16 present test results. In the figures, the pressure drop has a lower limit of 40 to 50 psi because, with the system accumulator pre-charged at this pressure, readings below this level are too noisy to be usefully measured. All the tests shown were performed under a room temperature of 69 deg. F without forced cooling of the fluid. Fluid temperature during the tests was between 80 to 90 deg. F. The reported pressure drop is the time-average value of the system pressure reading, assuming no return-line pressure loss. A previous reported fluid, fluid 2, was used[3].

The flow rate was fixed at 18.93 ml/s (0.3 gpm) for Figs. 13 and 14 while the electric excitation frequency and the electric field strength were adjusted. Figure 13 shows that the pressure drop increases with the electric field strength. When the voltage is at 4.5 kV or the field strength is at 4.5 kV/mm, the pressure drop is almost flat for electric excitation frequencies beyond 50 Hz. Fluid 2 has a yield shear stress between 551.6 to 689.5 Pa (0.08 to 0.1 psi) at 4.5 kV/mm and a density of about 1000 kg/m3. One obtains a flow pulsating frequency band limit from 92 to 116 Hz from Eq. 74. The corresponding electric excitation frequency is from 46 to 58 Hz, which is close to the experimental dynamic limit of 50 Hz. The cut off frequency decreases with the voltage thus the yield shear stress (Fig. 13), as predicted by the analysis (Eq. 74).

Figure 14 depicts the electric current through the valve under the same conditions. In general, the current increases initially with both electric field strength and frequency. The frequency-related component is dominated initially by the capacitive effect, where peak current follows frequency linearly, but then shows a saturation due apparently to inherent changes in the fluid's impedance. It is interesting, but unexplained, that current begins to saturate in the same frequency region where the pressure curves become flat in Fig. 13. This current saturation is unrelated the power supply. Clearly, since the current level drives, directly or indirectly, the basic ER response, any increase in fluid impedance with frequency will cause the shear stress response to decline, while, conversely, capacitive effects cause increased currents that result in greater shear stresses.

The electric excitation frequency was set at 1 Hz for Figs. 15 and 16 while the flow rate and electric field strength were varied. The curve pattern in Fig. 15 looks similar to a typical shear stress vs. shear rate plot for a Bingham fluid. Since the pressure drop is proportional to the shear stress and the flow rate is proportional to the shear rate, one can conclude that the flow in a parallel-plate valve is similar to the flow mode in a shearing viscometer. Finally, Fig. 16 reveals that the electric current increases with the field strength and is almost independent of the flow rate.

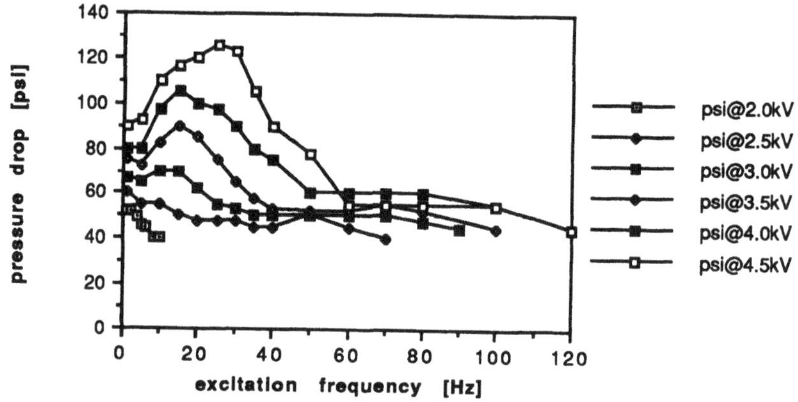

Fig. 13. Average pressure drop through an ER valve versus electric excitation frequency at various electric field strengths (with fluid 2, average flow rate = 18.93 ml/s, fluid temperature = 80-90 F).

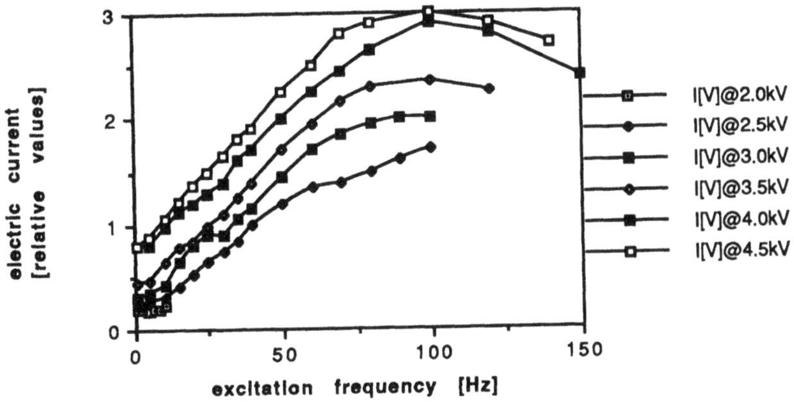

Fig. 14. Average electric current through an ER valve versus excitation frequency at various levels of field strength (with fluid 2, average flow rate = 18.93 ml/s, fluid temperature = 80-90 F).

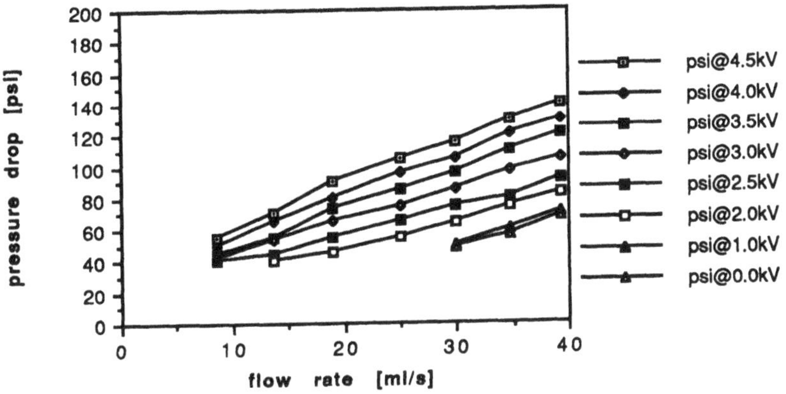

Fig. 15. Average pressure drop through an ER valve versus flow rate at various electric field strengths (with fluid 2, electric excitation frequency = 1 Hz, fluid temperature = 80-90 F).

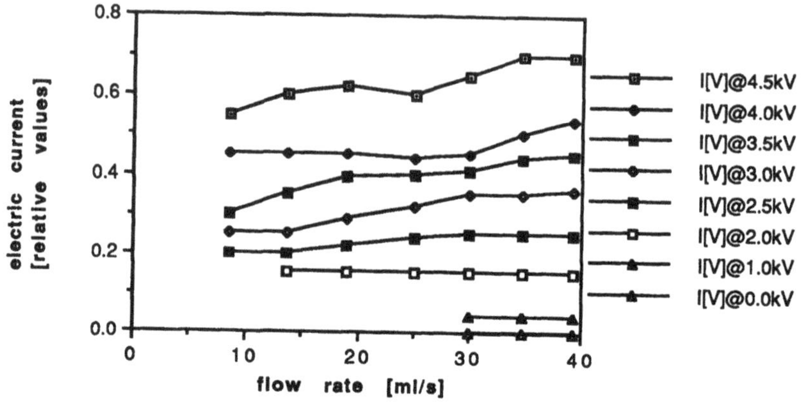

Fig. 16. Average electric current through an ER valve versus flow rate at various electric field strengths (with fluid 2, electric excitation frequency = 1 Hz, fluid temperature = 80-90 F).

9. Experimental Characterization of an ER Bridge

Tests were performed on an ER bridge, each arm of which has an ER valve of a gap of 1 mm, a width of 1 cm and a length of 1 meter. Fluid 2 was used, and the fluid temperature was around 80 to 90 deg. F. A special electric circuit with an d.c. supply and an high voltage a.c. amplifier was used to charge the bridge[1]. The electric field strength in valves 1 and 3 is

$$E_{13} = E_{dco} + E_{aco}\sin \omega t \qquad (75)$$

while that in valves 2 and 4 is

$$E_{24} = E_{dco} - E_{aco}\sin \omega t \qquad (76)$$

where E_{dco} and E_{aco} are two constants.

For the test in Fig. 17, the flow rate was 44.2 ml/s (0.7 gpm), and both E_{dco} and E_{aco} are 2 kV/mm such that the peak-to-peak amplitude of the final electric field strength on each arm was 4 kV/mm. P_{gage} is the ER fluid supply system pressure while P_{1os_pp} and P_{2os_pp} are peak-to-peak bridge output pressure readings for P_1 and P_2, respectively, on an oscilloscope screen. It is obvious from the data that the behavior of the bridge is frequency dependent in a way that cannot be fully explained. If the bridge were symmetric, of course, P_{1os_pp} and P_{2os_pp} would be identical in their values rather than exhibiting the very different frequency responses seen in Fig. 17, which will be discussed in the next section.

The results shown in Fig. 18 were obtained when the frequency was fixed at 10 Hz and the system flow rate at 25.2 ml/s (0.4 gpm). The peak-to-peak pressure readings increase with the input signal amplitude.

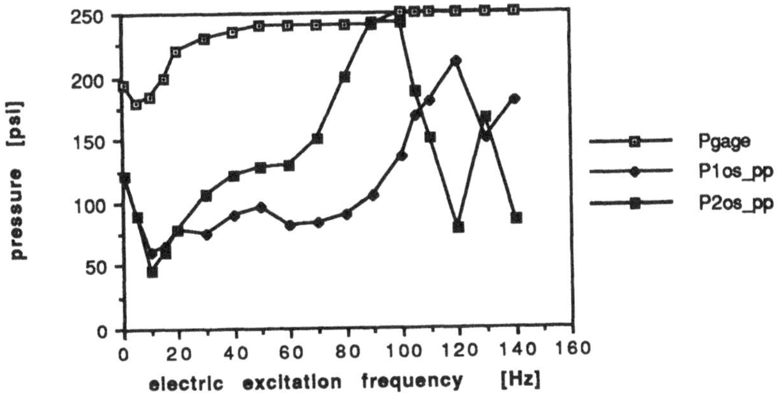

Fig.17. Pressure readings versus electric excitation frequency for an ER bridge ($E_{dco} = E_{aco} = 2$ kV/mm, Qs = 44.2 ml/s).

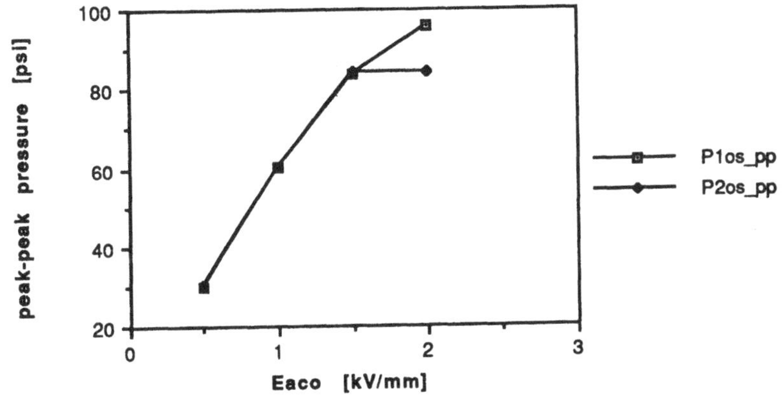

Fig. 18. Pressure readings versus electric field strength for an ER bridge ($E_{dco} = 2$ kV/mm, electric excitation frequency = 10 Hz, Qs = 25.2 ml/s).

10. Asymmetry in the ER Bridge

For an ideal ER bridge (Fig. 4), there is no flow resistance in any of its connecting passages. The only fluid impedances are located within the four electrorheologically-controlled resistors: R_1, R_2, R_3, and R_4.

A real or practical ER bridge, however, is more appropriately depicted by Fig. 19, where many more passive resistance elements exist within the connecting passages of the device. Even Fig. 19 is still somewhat simplified because resistors are shown as lumped elements instead of the distributed form that actually prevails. P_1 and P_2 in the figure are the actual locations of pressure sensors.

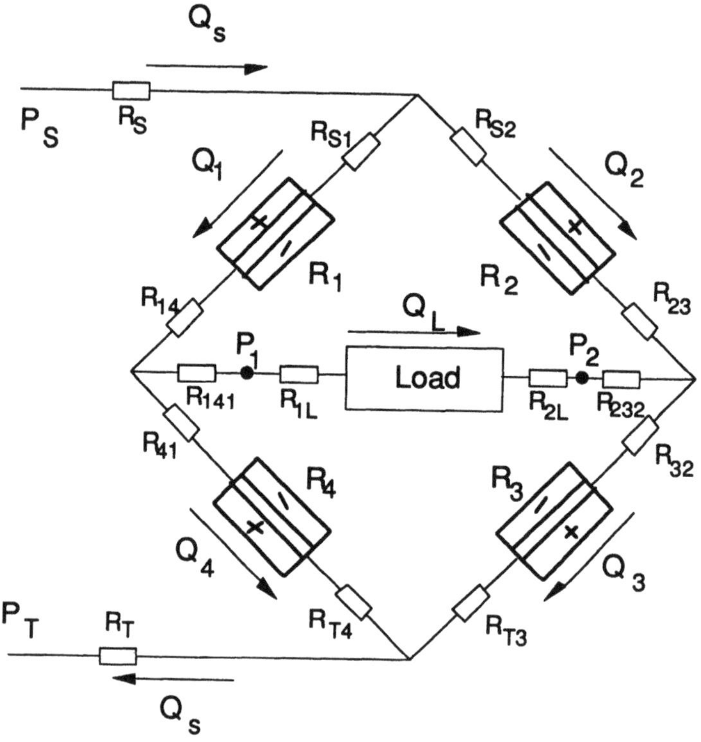

Fig. 19. An actual ER bridge.

In a steady-state flow, the pressure drop or loss due to friction over a pipe of length L is

$$\Delta P_{s.s.} = 8\pi\mu Q\left(\frac{L}{A^2}\right) \qquad \text{in a circular pipe} \qquad (77)$$

where μ is the viscosity, Q the flow rate and A the cross section area. In a channel, however, the pressure drop is

$$\Delta P_{s.s.} = 8\pi\mu Q(\frac{3L}{2\pi Wh^3}) \quad \text{in a channel} \tag{78}$$

With the same flow rate and fluid viscosity, the frictional pressure loss for a steady-state flow is proportional to (L/A^2) for a circular pipe or $(3L/2\pi Wh^3)$ for a channel.

For an unsteady flow, the pressure loss due to the inertia term, for both a circular pipe and a channel, is

$$\Delta P_{inertia} = \rho\frac{dQ}{dt}(\frac{L}{A}) \tag{79}$$

With the same flow rate and fluid density, the inertial pressure loss is proportional to (L/A). In an unsteady situation, the frictional loss is much more complicated and cannot be calculated using the equations for steady-state flow.

In the design of an ER bridge, it is preferable to make the connecting passages as short as possible and their cross sectional areas as large as possible, relative to those of the ER-controllable resistors, in order to reduce the pressure losses in the connecting circuit. However, passage lengths must obviously accommodate the physical arrangement and cross sectional areas will be limited by the need for compactness in the whole assembly.

Another important design criterion is to insure the symmetry of the bridge. The first step towards that goal is to design symmetric or identical controllable resistors: R_1, R_2, R_3, and R_4. The second step is to have symmetry in the ancillary passages. To do that, elements labelled in Fig. 19 as R_{S1}, R_{S2}, R_{T3}, and R_{T4} should be identical; R_{14}, R_{41}, R_{23}, and R_{32} should be identical; R_{1L} and R_{2L} should be identical; and R_{141} and R_{232} should be identical. We observe that it is difficult to obtain perfect symmetry in these various connections. However, their influence on the symmetry of performance for the entire system can be reduced by cutting their absolute resistance values.

Table 1 shows the parametric values pertaining to the various resistor segments in the ER bridge used in this study. For those resistors that have pipes or passages whose cross section varies along their length, they are divided into sub-resistors: (a), (b), etc., and the total resistance is taken as the summation of sub-resistances. Sub-resistor $R_{T(c)}$, for example, is a heat exchanger consisting of 25 small pipes in parallel. The total resistance is taken to be the equivalent of a parallel network of 25 individual resistors. L/A is the inertia resistance coefficient while L/A^2 is the steady-state frictional resistance coefficient. Their normalized values, against the controllable resistor's values, are listed in the column with a superscript "*." It is obvious that, in the actual bridge constructed for this project, {R_1, R_2, R_3, and R_4} and {R_{1L} and R_{2L}} are symmetric while {R_{S1}, R_{S2}, R_{T3}, and R_{T4}}, {R_{14}, R_{41}, R_{23}, and R_{32}} and {R_{141} and R_{232}} are not. The steady-state frictional resistances of the uncontrollable resistors is seen to be negligible—much less than 1%—compared with those of the ER-controllable elements. Thus, under steady state flow, the bridge appears virtually symmetric.

Table 1. Resistors in the ER bridge.

resistor	D	L	A	L/A	(L/A)*	L/A^2	(L/A^2)*
R1		40.000	0.0150	2667	1.000	795774	1.0000
R2		40.000	0.0150	2667	1.000	795774	1.0000
R3		40.000	0.0150	2667	1.000	795774	1.0000
R4		40.000	0.0150	2667	1.000	795774	1.0000
RS(a)	0.5000	13.000	0.1963	66	0.025	337	0.0004
RS(b)	0.3125	13.000	0.0767	169	0.064	2210	0.0028
RS				236	0.088	2547	0.0032
RT(a)	0.5000	12.000	0.1963	61	0.023	311	0.0004
RT(b)	0.3125	20.000	0.0767	261	0.098	3400	0.0043
RT(c)	0.2130	17.000	0.0356	478		13414	
25RT(c)				19	0.007	537	0.0007
RT				341	0.128	4248	0.0053
RS1	0.2500	1.500	0.0491	31	0.011	623	0.0008
RS2	0.2500	3.500	0.0491	71	0.027	1453	0.0018
RT4	0.2500	1.200	0.0491	24	0.009	498	0.0006
RT3	0.2500	3.000	0.0491	61	0.023	1245	0.0016
R14		0.000		0	0.000	0	0.0000
R41	0.2500	3.500	0.0491	71	0.027	1453	0.0018
R23	0.2500	3.500	0.0491	71	0.027	1453	0.0018
R32		0.000		0	0.000	0	0.0000
R141	0.2500	4.000	0.0491	81	0.031	1660	0.0021
R232	0.2500	4.500	0.0491	92	0.034	1868	0.0023
R1L(a)	0.2500	1.000	0.0491	20	0.008	415	0.0005
R1L(b)	0.1250	0.750	0.0123	61	0.023	4980	0.0063
R1L				81	0.031	5395	0.0068
R2L(a)	0.2500	1.000	0.0491	20	0.008	415	0.0005
R2L(b)	0.1250	0.750	0.0123	61	0.023	4980	0.0063
R2L				81	0.031	5395	0.0068

In a dynamic situation, the combined effect of R_S and R_T is as high as 20%, i.e. about 20% of the system pressure is lost along the pipes leading to and from the bridge. The losses from other resistors are also relatively significant. The total asymmetry deriving from corresponding resistors, {R_{14}, R_{41}, R_{23}, and R_{32}}, is 5.4%. This conclusion from analysis is consistent with our experimental observations that asymmetry is more prominent at higher frequencies, where inertial pressure losses are significant.

The above analysis of discrete resistances is valid for nominal, simplified, flow oscillation. A complete and precise analysis would take into account other factors such as the phase relation between the frictional and inertial losses, their interference, flow-developing regions, etc. Furthermore, possible asymmetry in electric circuit elements may also be important.

11. Summary

The main results from this study are:

• The flow characteristic of a single ER valve has been derived based upon a Bingham plastic fluid model. For a constant electric field strength, the flow rate is nonlinearly related to the pressure drop. A no-flow zone exists in the low pressure region for a single ER valve, which is translated into a load flow-saturation zone at low load pressure for an ER bridge.

• For an ER bridge, its volume and power efficiencies and the working range are related to the dimensionless load pressure and the dimensionless Bingham stress. The working range and efficiencies increase with the dimensionless Bingham stress. The efficiencies drop as the dimensionless load pressure rises.

• With an ER fluid whose Bingham stress limit is about 6.895 kPa (1 psi), and with ER valves whose internal shear length is 1.5 meters, a highly efficient ER bridge can be built for a 206.85 bar (20,685 kPa or 3,000 psi) fluid power system.

• Due to inertial forces accompanying oscillatory flow conditions, the dynamic response of an ER valve is constrained by, and linearly related to, the shear stress limit of an ER fluid. With current ER fluids, this effective bandwidth is much lower than the nominal bandwidth of the ER effect, itself (whereby rheological properties respond to time-variant electrical field strength.) The dynamic limit was verified experimentally in an ER valve.

• Experimental results of a particular ER bridge show good qualitative agreement with the analysis. Some substantial asymmetry in bridge performance at high frequencies is observed, and it can be caused by asymmetric design in conneting passages.

12. Acknowledgements

The authors are happy to acknowledge the advice and comments from Mr. Christopher Winkler and technical support from Ms. Theresa Filisko, Mr. Thomas Husen, Mr. John Koch and Mr. Michael Campbell. This work was partially supported by the U.S. Air Force through a contract with Dynamic Controls, Inc. It was also supported by the University of Michigan Industrial Affiliates Program in Electrorheology, which included the following sponsors: Aeroquip Corporation, Chrysler Corporation, Ethyl Corporation, Ford Motor Company, Monroe Automotive Equipment Co., Naval Underwater Systems Center, Nissan Motors, and Tremec Trading Company.

13. References

1. Z. Lou, R.D. Ervin and F.E. Filisko, Electrorheological fluids for aircraft flight control: Feasibility of a position servomechanism. UMTRI-90-10. (University of Michigan Transportation Research Institute, Ann Arbor, 1991).

2. Z. Lou, R.D. Ervin and F.E. Filisko, An electro-rheological servo position control system. Conference on Electrorheological (ER) Fluids. Southern Illinois University at Carbondale, Illinois, Oct. 15-16, 1991.

3. Z. Lou, R.D. Ervin and F.E. Filisko, The Feasibility of using electro-rheological fluids in aircraft flight controls—Phase 1 report. UMTRI-91-30. (University of Michigan Transportation Research Institute, Ann Arbor, 1990).

TESTING ON A HIGH SPEED ELECTRO-RHEOLOGICAL CLUTCH

by

A. R. Johnson,
Department of Mechanical and Process Engineering
University of Sheffield
Mappin Street
Sheffield S1 3JD
Tel: (0742) 768555 Extn. 5151
Fax: 0742 753671

W. A. Bullough, R. Firoozian, A. Hosseini-Sianaki, J. Makin, S. Xiao,
Department of Mechanical and Process Engineering
University of Sheffield

ABSTRACT

The design of a prototype ER clutch and its associated test facility have been described and sample results from the initial test programme are presented. The ER clutch shows considerable promise and outperforms various magnetic devices. Further clutch and ER fluid developments should increase these advantages.

The test facility described provides a quick method for obtaining a wide range of important fluid properties from a single small quantity of fluid in a single test facility operating under realistic engineering conditions. The results reported emphasise the need for continuing ER fluid development to increase the ER shear stress, to reduce the viscosity of the ER fluid, to reduce the electronic to ER shear stress time delay, and to improve the durability of ER fluids.

Nomenclature

E	-	electric field intensity (V/mm)
h	-	radial inter electrode spacing (mm)
i	-	excitation current (mamp)
I_L	-	load moment of inertia (kg m^2)
I_R	-	rotor moment of inertia (kg m^2)
I_Σ	-	total moment of inertia to be accelerated (I_R + I_L)
R	-	radius of clutch rotor (mm)
t^*	-	electronic to shear stress time delay (ms)
t^{**}	-	mechanical run up duration (ms)
T_e	-	Electro-Rheological torque (Nm)
T_f	-	friction torque (Nm)
T_o	-	viscous torque (Nm)

T_Σ	-	total clutch torque ($T_o + T_e + T_f$)
V	-	excitation voltage
$\dot{\gamma}$	-	fluid shear rate (sec^{-1})
τ_e	-	Electro-Rheological shear stress (Pa)
ϕ	-	fluid temperature (°C)
ω_i	-	angular speed of input member (rpm)
ω_o	-	angular speed of output member (rpm)

1. Introduction

The use of Electro-Rheological (ER) fluids to construct useful controllable devices such as clutches and catches[1,2] is showing considerable promise. The development of these devices have however highlighted the need for the development of ER fluids to be done on the basis of steady, time and frequency response. This generally involves the determination of steady, step response and frequency performance of the ER fluid from a variety of tests.

In a contribution to the speeding up of the ER fluid developmental test process a test facility is described in this paper based on an ER clutch from which all of these measurements can be taken from one fluid sample operating under realistic engineering conditions. Typical results are given and problem areas in ER fluid properties, test rig equipment and procedures are highlighted.

The development of a high speed ER clutch[1-7], for example, illustrates that the following information about its and the E. R. fluid performance are required.

a) What is the level of the ER fluid shear stress τ_e, and is it high enough for the clutch application?

b) Will the clutch pick up and drop a load satisfactorily at speeds up to 3000 rpm? (The maximum shear rate $\dot{\gamma}$ is 19,000 s^{-1} in the present case).

c) Will τ_e persist at rotational speeds of 3000 rpm, that is will centrifugal effects and/or high shear rates severely affect the ER effect?

d) What are the time constants and delays associated with the ER effect and what are the consequential effects on controllability?

e) Does the ER effect persist over an adequate frequency range?

f) Are there any hysteresis and non-linear effects which would affect the performance and control of the clutch?

g) Is there any ER fluid degradation under these severe operational conditions?

At present much of this information is obtained from separate test rigs, and on several different fluid samples. Further some of this information is little known or understood or even unavailable. The clutch test rig described in this paper can provide information on all these questions. It therefore not only provides essential developmental information for the clutch application but also very useful data for the ER fluid developer. Indeed the test facility and associated data analysis expertise could speed up fluid development by enabling data to be obtained quickly from one possibly automated test rig and one sample of ER fluid.

A brief description of the clutch and a fuller one of the test facility will now be given. This is followed by indications of the variety of results which have been obtained from this test facility.

2. Test clutch

Two basic configurations are feasible for an ER Clutch, namely radial plates and concentric cylinders[1]. For the test facility described the concentric cylinder form shown in figure 1 was adopted. The concentric cylinder form is essentially a high torque/inertia ratio geometry. The rotor, which is in effect the output member of the clutch, needs to have a low inertia so that the speed of response of the clutch will be as fast as possible when the ER fluid is energised causing torque to be transmitted from the rotating outer electrode. The rotor inertia can be made low by using a lightweight non-metallic material which carries only a thin conducting film on its outer surface. It should be emphasised that the outer electrode is connected to the driving motor which is of high inertia and running at a constant speed[1], thus the inertia of the outer electrode is not of any significance.

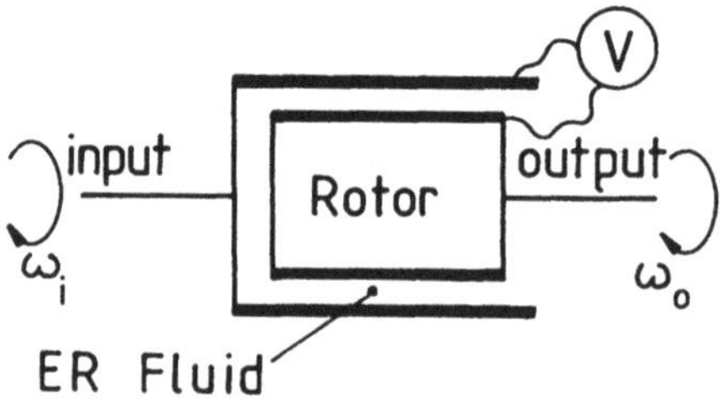

Fig. 1 Concentric cylinder ER clutch configuration

Importantly the adoption of the concentric cylindrical form enables the inter electrode gap, h, to be made small compared with the radius of the inner electrode R. However, the viscous torque, and hence the generation of heat in the ER fluid, is proportional to the cube of the radius per unit length of the clutch whilst the cooling surface area depends only on the radius per unit length of the clutch. This imposes an upper limit on the clutch radius and thus the h/R ratio possible. In the clutch tested the gap h = 0.5 mm (with a further rotor having a gap h = 0.75 mm) and the electrode radius R = 30 mm. This fairly large ratio R/h has the advantage that the electric field intensity E (V/h), and the fluid shear stress, $\dot{\gamma}$ ($([\omega_i - \omega_o]R/h)$), vary little across the ER fluid in the gap, that is radial effects are small[1,3]. Thus the data obtained from the test is not significantly dependant on the geometry and in fact approximates to that which would be obtained between moving flat plates. The data obtained should therefore be useful for quantifying fluid properties, advancing knowledge of the ER mechanism as well as providing design data for clutches and other ER devices.

The size of the test clutch was determined by the nominal torque (2 Nm) required and the properties of the ER fluid used[8-11]. The length was made equal to the diameter (60 mm) in an attempt to avoid any hydrodynamic instability, to minimise electrode end effects, and to give an acceptable high R/h ratio. Whilst it is possible to envisage a very light material rotor with a very thin conducting coating for the electrode the test clutch had a 3 mm stainless steel sleeve pressed onto a glass reinforced nylon rotor. This was easy to manufacture accurately but had the disadvantage that about 80% of the rotor inertia of 4×10^{-4} kgm^2 was due to the stainless steel sleeve. Full details of the philosophy and design of the test clutch are given elsewhere[1,3,7].

3. Experimental test facility

The overall philosophy behind the design of the experimental test facility was that the data obtained should be of sufficiently high quality to enable identification studies to be performed. This was for two main reasons: firstly, to enable the dynamic performance and characteristics of the clutch to be fully observed, analysed, understood and later to be modelled. Secondly, to enable further insights to be made into the basic ER mechanism, and also the properties of the ER fluid.

The test facility was also required to be flexible to enable a wide range of clutch geometries, excitation and loading conditions to be investigated. For example, envisaged excitations included steady state, step inputs, sine waves, other periodic functions and also possibly random inputs. Anticipated loadings included steady, step and periodic changes in torque, and also inertial loading.

In order to achieve the above aims the test clutch, motor and instrumentation could be mounted in various configurations, yet precisely aligned, on a rigid steel girder bed assembly. Care was also taken to choose the best appropriate instrumentation in order to provide the highest quality data, with the instrumentation having the least effects on the dynamics of the system. The basic measurements required were the clutch excitation voltage (V) and current (i), the input speed (motor speed) (ω_i) and the output speed of the clutch (ω_o), the torque transmitted by the clutch (T_Σ) and the various temperatures of the ER fluid (ϕ) within the clutch. Each measurement would be required simultaneously and, in view of the anticipated fast response, also at a fast sampling rate in order to achieve the aim of using the results for comprehensive identification studies. To achieve this a high quality computer controllable data logging system was chosen. This was capable of capturing the six channels simultaneously, to 12 bit resolution, at sampling rates of up to 1 MHz. Further, the system was capable of expansion and other enhancements to meet future needs.

A schematic layout of the apparatus is shown in figure 2. As mentioned previously an important conceptual idea behind the electrorheological clutch was that it should latch the load on to a constant speed drive. To this end an oversized (8 kw) speed controllable (0-3000 rpm) dc electric motor (1) was used, the effects on motor speed of clutching in typical loads was later shown to be negligible. The clutch input member (2), supported in two rolling element bearings (3), was connected to the motor by a torsionally stiff flexible coupling (4) to eliminate unwanted dynamic effects. The speed of the motor and clutch input member was measured with a small tacho-generator (5). High quality silver/silver-graphite slip-rings and brushes (6) were attached to the input member and enabled the clutch outer casing to be earthed, and the signals from the thermocouples measuring the fluid temperatures to be transmitted to the stationary measuring instrumentation with the minimum amount of signal noise.

The output member of the clutch (7) carried a further set of silver/silver graphite slip-rings and brushes (8), which were used to transmit the high voltage excitation and associated current to the inner gap electrode. The high voltage power supply, to provide the clutch excitation, was capable of producing 2 kV at a maximum current of 50 mamps. It was capable of generating step voltages with negligible voltage overshoot and time delay for the maximum predicted current range for the clutch of 0 to 50 mamps.

It is on the output side of the clutch (area A in figure 2a) that the flexibility of the test facility is available as a wide range of instrumentation/loading arrangements are possible. The four main configurations are shown in figure 2b-e. The static torque to cause the stationary elements of the clutch to slip can be measured using the torque arm (9) shown in figure 2b. Figure 2c shows a 0-4 Nm dynamic torque transducer (10) to measure the torque transmitted through the E.R. clutch when the output member is held stationary (achieved by

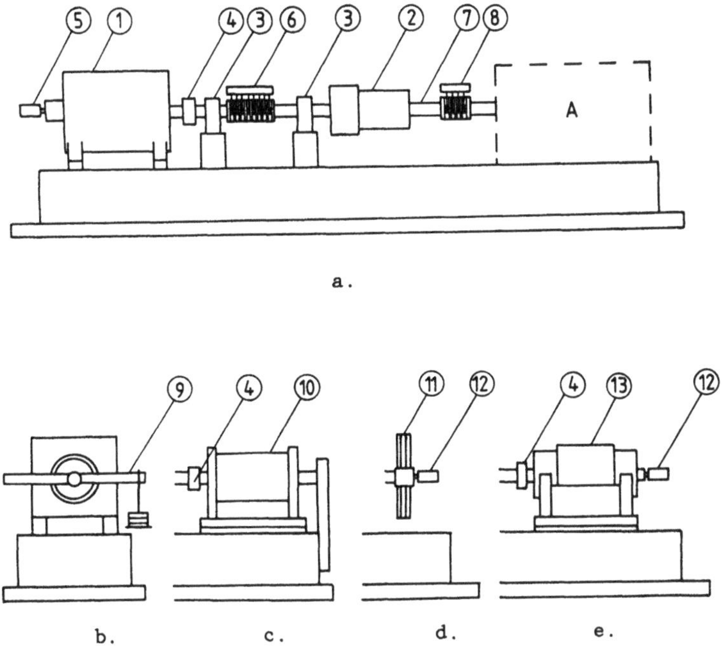

1 - Variable Speed Motor
2 - Clutch Input Member
3 - Rolling Element Bearings
4 - Flexible Coupling (Torsionally Stiff)
5 - Tacho-Generator
6 - Instrumentation and Earthing Slip-rings
7 - Clutch Output Member
8 - High Voltage Slip-ring
9 - Static Torque Arm
10 - Dynamic Torque Transducer
11 - Inertial Load Discs
12 - Tacho-Generator
13 - Powder Brake Dynamometer
A - Area shown in b,c,d and e

Fig. 2 Schematic View of Experimental Test Facility
a. side view of test facility
b. end view of static torque measuring configuration
c. side view of dynamic torque measuring configuration
d. side view of inertia loading configuration
e. side view of dynamometer loading configuration

holding the output shaft of the torque transducer stationary). Unfortunately, for a torque transducer to be sensitive it is necessary for it to be torsionally flexible. In this case this meant that the torque transducer has a natural frequency of 560 Hz. Worse, when it is connected to the clutch output rotor the resulting natural frequency is 110 Hz well within the operating torque frequency band width of the clutch. The actual response thus needs to be estimated by implementing an identification techniques and digital simulation software[5].

Figure 2d shows the configuration where inertia discs (11) could be added to vary the load inertia, I_L, from 0 to 16 x 10^{-4} kgm^2 in steps of 4 x 10^{-4} kgm^2. The total inertia to be accelerated was thus that of the clutch rotor (I_R = 4 x 10^{-4} kgm^2) added to the load inertia (that is $I_\Sigma = I_R + I_L$). A catch (not shown) holds the output member stationary whilst the input member runs up to full speed. The catch can then be released, also energising the clutch at the same instant. The increasing speed of the output member is measured by the small tachogenerator (12).

The final main configuration is shown in figure 2e where a 0-10 Nm powder brake dynamometer (13) can be used to apply various loads to the clutch. This type was chosen since it has a low inertia and can apply constant loads from stationary through increasing speed of rotation. A disadvantage is that the time constant for its response is about 200 ms, thus it can only apply a load torque and measure the torque in a quasi-stationary manner. The major problem with this type of dynamometer is that because of powder alignment restrictions it must be horizontal. This precluded having the test rig vertical and necessitated extra seals in the clutch which increased the unwanted parasitic frictional drag and increased the detailed design complexity. It also precluded the use of a water bath surrounding the outer electrode for temperature control/cooling duties, since this would require additional seals so adding to the parasitic frictional drag.

4. Test procedures and results

The simplest test on the clutch was to find the static torque to cause slipping of the clutch using the test configuration shown in figure 2b, where both the input and output clutch members are originally stationary. Whilst the test provides no information as to clutch or ER fluid performance at high rotational speeds the static shear stress obtained was equivalent to static ER shear stress values obtained from previous Couette viscometer tests. Further it enabled the adequate functioning of the test facility and instrumentation to be quickly assessed. Importantly it also provided a quick check for any degradation of the ER fluid, this will be discussed later.

The transmission of torque by the ER clutch at speeds of up to 3000 rpm could be tested using the configuration shown in figure 2c. The clutch input member is running at constant speed whilst the output member is held stationary. The dynamic torque transducer is used to measure the torque transmitted by the clutch. Initially no voltage is applied across the clutch so that the only torque transmitted is the parasitic torque due to fluid viscosity T_o and seal and slip-ring friction T_f as shown in figure 3. The excitation was switched on producing a step increase in voltage. A very rapid rise in transmitted torque was measured followed by a slow rise over a much longer time period up to a constant level which agreed closely with the slipping torque derived using ER shear stress values obtained from the earlier valve tests[9-11]. These earlier valve test values had been used for the initial basic design of the clutch[1,3]. The test therefore confirmed that ER torque would be transmitted at rotational speeds up to 3000 rpm. The severe vibrations seen in figure 3 about the mean response (dashed line) are due to the low natural frequency of the torque transducer. This necessitated the use of identification and digital simulation techniques to enable the torque response of the clutch to be evaluated[5]. The analysis showed that following the application of the excitation voltage there was an initial delay t* of about 0.5 milliseconds before the ER torque started to increase. The increase in ER torque was then very rapid with a time constant of about one millisecond. This rapid response was then followed by the slow rise to the steady value shown in figure 3. These features had been noted in previous valve tests[9-11].

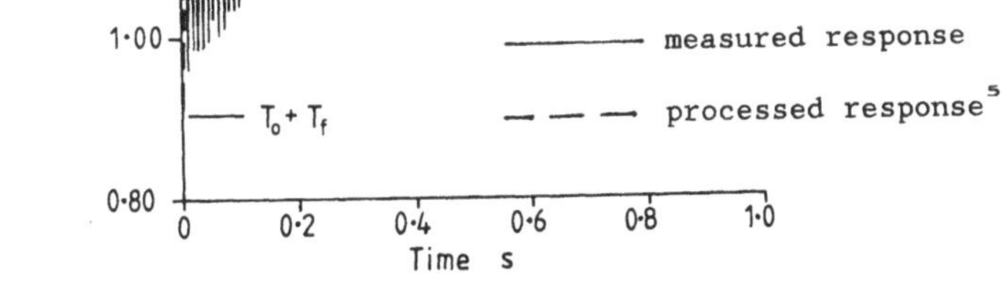

Fig. 3 Typical torque response of the ER Clutch to a step increase in the excitation voltage, also shows the increase in ER torque for time periods in excess of 200 ms. Excitation voltage of 1200 volts, input member speed of 3000 rpm, h = 0.5mm

Similar tests on a clutch with a 0.75 mm gap rather than the 0.5 mm gap previously used indicate that the time delay t* and the time constants are both similar. These tests were done using the same fluid shear, rate $\dot{\gamma}$, and electric field intensity across the ER fluid, E. The detailed findings will be reported shortly. Similar results to these from the step excitation test shown in figure 3 have recently been obtained from biased sine wave excitations at various frequencies. Again the raw results are noisy due to the low natural frequency of the torque transducer. Full results will be presented in a future paper, however, the frequency response data can also be dramatically improved by similar processing techniques to those for the step excitation[5]. For example, the typical noise on the raw data for the step excitation is clearly seen in figure 3, and similar noise is seen on the biased sine wave excitation raw data. However, a typical processed biased sine wave result is shown in figure 4, clearly showing the dramatic improvement. Nevertheless further improvements need to be implemented before the whole range of biased sine wave data can be presented and the corner frequency established by this method. One feature to note from figure 3 is the high level of the parasitic drag, $T_o + T_f$. Elimination of seal friction would be possible with a vertical test facility. However, the need for a reduction in fluid viscosity is a pointer to the need for further ER fluid development.

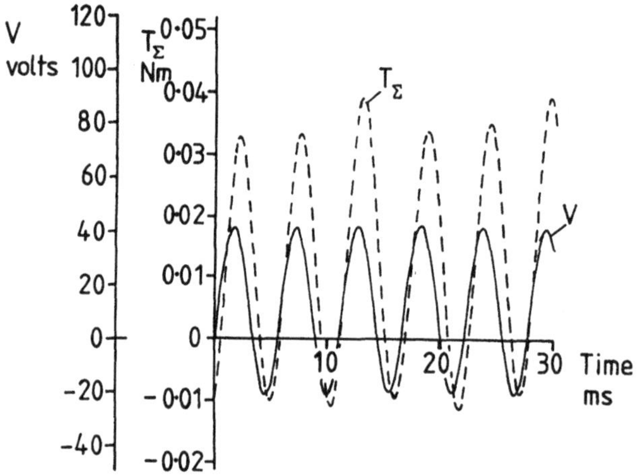

Fig. 4 Typical processed results from the biased sine wave tests for a excitation voltage frequency of 180Hz, h = 0.75mm, bias voltage = 1500V, $\dot{\gamma}$ = 3142 s^{-1}

The performance of the clutch in picking up a load could be evaluated using the test configuration shown in figure 2d, and the performance in dropping a load by use of that shown in figure 2e. For picking up a load the output member was held stationary whilst the input member was run up to speed. The output member was then released at the same instant that the excitation voltage was applied. The typical increase in speed of the output member is shown in figure 5a. Further tests of this type where undertaken with added load inertias up to four times the $4x10^{-4}$ kgm^2 of the output member alone[7]. The results are similar except for the longer run up times due to the larger inertia needed to be accelerated. The total run up times until clutch slipping ceased, t* + t**, are shown in figure 6 for a variety of speeds, excitation voltages and inertias. An extrapolation to zero inertia provides a rough indication of the initial time delay t*, and confirms that it is of very short duration, that is consistent with the value of t* = 0.5 ms obtained from the identification techniques[5]. The drop load tests using test configuration 2e were conducted by applying the excitation voltage to the clutch and running the locked input and output members up to speed. The dynamometer torque was then set to just below the level which would cause the clutch to slip. The excitation was then switched off and the clutch output member ran down to zero speed, a typical result is shown in figure 5b. The resulting drop load times were consistent with the pick up times after allowing for the additional inertia of the dynamometer[1]. Thus if the pickup tests can be considered as indicative of drop load tests then the dynamometer can be dispensed with. The test rig could then be vertical, seal friction could be reduced or eliminated and a cooling/temperature controlled bath introduced if required for accurate ER fluid property investigations.

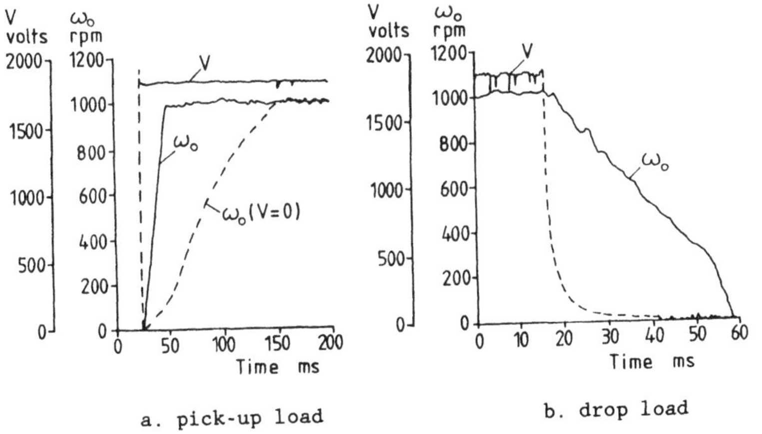

Fig 5. Typical pick-up and drop-load results for an input member speed of 1000rpm, h = 0.5mm

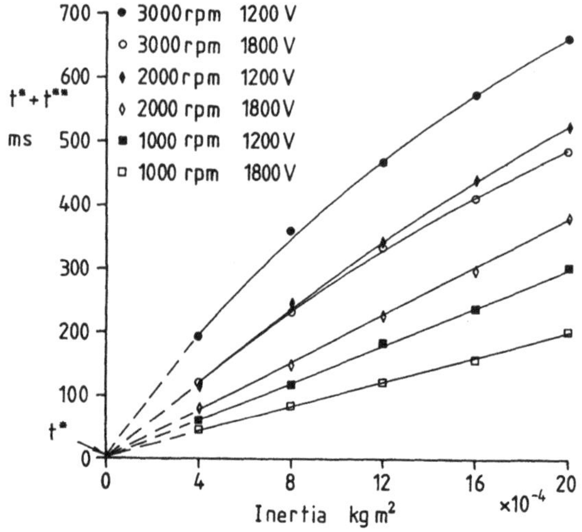

Fig. 6 Total acceleration times ,t^{*} + t^{**}, for various inertias at different excitation voltages, h = 0.5mm

An interesting and important consideration for the design and control of an ER clutch, but also for ER devices in general, is whether the voltage or the current is the better controlling parameter. Recent investigations on the ER clutch have suggested that sometimes current rather than voltage might be the controlling parameter[4-6]. In an attempt to investigate this query two tests were conducted using the test configuration shown in figure 2e with the output member held stationary by the dynamometer which was also used to measure the torque, this was possible since the torque would vary slowly in quasi-stationary fashion. In the first test the excitation voltage was held constant resulting in an essentially constant torque as shown in figure 7a, with the current rising as the temperature rose due to the shearing of the ER fluid. This appeared to confirm that voltage was indeed the main controlling parameter. However, to confirm this a second test was conducted in which the current was kept constant by reducing the voltage as the ER fluid temperature increased, the result is shown in figure 7b where clearly the torque falls as the voltage falls. These tests show that at variable temperature operation the control excitation would involve a complex function of current, temperature, and voltage.

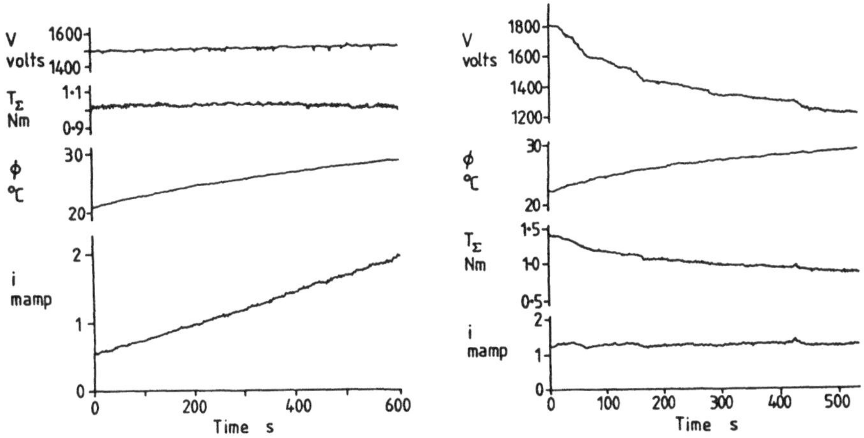

a) excitation voltage is kept constant

b) excitation current is kept constant

Fig 7. Variation of torque, fluid temperature, excitation current and excitation voltage, for a clutch with h = 0.5mm

For a controllable ER fluid clutch any hysteresis effects must be fully known. Using the test configuration shown in figure 2e tests were done by noting the effect on the output member speed of varying the voltage, the dynamometer being set to provide a torque just below that required to cause slip. Providing the clutch plates never locked the output member speed could be changed by varying the voltage in the region A to B in figure 8. However, if the voltage was raised so that the plates locked for a time, that is at C, then the voltage had to be reduced to the level D before the plates subsequently disengaged causing the output member to stop, that is at E. The voltage had then to be increased to A before the output member started to rotate again. This hysteresis loop clearly affects any control strategy and can probably be best dealt with using digital pulse width (0 to max. to 0 volts) modulation techniques. Further tests were done in a similar fashion to investigate the effects of prolonged plate lock up, with a typical result shown in figure 9. Following the prolonged plate lock up, X to Y, the voltage had to be reduced to a lower level before the plates disengaged, subsequently the disengagement voltage returned to the higher level unless the plates were again locked for a prolonged period. Clearly some time dependent feature is involved. Again this will influence control strategies[12].

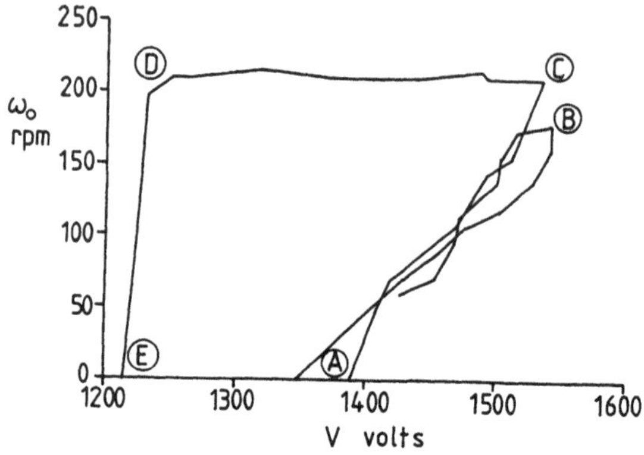

Fig 8. Variation of output member speed with the applied voltage for an input member speed of 210 rpm, h = 0.5mm

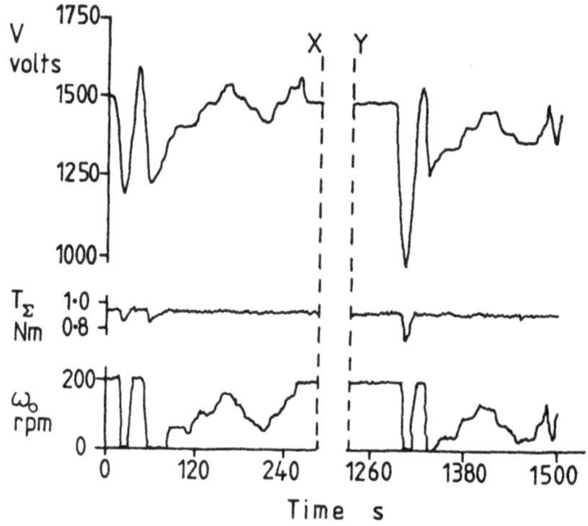

Fig 9. Pseudo-hysteresis effect of a prolonged lock up of the clutch, h = 0.5mm

There appears to be some problems with fluid degradation in the prototype clutch. Load pick up times have increased by about 25% over the six months that the fluid has been in the clutch[1,4,6]. Further, more recent static slipping torque tests on new samples of fluid indicates that this fall in performance occurs most rapidly in the first few weeks, followed by a much slower fall off in performance. The reasons for this apparent degradation have yet to be fully investigated and understood, but will obviously affect any control strategy; it had not shown itself in prolonged valve tests[10]. Further, it again demonstrates the importance, and wide requirement, of continued ER fluid development and testing in appropriate test facilities.

5. Comparison with Electro-Magnetic Devices

The performance in accelerating inertial loads for the prototype ER clutch is shown in figure 10 compared to that for an equivalent sized rare earth magnet dc motor with pulse-width modulation drives. These motors are among the fastest available[13]. Full details have been given elsewhere[7] but it is clear from figure 10 that for low inertia loads the ER clutch is significantly faster. Known improvements in the rotor design to reduce its inertia will produce the worthwhile improvements shown in figure 10. The largest improvement shown in figure 10 is when the ER shear stress is increased from the present 3 to 4 kPa maximum fluid to a 10 kPa fluid[14]. The ER clutch then outperforms the dc motor over a wide range of loads for high speed applications. Since the dc motors are highly developed dramatic improvements in the inertia or torque are unlikely, whereas for the ER clutch the inertia can be reduced and much higher yield shear stress ER fluids are in prospect. Thus there are clear aims for the fluid developer, namely to increase the shear stress, reduce parasitic viscous drag, improve fluid durability and reduce the time delay t*. Further, these new improved fluids need to be tested in appropriate test facilities giving realistic operational and not idealistic laboratory test conditions.

Theoretically, given the low inertia geometric form available to both the rare earth magnet dc motor and the ER clutch, an increase in torque should normally be gained from an increase in length of the rotor. From this point of view the ER clutch should always outperform the motor, that is given a satisfactory level of shear stress and utilising the rotor weight advantage. There is also no equivalent to the L/R delay inherent with the dc motor in the ER clutch where the excitation current i leads the excitation voltage V[12].

In a previous comparison with controllable hysteresis and powder filled friction brakes t* for the ER clutch was seen to be at least an order faster than the L/R values encountered for both. Additionally the inertia of the clutch rotor is much smaller than the iron utilising machines. A further comparison this time with a solenoid activated magnetic clutch[17] again shows t* for the ER clutch prominently advantageous against the magnetic time constant. By way of example a 1.1 Nm solenoid operated friction clutch reaches maximum torque some 25 ms after trigger and similarly takes 18 ms to drop the torque from the rated value to zero. Further, heating problems limit the on-off frequency to one hundred times per minute. The parasitic driven inertias in both the ER clutch and the solenoid magnet clutch are similar.

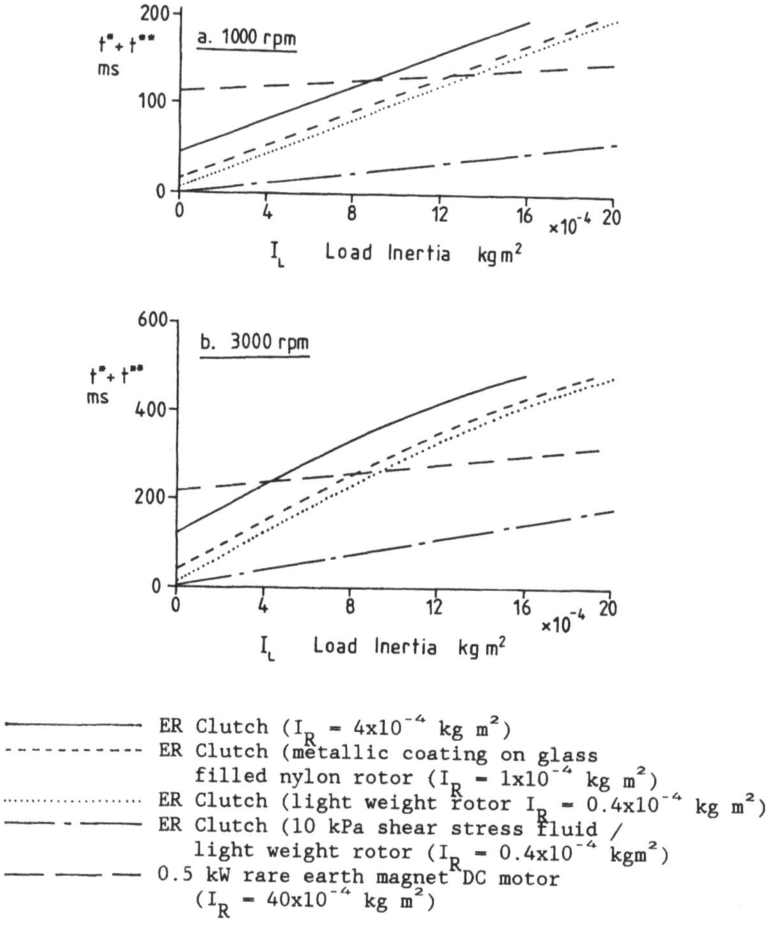

Fig. 10 Comparison of performance, in accelerating from 0 to 1000/3000 rpm, of the ER clutch with a 0.5 kW (1.6 Nm) rare earth magnet dc motor with pulse width modulation drive

6. Development of Test Facility

An improved torque transducer, that is one having a higher natural frequency, would improve the testing procedures considerably by making many readings more direct and reducing the need for identification and computer simulation techniques[5]. Such a transducer is available to the investigators[16], however, it cannot be used in a large diameter clutch rig that rotates. Recently results similar to those shown in figure 3 have been obtained with this transducer for a step increase in voltage and confirm the accuracy of the transform technique[5]. (See elsewhere in these proceedings). Similar tests for the frequency response are nearing completion and will be reported shortly.

Detailed analysis of the results indicate that the pick up results shown in figure 5a are indicative of the drop load results shown in figure 5b. If fully confirmed this would eliminate the need to use the dynamometer as a brake. The test rig could then be vertical, unwanted seals with their parasitic frictional drag could be removed, and a cooling bath incorporated for precise temperature control during long term testing. It is thus anticipated that a test facility would be available that could give ER fluid developers full and realistic data on the fluid performance from one small sample of fluid even quicker than using the existing facility.

7. Conclusions

A prototype ER clutch and its associated test facility have been described together with some results from its initial test programme. The ER clutch shows considerable promise and outperforms various magnet devices. Further clutch and ER fluid developments should increase these advantages.

The test facility described provides a quick method for obtaining a wide range of important fluid properties from a single small quantity of fluid in a single test facility operating under realistic conditions. The results reported emphasise the need for continuing ER fluid development. An automation of this method could give enhancement of data for fluid development purposes and greatly increase the speed and convenience of achieving it.

8. Acknowledgements

The authors are grateful to the Science and Engineering Research Council for their support of this work through the Specially Promoted Programme : The Design of High Speed Machines. Thanks are also due to Mr. D. Butcher who constructed the prototype clutch and test facility.

9. References

1. W. A. Bullough, et al. *The Electro-Rheological Catch/Latch/Clutch, Implementation and Prospects*, Paper **C414/070**, Proc. I.Mech.E., Eurotech Direct '91 Conf. NEC, Birmingham (1991) p. 129-136.

2. W. A. Bullough, et al., *Quick-Acting Drive Connections*, Patent Application.

3. R. Atkin, Xiao Shi and W. A. Bullough, *The Flow of an Ideal Electro-Rheological Fluid in a Rotational, Cylindrical Clutch*, Proc. Atuator 90 Congress, VDI/E. (German Assn. of Engrs), Bremen, (June 1990) p. 211-214.

4. W. A. Bullough et al, *Aspects of Steady, Quasi-Steady and Fast Time Response in a High Speed Electro-Rheological Clutch*. Preprint **N15**, Intro Seminar on Rheophysics and Thermophysics of Non-equilibrium Systems, Byelorussian Academy of Sciences, A.V. Luikov Heat and Mass Transfer Institute, Minsk, (1991).

5. R. Firoozian, et al. *Dynamic Torque Response of an Electro-Rheological Clutch*. ASME, DE **35** Rotating Machinery and Vehicle Dynamics, Miami (1991) p. 295-301.

6. A. Hosseini-Sianaki, et al. *Operational Considerations in the Use of an Electro-Rheological Catch Device*, Proc. China Fluid Power Society, 1st Fluid Power Trans. and Control Symposium, Beijing, (1991) p. 591-595.

7. A. R. Johnson, et al, *Electro-Rheological Clutch under Inertial Loading*, Proc. MPT, **91**, Int. Conf. of Motion and Power Transmissions, JSME, Hiroshima (1991).

8. J. E. Stangroom, *Improvements (further improvements) in or relating to Electric Field Responsive Fluids*, UK pat. 1, 570, 234, (25 June 1980).

9. W. A. Bullough, and D. J. Peel, *Progress Towards a Hydraulic Semiconductor for Vehicle Application*, paper 881786, Society of Automotive Engineers, Inc., (1988).

10. W. A. Bullough, and D. J. Peel, *The Field Controlled Liquid State*. Proc. I.Mech.E., Int. Conf. Mechatronics-Designing Intelligent Machines. Cambridge, (Sept. 1990) p. 171-178.

11. R. Firoozian, D. J. Peel, and W. A. Bullough, *Magnetic Effects in an Electro-Rheological Controller*. Proc. I.Mech.E., Int. Conf. MechatronicsDesigning Intelligent Machines, Cambridge, (Sept. 1990) p. 231-238.

12. R. Firoozian, et al., *Control Techniques for Electro-Rheological Fluids*, Proc. 24th ISATA Int. Symp. on Automotive Technology and Automation, Florence (1991) p. 479-487.

13. R. Firoozian, and K. Foster, *The Choice of Servo Motor for a Specific Application*, Proceedings of I.Mech.E. Conf. on Electric versus Hydraulic versus Pneumatic Drives, (1985) p. 67-76.

14. D. Brooks, *Fluids Get Tough*, Physics World, **2** (1989) p. 35-38.

15. W. A. Bullough, *Integrated Electron-Hydraulics*, Proc. of Mech 88. World Congress of the Australian Bicentennial Exposition, Institution of Engineers, Brisbane, Australia. (1988) p. 18-23.

16. J. E. Stangroom, *The Measurement of the Speed of Response of ER Fluids*, Proc. 1st Int. Symp. on Electro-Rheological Fluids, 18th Annual Meeting of the Fine Particle Society, Boston, (1987) p. 81-98.

17. *Clark Model 175 Clutch*, Technical Catalogue, Clark Electric Clutch & Controls Ltd., Surrey, England.

A LOW LOSS TRANSMISSION SYSTEM BASED ON ER FLUIDS

JAMES E. STANGROOM
ER Fluid Developments Ltd, Brough, Sheffield S30 2HG, U.K.

ABSTRACT

A standard fixed-displacement piston pump can be converted to operate as a variable displacement pump by the application of ER control to the one-way valve at the inlet port. The resulting device is considerably simpler than conventional variable displacement pumps, and can be directly electronically controlled. Combined with a conventional hydraulic motor, such a pump could provide a continuously variable, low-loss drive which could be used in many applications.

1. Introduction

Almost the only aspect of ER fluids on which there is universal agreement is that they are dissipative, i.e. they convert mechanical energy into heat under control, like viscous and frictional devices. All of these suffer from the same problems:- they are wasteful, since mechanical energy is costly to produce, and the low-grade heat generated is a positive embarrassment. Provision for cooling is a major consideration in all dissipative systems: without it, as in a car clutch, they can be used only intermittently.

Dissipative systems are used continuously only when there is no other way of achieving the desired control, and the waste heat can be easily removed. In much heavy engineering, the prime source of mechanical energy, e.g. constant speed a.c. motors, cannot be readily controlled. Conventional hydraulics provides the answer. The motor runs continuously, producing a constant flow of high-pressure oil from a pump. As shown in Fig 1, the speed of the actuator (shown as a simple ram) is regulated by bleeding the excess flow back to the low pressure reservoir: the energy used to compress the oil appears as heat. The great strength of hydraulic systems lies in the fact that the working liquid also acts a coolant, carrying the heat away. This allows the local power densities to be very high indeed. However, this merely shifts the problem; it does not solve it. The heat must eventually be removed, either through a radiator, or through the walls of an oversize reservoir.

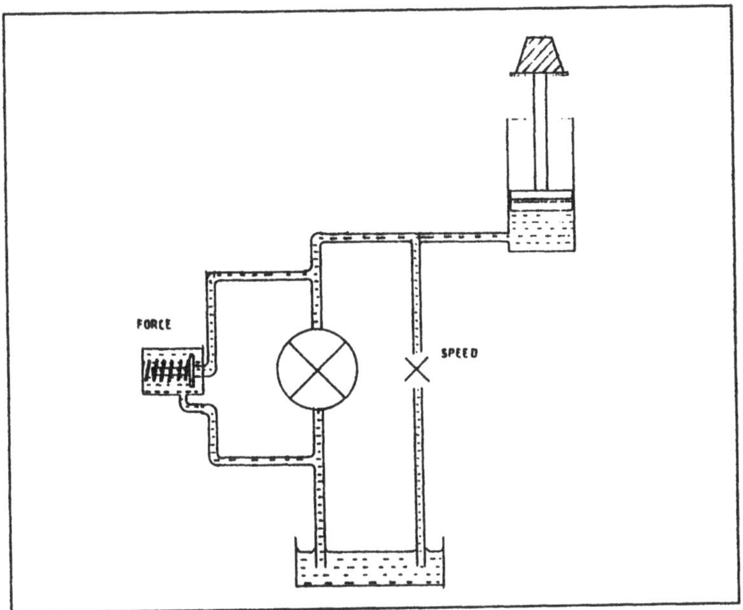

Figure 1 –Conventional Hydraulic Control

An ideal transmission system would decouple the speed of the actuator from that of the prime mover without loss of energy. Conventionally, this is done in three main ways:-

a) Gearboxes. These are very efficient, but they are normally restricted to fixed ratios and become very complex and costly when a large number of ratios are required. Some continuously variable gearboxes have been proposed but they have not achieved general acceptance.

b) Friction Drives. In the very early days, the speed of machines was varied by moving a flat belt over a pair of cone pulleys. This principle has re-appeared in various forms, such as the variable diameter pulleys (van Doorn drive) used in the original Daf car, and has been widely researched by the automobile industry. However, the basic problems of wear and limited torque capability remain to be solved.

c) Hydrostatic Drives. In these, the prime mover drives a variable displacement pump which in turn drives a fixed displacement hydraulic motor. Such drives are effective and reliable, and are widely used on heavy vehicles, such as agricultural and earthmoving machinery, but in spite of their advantages, they are little used in other areas. [1]

This Paper describes the application of ER to variable displacement pumps and hence to hydrostatic drives.

2. Conventional Variable Displacement Pumps.

Fig 2 shows the main features of a conventional variable displacement pump.

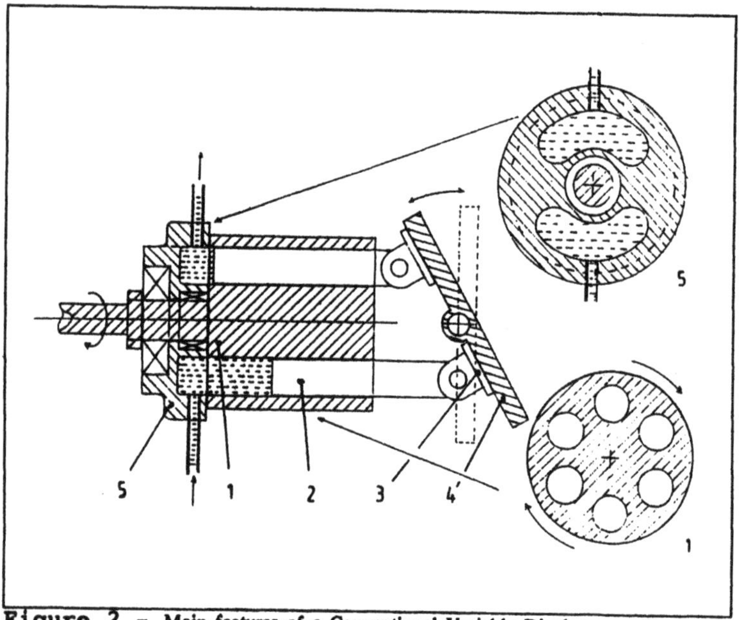

Figure 2 - Main features of a Conventional Variable Displacement Pump.

[1]. The writer is indebted to Mr.W.A.Bullough, of Sheffield University, Department of Mechanical Engineering, for drawing his attention to some early trials in which a dumper truck was fitted with a hydrostatic drive, and out-performed all competing systems.

The prime mover turns a solid steel cylinder block (1) in which a series of axial holes are bored. Close-fitting pistons (2) slide in these holes. The outer ends of these carry "Slipper Pads" (3) which run against a flat steel "Swash Plate" (4). The swash plate can be tilted externally, thereby varying the stroke of the pistons. The back of the cylinder block bears against a ground steel plate (5) carrying the bearings. Two curved ports are cut in this plate; these are so located in relation to the axis of tilt of the swash plate that they act as inlet and outlet ports respectively.

Although the principles of the device illustrated in Fig 2 are simple, reliability and efficiency demand first-class precision engineering. Pistons and cylinders are routinely fitted together (e.g. diesel injector pumps) - but achieving a good seal between the cylinder block (1) and the back-plate (5) when the oil pressure is driving them apart is more difficult. To prevent wear, the slipper pads (3) must not come into direct contact with the swash-plate (4); this is avoided by bleeding a small amount of high-pressure oil onto the slipper pads so that they run on an oil film rather than metal. Finally, the loads on the swash-plate (4) are so heavy that an auxiliary hydraulic jack is usually needed to move it. Not only does this further increase the complexity of the pump, but the power available to operate this jack limits the speed of response. Overall, conventional variable displacement pumps are complex and highly stressed, requiring very sophisticated materials and machining. As a result, they are costly, and uneconomic in many potential applications.

3. An ER-Operated Variable Displacement Pump

4.1 Conventional Fixed Displacement Piston Pumps

Fig 3 shows the main features of a conventional, fixed displacement piston pump.

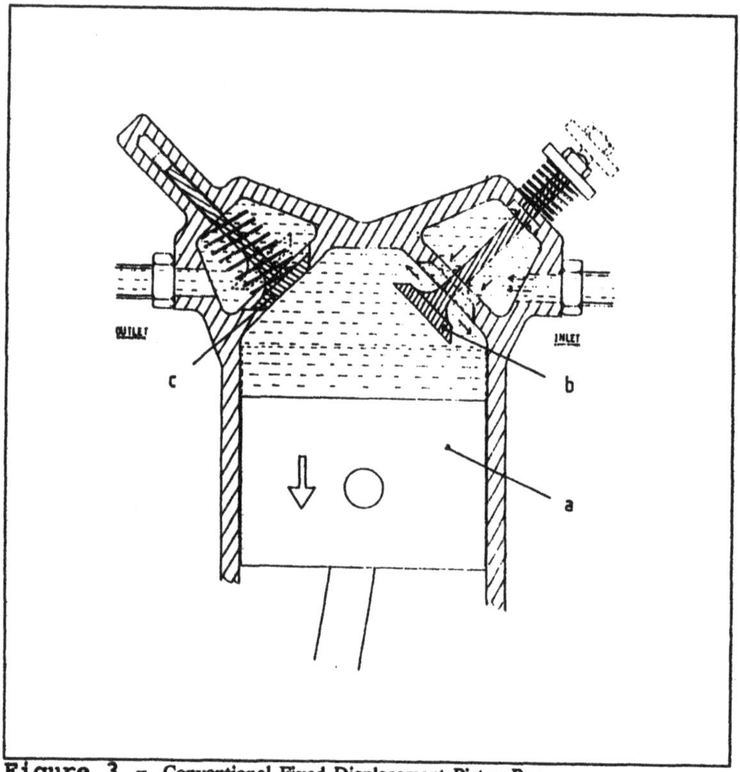

Figure 3 - Conventional Fixed Displacement Piston Pump.

The piston (a) moves up and down a fixed distance in the cylinder. As it descends, the pressure in the cylinder falls below that in the low pressure reservoir, so the inlet valve (b) opens, allowing oil to flow in. When the piston starts to ascend, oil flowing back into the low-pressure reservoir creates a pressure drop across the open valve (b) which tends to close it. Once the valve starts to shut, the pressure drop increases, so it closes more rapidly, so that although a closing spring is shown, it is often omitted in practice. With valve (b) fully closed, the cylinder pressure in the cylinder rises; this pressure acts on the valve face, exerting a force on the seat. Since the area of the latter is much less than that of the face, the local pressure is much higher than cylinder pressure and the valve seals. When the cylinder pressure has risen sufficiently above that in the outlet tube to overcome the closing spring on the outlet valve (c), the latter opens and the oil is expelled into the outlet pipe. At the end of the piston stroke, valve (c) is closed by the spring assisted by the high pressure oil in the outlet pipe.

The fixed displacement pump is clearly a relatively simple mechanism. As before, the piston must be a good fit in the cylinder, but this is not difficult to achieve. Since there is essentially a servo-action on the valves, i.e. the higher the pressure, the more tightly they are closed, these do not present anything like the same constructional problem as the face-seal on the variable displacement pump. As a result, this pump is much cheaper to make. If it were possible to adapt such a pump relatively simply and cheaply to give a variable flow, the resulting device would have very wide potential application, particularly if it could be controlled directly electronically.

4.2 Basic Principle

If the inlet valve (b) in Fig 2 jammed open, there would be no net flow at the outlet: the oil would merely flow in and out through the inlet port, without being compressed, and the energy losses, due purely to viscosity, would be small. If the inlet valve closed halfway through the delivery stroke of the piston, only the oil remaining in the cylinder would be driven through the outlet - the rest would have been expelled through the inlet. Thus, the delivery of the pump can be varied from zero to full swept volume merely by delaying the closure of the inlet valve. As shown in Fig 3, the former corresponding to the inlet valve being held open permanently, and the latter to the valve working normally. Intermediate delivery can be obtained by releasing the inlet valve at a pre-determined point during the delivery stroke of the piston.

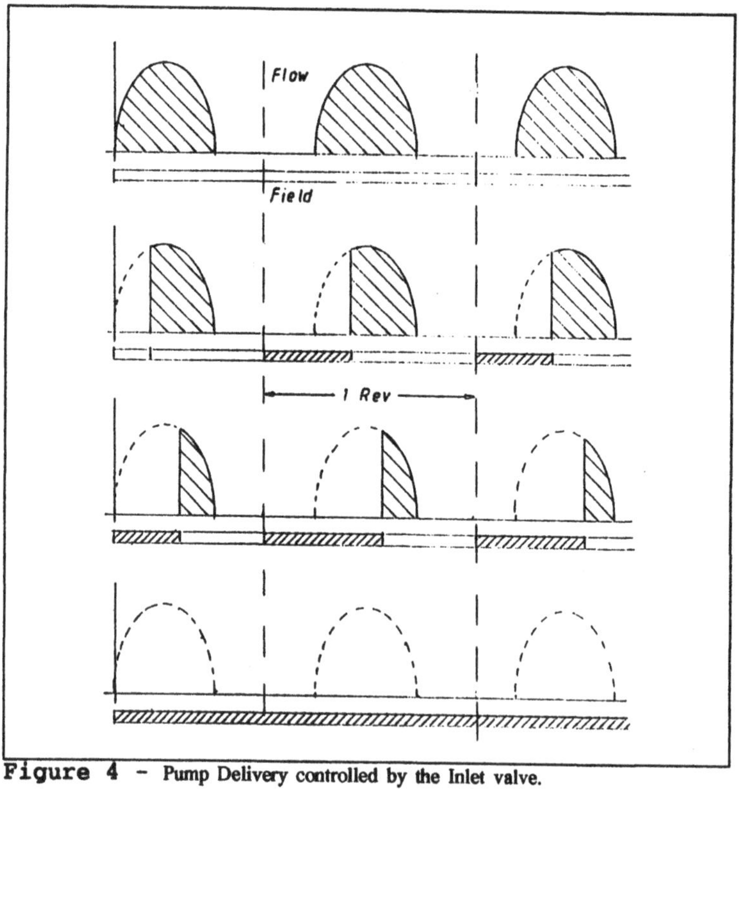

Figure 4 - Pump Delivery controlled by the Inlet valve.

Controlling the output of a pump by modifying the operation of the inlet valve has been previously proposed in various forms [1,2,3]; the novel element is to achieve this by ER technology. Flow control relies on the timing of the release of the inlet valve in relation to the movement of the piston. This will require high speed of response which ER fluids can provide - other aspects, such as timing, are well within the capabilities of modern electronics. The ER sub-system will handle very little power in relation to the power of the pump so it can be small and self-contained and will need only a small high voltage power supply. Little heat will be generated, particularly since the ER system will work in essentially a "Mark-Space" mode.

4.3 Noise and Shock Loading

It has been suggested that the final pump will be very noisy, and will be subjected to intolerable shock loading. In a conventional, fixed displacement pump, the inlet valve ((b) in Fig 3) closes at the start of the delivery stroke; the piston is moving relatively slowly, so the flow forces closing the inlet valve are moderate and the valve closes relatively gently. In the pump proposed, if the valve is released in the middle of the delivery stroke, flow forces will be at their maximum and the valve will close very violently. This will generate noise, which will be transmitted throughout the whole installation through the pipework, and subject the pump itself to a severe shock load.

This objection would be fully justified if the control on the inlet valve was a simple mechanical lock or solenoid, but it does not apply to an ER controlled system. Given the speed of response of the latter, there is no fundamental difficulty in programming the closure of the inlet valve, so that it slows down as it approaches the fully-closed position. Clearly, this is a second stage of refinement, but it underlines the advantages of using ER in this application.

4.4 Design Possibilities

As pointed out in the original Patent [4], the actual ER valve control can take many forms. A simple developed pressure unit is probably the simplest, since the problem of sliding contacts is avoided. The high viscous damping of such a system may well be advantageous in terms of reducing noise; on the other hand, excessive damping could limit the speed of operation of the pump, so fast pumps may require the inlet valve control in the form of a linear brake. Complete control of the inlet valve would be achieved by operating it with a powered actuator, with its own ER circuit. In a multiple piston pump - most practical pumps have several cylinders to reduce fluctuations - such a powered system may well prove economic compared with individual dampers since they could be operated by a single pump.

5. Initial Prototype

A prototype has been built and is now being tested. Since this is intended primarily to prove the concept and provide basic information, no attempt has been made to operate at the pressures and speeds needed for a realistic hydraulic pump. The prototype was therefore designed to be driven by a 3kW electric motor at up to 100 r.p.m. and generate pressures up to 20Bar. The crank, piston and cylinder and the output valve are conventional parts bought off the shelf. The inlet valve assembly is shown in Fig 5 overleaf.

The unit is basically a developed-pressure damper. As the main piston descends, pressure in the inlet port (i) opens the valve (ii), the stem of which slides in a hole in the steel mounting block (iii). A metal cylinder (iv) on the end of the valve stems acts as the "piston" of a rolling diaphragm, so that as the valve opens, ER fluid flows towards it. Three paths are provided for this flow, namely two annular gaps inside and outside the central "live" electrode (v) and a central hole with a ball-valve. When the main valve closes, the ball valve closes off the central hole, so the ER fluid is forced to flow back past the live electrode (v). This is insulated from the rest of the unit, which is at earth potential by "Teflon" rings (vi and vii). Applying a voltage to this electrode (contact not shown) therefore prevents flow and holds the valve open. Notice that the live electrode (v), which is in the form of a tube, has a series of radial holes at either end to allow the ER fluid to flow down both surfaces.

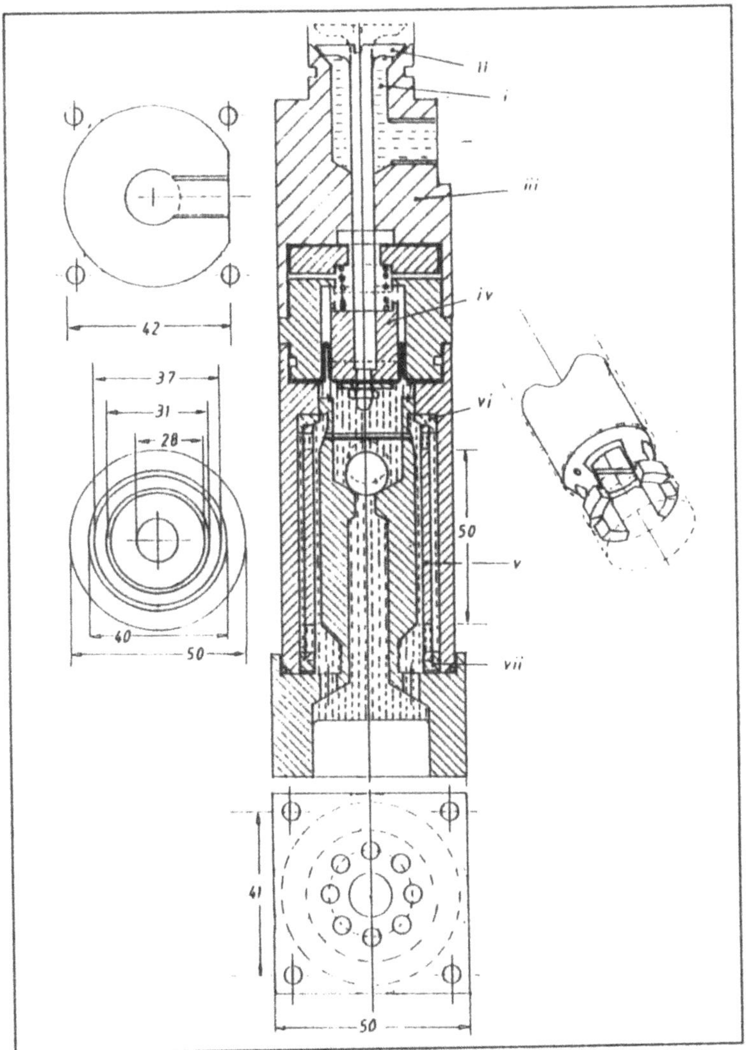

Figure 5 - ER Control Unit for Inlet Valve
(Dimensions - mm)

The rolling diaphragm provides an absolute seal, and avoids the normal friction of a lip-seal. However, the range of materials available for such seals is limited so it is essential to use an ER fluid with a base liquid which does not attack them. Fluorosilicone oil fulfils this requirement, and since only a small amount of ER fluid is used, its price is not a severe disadvantage.

The ball valve in the central channel is primarily to allow the main valve to open smartly, although the force available from the main low-pressure reservoir is small. However, it also imposes an overall circulation on the ER fluid, which will tend to come in down the central channel and go out through the ER valves. Experience with other ER devices suggests that it is important to avoid purely tidal flow into and out of a closed end, since deposition of solid tends to occur at these points. If some direction is imposed on the flow, this is entirely avoided.

At the time of writing (November, 1991), the unit has been tested and shown to function qualitatively. Numerical data is being collected to quantify the performance.

6. Conclusion

There is every indication that ER control applied to the inlet valve of a fixed displacement piston pump can provide variable displacement capability. The resultant unit should be economically attractive in a wide range of transmission applications.

Acknowledgements

We gratefully acknowledge the financial assistance we have received by way of a SMART Award (Small firms Award for Research and Technology) from the Department of Trade and Industry of the United Kingdom in the construction of the prototype.

References

1. J.E.Miller. *U.S. Patent No. 3459363.* 5th August 1969

2. Tron-Halvard Fladby. *U.K Patent No. 1565878.* 1st July 1977

3. *"Fluid Power International".* December, 1965, 414

4. J.E.Stangroom. *U.K. Patent No. 8913343.3.* 9th. June 1989

AN ELECTRO-RHEOLOGICAL SERVO POSITION CONTROL SYSTEM

ZHENG LOU*, ROBERT D. ERVIN* AND FRANK E. FILISKO**
**Transportation Research Institute*
***Material Science and Engineering Department*
The University of Michigan, Ann Arbor, MI 48109

Abstract

An electrorheological (ER) servo position control system has been designed, fabricated and assembled to examine the feasibility of using ER fluids in hydraulic control systems. To deal with current limitations in the energized shear strength of ER fluids, the servo device employs ER fluid only in a pilot stage of control, with a higher-pressure conventional fluid supplied separately to the power stage. The ER fluid supply system incorporates a diaphragm pump and a fluid tank with porting to assure suitable fluid mixing, in steady state operation.

The pilot stage is an ER bridge with four arms, each of which employs a parallel-plate ER valve. A special electric circuit, utilizing the combination of a high voltage linear amplifier and a fixed high voltage supply, was devised to control field voltages in the ER bridge. This scheme achieves "push-pull" energization of opposing arms of the bridge by means of only one linear amplifier. It changes the system stiffness from zero to a finite value, which is fundamentally important in control strategy. The ER bridge was seen to achieve pressure differentials up to 250 psi for driving the main-stage spool of a modified servo valve.

Although the ER module provided good control of differential pressures, shortcomings in the main-stage assembly prevented complete characterization of closed-loop performance during this study.

1. Introduction

This study is to test the feasibility of an electrorheological servo position control system (Fig. 1). This package is to control the position of a hydraulic cylinder by means of a two-fluid system. An ER fluid is employed in the pilot stage of a servovalve device while a conventional hydraulic fluid is supplied to the power stage of the valve. This hybrid design approach was prompted both by the problems of handling ER particulates in a hydraulic cylinder (with sliding seals) and by current limitations in the shear strength of ER fluids. Although the system is to demonstrate current ER capabilities for hydraulic control application, the package is not intended as a practicable application, in and of itself.

The pilot stage is an ER pressure controller. It is configured as a diamond array, or bridge, of four individual valve elements whose flow resistance is electrorheologically-controlled. The ER bridge produces a pressure differential across the load in a manner that is analogous to the electrical Wheatstone bridge that amplifies resistance changes in the arms—generating a voltage differential across its diagonal.

The actuator, servo valve and ER bridge are physically connected through an integral manifold, as shown in Figure 2. The supply of ER fluid to the bridge device is provided through a diaphragm pump.

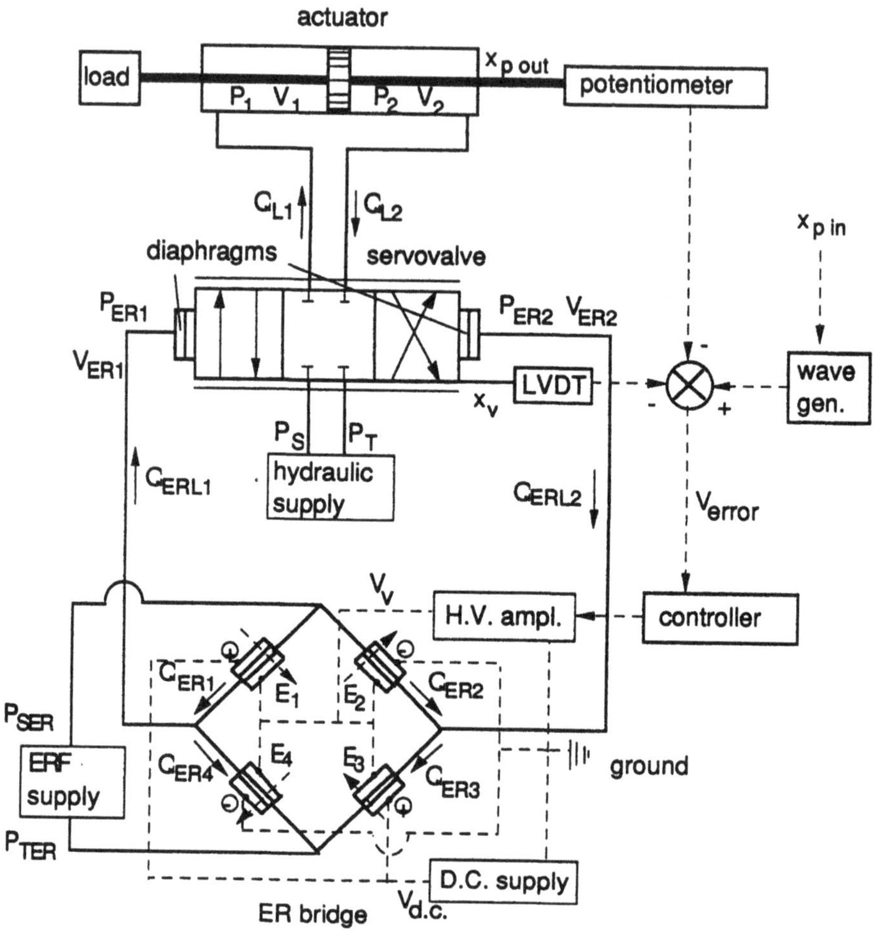

Fig. 1 The schematic of an electrorheological servo position control system.

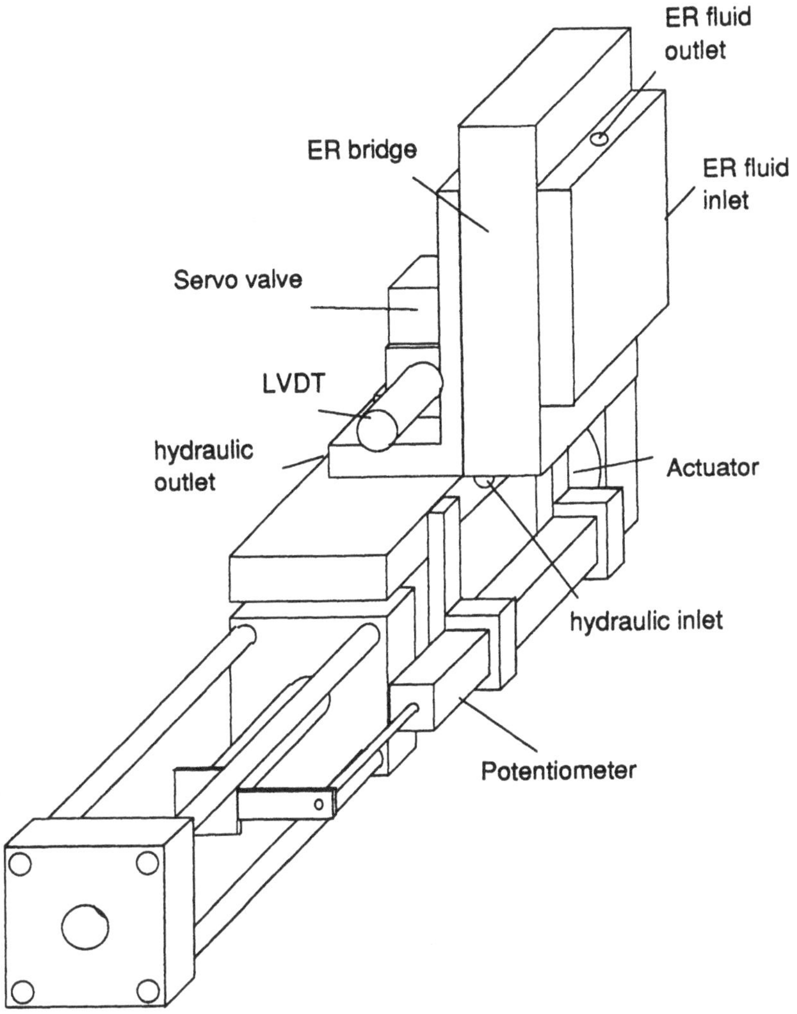

Fig. 2. Arrangement of components on the integrated package, including the servo valve, ER bridge, and actuator.

The load for the ER bridge is a floating spool that comprises the power stage of a modified servo valve. A pressure differential appearing across the diagonal of the ER bridge is exerted across the spool. Movement of the spool allows flow of conventional fluid, at high pressure, to the hydraulic cylinder. Small diaphragms are employed at both ends of the spool to separate the two fluid systems. Ideally, such a spool element should have a critical center or slight under-lap; however, the valve employed in this package turned out to be widely underlapped, thus rendering various problems in the system performance.

The bridge is excited in the push-pull fashion needed for control of differential pressure across the diagonal. The ER fluid samples are a mixture of kerosene and zeolite, which offers a low base viscosity, about 40 cP.

Both an individual, parallel-plate, ER valve and the four-valve ER bridge have been experimentally characterized as stand-alone items[1]. The complete servo control system was later experimentally characterized.

2. ER Bridge

As stated in the previous section, an ER bridge has four identical arms or valves, each of which can be electrorheologically controlled. A detailed study of the characteristics of ER valves and bridges has been performed by the authors[1]. For a bridge to deliver a load pressure as high as the ER system pressure, P_{SER} (assuming $P_{TER} = 0$), its flow resistors must be designed per the relationship,

$$\frac{L}{h} = \frac{P_{SER}}{2\tau_{y\,max}} \tag{1}$$

where L and h are the length and gap of each arm, respectively, and τ_{ymax} is the Bingham shear stress limit for an ER fluid. If $\tau_{ymax} = 0.1$ psi and L = 1 meter, the bridge is able to deliver a maximum lock-up pressure of 200 psi. The bridge design has also to accommodate the four arms, each with relatively long valve length, within a compact volume.

The ER bridge shown in Fig. 3 has been designed and fabricated for this purpose. On each gallery plate, slotted passages are provided to comprise the four valves of the bridge. All the plates (gallery plates, electrodes and insulators) are clamped between the top and base blocks. A problem encountered in this assembly involved some fluid leakage due to insufficient distribution of clamping forces, given the high stiffness of the nylon insulation plates. More distributed fastening would be recommended for any future design.

The top and base blocks serve as manifolds that afford the necessary connections among the four individual ER valve sections, the ER fluid and hydraulic fluid supply ports, and the ports on the servo (spool) valve. The base block also provides two taps from which the ER pressure differential was transduced.

Design criteria for the slotted passages that formed each ER valve included the following: (1) a minimum total fluid volume to reduce the amount of ER fluid needed in the system; (2) a minimum base flow resistance or pressure loss; (3) a symmetric geometric layout so that the bridge response is symmetric over the full range of dynamic operation[1]; (4) a faster dynamic response[1].

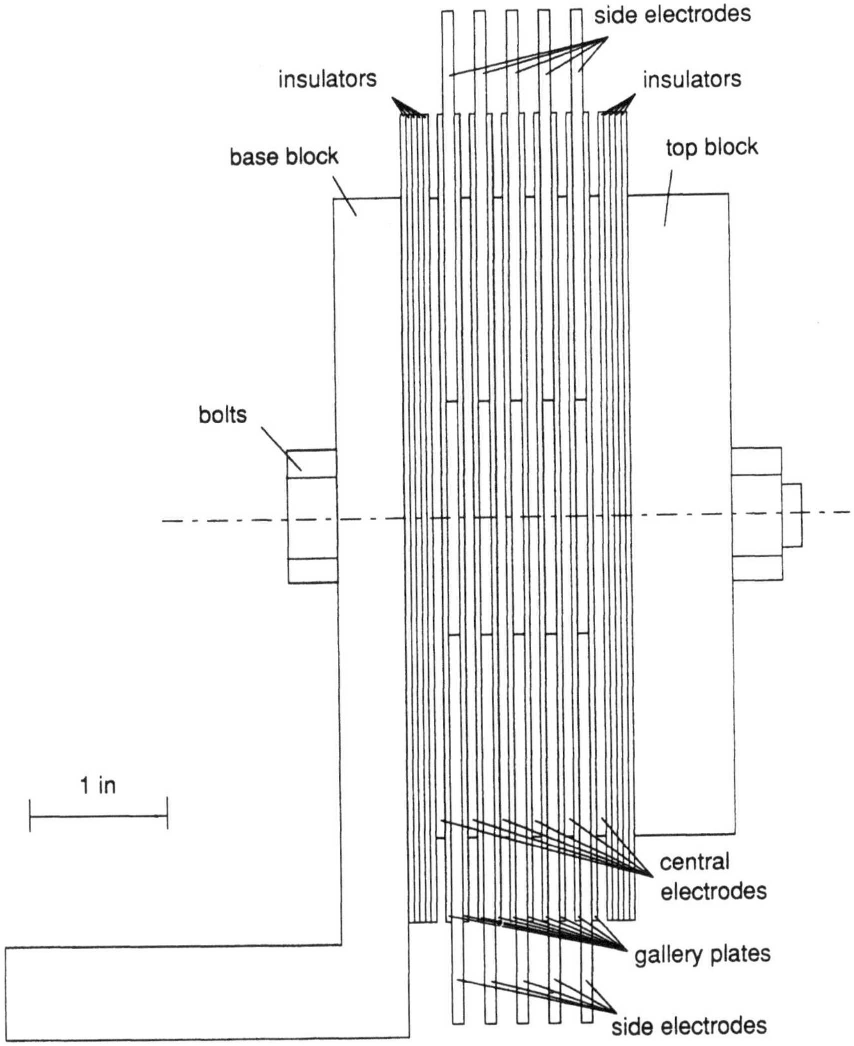

Fig. 3. An ER bridge.

3. Electric Circuit Activating the ER Bridge

A d.c. supply with a fixed voltage of $2V_{dco}$ is combined with a linear high-voltage amplifier whose output is offset by a value of V_{dco}, giving a peak-to-peak response of $2V_{aco}$ (Fig. 4). The limit of V_{aco} is V_{dco}. When the value of the reference input signal is zero, both valve pairs (valves 1 and 3 and valves 2 and 4) are charged with a uniform voltage differential of V_{dco}. As a sinusoidal signal changes from zero to the positive peak, the potential difference between points a and b decreases from V_{dco} toward $(V_{dco} - V_{aco})$ while that between points b and c increases from V_{dco} toward $(V_{dco} + V_{aco})$.

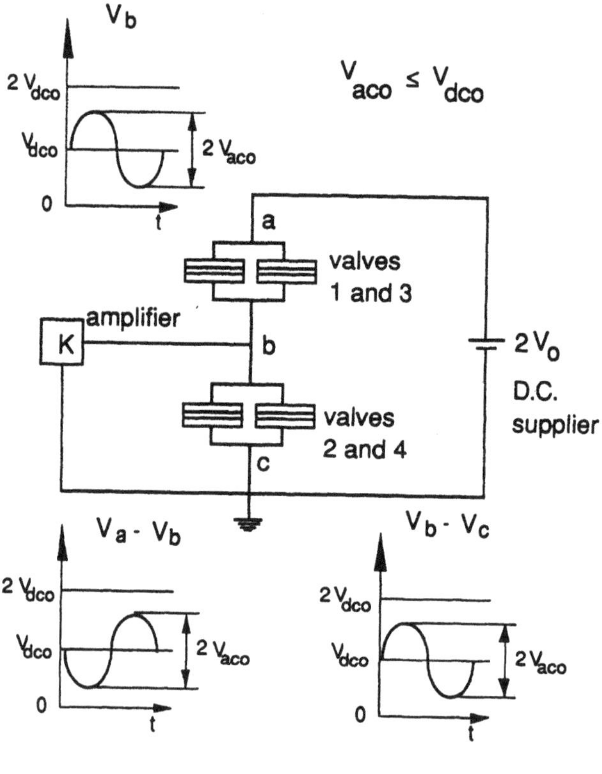

Fig. 4. Electric circuit for energizing the ER bridge.

The electric field strength in valves 1 and 3 is

$$E_{13} = E_{dco} + E_{aco}\sin \omega t \qquad (2)$$

while that in valves 2 and 4 is

$$E_{24} = E_{dco} - E_{aco}\sin \omega t \qquad (3)$$

where $E_{dco} = V_{dco}/h$ and $E_{aco} = V_{aco}/h$.

This arrangement enables the balanced push-pull exercise of the four-arm bridge using only a single linear amplifier. Perhaps what is more important, the biased voltage differential retained on each valve arm provides for an elevated stiffness in the ER bridge response about the zero reference operating point. This feature avoids the nearly zero gain of typical ER valves around zero voltage or electric field strength (Fig. 5).

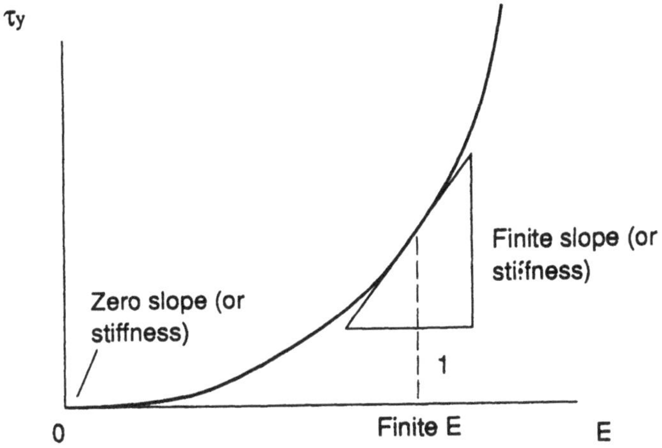

Fig. 5. Stiffness of a typical ER fluid.

4. ER Fluid Supply

Studies were performed to find an effective way to pump ER fluids, whose solid phase poses various problems to conventional hydraulic pumps. A commercial centrifugal pump was tested and a special pumping mechanism was designed before a commercial diaphragm pump having suitable output pressure was found. This diaphragm pump (L-07138-74, Cole Parmer) is capable of delivering ER fluid at a flow rate of 0.8 gpm and a pressure of 225 psi or higher. It was calibrated for both kerosene-based ER fluid (Fig. 6) and pure kerosene. The output flow rate is roughly linear to the dial value and is quite insensitive to the load pressure. Nevertheless, a dead zone exists near the zero value on the flow-control dial. Some deviation in the calibrations was seen between that obtained with pure kerosene and the higher viscosity ER fluid. Although this pump proved to be quite reliable over the period of the study, it does exhibit pulsations in both output flow rate and pressure due to the reciprocating nature of its design.

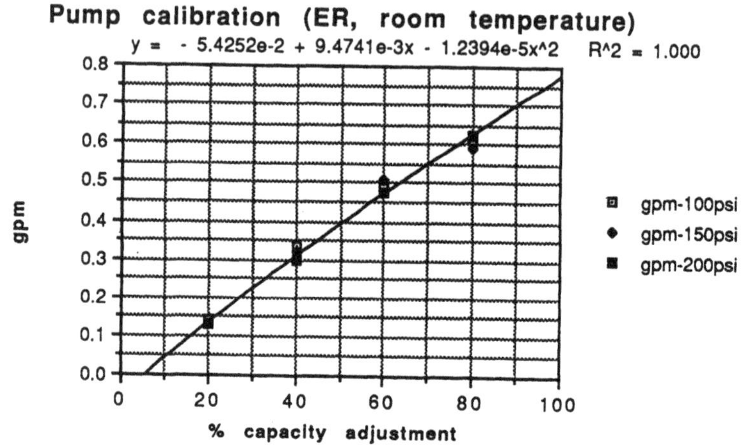

Fig. 6. Calibration of the diaphragm pump with an ER fluid.

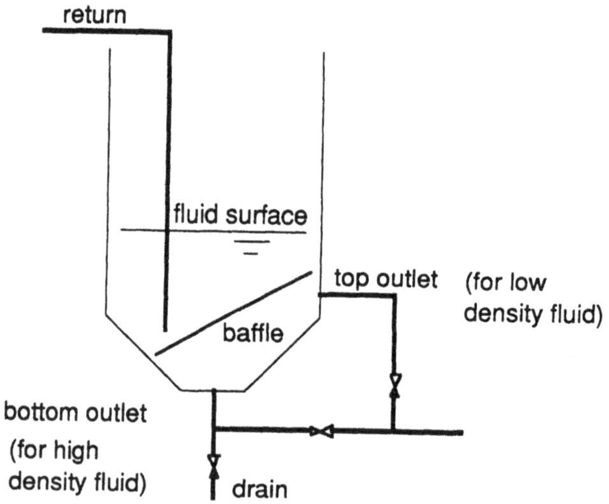

Fig. 7. ER fluid tank.

A specialized tank or reservoir device was designed (Fig. 7) for storing ER fluid in the bridge-charging system The fluid return line to this tank is immersed beneath the fluid surface to reduce or eliminate bubble generation. The tank was provided with two outlets. The pump draws relatively low or high density fluids from the upper or lower of these outlets. As a procedural convenience, it was occasionally necessary to pump the lower-density fluid into the system to (1) minimize the amount of ER particulate that would be lost during a partial disassembly of the apparatus, (2) to facilitate the bleeding of air bubbles, and (3) more readily flush accumulated particulate from portions of the system where insufficient flow occurs to maintain the suspension.

Alternatively, the normally high-density ER fluid is needed for a controlled operation. In these cases, the pump draws from the bottom tank port, accessing a relatively high solid content in the ER fluid that is mixed reasonably well due to flow agitation.

5. Modified Servo Valve

5.1. General Scheme

A conventional servo valve was modified to eliminate the electromagnetic torque motor element from its pilot stage. Fig. 8 illustrates the components of the valve that were provided for the hybrid operation, involving ER fluid and conventional hydraulic fluid. ER fluid is introduced to both ends of the spool through two ports. Diaphragms are used to prevent the ER fluid from entering the sliding clearance between the spool and sleeve. The forces from the diaphragms are transferred to the spool through two connectors, which slide within the cylinder sleeve. Two leak passages are drilled to channel leaking oil from the ends of the spool to the return port. Two sealing screws are used to separate the returning ports for the respective fluids.

5.2. Forces on the Spool

To scale the level of differential pressure needed across the diagonal of the ER bridge, it is necessary to compute the peak forces required to move the power spool at the maximum mechanical frequencies of interest.

For a four-way open-center spool valve with a linear area gradient, no net damping length, no flow force compensation, and no reaction of the nozzle flow forces, the momentum equation is[2]

$$P_{LP}A_v = M_v \frac{d^2x_v}{dt^2} + 0.86w(P_s - P_L)\, x_v \,. \tag{4}$$

where P_{LP} is the pilot valve load pressure ($= P_{ER1} - P_{ER2}$) (Fig. 1), A_V the end area of the spool, M_V the spool mass, x_V the spool displacement, t the time, P_S the hydraulic system pressure, P_L the load pressure, and w the spool wet perimeter or the area gradient. The term on the left hand side of the equation represents the driving force. On the right hand side of the equation, the first term depicts the inertia force, and the second one is the steady-state flow force. The steady-state flow force, like a spring, always tries to push the spool back to center or zero position, countering the driving force.

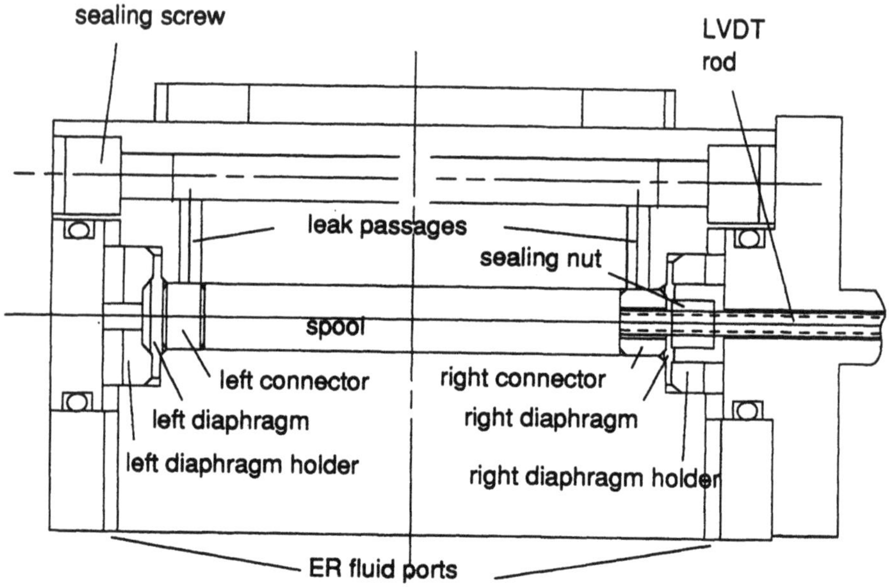

Fig. 8. Servo valve modifications

The valve used in this study happened to be an open-center valve. For example, when the actuator is locked, the entire output flow from the hydraulic pump (about 0.93 gpm) goes through the servo valve with a system pressure of about 400 psi. For an oil specific gravity of 0.9 and a discharge coefficient of 0.61, one gets

$$wU = 0.8677 \times 10^{-6}\, m^2 \tag{5}$$

where U is the underlap. The underlap was also tested experimentally with a pressurized air line, and it was found to be slightly over 1.016 mm (40/1000 inches). If U = 1.016 mm, then one obtains

$$w = 0.854 \text{ mm.} \tag{6}$$

However, the maximum w value for the spool in question, whose diameter is 4.7625 mm (3/16 inches), considering that the ports cover the full periphery of the spool, is 14.962 mm. Recognizing that it is difficult to measure w directly, we might only say that the real value of w is somewhere between 0.854 and 14.962 mm.

At $P_{LP} = 250$ psi, the driving force is 6.9 lbf. At $P_s = 3000$ psi and $P_L = 0.0$ and with $wx_v = wU = 0.8677\ mm^2$, the steady-state flow force is 3.47 lbf. The inertia force from the spool mass is less than 1 lbf at 200 Hz.

This analysis shows that a 250 psi pressure differential should be sufficient to drive the spool up to 200 Hz. We should also note that, in the modified servo valve, there is also a nonlinear spring force from the diaphragms, which has not been characterized, but which acts equally on both ends of the spool.

5.3. *Diaphragm and LVDT Problems*

When the valve was initially modified in the manner depicted in Fig. 8, there was an unbalanced fluid force, which tended to move the spool towards the right. It was suspected that internal hydraulic pressures at the two ends of the spool were causing an unbalanced spool force because of the asymmetry afforded by the LVDT installation through the right-side diaphragm. Also, the small hole in the right diaphragm, which was necessary to accommodate the LVDT rod, was a source of leakage. Therefore, the entire LVDT assembly was removed in the final version of the system, yielding a symmetric containment of the spool. Removal of the LVDT assembly strongly degrades the closed-loop system performance, however, such that any future system of this type should incorporate an alternative spool-displacement transducer, perhaps by strain-gaging the original leaf-type feedback spring.

6. ER Fluid Stability

A zeolite-based ER fluid, fluid 2, from another study[3] was prepared during summer of 1989 and utilized throughout the testing activities of the current study. It was frequently replenished with commercially-available kerosene to recover from kerosene leakage in the system. Further pollution of this fluid sample occurred when diaphragms isolating the power spool ruptured, admitting conventional hydraulic oil into the ER fluid circuit. Even after this extent of mixing with the parent fluid sample, however, the polluted ER fluid still performed essentially as seen in the initial rheometry measurements. Perhaps this experience suggests that at least some ER fluids are rather robust.

7. System Performance

The system did function in a closed-loop manner, responding to set-point position commands and achieving zero steady-state errors. Because of removal of the LVDT as a spool-position transducer, however, and the inability to develop high fluid pressures across the extremely under-lapped spool, system dynamic performance was minimal and was not explicitly measured.

One incident worth noting, for its experiential value, was a short circuit that appeared in the ER bridge during these tests. It was discovered that a virtual short (100 ohms resistance between adjacent valve electrodes) was derived from an inadvertent over-voltage excitation of the bridge that cause repeated and prolonged arcing between two plates within the bridge assembly. In the electronic control circuit, the error signal limit had been accidentally set at 8 kV, instead of the intended 4 kV level. The nylon insulation plate between two electrodes had become carbonized due to the arcing right at its edge, thus producing a failed, conductive section, approximately 10 mm in diameter. This observation prompts the view that since (a) inadvertent arcing may be inevitable, eventually, then (b) it should be designed-for. That is, we should seek insulation materials that, if subjected to intense local heating due to nearby arcing, will fail in a manner that does not provide a low-resistance electrical path. Otherwise, an unstable condition is possible, by which a temporary short causes resistance to decline, thus locally-elevating the current to thereby cause further heating, breakdown, and finally low levels of sustained resistance.

8. Conclusions

An ER servo position control system has been designed, fabricated and assembled to test the feasibility of using ER fluids in hydraulic control systems. In a pilot-valve stage of this system, an ER bridge has been shown to be successful in producing a controlled pressure differential up to 250 psi for driving the spool of a power-stage valve.

A special electric circuit, utilizing the combination of a high voltage linear amplifier and a fixed high voltage supply, was devised to control field voltages in the ER bridge. This scheme achieves "push-pull" energization of opposing arms of the bridge by means of only one linear amplifier. It changes the system stiffness from zero to a finite value, which is fundamentally important in any control system.

Upon closing the loop around the position of a hydraulic cylinder, the assembled system was able to control the motion of the actuator in slow-speed scenarios. Lack of refinement of certain details in the power-stage valve prevented the system from achieving any dynamic performance levels worth measuring.

Throughout the process of system testing, ER fluid samples were seen to be stable and consistent in performance.

9. Acknowledgements

The authors would like to acknowledge the advice and comments from Mr. Christopher Winkler and technical support from Ms. Theresa Filisko, Mr. Thomas Husen, Mr. John Koch and Mr. Michael Campbell. This work was partially supported by the U.S. Air Force through a contract with Dynamic Controls, Inc. It was also supported by the University of Michigan Industrial Affiliates Program in Electrorheology, which included the following sponsors: Aeroquip Corporation, Chrysler Corporation, Ethyl Corporation, Ford Motor Company, Monroe Automotive Equipment Co., Naval Underwater Systems Center, Nissan Motors, and Tremec Trading Company.

10. References

1. Z. Lou, R.D. Ervin and F.E. Filisko, Behaviors of electrorheological valves and bridges. (Southern Illinois University at Carbondale, Illinois, Oct. 15-16, 1991).
2. H.E. Merritt, *Hydraulic Control Systems*. (John Wiley & Sons, New York, 1967).
3. Z. Lou, R.D. Ervin and F.E. Filisko, The Feasibility of using electro-rheological fluids in aircraft flight controls—Phase 1 report. UMTRI-91-30. (University of Michigan Transportation Research Institute, Ann Arbor, 1990).

The Research in the Applications of ER Technology in Engineering

Wei Chenguan, Wei Aixia
Vehicle Engineering Department
Beijing Institute of Technology
Beijing, 100081,P. R. China

ABSTRACT

In this paper, the authors introduce the results of research on the basis of practical experience: 1. the development of ERF in the engine-ering application, the requirement of the charactericstics of practically used ERF and the problems still existing; 2.some application examples of ER technology; 3. the highlighting of ER technology in the engine-ering application.

1.Introduction

ER effect is a great discovery made in the 1940's. Before it was disclosed by Winslow, many scientists had made a great many of experiments on the variation of fluid viscosity on application of an electric field. But there is an essential distinction between the experiments made by Winslow and that made by them. What they used were pure liquids or monophase materials, such as polar liquid and norpolar liquid. It was shown that the viscosity of norpolar liquids doesn't change with the application of electric field and that of polar liquids do. But the range of variation does not exceed 200%. Because these liquids are not sensitive to the electric field, they were only studied in the laboratories. In 1940's, for the first time, Winslow used the mixture of dispersed phase and dispersing phase whose yield stress could increase to several hundred dyne/cm^2. It is the important change that put the ER effect into the stage of industrial application and stimulated the interest of many scientists in this field.

The following are some distinguished features of ER effect:

1. The shear stress increases with increasing the field strength and decreases with decreasing the field strength. The range of variation should be wide.
2. The variation of shear stress should be stepless and reversible under the control of electric field.
3. With the control of electric field, the change from one state to another should be very fast. That means the frequence response is very high.

Because of the special advantages of ER fluids, the ER effect is very promising in the following fields:

1.1 Transmission technology.

According to the ER effect, we can design ER cluches of high performance. The torque transmitted and the output speed can be controlled by the appplied field above the "no-field" condition to meet the special need of load. But the yield stress of present ER fluids is not high enough. There is still a long way for ER clutches to become practical.

1.2 Control technology.

A kind of ER valve can be designed in terms of ER effect. The pressure drop can be controlled by the applied field. The fast response of ER fluids could make the bandwidth exceed 1kHz. But this kind of devices need a great amount of ER fluid. It is not practical to use the expensive ER fluid instead of modern hydrolic system at present.

1.3 Sealing technology

It is well known that magnetic fluids have achieved great successes in sealing gas. The shear stress of magnetic fluids is not very high, so the sealing pressure is limited. With the similar principle, ER fluids could be very useful in sealing gas and liquid.

1.4 Clamping and position setting technology

This is another potential application area for ER effect. It could used in the locating of workpieces, especially for thin wall cylinders during the machining. As a result, the prisision and the efficiency could be increased greatly. This is a new method to develop new kinds of clamp appratus. Another case is microelectronic motor in which the piezo-ceramic could be located by ER fluids of high performance. In this way, a series of advanced motors could be produced.

1.5 Damping technology

A kind of dampers can be designed in terms of ER effect. The damping force can be controlled in response to road surface conditions, Automobiles with this kind of dampers will be more comfortable.

2. ER fluids

Since Winslow disclosed the ER effect, many investigators have studied the ER fluids. The fluids have progressed considerably, being stronger and requiring less control power than the earlier fluids. So far there have been the following ER diperse phases:

1. Calcium titanate — Carboxymethyl dextran
   Cellulose — Ion exchange resins
   Iron oxides — Piezo-ceramic
   Poly(acrylate) salts — Poly(methacrylate) salts
   Poly(vinylalcohol) — Silica
   Sodium carboxymethyl cellulose — Starch
   Tin oxide — Titanium dioxide
2. Poly(acenequinones)

There must be some surfactants, mainly water in the first class of disperse phases to increase the ER effect. This class of ER fluids are called wet fluids.

The second class of disperse phases are quite different. They do not need the water to increase the ER effect. This calss of ER fluids are called water-free fluids.

There are not special restrictions on the dispersing phase of ER fluids. The dielect constant must be high, the viscosity low. Silicone oil, kerosine, transformer oil and some heavy synthesized oil are commonly used.

Untill now, many investigators still use wet ER fluids for experiments.

Chinese investigators began to study in this field in 1988. To meet the need of the industrial application, we developed several kinds of ER fluids based on silicone oil. One of the disperse phase is semi-conductor. The yield stress(weight fraction is 40%) is 4.6KPa at E=3.0KV/mm. The current density is less than $4uA/CM^2$. It could operate at 120℃.

Considering the performance of the present ER fluids, we think they are not suitable for the situation of high shear rate, high temperature and large demand. Now our research areas are limited in clamping, microelectronic motors, dampers and clutches.

The factors which influence the application of ER effect are the yield stress, response time of ER fluids. The yield stress is relatc to the field strength, the viscosity of the base oil and the fraction of the dispersed phase. In addition, the power should be reduced and stability improved. A ER fluid(weight fraction is 0.4) whose yield stress could reach 10KPa at 3000V/mm, the yield stress with no field is 0.5− −1.0KPa and the physical and chemical properties are stable will play a very important role in the clamping and position setting technology.

3. ER Dampers

Dampers can absorb the vibration excited by many kinds of stimulations. They are widely used in all kinds of automobiles. Dampers are divided into two catagories of mechanical and hydraulic in which the damping force is unchangeble. ER dampers can be designed that have a variable damping force characteristics so a vehicle surpension can automatically compensate for loading conditions and adverse road conditions. Thus make the automobiles more comfortable. In our opinion, the modification of the present dampers should be as small as possible to make it easier to populized the ER dampers. We use the spark plug to apply the electric field. Once it is successful, there will be no need to change the production line greatly. The ER dampers will be produced more easily to replace the mechanical and hydrolic ones.

4. ER clutch

At present, there are many kinds of clutches. The most commonly used are electromagnetic ones. Its disvantage is that there is a very sudden action during the engagement and is not reliable for precision instruments. A ER clutch can be controlled by rapid and continuous electrical signal variation and can even form a servosystem. It certainly has advantages over other kinds of clutches.

ER clutches can be cylindrical or discal ones. Each of them has been developed by us. To meet the requirement of torque and power transmitted, they could be muti−cylindrical or muti−discal. During the operation the heat generated from the slip should be conducted to avoid the failure of the whole device. But it is very difficult for the ER fluid to circulate in the very sticky state. So circulators are needed for high power clutches and make the system more complicated and expensive. Only if the performance of ER fluids are improved could the ER clutched be applicatable to industry.

5. Highlight of ER technology

ER technology is a promising technology. Many kinds of devices exploiting ER effect have been produced in the laboratories. The priority should be given not only to the development of ER fluids of higher yield stress, lower viscosity without field, the greater operating temperature range but also to the practical inventions based on the ER effect. The study of the mechanisms of ER effect is as important as that of engineering application. A new era will come when the performances of ER fluids are considerably improved, the cost reduced, the size of high voltage power minimized.

BALANCING ROTARY MACHINERY WITH ELECTRORHEOLOGICAL FLUIDS

MICHAEL KAZZ
Department of Agricultural and Biosystems Engineering
The University of Arizona, 507 Shantz Building #38
Tucson, AZ 85721 U.S.A.

DENNIS L. LARSON
Department of Agricultural and Biosystems Engineering
The University of Arizona, 507 Shantz Building #38
Tucson, AZ 85721 U.S.A.

ABSTRACT

A machine balancer-analyzer system could be merged with an ER Fluid balancing device to automatically balance rotating machinery. The ER Fluid system is used for correction weight placement within a balancing device. The authors report on an ER Fluid flow model used to calculate the precision of ER Fluid apportionment. Design of the balancer device is discussed along with a strategy for the electronic control system. An advantage is ER Fluid use could permit electronic control of hydraulic-like systems without complex moving parts.

1. Introduction

Balancing of rotary machinery applies to many manufacturing operations. Some examples include paper and printing machines, turbines, hammermills, conveyors and spindles. Some of these machines play a critical role in the quality of the manufactured product. Consider a paper making machine, perhaps the heart of a paper mill. The paper machine's roller must be precisely balanced within a certain tolerance or the quality of the paper will be reduced and critical components will wear prematurely. Components also can be expensive, e.g., a roller bearing for a large paper machine can cost $20,000.

Rotary machines must operate within a certain range of balance tolerance. Critical components, like the paper roller, are often continuously monitored. In order to monitor a rotor, sensors such as accelerometers and velocity meters are mounted perpendicular to the axis of the rotor against a secure frame. These sensors provide vibration data for the balancer - analyzer which monitors one or more of a machine's rotors. By adopting this "predictive maintenance" approach, the user has a means to locate a potential fault in a rotating machine long before tolerance levels have approached the critical state of imbalance.

When a rotor creates excess vibration, an analyzer-balancer provides the balance correction data. A technician uses the data to perform an "in place" balance on the rotor(s). To do this, the technician must stop the machine and apply correction weights to the rotor, resulting in downtime. The technician must be trained to do balancing, and the correction weights are available only in discrete sizes.

A technician would not be required to operate the proposed device. Furthermore, the device could balance in real time without stopping the machine. Current systems capable of this technology are very expensive. Schenck Trebel Corporation produces an automated balancer that uses a laser to remove weight from an externally mounted rotor. The cost of this balancing device is over $1,000,000[1]. A simple system with few moving parts, feasible construction and compatibility with current balancer-analyzer systems is the desired goal of the proposed device.

1.1 Existing Balancing Technology and ER Fluid Applications

Most of the research dedicated to balancing technology has been oriented towards analyzing rotors while in operation. This research has provided engineers with programs that schedule maintenance operations for many kinds of machinery.[2] Other new developments include fully automated balancing systems.

Machine balancing systems are designed so many rotors can be monitored by one central balancer-analyzer. The missing link is the act that the balancer-analyzer cannot do -- balance. The balancer-analyzer only provides correction data for use in alleviating vibrations and imbalances. To physically correct an imbalance, a technician normally must gain access to the rotor and add correction weights. Critical rotary machine operations could be improved by having a balancing system integrated into the machine.

A balancing system utilizing ER Fluids instead of conventional hydraulic fluids could save space, perform precision balancing and reduce balancing costs. ER Fluid valves can be simpler and lighter than solenoid valves. The principle cost of an ER Fluid balancing system would be the electronic components and the ER Fluid. The balancer skeleton, including the valves, could be made of injection molded plastic, fitted to the dimensions of the rotor.

2. Development of an ER Fluid Balancer

The balancing system uses a series of ER Fluid valves that apportion ER Fluid into reservoirs or cells where the ER Fluid serves as the correction weight. The amount of correction weight added to various cells depends upon the balancing requirements of the rotor. The balancing system is to be integrated with an existing analyzer system that senses imbalances and computes balance correction requirements.

The ER Fluid balancing system consists of containment cells, an ER Fluid movement system, and a computer controlled fluid apportionment system (Figure 1). For each cell, an ER Fluid valve controls ER Fluid flow into and out of the cell. ER fluid is pumped into the cells when the by pass valve is opened and drawn out from the cells when the valve is closed (Figure 2). The remainder of the ER Fluid movement system consists of a pump, selected for low flow rates with a high viscosity fluid and a reservoir (sump).

An important component of the cell is a flexible Viton diaphragm that permits fluid to be drawn out of the cell without creating a partial vacuum and provides an ER Fluid to air barrier. On the external side of the diaphragm is a space that permits air entrance from an outside "breathing" hole (Figure 3). When fluid is pumped in, the cell expands to the outside walls of the box and when fluid is drawn out, the flexible diaphragm is pulled inward.

The present balancer test model is composed of eight cells. Each valve connects a cell with the spider which is connected to a main line. ER fluid is pumped to the main line by a stationary pump. The ER Fluid passes through the main line from the pump to a swivel, located under the spider, that rotates with the balancer. This permits distribution of ER Fluid to the cells while the balancer and rotor are rotating.

The control system for the balancer is composed of two parts, an electronic controller mounted inside the balancer and a digital timing circuit. The digital timing circuit is linked to the electronic controller by a multiple slip ring assembly containing the needed connections for high voltage supply and low voltage service.

The ER Fluid flow is controlled by high voltage electronics, which link digital control with a high voltage power supply. One solution is use of high voltage operational amplifiers with digital or analog control which permits up to 10,000 volts with a 10 volt input. However, each operational amplifier could control only one ER Fluid valve. Thus, a second alternative is use of a multiplexing high voltage operational amplifier.

These solutions are too costly for this particular application. High voltage relay technology permits control of all eight valves at low cost. High voltage relays can release 50,000 volts with a control voltage less than 25 volts. The particular relays selected for this application were latching relays supplied by Kilovac Corporation rated at 10,000 volts.

In operation, the controller receives digital signals to open the ER Fluid valves for specified periods. A series of low voltage pulses are sent through the slip ring into an asymmetric transmitter receiver located in the balancer which controls the high voltage latching relays. The timing circuit sets the period for which the valve is opened or closed. Figure four depicts the control strategy of the system.

2.1 System Evaluation

An ER Fluid flow model was developed to predict ER Fluid flow into and out of cells under various operating conditions (voltage/mm, pump pressure, flow rate).[3] This model was used to size flow and valve components. Fluid mass inflow and outflow from the cell then were measured to compare mass apportionment with estimates made by the ER Fluid flow model. The following equation reveals a linear relationship of mass apportionment to time.

The statistically derived equation predicting ER Fluid flow apportionment is:

$$Y = 1.07(t) + 6.75(R^2 = 0.992)$$

Where Y is mass(g), t=time (s).

3. Discussion

The balancing system described above has been evaluated for rotational speeds of less than 1000 r.p.m. The tests provided information on design limitations caused by separation of ER Fluid materials subjected to large centrifugal forces. The separation of ER Fluid particles within the cells of the balancer results in formation of a sediment which is difficult to remove by pumping. One means of reducing particle build up is use of low particle concentration ER Fluid, so that there is less accumulation of sediment in the walls of the balancer. Current experimentation is attempting to resolve the sedimentation problem by inducing temporary vibrations in the cells to agitate the centrifuged ER Fluid and restore the fluid suspension.

The initial phase of testing reveals production of a high precision automatic balancer is possible. However, additional development and testing are required to provide more adequate concept evaluation.

An ER Fluid based balancing system appears to have distinct advantages over a hydraulic or conventional liquid balancer. The cost could be low due to the simplicity of the ER Fluid valves and an ER Fluid balancer is potentially reliable since the only moving parts are the ER Fluid pump and the swivel assembly.

4. Acknowledgements

This research was supported by a grant from the National Science Foundation's "Creativity Awards in Undergraduate Engineering." The author would like to thank NSF for support during his graduate program. Additional thanks to Doug Brooks of Advanced Fluid Systems, London, for his cooperation in supplying an ER Fluid used in this research. Special thanks to John Tiss of The University of Arizona for his expert technical advice.

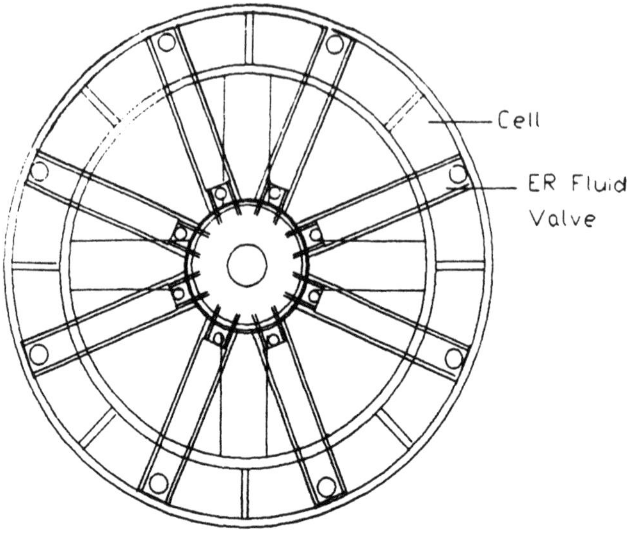

Figure 1. Balancer Skeleton

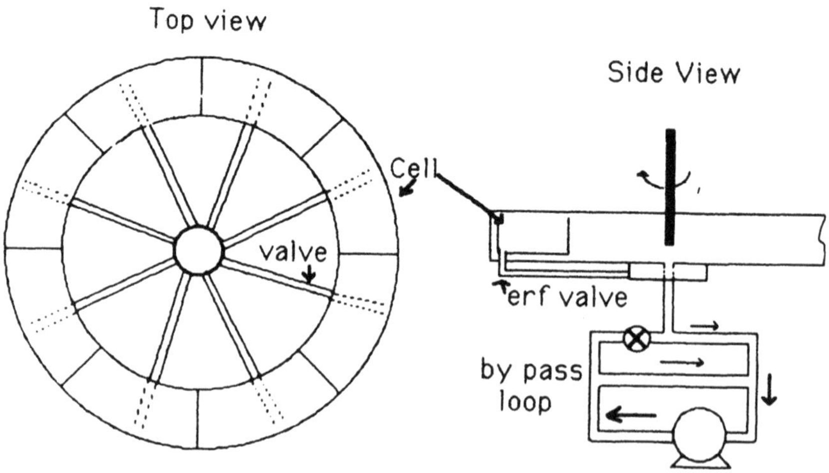

Figure 2. ER Fluid Movement System.

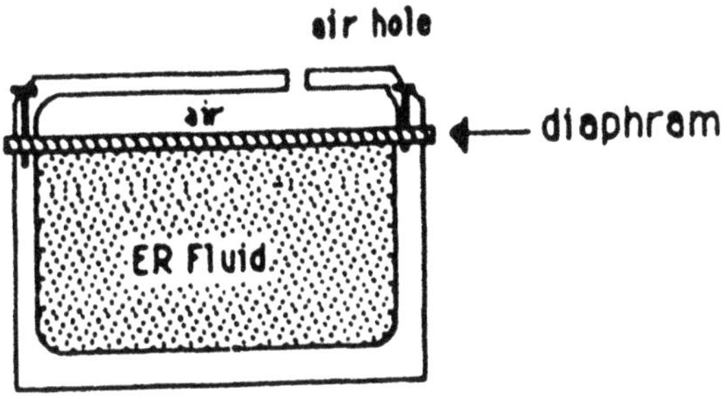

Figure 3. Cross-sectional view of a cell.

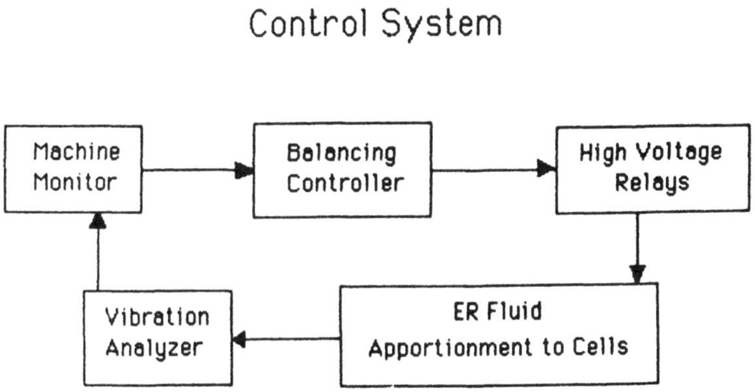

Figure 4. Automatic balancer's control system.

5. References

1. J. Harrel. Personal Communication. IRD Mechanalysis Inc. 4130 E. Palo Brea Ln., Cave Creek, Arizona 85331. August 1991.

2. M. Valenti. "Maintenance Software Keeps Machines Up and Running." Mechanical Engineering. V 113/NO. 11. pp. 63-65. November 1991.

3. M. Kazz and D. L. Larson. "Development of a System for Electrorheological Fluid Flow Control." American Society of Agricultural Engineers. Paper 913046. June 1991.

DISTRIBUTED ANECHOIC SYSTEM USING ELECTRICALLY ACTIVE FLUID

By

D. K. BHADRA

General Atomics
San Diego, CA 92186, U.S.A.

and

W. B. THOMPSON

University of California
San Diego, CA 92093, U.S.A.

ABSTRACT

A novel application of electrorheological fluid as a potential candidate for externally controllable anechoic material is discussed. The mechanism postulated is a resonant interaction of the incident compressional waves with the subsonic shear modes of the system. The energy dissipation is expected due to the existence of a loss modulus of the system.

1. INTRODUCTION

Most Present-day anechoic materials or surface treatments suffer from a lack of control of the acoustic properties of the material, generally in the form of a constraint on the frequency range over which the material can be utilized. We wish to present the potential likelihood of a smart material using electrorheological (ER) fluid for acoustic applications that may allow control of the response of the material.

For the sake of simplicity, we will consider an isotropic viscoelastic system and proceed to show that the ER-fluid as a viscoelastic substrate will provide for significant decrease in the reflection coefficient for compressional oscillations. Existence of a free parameter, such as the applied electric field for ER fluids, might allow for the control of the dynamic response of the viscoelastic matrix.

ER fluids offer promise as an ingredient in smart structure, due to their rapid response, occurring within milliseconds, and is completely reversible. This speed and reversibility allows for applications of a smart nature on acoustic time scales. Because an electric field controls the electro-viscous properties, they naturally lend themselves to automation of control. An external signal can directly modify the function or response of an ER fluid coating material. The electronic circuitry required is rather routine.

2. ELECTRORHEOLOGICAL FLUIDS AS VISCOELASTIC MEDIA

2.1 Relevant Properties of ER-Fluid

There are a number of common attributes that characterize ER-fluids. All are comprised of dispersed polarizable or conducting materials and it is commonly agreed that the ER phenomenon arises from some form of induced polarization originating within or around the particulates. However, there is less agreement about what occurs subsequently to induce the rheological effects. The ER phenomenon involves primarily the sciences of rheology and dielectrics and, for this reason, the variables influencing the phenomenon are many, including variation of shear stress τ with electric field strength and frequency, temperature, composition in terms of volume fraction of particles, the amount of water or other ingredient, dielectric constants and viscosities of the dispersing oils, and of course the shear rate $\dot{\gamma}$.

ER material behavior can be divided into pre-yield and post-yield stages, with these stages defined as those in which the material is strained below and above a critical yield strain (generally, the value of critical yield strain is about 1% for many ER formulations). The controllable viscoelastic behavior of ER-materials in the pre-yield phase is of our interest for the applications. Generally, the complex shear modulus G^* increases with increasing applied electric field[1]. Similar increases occur in pre-yield mechanical moduli when ER materials are loaded in tension along a direction parallel to the applied electric field. This controllable pre-yield mechanical moduli of ER materials can be used in a distributed fashion to obtain controllable elastodynamic behavior of structures.

The experimental data for the pre-yield complex shear modulus of a particular fluid is presented in Fig. 1 as a function of frequency and electric field. The results are presented in the form of shear storage modulus G' and shear loss factor $\delta = G''/G'$, where $G^* = G' + iG''$.

The unique feature of ER materials in pre-yield situations is the controllability of material properties within certain ranges, as shown in Fig. 1. For each frequency, the complex shear modulus

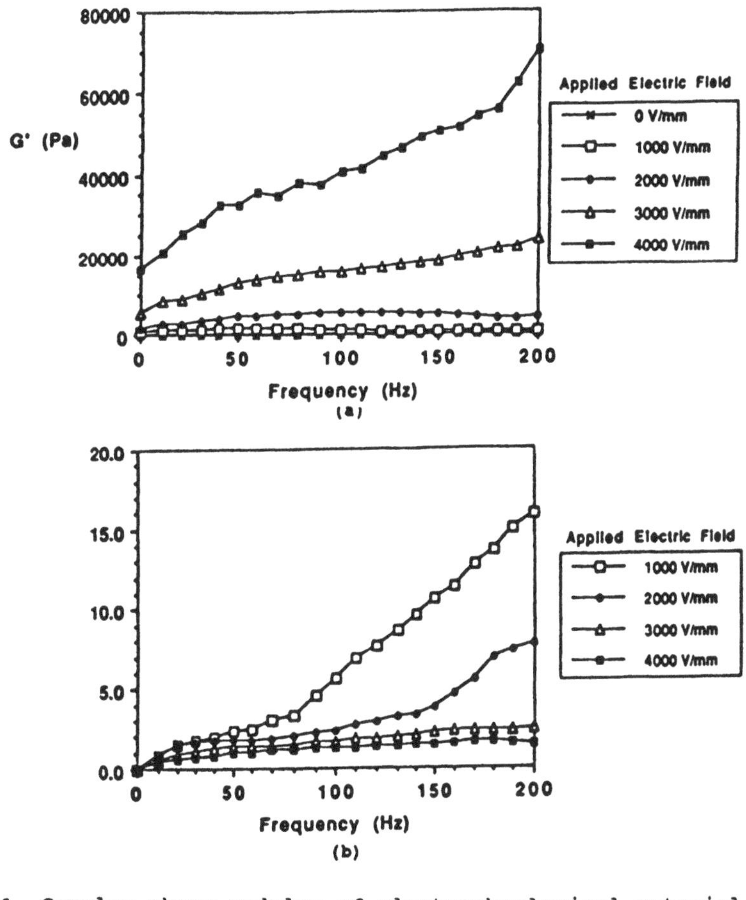

Fig. 1 Complex shear modulus of electrorheological material Dow Corning ERX-03-145[1]. (a) Storage modulus, (b) Loss factor.

can be controlled within the minimum and maximum values by adjustment of the applied electric field. This control can be exercised over time increments of several milliseconds, which is the response time of present ER materials to changes in electric field strength. Such strong dependence of ER-fluid dynamic moduli on the strength of the imposed electric field has also been predicted from theoretical micro-structural models. In addition to the external electric field, both the storage and loss moduli depend on the nature of the suspended particles and their concentration.

2.2 Shear Flow Development

It is helpful, at this point, to note how acoustic tiles work. Such tiles are characterized by pores in a mildly viscoelastic matrix. In the presence of oscillations, these pores induce fairly high shear which ultimately leads to viscous damping. The development of such shear flow is also possible in a fluid-loaded compliant surface, as shown schematically in Fig. 2. The added

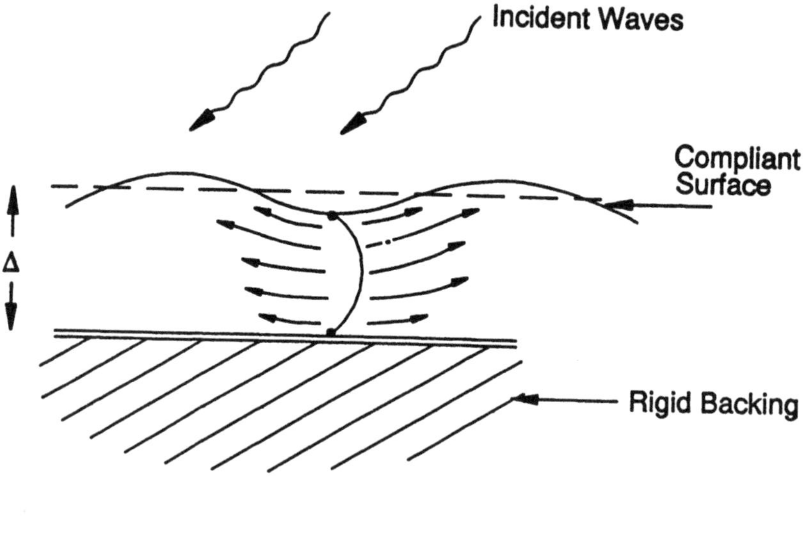

Fig. 2 Development of shear flow

feature of loading such matrix with ER-fluid is the potential for parametric control of visco-elastic properties of such distributed configurations.

3. A DISTRIBUTED SYSTEM USING ER-FLUID

A distributed system, for our application, may be defined as a configuration where the acoustic absorption and reflection properties are effective over an extended surface and such properties do not necessarily depend on a characteristic frequency as in a lumped-element system. An example of such a distributed system may be a flexible layer of soft silicone rubber covered with a thin protective layer of latex rubber. Acoustic characteristics of such a system may include a significant amount of sound energy absorption over a fairly wide band of frequencies together with a low coefficient of reflection. Figure 3 shows a possible schematic for such a configuration. Inside the softer matrix region (very soft silicone rubber in our example), one may insert some aligned fibers or aligned structures with variable stiffness (and possibly also variable damping). The electrically controllable properties of the ER fluid provide a potentially novel means for developing such structures with variable stiffness.

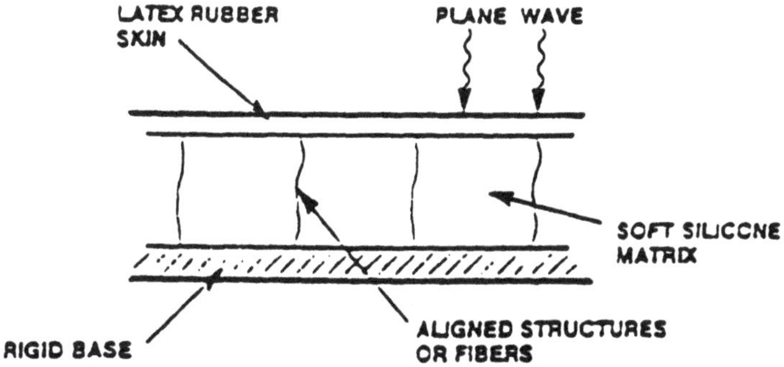

Fig. 3 A distributed configuration

The physics of interaction of compressional waves with a compliant distributed configuration is not completely developed at the present time. Obviously, the mechanics of the distributed configuration is the most important element in this interaction process. The basic model should be a volume-based one where the equations of motion for the configuration depend on the coordinates both parallel and perpendicular to the surface. The principle behind the operation of such a distributed system is to generate motions in the configuration so as to minimize the reflection of compressional waves. This is equivalent to the statement that the configuration presents an effective acoustical impedance to the environment very similar to that of the environment itself over a fairly broad range of frequencies. The transmitted sound energy may then be absorbed inside the visco-elastic matrix. Some of the conventional elastomeric surfaces do provide some of these functions though in limited frequency ranges and without any real means of external control. Such elastomeric configurations are basically passive elements and their utility is rather limited.

In order to understand the nature of the interaction of acoustic perturbations with such ER-fluid matrix configurations, one needs to formulate the elastodynamic equations for an anisotropic visco-elastic medium in the presence of an oscillatory compressional pressure pulse.

There are five competing physical mechanisms that need to be included in the modeling: the structural stiffness of the panel, its inertia, fluid pressure, fluid compressibility and fluid inertia. Depending on the balance between the various mechanisms, a wide range of phase and amplitude of surface displacements can be obtained. In our treatment, we neglect the structural stiffness of the panel and its inertia.

4. ANALYSIS OF DISTRIBUTED ELASTODYNAMIC SYSTEM

The equations for an anisotropic viscoelastic system are extremely complex. The existence of anisotropy also introduces the added complexity that the compressional and shear waves modes are not separable. For the sake of simplicity, we will consider only an isotropic viscoelastic system and proceed to show that the ER-fluid as a visco-elastic substrate will provide for significant decrease in the reflection coefficient for compressional oscillations. Existence of a free parameter, like the external electric field, will allow for the potential to control the dynamic response of visco-elastic matrix. Under certain conditions, an anechoic condition for the compliant surface may be obtained. The absorption of compressional energy inside the matrix, driven by the electrically-controlled loss modulus of the medium, is expected to occur but needs to be demonstrated in a controlled laboratory situation.

4.1 Derivation of Elastodynamic Equations

Referring to Fig. 4, we have defined the coordinate z to be in the plane of the paper (and perpendicular to the substrate), and x, y define the coordinates perpendicular to z; however, for the sake of simplicity, we choose the coordinate x as the perpendicular one and quantities in that direction are denoted in the corresponding directions we denoted by ξ_i (i = z, x). The strain tensor is defined as

$$\varepsilon_{ij} = \frac{1}{2}\left(\frac{\partial\xi_i}{\partial x_j} + \frac{\partial\xi_j}{\partial x_i}\right) \quad , \quad (i,j = z,x) \tag{1}$$

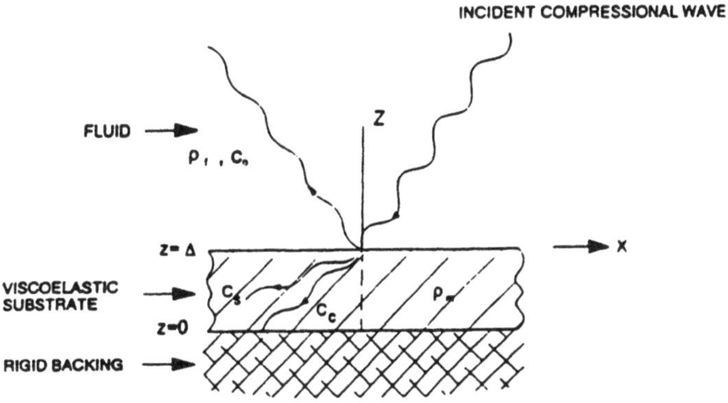

Fig. 4. Coordinate system for the configuration

Using Hooke's law, one can obtain the stress tensor[2]

$$T_{ij} = -\mu\, \varepsilon_{ij} - K\, \delta_{ij}\, \vec{\nabla}\cdot\xi \tag{2}$$

where

K = bulk modulus = $\lambda + \frac{2}{3}\mu$

λ and μ being the Lame constants.

A widely used model for introducing losses in a visco–elastic medium is the so–called Voigt solid. This model has the advantage that it yields a linear wave equation which can be solved for arbitrary time–dependence. The assumption is made that stresses are directly proportional to strain rates, as well as to the components of strains themselves. The Hookes' law is then modified, including strain–rate terms:

$$T_{ij} = -\mu \, \varepsilon_{ij} - \mu' \frac{\partial \varepsilon_{ij}}{\partial t} - K\xi_{ij} \, \vec{\nabla}\cdot\xi - K'\delta_{ij}\cdot\vec{\nabla}\cdot\frac{\partial \xi}{\partial t} \tag{3}$$

In the frequency domain, it is now equivalent to the statement that the elastic modulü are complex with the primed quantities denoting the imaginary component, e.g., μ now becomes $\mu + i\omega\mu'$, etc. The dynamical equations of motions are obtained from

$$\rho_m \frac{\partial^2 \xi}{\partial t^2} = \vec{\nabla}\cdot\vec{T} \tag{4}$$

After some algebra, one can show that:

$$\begin{aligned} -\omega^2\rho_m\xi_z &= (\hat{\lambda} + 2\hat{\mu})\frac{\partial^2 \xi_z}{\partial z^2} + \hat{\mu}\frac{\partial^2 \xi_z}{\partial x^2} + (\hat{\lambda} + \hat{\mu})\frac{\partial^2 \xi_x}{\partial x \partial z} \\ -\omega^2\rho_m\xi_x &= (\hat{\lambda} + 2\hat{\mu})\frac{\partial^2 \xi_x}{\partial x^2} + \hat{\mu}\frac{\partial^2 \xi_x}{\partial z^2} + (\hat{\lambda} + \hat{\mu})\frac{\partial^2 \xi_z}{\partial z \partial x} \end{aligned} \tag{5}$$

where we assumed that quantities $\sim e^{i\omega t}$ and $\hat{\lambda}$, $\hat{\mu}$ are now the complex Lamé coefficients as discussed above. The relationships between $\hat{\lambda}$, $\hat{\mu}$ and the dynamical moduli are well-known and are not repeated here (e.g., $\mu \to G'$ and $\omega\mu' \to G''$).

For briefness, we would omit the detailed mathematical steps in the calculation but only retain some significant equations relevant for our purpose.

Solutions of the above equations may be obtained by expressing the components of ξ in terms of potentials ϕ and Ψ (in the two-dimensional case). Explicitly

$$\xi_z = \frac{\partial\phi}{\partial z} + \frac{\partial\psi}{\partial x}$$
$$\xi_x = \frac{\partial\phi}{\partial x} - \frac{\partial\psi}{\partial z} \tag{6}$$

Inserting these expressions in the equations of motion above implies that each of the potential obeys a two-dimensional wave equation, ϕ being associated with a compressional wave and Ψ with a shear wave with velocities

$$C_c = [(\lambda + 2\ \mu)/\rho_m]^{1/2}$$
$$C_s = (\mu/\rho_m)^{1/2} \tag{7}$$

respectively.

In order to treat the pressure fluctuations, one readily assumes:

$$\phi = (a_+ \cosh lz + a_- \sinh lz)\, e^{i\omega t - ikx} \tag{8}$$

and

$$\psi = (b_+ \cosh mz + b_- \sinh mz)\, e^{i\omega t - ikx} \tag{9}$$

as the double fourier transform for the potentials valid for $o \le z \le \Delta$.

The separable wave equation's for ϕ and Ψ then yield:

$$(k^2 - l^2)\, c_p^2 = \omega^2$$
$$(k^2 - m^2)\, c_s^2 = \omega^2 \tag{10}$$

where

$$C_p^2 = C_c^2\,(1 + i\omega\,\tau_\ell)$$
$$C_e^2 = C_s^2\,(1 + i\omega\,G''/G') \qquad (11)$$

and

$$C_p \to C_c \;,\quad c_e \to c_s \quad when \quad \lambda' = \mu' = 0.$$

The quantity τ_ℓ is related to the relaxation time of the medium with respect to compressional disturbances.

The determination of the unknown coefficient a_+, a_-, b_+ and b can be achieved by taking into account boundary conditions at the interface $z = o$ and $z = \Delta$. Rigid backing at lower interface imposes the vanishing of both the normal and the tangential components of the displacement at $z = o$. On the other hand, the two remaining boundary conditions at the fluid—matrix interface are given by:

$$p_{zz} = -p_{ext} \;(continuity\ of\ normal\ stress)$$
$$p_{xz} = 0 \;(vanishing\ of\ tangential\ stress) \qquad (12)$$

where $p_{xz} = p_{zx}$ is the tangential stress and p_{zz} is the normal component of stress; both these quantities can eventually be expressed in terms of the potentials ϕ and Ψ. The quantity p_{ext} denotes the pressure due to an externally imposed sound wave. This is obtained form the two—dimensional wave equation for the propagation of sound in that medium:

$$\frac{\partial^2 p_{ext}}{c_o^2 \partial t^2} = -\frac{\partial p_{ext}}{\partial z} \qquad (13)$$

where c_o is the speed of sound in the fluid medium with density ρ_f. At $z = o$, one then also has to satisfy the following equation:

$$\rho_f \frac{\partial^2 \xi_z}{\partial t^2} = - \frac{\partial p_{ext}}{\partial z} \text{ at } z = 0 \tag{14}$$

For the external displacement field, one assumes

$$\xi_z^{ext} = \frac{\partial \Phi_e}{\partial z} \tag{15}$$

where

$$\Phi_e = A_e(e^{ik_z z} + Re^{-ik_z z})e^{i\omega t} \tag{16}$$

R being the reflection coefficient.

4.2 Evaluation of the Reflection Coefficient

After considerable amount of algebra, one can show that

$$\frac{ik_z}{\ell} \frac{\rho_m}{\rho_f}(1 + \bar{A}) \left(\frac{1 - R}{1 + R}\right) = - \frac{H_1}{H_2 + H_3} \tag{17}$$

where

$$H_1 = \frac{m^2 - k^2}{2k^2} \left(G_1 - \frac{k^2}{m\ell} G_2\right) \tag{18}$$

$$H_2 = \frac{m^2 + k^2}{2k^2} \left(G_3 - \frac{k^2}{m\ell} G_4\right) \tag{19}$$

$$H_3 = \bar{B}\left(G_3 - \frac{m\ell G_4}{k^2}\right) - \left(1 + \frac{m^2 + k^2}{2k^2}\bar{B}\right) \tag{20}$$

$$\begin{aligned} G_1 &= \sinh \ell\Delta \; Cosh\, m\Delta \\ G_2 &= Cosh\, \ell\Delta \; Sinh\, m\Delta \\ G_3 &= Cosh\, \ell\Delta \; Cosh\, m\Delta \\ G_4 &= Sinh\, \ell\Delta \; Sinh\, m\Delta \\ \bar{A} &= -\frac{k^2 C_e^2}{\omega^2} \\ \bar{B} &= \frac{\bar{A}}{1+\bar{A}} \end{aligned} \tag{21}$$

We make the following assumptions:

$$k_z \Delta < 1$$
$$\ell\Delta < 1$$

and rewrite Eq. (10) in the following form:

$$m^2 = k^2 - \frac{\omega^2}{C_e^2} \tag{23}$$

When

$$\frac{\omega}{k} < C_e \quad , \quad m^2 \gtrsim 0 \quad ,$$

and when

$$\frac{\omega}{k} > C_e \quad , \quad m^2 < 0 \quad \textit{but large.}$$

The latter result is more appropriate for ER-fluids when one considers its not too high shear-bearing capacities. Under these assumptions, one can show that;

$$ik_z \frac{\rho_m}{\rho_f}\left(\frac{1-R}{1+R}\right) = \ell^2\Delta\left[1 - \frac{k^2}{\ell^2}\ \frac{\tanh m\Delta}{m\Delta}\right] \tag{24}$$

Realizing that the right hand side in the above equation is complex for $m^2 < 0$, the requirement that the complex quantity $R \to 0$ leads to the condition that the real part of R be close to unity whereas its imaginary part be close to zero. To explore what restrictions these conditions would impose on the right hand side of Eq. (24) above, for convenience one replaces the complex quantity m by another complex quantity $\bar{m}$ where $m = i\bar{m}$. It is evident from Eq. (24) that, for real values of the quantity $m\Delta$, the second term inside bracket on the right hand side has singularities at $\bar{m}\Delta = (n+\frac{1}{2})\pi$ where n = 0, 1, 2, ... etc. These singularities correspond to poles in the complex $\bar{m}$-plane and, for small values of the quantity Im$\bar{m}$, these poles are very close to the real axis as shown schematically in Fig. 5(a). If Im$\bar{m}$ = 0, then the real part of the second term on the right hand side of Eq. (23) is shown in Fig. 5(b) as solid lines. We also show the horizontal lines intersecting the solid lines for the case when $\ell = k$, $\ell > k$ and $\ell < k$. The singularities actually do not go to infinity due to the fact that $\bar{m}$ is not completely real, it actually has a small imaginary part. This results in limiting the values of the second term on the right hand side to finite values (shown schematically by cross-hatched areas in Fig. 5(b) at the points $(n+\frac{1}{2})\pi$. The reason why Im $\bar{m}$ is small can be shown ultimately to be related to the fact that, for the material medium we are considering, we have assumed G'' to be smaller than G_e'.

It can then be shown that $R \to 0$ if

$$m\Delta = i\bar{m}\Delta = X + iY \tag{25}$$

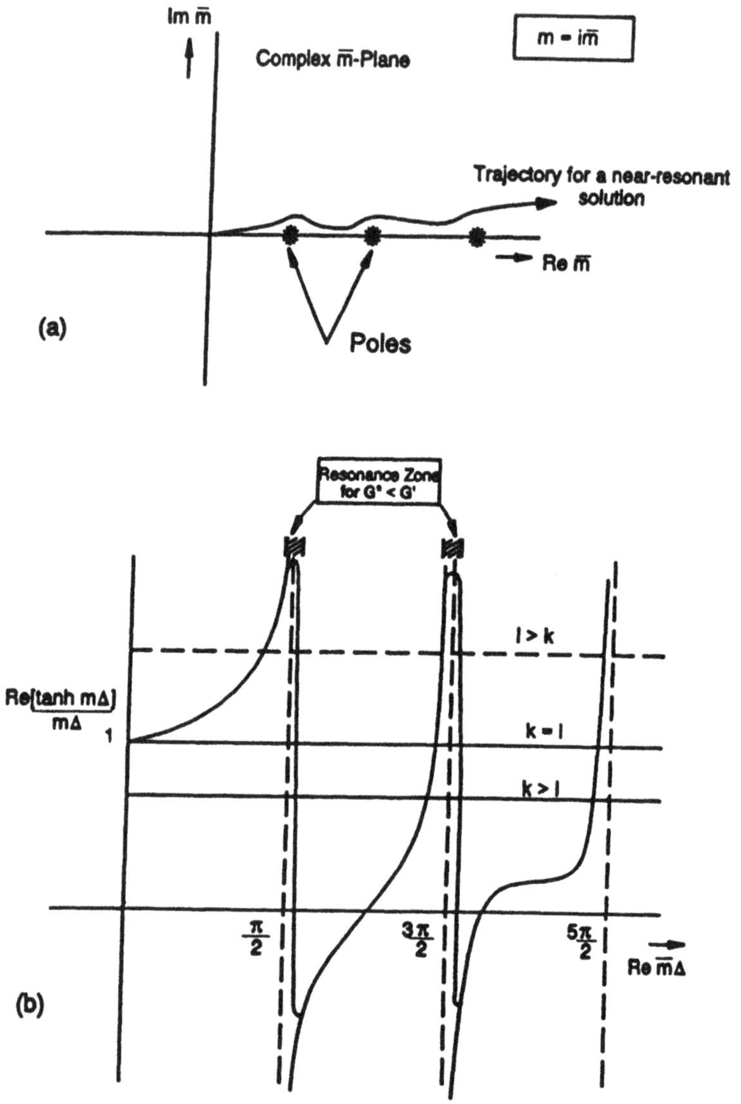

Fig. 5 (a) Poles in the complext $\dot{m}$-plane
(b) Singularities of the function tanh mΔ/mΔ

where

$$Y = (n + \tfrac{1}{2})\pi \quad , \quad n = 0,1,2,-- \quad . \tag{26}$$

$$\frac{X}{Y} = \frac{2\pi\Delta}{\lambda_w} \frac{\sin^2\theta}{\cos\theta} \quad . \tag{27}$$

In the above expression, λ_w is the wavelength of compressional disturbances in the external medium and θ is the angle of incidence of the plane wave (as shown in Fig. 2 earlier).

4.3 Maximization of Shear Flow

It is evident that the quantities X and Y are determined by the properties of the matrix containing ER-fluid and such properties could be controlled by an external electric field. The condition Y can essentially be viewed as a resonance condition, viz.,

$$\frac{\omega}{k} \sim V_{ph} \sim mC_e \quad , \tag{28}$$

realizing that $V_{ph} >> C_t$ but m >> k. This is the resonance of the external perturbation as it interacts with the shear motions inside the viscoelastic matrix. As we have mentioned earlier, the incident compressional wave is converted both into compressional and shear waves inside the medium. The above mentioned resonance condition maximizes the shear flow when $G'' < G'$. When G'' is strictly zero, these modes correspond to the natural modes of a theoretically undamped viscoelastic layer. One may identify them as subsonic shear modes. Such a mode satisfies the vanishing boundary conditions both at the rigid backing and also at the interface and the lowest harmonic has a displacement pattern as shown schematically in fig. 2. However, $G'' \neq 0$, and as a consequence, such modes will be damped and the specific properties of the layer will determine the nature of such damping. Similarity of such modes to the normal oscillations of a fluid-loaded viscoelastic layer (with rigid backing) is to be noted[3].

The condition on X/Y, as shown in Eq. (27) above, may be termed as a "tuning" condition. Adjustments of this ratio could be achieved via altering the dynamic moduli of the ER-fluid with the application of an external electric field.

5. Design of a Distributed System

Conceptually, one may design such an acoustically active structure by imbedding ER-fluid in a visco-elastic matrix. For such a matrix, the modulus G′ is mostly likely be determined by matrix or the skeleton itself whereas the presence of ER-fluid will provide controllable variability of the other dynamic modulus G″ of the system. It is possible that such panels could be modeled in terms of what may be called as a synthetic "Biot" model of a porous medium. Some of the major elements of the medium may consist of: (i) a skeleton or aggregate which is statistically isotropic, and (ii) a fluid which fills interconnected pores. Such pores may be random or organized in a periodic or geometric fashion. The response of the fluid may be in the inertial range or viscous range, depending on the various parameters of the system. Figure 6 shows a possible embodiment of a distributed porous configuration with ER-fluid embedded in it. The skeleton could be come form of diluted soft silicone rubber with voids in it, these voids being filled with ER-fluid. These voids could be random and a set of wires with alternate psotive and negative polarity may be all that would be necessary to activate the porous material.

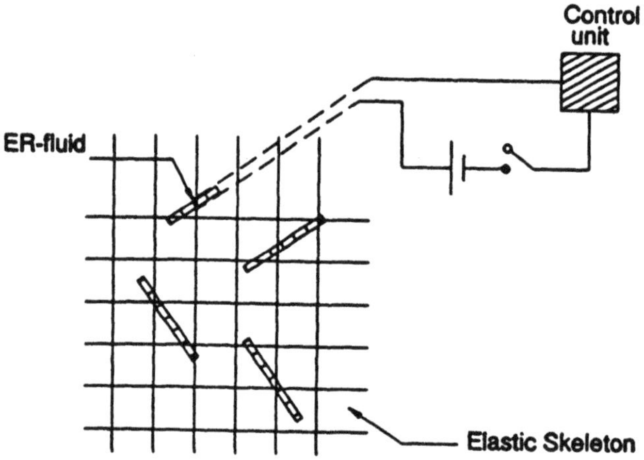

'Pores' Imbedded in an Elastic Skeleton

Figure 6. Synthetic "Biot" solid model

6. Conclusions

In this paper, we have explored the dynamic of oscillations of a distributed configuration with ER-fluid imbedded in it. We have indicated that such a system, under suitable conditions, may behave like a near-anechoic surface with respect to incident plane compressional waves. The mechanism that we have postulated is a resonant interaction with the subsonic shear modes of the system. The energy dissipation is expected due to the existence of a loss modulus of the system.

An architecture with ER-fluid elements imbedded in an elastic matrix is an example of the embodiment of such a configuration. In a more general sense, such configurations form the basis of a class of "smart" surfaces or "intelligent" materials. This is because:

(i) The system is able to sense changes in its environments in some dynamic modality,

(ii) Some associated functions (controlled by the electric field) allows response to these changes,

(iii) Such a link could be governed by an acceptable algorithm which would provide the level of "intelligence".

7. REFERENCES:

1. J. P. Coulter and T. G. Duclos, in Electrorheological Fluids, ed. J. D. Carlson, A. F. Sprecher and H. Conrad (Technomic Pub. Co., Lancaster, Pa., 1990) p. 300.

2. L. D. Landau and E. M. Lifshitz, Theory of Elasticity (Pergamon Press, New York, 1986), p. 11.

3. D. V. de La Croix, in Progress in Underwter Acoustics, Ed. H. M. Merklinger (Polenum Press, New York, 1987), p. 681.

PASSIVE TUNEABLE FIBERS AND MATRICES

CAROLYN DRY

School of Architecture, University of Illinois, 608E. Lorado Taft Drive

Champaign, Illinois 61820, USA

ABSTRACT

Active smart electrorheological fluids in building materials are expensive, need to be widely dispersed, and require large amounts of electricity. These practical considerations led to the consideration of ways of having similar effects in a matrix but using different materials. Tuneable fibers is an inexpensive dispersed passive system, requiring no electricity. The modulus of elasticity of a solid filled fibers is very different from a liquid or gas-filled fiber or a hollow fiber. The ability to empty and fill fibers with a material which could change phase in the fiber or in the matrix in response to some stimulus or signal from the environment is promising. In the laboratory we have used filled fibers which empty their contents into the matrix in response to a change in the environment. In this case as a substitute for ER fluid, a blast, vibration, or shaking motion could release a catalyst from flexed fibers. It would cross-link with a resin either in the porous fiber or in the matrix. Passive tuneable fibers could also be used for nonbuilding applications such as radar evasion due to changing modulus profile.

Active smart electrorheological fluids in building materials require large amounts of electricity and need to be widely dispersed. These practical considerations led to the consideration of ways of having similar effects in a matrix but using different materials. Tuneable fibers is an inexpensive dispersed passive system, requiring no electricity. The modulus of elasticity of a solid filled fiber is very different from a liquid or gas-filled fiber, an emptied hollow fiber, or broken fiber (see Figure 1). Very preliminary results as in Figure 2 verify this. The ability to empty and fill fibers with a material which could change phase in the fiber or in the matrix in response to some stimulus or signal from the environment is promising. In the laboratory we have used filled fibers which empty their contents into the matrix in response to a change in the environment (see Figure 3). For example, excessive structural loading causes fibers to debond, the coating is stripped off in the process, or the fibers are simply broken; the chemical released could cause the fiber t rebond to the matrix (see Figure 4). These cases are used as a substitute for E.R. fluid. A blast, vibration, or shaking motion would release the catalyst, polymer, or adhesive from the fibers. Not only would the dynamic modulus be altered, but the fibers could be reattached (see Figure 5). If a phase change material sensitive to motion could be found, we would try this also (see Figure 6).[1] In our work on smart materials we have researched the release of chemicals into cements and ceramics by fiber coating removal due to heat melting, degradation in alkalinity change, to improve such properties as resistance to corrosion, cracking, freeze/thaw damage, or permeability (see Figure 7).

---

[1] M. F. Asoro, A. T. Gaynor, S. Hettiarachelic, *Electrochemical Chloride Removal and Protection of Concrete Bridge Components (Injection of Synergistic Corrosion Inhibitors)*, Strategic Highway Research Program (October, 1990).

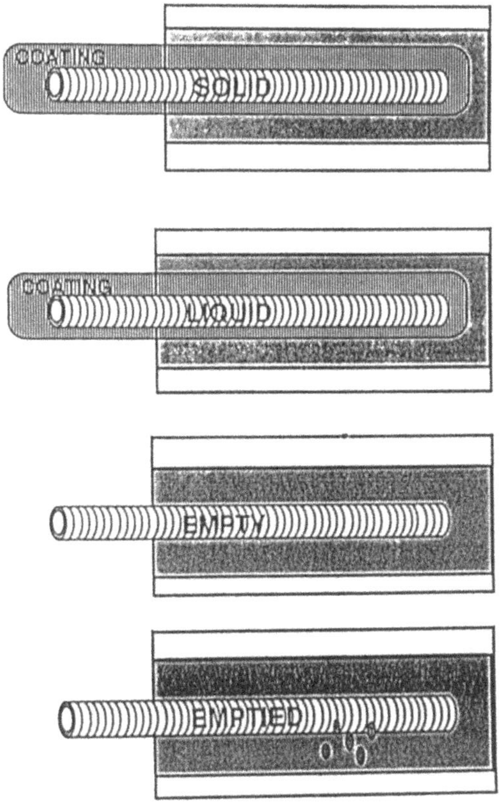

Figure 1. Dynamic Modulus of Elasticity Varies for Each Condition Based on the Contents of the Hollow Porous Fibers

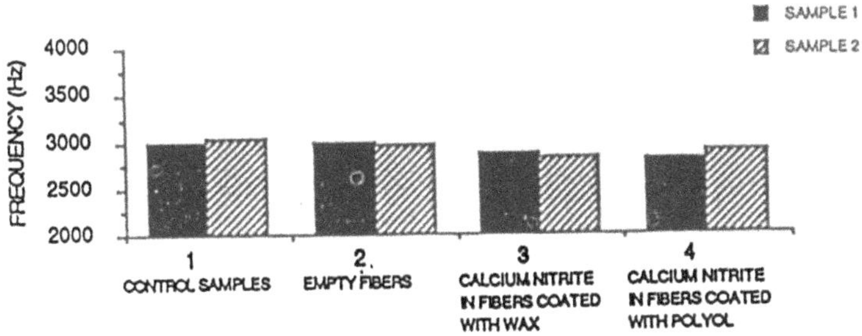

Figure 2. Resonant Frequency of Samples Containing Empty Fibers, Fibers containing Calcium Nitrite Coated with Wax, and Fibers containing Calcium Nitrite Coated with Polyol

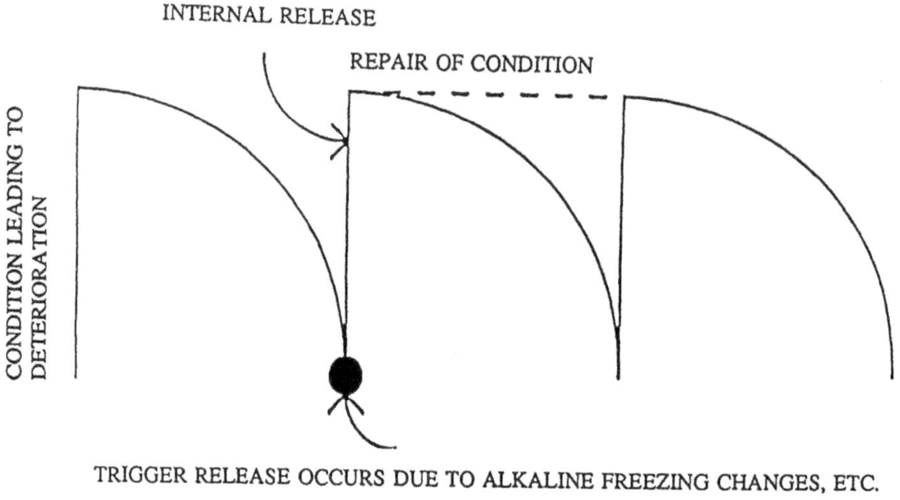

Figure 3. Release in Direct Response to Naturally Timed Intrusion of the Environment

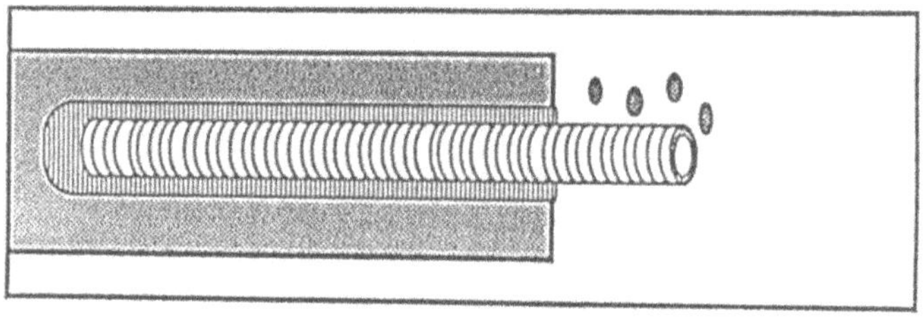

Fiber Yields and Releases Chemical into Matrix

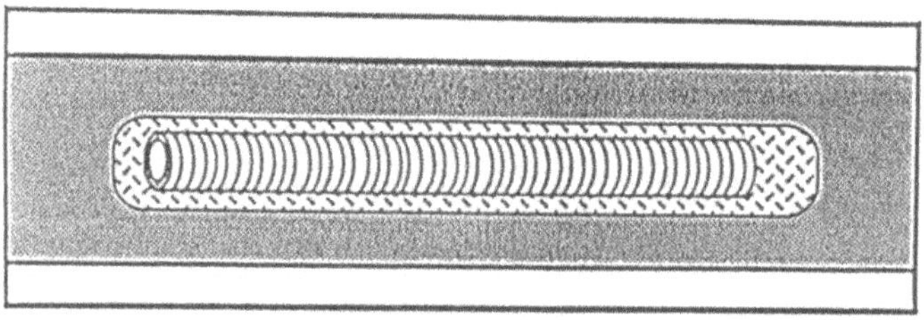

Chemical Released from Fiber Reattaches or Rebonds the Fiber to Matrix

Figure 4.

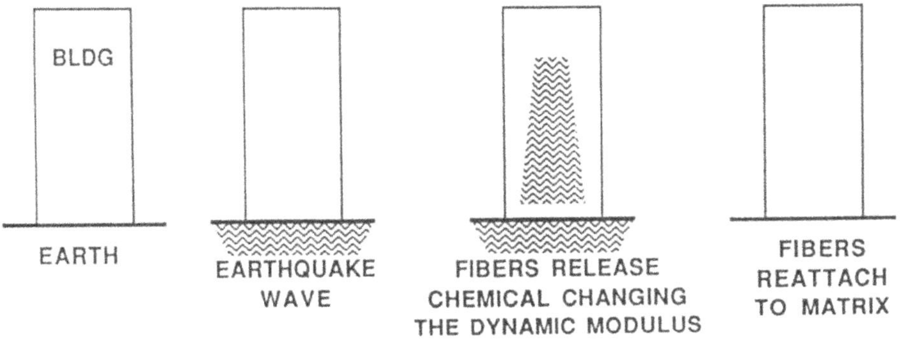

Figure 5.

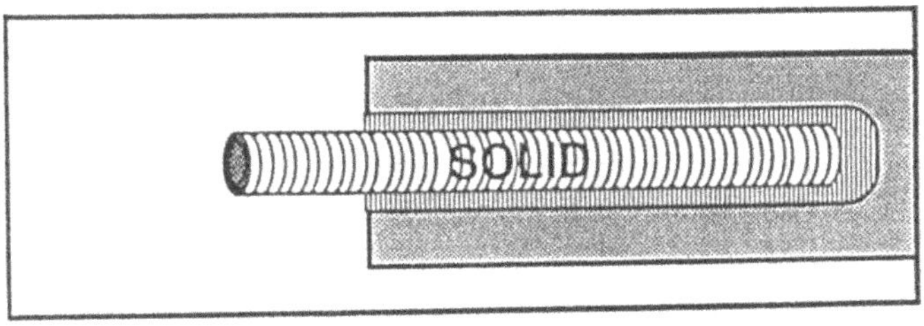

Hollow Fiber Contains Vibrationally Phase Changing Material

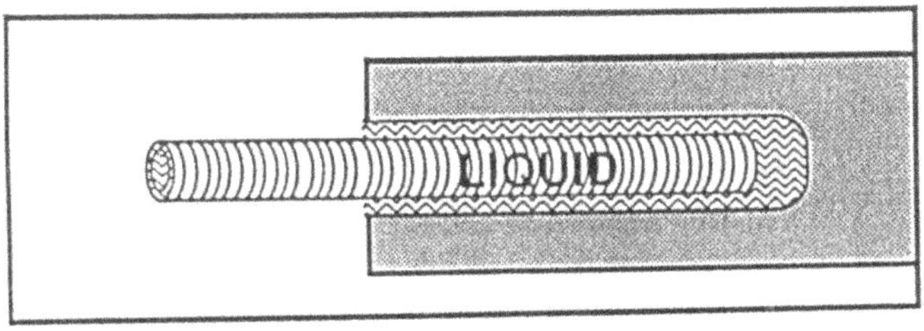

Phase Change Material Changes to Liquid Upon Vibration and Alters the Dynamic

Figure 6.

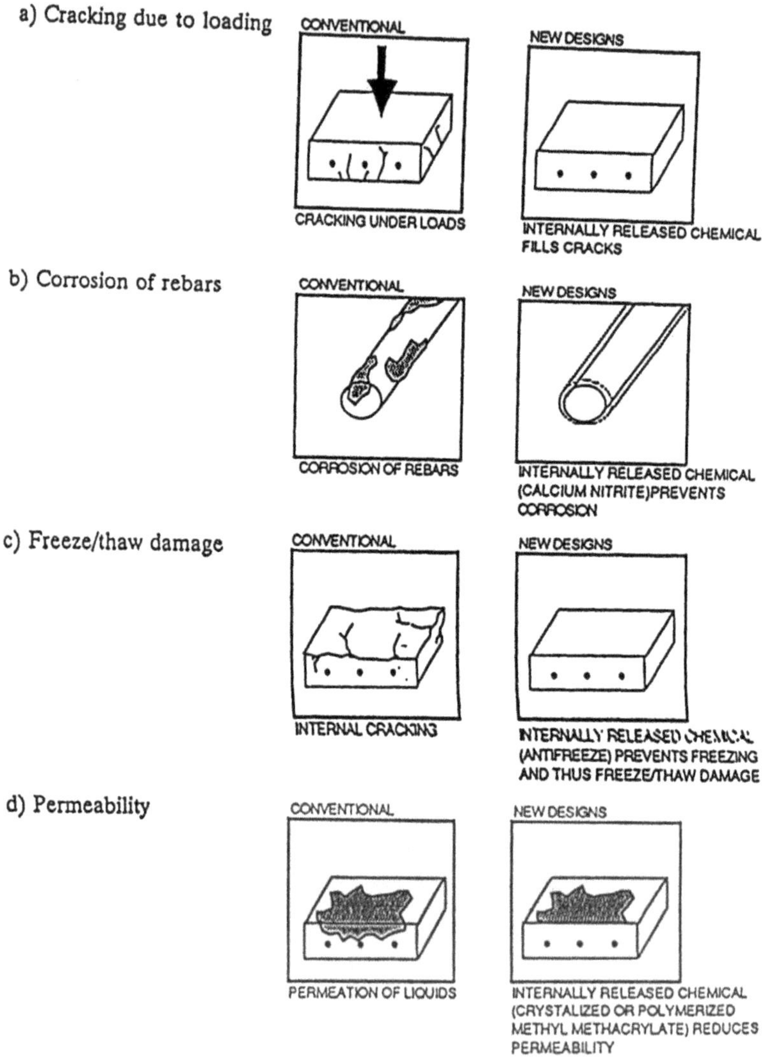

Figure 7. Alleviating Environmental Distress

INDEX

www.ingramcontent.com/pod-product-compliance
Lightning Source LLC
LaVergne TN
LVHW082253241026
850861LV00008B/39
* 9 7 8 9 8 1 0 2 0 9 0 2 5 *